BASICS OF ELECTRICAL MACHINES: THEORY AND PRACTICALS

AUTHOR (S)

Prof. D. K. Palwalia
Professor
Department of Electrical Engineering, Rajasthan Technical University, Kota, India

Dr. U. K. Kalla
Associate Professor
Department of Electrical Engineering, MANIT, Bhopal, India

Dr. R. K. Kumawat
Assistant Professor
Department of Electrical Engineering, I K Gujral Punjab Technical University Amritsar, India

REVIEWER

Prof. R. K. Srivastava
Professor
Department of Electrical Engineering,
Indian Institute of Technology (BHU) Varanasi, India

All India Council for Technical Education
Nelson Mandela Marg, Vasant Kunj, New Delhi, India

Dear Readers,

To prevent the piracy, this book is secured with HIGH SECURITY HOLOGRAM on the front title cover. In case you don't find the hologram on the front cover title, please write us to at contact@khannabooks.com or whatsapp us at +91-99109 09320 and avail special gift voucher for yourself.

Specimen of Hologram on front Cover title:

Moreover, there is a SPECIAL DISCOUNT COUPON for you with EVERY HOLOGRAM.

How to avail this SPECIAL DISCOUNT:

Step 1: Scratch the hologram

Step 2: Under the scratch area, your "coupon code" is available

Step 3: Logon to www.khannabooks.com

Step 4: Use your "coupon code" in the shopping cart and get your copy at a special discount

Step 5: Enjoy your reading!

ISBN: 978-93-5538-712-7

Book Code: UG410EN

Basics Of Electrical Machines: Theory And Practicals

by Dr. D. K. Palwalia, Dr. U. K. Kalla and Dr. R. K. Kumawat

First Edition: 2026

Originally Published by:

All India Council for Technical Education, Ministry of Education, GOI

Exclusively Published & Distributed by:

Khanna Book Publishing Co. (P) Ltd.
4C/4344, Ansari Road, Daryaganj,
New Delhi (INDIA)

Visit us at: **www.khannabooks.com**
Write us at: **contact@khannabooks.com**

To view complete list of books, Please scan the QR Code:

Printed in Bharat.

प्रो. टी. जी. सीताराम
अध्यक्ष
Prof. T. G. Sitharam
Chairman

सत्यमेव जयते

अखिल भारतीय तकनीकी शिक्षा परिषद्
(भारत सरकार का एक सांविधिक निकाय)
(शिक्षा मंत्रालय, भारत सरकार)
नेल्सन मंडेला मार्ग, वसंत कुंज, नई दिल्ली-110070
दूरभाष : 011-26131498
ई-मेल : chairman@aicte-india.org

ALL INDIA COUNCIL FOR TECHNICAL EDUCATION
(A STATUTORY BODY OF THE GOVT. OF INDIA)
(Ministry of Education, Govt. of India)
Nelson Mandela Marg, Vasant Kunj, New Delhi-110070
Phone : 011-26131498
E-mail : chairman@aicte-india.org

FOREWORD

Engineers are the backbone of any modern society. They are the ones responsible for the marvels as well as the improved quality of life across the world. Engineers have driven humanity towards greater heights in a more evolved and unprecedented manner.

The All India Council for Technical Education (AICTE), have spared no efforts towards the strengthening of the technical education in the country. AICTE is always committed towards promoting quality Technical Education to make India a modern developed nation emphasizing on the overall welfare of mankind.

An array of initiatives has been taken by AICTE in last decade which have been accelerated now by the National Education Policy (NEP) 2020. The implementation of NEP under the visionary leadership of Hon'ble Prime Minister of India envisages the provision for education in regional languages to all, thereby ensuring that every graduate becomes competent enough and is in a position to contribute towards the national growth and development through innovation & entrepreneurship.

One of the spheres where AICTE had been relentlessly working since past couple of years is providing high quality original technical contents at Under Graduate & Diploma level prepared and translated by eminent educators in various Indian languages to its aspirants. For students pursuing 2nd year of their Engineering education, AICTE has identified 87 books, which shall be translated into 12 Indian languages - Hindi, Tamil, Gujarati, Odia, Bengali, Kannada, Urdu, Punjabi, Telugu, Marathi, Assamese & Malayalam. In addition to the English medium, books in different Indian Languages are going to support the students to understand the concepts in their respective mother tongue.

On behalf of AICTE, I express sincere gratitude to all distinguished authors, reviewers and translators from the renowned institutions of high repute for their admirable contribution in a record span of time.

AICTE is confident that these outcomes based original contents shall help aspirants to master the subject with comprehension and greater ease.

T. G. Sitharam

(Prof. T. G. Sitharam)

ACKNOWLEDGEMENT

I am grateful to the authorities of AICTE, particularly Prof. T. G. Sitharam, Chairman; Dr. Abhay Jere, Vice-Chairman; Prof. Rajiv Kumar, Member-Secretary, Dr. Ramesh Unnikrishnan, Advisor-II and Dr. Sunil Luthra, Director, Training and Learning Bureau for their planning to publish the books on ***Basics of Electrical Machines: Theory and Practicals***. We sincerely acknowledge the valuable contributions of the reviewer of the book Prof. R. K. Srivastava, Professor, Department of Electrical Engineering, Indian Institute of Technology, Banaras Hindu University Varanasi for making it students' friendly and giving a better shape in an artistic manner.

This book is an outcome of various suggestions of AICTE members, experts and authors who shared their opinion and thought to further develop the engineering education in our country. Acknowledgements are due to the contributors and different workers in this field whose published books, review articles, papers, photographs, footnotes, references and other valuable information enriched us at the time of writing the book.

Prof. D. K. Palwalia,
Dr. U. K. Kalla
Dr. R. K. Kumawat

PREFACE

The book "Basics of Electrical Machines: Theory and Practical," is a comprehensive guide designed to introduce the fascinating world of electric machines. This book aims to provide a solid foundation in the fundamental principles and practical applications of electric machines, focusing on magnetic circuits, electro-mechanical energy conversion, DC motors and generators, and transformers.

Electric machines are at the heart of numerous technological advancements and are vital components in a wide range of industries. Electric machines play a pivotal role in various industries, driving technological advancements and enabling the efficient conversion and transmission of electrical energy. A firm grasp of the underlying principles is crucial for anyone seeking to navigate this dynamic field and contribute to its continued growth.

This book endeavour to present the complex concepts of electric machines in a clear and accessible manner. It adopts a systematic approach to gradually introduce the key topics and provide a strong conceptual framework for understanding the underlying principles.

The journey begins with an exploration of magnetic circuits, where the fundamental principles of magnetic fields and their behaviour in various materials. Through this understanding, one will gain insights into the properties of magnetic materials, magnetic flux, and the concept of reluctance.

Building upon this foundation, we investigate electro-mechanical energy conversion, which lies at the core of electric machines. By comprehending the interaction between electrical and mechanical systems, one will grasp the principles of energy conversion and the operation of electric machines.

The subsequent chapters focus on direct current (DC) motors and generators. The construction, working principles, and applications of these machines will enable us to understand their fundamental operation and practical considerations. Topics such as armature winding, commutation, torque production, and speed control are explored to equip you with a comprehensive understanding of DC machines.

The Transformers are essential devices for electrical energy transmission and distribution. You will learn about the principles of electromagnetic induction, transformer construction, transformer types, and the theory behind their operation. Practical aspects, including transformer performance, efficiency, and considerations in their application, are also explored.

Throughout this book, we have strived to strike a balance between theoretical concepts and practical applications. Numerous examples, illustrations, and real-world case studies are included to help bridge the gap between theory and practice. Additionally, problem-solving exercises and numerical examples are provided to reinforce understanding and allow to apply the concepts learned.

It is sincerely hoped that this book serves as a valuable resource for students, professionals, and anyone interested in developing a strong foundation in the field of electric machines. By combining theory with practical insights, it aims to empower to tackle the challenges and embrace the opportunities presented by electric machines.

The authors express gratitude to the countless researchers, engineers, and educators whose work has contributed to the knowledge and understanding in this field. Their dedication and innovation have paved the way for the remarkable advancements we witness today.

The authors invite you to embark on a captivating journey through the "Basics of Electric Machines: Theory and Practical." It is hoped that this book sparks your curiosity, expands your knowledge, and ignites your passion for electric machines.

Prof. D. K. Palwalia,
Dr. U. K. Kalla
Dr. R. K. Kumawat

OUTCOME BASED EDUCATION

For the implementation of an outcome-based education the first requirement is to develop an outcome-based curriculum and incorporate an outcome-based assessment in the education system. By going through outcome-based assessments evaluators will be able to evaluate whether the students have achieved the outlined standard, specific and measurable outcomes. With the proper incorporation of outcome-based education there will be a definite commitment to achieve a minimum standard for all learners without giving up at any level. At the end of the programme running with the aid of outcome-based education, a student will be able to arrive at the following outcomes:

PO1. Engineering knowledge: Apply the knowledge of mathematics, science, engineering fundamentals, and an engineering specialization to the solution of complex engineering problems.

PO2. Problem analysis: Identify, formulate, review research literature, and analyze complex engineering problems reaching substantiated conclusions using first principles of mathematics, natural sciences, and engineering sciences.

PO3. Design / development of solutions: Design solutions for complex engineering problems and design system components or processes that meet the specified needs with appropriate consideration for the public health and safety, and the cultural, societal, and environmental considerations.

PO4. Conduct investigations of complex problems: Use research-based knowledge and research methods including design of experiments, analysis and interpretation of data, and synthesis of the information to provide valid conclusions.

PO5. Modern tool usage: Create, select, and apply appropriate techniques, resources, and modern engineering and IT tools including prediction and modeling to complex engineering activities with an understanding of the limitations.

PO6. The engineer and society: Apply reasoning informed by the contextual knowledge to assess societal, health, safety, legal and cultural issues and the consequent responsibilities relevant to the professional engineering practice.

PO7. Environment and sustainability: Understand the impact of the professional engineering solutions in societal and environmental contexts, and demonstrate the knowledge of, and need for sustainable development.

PO8. Ethics: Apply ethical principles and commit to professional ethics and responsibilities and norms of the engineering practice.

PO9. Individual and team work: Function effectively as an individual, and as a member or leader in diverse teams, and in multidisciplinary settings.

PO10. Communication: Communicate effectively on complex engineering activities with the engineering community and with society at large, such as, being able to comprehend and write effective reports and design documentation, make effective presentations, and give and receive clear instructions.

PO11. Project management and finance: Demonstrate knowledge and understanding of the engineering and management principles and apply these to one's own work, as a member and leader in a team, to manage projects and in multidisciplinary environments.

PO12. Life-long learning: Recognize the need for, and have the preparation and ability to engage in independent and life-long learning in the broadest context of technological change.

COURSE OUTCOMES

After completion of the course the students will be able to:

CO-1: Understand the concepts of magnetic circuits.

CO-2: Understand the operation of DC machines.

CO-3: Analyse the differences in operation of different DC machine configurations.

CO-4: Analyse single phase and three phase transformers circuits.

Course Outcomes	Expected Mapping with Programme Outcomes *(1- Weak Correlation; 2- Medium correlation; 3- Strong Correlation)*											
	PO-1	**PO-2**	**PO-3**	**PO-4**	**PO-5**	**PO-6**	**PO-7**	**PO-8**	**PO-9**	**PO-10**	**PO-11**	**PO-12**
CO-1	3	2	2	1	1	-	-	-	-	-	-	-
CO-2	3	3	2	2	1	-	-	-	-	-	-	-
CO-3	3	3	2	1	1	-	-	-	-	-	-	-
CO-4	3	3	3	2	1	-	-	-	-	-	-	-

GUIDELINES FOR TEACHERS

To implement Outcome Based Education (OBE) knowledge level and skill set of the students should be enhanced. Teachers should take a major responsibility for the proper implementation of OBE. Some of the responsibilities (not limited to) for the teachers in OBE system may be as follows:

- Within reasonable constraint, they should manoeuvre time to the best advantage of all students.
- They should assess the students only upon certain defined criterion without considering any other potential ineligibility to discriminate them.
- They should try to grow the learning abilities of the students to a certain level before they leave the institute.
- They should try to ensure that all the students are equipped with the quality knowledge as well as competence after they finish their education.
- They should always encourage the students to develop their ultimate performance capabilities.
- They should facilitate and encourage group work and team work to consolidate newer approach.
- They should follow Blooms taxonomy in every part of the assessment.

Bloom's Taxonomy

Level	Teacher should Check	Student should be able to	Possible Mode of Assessment
Create	Students ability to create	Design or Create	Mini project
Evaluate	Students ability to justify	Argue or Defend	Assignment
Analyse	Students ability to distinguish	Differentiate or Distinguish	Project/Lab Methodology
Apply	Students ability to use information	Operate or Demonstrate	Technical Presentation/ Demonstration
Understand	Students ability to explain the ideas	Explain or Classify	Presentation/Seminar
Remember	Students ability to recall (or remember)	Define or Recall	Quiz

GUIDELINES FOR STUDENTS

Students should take equal responsibility for implementing the OBE. Some of the responsibilities (not limited to) for the students in OBE system are as follows:

- Students should be well aware of each UO before the start of a unit in each and every course.
- Students should be well aware of each CO before the start of the course.
- Students should be well aware of each PO before the start of the programme.
- Students should think critically and reasonably with proper reflection and action.
- Learning of the students should be connected and integrated with practical and real life consequences.
- Students should be well aware of their competency at every level of OBE.

ABBREVIATIONS AND SYMBOLS

List of Abbreviations

General Terms			
Abbreviations	**Full form**	**Abbreviations**	**Full form**
PM	Permanent Magnet	EHV	Electric Hybrid Vehicles
DC	Direct Current	MNA	magnetic neutral axis
AC	Alternating Current	GNA	geometric neutral axis
Emf	electromotive force	CRGO	Cold-Rolled Grain-Oriented
NdFeB	Neodymium magnet	LPF	Low Power Factor
ICs	Integrated Circuits	*hv*	High Voltage
PMBL	Permanent Magnet Brush Less	*lv*	Low Voltage
SRM	Switched Reluctance Motor	AN	Air Natural cooling
EM	Electromagnet	ON	Oil immersed Natural cooling
AT	Ampere-Turn	OFN	Oil immersed Forced oil circulation Natural cooling
Mmf	Magneto Motive Force	OFB	Oil immersed Forced oil circulation with air Blast cooling
TV	Television	OB	Oil immersed air Blast cooling
CRT	Cathode Ray Tube	AB	Air Blast cooling
MEMS	Micro Electromechanical System	OW	Oil immersed Water cooling
PV	Photovoltaic	OFW	Oil immersed Forced oil circulation with Water cooling
EV	Electric Vehicles	MOG	Magnetic Oil Gauge
OCC	Open circuit characteristic	RTCC	Remote Tap Changer Control Panel
BHP	Brake Horse Power	OTI	Oil Temperature Indicator
NVRC	No Volt Release Coil	WTI	Winding Temperature Indicator
OLRC	Over Voltage Release Coil		

List of Symbols

Symbols	Description	Symbols	Description
I	Electric current	W_{mo}	Mechanical energy output
ϕ	Flux	W_{es}	Energy stored in the magnetic field
Wb	Weber	W_{ms}	Energy stored in the mechanical system
ℓ	Mean length	dW_{elec}	Differential electrical energy input to coupling field
A	Cross-sectional area	dW_{mech}	Differential mechanical energy output
μ_r	Relative permeability	dW_{fld}	Differential change in energy stored in the coupling field
B	Flux density	F_e	Magnetic force
H	Magnetic field intensity	L_s	Self-inductance of the stator winding
N	Number of turns	L_r	Self-inductance of the rotor winding
$\Re$	Magnetic reluctance	M_{sr}	Mutual induction between stator & rotor windings
F	Mmf	N_C	Number of coils
A	Ampere	Y_A	Average pitch
T	Torque	Y_C	Commutator pitch
V	Volt	Y_R	Resultant pitch
m	Meter	Y_B	Back pitch
J	Joule	Y_F	Front pitch
s	Second	E_g	Armature emf
H	Henry	E_b	Back emf
$M.x$	Maxwell	P	Number of poles
μ_0	Permeability of free space	ϕ_a	Armature flux
P	Magnetic permeance	ω_m	Electromagnetic torque
L	Inductance	ϕ_R	Resultant flux
μ	Absolute permeability	Z_a	Number of armature conductor/pole
Ω	Ohms	Z_c	Number of compensating conductor/pole face
σ	Conductivity	I_f	Field current
δ	Current density	V_t	Terminal voltage
E	Electrical field intensity	η	*Efficiency*

Symbols	Description	Symbols	Description
v	Velocity	k	Coefficient of coupling
Hc	Coercive force	I_{μ}	Magnetizing current
EI	Energy or power	P_h	Hysteresis loss
ωT	Mechanical energy or power	P_e	Eddy current loss
W_{ei}	Electrical energy input from the supply means	P_i	Iron loss
v_b	Brush voltage drop	P_{Cu}	Copper loss
P_g	Power developed	η_m	Maximum efficiency
P_L	Power delivered to the load	η_{fl}	Full load efficiency
I_{se}	Series field winding current	$\cos\theta$	Power factor
I_{sh}	Shunt field winding current	P_{cfl}	Full-load copper losses
R_h	Rheostat	V_{Ph}	Phase voltage
T_{sh}	Shaft torque	V_L	Line voltage
T_n	Net torque	V_{LP}	Primary line voltage
T_f	Fractional torque	V_{LS}	Secondary line voltage
T_w	Windage torque	V_{phP}	Primary phase voltage
P_c	Constant loss	V_{phs}	Secondary phase voltage

LIST OF FIGURES

Unit 1 Magnetic Fields and Magnetic Circuits

Unit 2 Electromagnetic Force and Torque

Unit 3 DC Machines

Unit 4 DC Machine-Motoring and Generation

Unit 5 Transformers

CONTENTS

1 Magnetic Fields and Magnetic Circuits

UNIT SPECIFICS

Through this unit, the following aspects have been discussed:

- *Magnets, magnetic field, and magnetic induction;*
- *Laws of magnetism;*
- *Magnetic circuits and determination of flux, flux density, field intensity, and reluctance;*
- *Review of Ampere's Law and Biot-Savart's Law;*
- *Visualization of magnetic fields produced by a bar magnet and a current-carrying coil through the air and a combination of iron and air.*
- *The analogy between magnetic and electric circuits*

The unit contains a detailed description of the topics. It further gives solved and unsolved numerical problems to illustrate the theoretical study. A number of multiple-choice questions as well as questions of short- and long-answer types are also given. A list of references and suggested readings are given in the unit so that one can go through them for practice. Some QR codes have been provided in different sections which can be scanned for relevant supportive knowledge.

RATIONALE

This fundamental unit on Magnetic fields and Magnetic circuits will help students to get primary knowledge of magnets, magnetic fields, and magnetic induction. It gives an understanding of magnetic circuits and the determination of flux, flux density, field intensity, and reluctance. It explains the basic laws in describing magnetic circuits. Further, the visualization of magnetic fields produced by magnets or by the magnetic circuit is explained with graphic diagrams, solved examples, and numerical problems.

PRE-REQUISITES

Mathematics: Vectors, Integrals, Differentiation, Algebra (Class XII)
Physics: Electricity and Magnetism (Class XII)

UNIT OUTCOMES

The list of outcomes of this unit is as follows:

U1-O1: *Understand the concept of Magnets, Magnetic fields, and Magnetic induction.*

U1-O2: *Understand the Magnetic circuits and determination of flux, flux density, field intensity, and reluctance.*

U1-O3: *Explain laws in describing magnetic circuits.*

U1-O4: *Visualization of magnetic fields produced by magnets or by magnetic circuits.*

U1-O5: *Apply magnetic relation to solving problems.*

Unit-1 Outcomes	***EXPECTED MAPPING WITH COURSE OUTCOMES*** *(1-Weak Correlation; 2-Medium correlation; 3-Strong Correlation)*			
	CO-1	***CO-2***	***CO-3***	***CO-4***
U1-O1	*3*	*1*	*1*	-
U1-O2	*3*	*1*	*1*	*1*
U1-O3	*2*	-	-	-
U1-O4	*1*	*1*	*1*	*1*
U1-O5	-	*1*	*1*	*1*

1.1 Introduction

All of us are familiar with a natural magnet. It is an ore of iron that attracts small pieces of ferromagnetic material like - iron, cobalt, and nickel toward it. It is usually an oxide of iron named magnetite or Lodestone (Fe_3O_4), which is a natural magnet. In 1820, Han Christian Oersted observed that a compass needle gets deflected when placed near a wire carrying an electrical current. Moving charges or current in a wire or coil produces magnetic lines of forces called magnetic flux in the surrounding space causing deflection of a magnetic needle. These lines of force are termed magnetic fields. The property of such current is called the magnetic effect of electric current.

The magnetic field can be obtained from a permanent magnet or by using an electromagnet. An electromagnet consists of a coil (winding) carrying electrical current. A Permanent Magnet (PM) can be replaced by a coil carrying a relatively small electrical current for obtaining magnetic field. The electric current could be either Direct Current (DC) or Alternating Current (AC). The electromagnet's open magnetic circuit is made of ferromagnetic (iron or its alloy) material through which the magnetic flux influences the iron parts to attract. Once the iron part is in close proximity to an electromagnet a closed magnetic circuit is formed. Ideally, the magnetic flux passes through the magnetic circuit as that electrical current conducts through a conductor. As the copper conductor provides a low resistance path to electrical current, so does the magnetic circuit provide a path to magnetic flux with low magnetic reluctance, which is analogous to electric resistance.

To understand the relationship between electromotive force (emf), electric current, and the mechanical force experienced by the ferromagnetic object in a magnetic circuit, it is necessary to study the fundamental concept of magnet circuits. In this chapter, the laws of magnetism, magnetic field, magneto motive force, flux, reluctance, inductance, Ampere's Law, Biot Savart's Law, series, and parallel magnetic circuit are discussed.

Magnetic flux cannot be completely directed to follow a given path in a magnetic material; but if the permeability of the magnetic material is high as compared to material surrounding it (such as free space), then most of the flux flows through highly permeable material and almost negligible flux conducts through surrounding low permeability medium. This flow of magnetic flux is similar to the flow of electrical current through a conductor. Therefore, the path followed by the flux in a magnetic material is called magnetic

Scan to know more about magnetic flux

1.2 Magnet and It's Properties

A magnet is a material that can produce a magnetic field. A magnet can attract ferromagnetic materials; attract or repel any other magnet due to its magnetic fields. A bar-shaped Permanent Magnet (PM) suspended through a string, always points in the North-South direction; the magnetic poles of PM are accordingly marked. A magnet always comes with a pair of magnetic poles, which cannot be separated. These are referred to as the "North or N pole" and the "South or S pole". Like poles repel one another whereas opposite poles attract.

Rare earth magnets such as Neodymium magnet (NdFeB) are finding their use in making modern high-power PM motors and generators. Some elementary used magnets are Bar Magnet, Magnetic Needles, and Horseshoe Magnet. Most of the magnets used in research and technology are artificially made into different sizes and shapes depending upon requirements. Based on the source of magnetism, Magnets are of two types, permanent magnets, and electromagnets. fig. 1.1 shows two types of magnets.

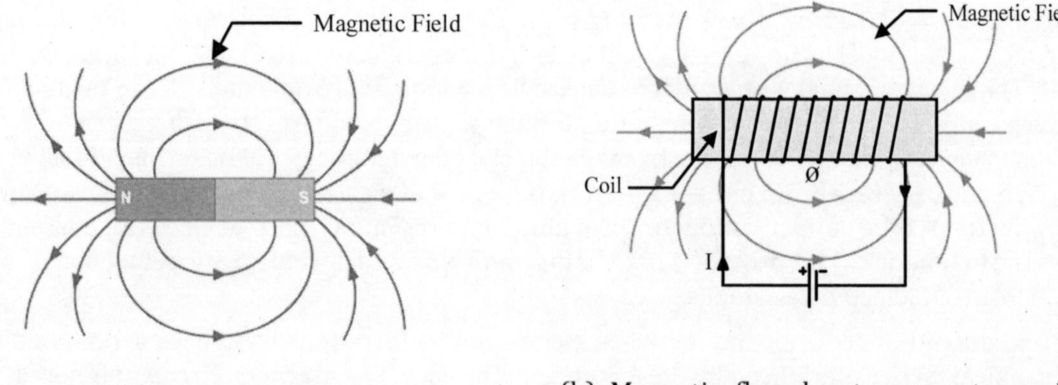

(a) Permanent magnet

(b) Magnetic flux due to current-carrying coil

Fig 1.1 Magnetic field due to permanent magnet and current-carrying solenoid

The permanent magnet has magnetism present and it does not require any electrical source. The electromagnet is formed by creating a magnetic field around a ferromagnetic core. Electromagnets are strong magnets, consisting of electrically insulated conducting wire in the form of a multi-turn coil embedded in an iron or ferromagnetic core. The core then acts as a magnet as long as the electric current is present. However, it loses its magnetic properties as soon as the current stops flowing. A ferromagnetic core may retain a feeble magnetism known as Residual Magnetism when the current is made zero. This property of ferromagnetic materials is used in the self-excitation of DC shunt generators, self-excited induction generators etc.

1.2.1 Properties of magnet

Magnets have distinctive and interesting properties as follows,

- *Attractive Property of Magnet:* A magnet attracts ferromagnetic materials like iron, nickel, and cobalt.

- *Directive Property of a Magnet:* If a magnet is suspended from rigid support such that it can rotate freely, the magnet always points in the North-South direction.
- *Poles of a Magnet:* Magnets have two poles, where the strength of the magnetic field is the strongest. Magnetic poles always exist in pairs. No matter how small a magnet is, it is impossible to separate one pole.
- Like poles always repel each other but opposite poles attract.
- The magnetic force (attraction or repulsion) between two objects is inversely squarely proportional to the distance between them. The force is stronger when the objects are close.
- There is no insulation for the magnetic field. The magnetic flux can also flow through the air.

1.2.2 Applications of magnet

Some of the applications of magnets are listed below:

- Speakers and microphones: Several speakers use a static magnet and a tiny coil that converts power (signal) into mechanical (noise-causing motion) resulting in audio output. Microphones also operate on a similar principle.
- Some electrical motors use a combination of electrical magnets and permanent magnets to strengthen the magnetic field resulting in higher output power. A generator converts mechanical energy into electrical energy by moving the conductor in a magnetic field.
- Magnets are widely used in daily life, science, and technology. Some applications are permanent magnets used in hard disk drives (HDD) and cooling fans of personal computers and laptops, cars, and motors.
- With the invention of high magnetic field Rare Earth Magnets like NdFeB Magnets, these are finding their applications in high-power PM motors and generators.

1.3 Magnetic Field

The magnetic field is defined as the region near a magnet within which the effect or influence of the magnetism is felt. The influence of a static magnet can be tested by using a simple compass needle, another magnet, and a Hall Effect sensor. The magnetic field is measured using the Hall Effect Gauss meter, though the alternating magnetic field is also measured using the search coil technique. Permanent Magnets mounted on the shaft and Hall Effect Sensor & Latch Semiconductor Integrated Circuits (ICs) are being used for instantaneous rotor position sensing in Permanent Magnet Brush Less (PMBL) motors and Switched Reluctance Motor (SRM) for their operation. PM with Hall effect sensor IC is also used in position-based feedback in control systems and robotics.

1.4 Magnetic Lines of Force

The magnetic lines of force used for representing the magnetic field of a magnet are shown in fig 1.1(a) and for two magnets of similar poles (N-N) and dissimilar poles (N-S) in fig. 1.2(a) and fig. 1.2(b) respectively. These lines are imaginary and do not have any physical existence. They were introduced by scientist Faraday to get the pictorial representation of the distribution of the magnetic field of a magnet. The magnet lines of force are also known as lines of magnetic field.

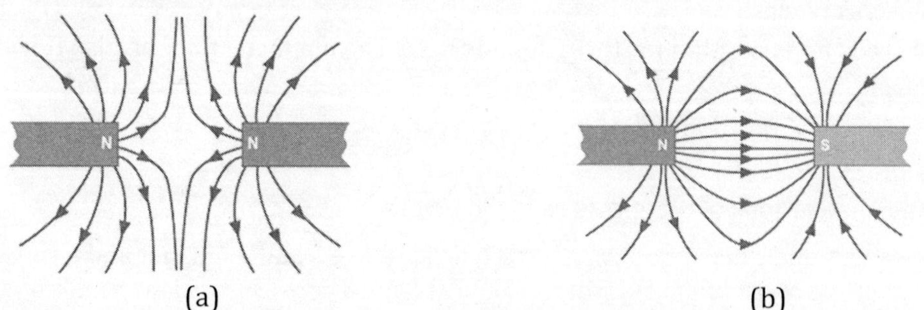

(a) (b)

Fig 1.2 Magnetic lines of forces due to (a) similar poles and (b) due to dissimilar poles

The magnetic force is caused by the magnet's magnetic field and points in the direction of the field lines. When two magnets are next to each other and their North poles are facing each other or their South poles are facing each other, then the field lines move away from each other, so repelling force appears between the two magnets as shown in fig. 1.2(a).

When the North pole of one magnet is next to the South pole of the other, then the field lines go straight from the North pole of the first magnet to the South pole of the second, and an attractive force appears between the two magnets as shown in fig. 1.2(b).

Permanent magnets with alternating poles when placed side by side are referred Halbach array.

1.5 Magnetic Induction

When a magnet is placed near a piece of magnetic material such as iron, then without any physical contact, magnetism gets transferred to the piece of iron. This is called magnetic induction. The presence of nearby ferromagnetic material or another current-carrying coil or another magnet drastically affects the line of forces.

1.6 Magnetic Circuit

Magnetic flux lines form a closed path. The closed path followed by magnetic lines of forces is called the magnetic circuit. In the magnetic circuit, magnetic flux or magnetic lines of force starts from a point and end at the same point after completing its path.

An electric circuit provides a path for the electric current (I) to flow, whereas a magnetic circuit provides a path for the magnetic flux (ø). The magnetic flux is generated by a magnet; it can be either a permanent magnet (PM) or an electromagnet (EM). A magnetic circuit is made up of magnetic materials having high permeability ferromagnetic material such as iron, soft steel, ferrite, etc. The study of the magnetic circuit is important in the design, analysis, and application of electromagnetic devices like electric motors, transformers, relays, generators etc. An elementary-level magnetic line of forces can be drawn on paper using a simple bar magnet and compass. The magnetic field can be also computed using software such as ANSYS, MAXWELL, JMAG etc. In these software numerical solutions of Maxwell's equations in 2D/3D are obtained using the Finite Difference Method, Finite Element Method, or Boundary Element Method.

Consider a solenoid having N number of turns wound on an iron core. The magnetic flux 'ø' Weber (Wb) sets up in the core when the current of I ampere is passed through the solenoid as shown in fig.1.3. This magnetic flux follows a closed path ABCDA and hence ABCDA is the magnetic circuit.

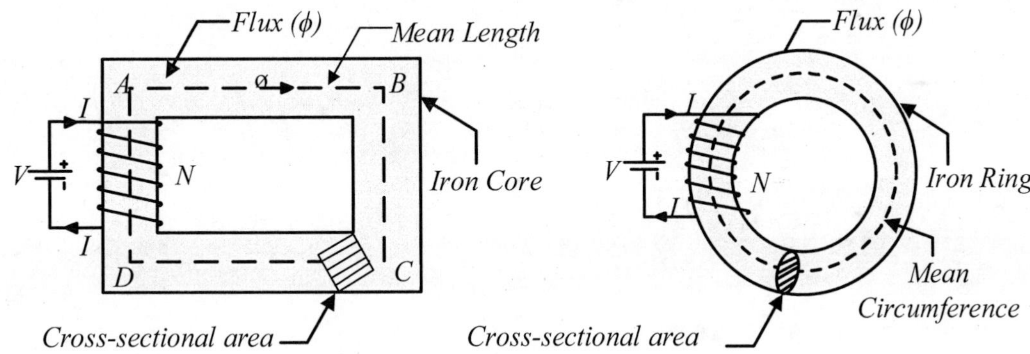

Fig.1.3 Simple magnetic circuit

Let, ℓ = Mean length of the magnetic circuit or path flux in meter
A = Cross-sectional area of the core in square meter
μ_r = Relative permeability of the core

Now the flux density in the core material

$$\boldsymbol{B} = \frac{\phi}{A}\ Wb/m^2 \tag{1.4}$$

Magnetizing force in the core, which induces the line of forces through the magnet is referred to as magnetic field intensity,

$$\boldsymbol{H} = \frac{\mathbf{B}}{\mu_0\mu_r}\ AT/m \tag{1.5}$$

$$\boldsymbol{H} = \frac{\phi}{A\,\mu_0\mu_r} AT/m \tag{1.6}$$

According to work law, the work done in moving a unit pole once round the magnetic circuit is equal to the ampere-turns (*AT*) enclosed by the magnetic circuit.

$$\boldsymbol{H}\ell = NI \tag{1.7}$$

$$\boldsymbol{H}\ell = \frac{\phi}{A\,\mu_0\mu_r} \times \ell = \text{NI} \tag{1.8}$$

$$\phi = \frac{NIA\,\mu_0\mu_r}{\ell} \tag{1.9}$$

Thus, a magnetic flux is directly proportional to the number of turns (*N*) and current (*I*). It shows that the flux increases if the number of turns or current increases and decreases when either of the two quantities decreases. *NI* is the magneto motive force (mmf). The magnetic flux is inversely proportional to magnetic reluctance,

$$\mathcal{R} = \frac{\ell}{A\,\mu_0\mu_r} \tag{1.10}$$

The lower the reluctance, the higher the flux, and vice-versa.

Types of magnetic circuit:

There are two types of magnetic circuits –

- Series magnetic circuit
- Parallel magnetic circuit

1.6.1 Series magnetic circuit

When the same magnetic flux (ϕ) flows through each part of the magnetic circuit, then the circuit is called a series magnetic circuit.

Consider a composite series magnetic circuit consisting of two different magnetic materials of different relative permeability as shown in fig. 1.4. Each part of this series magnetic circuit will offer reluctance to the magnetic flux (ϕ). Since the different parts of the magnetic circuit are in series, the total reluctance is equal to the sum of the reluctances of individual parts.

$$Total\ reluctance = \mathcal{R} = \frac{\ell_1}{A_1\mu_0\mu_1} + \frac{\ell_2}{A_2\mu_0\mu_2} \tag{1.11}$$

Total mmf= Magnetic flux × Total Reluctance

$$Total\ mmf = \phi\left(\frac{\ell_1}{A_1\mu_0\mu_1} + \frac{\ell_2}{A_2\mu_0\mu_2}\right) \tag{1.12}$$

$$Total\ mmf = \left(\frac{\boldsymbol{B_1}}{\mu_0\mu_1}\right)\ell_1 + \left(\frac{\boldsymbol{B_2}}{\mu_0\mu_2}\right)\ell_2 \tag{1.13}$$

$$Total\ mmf, \mathcal{F} = \boldsymbol{H}_1\ell_1 + \boldsymbol{H}_2\ell_2 \tag{1.14}$$

Therefore, the total mmf ($\mathcal{F}$) required to set up the magnetic flux in a series magnetic circuit is the sum of mmf required by the individual parts of the magnetic circuit.

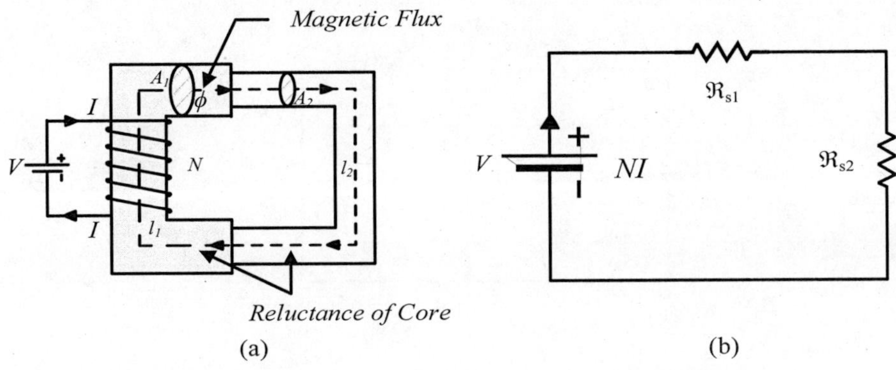

Fig.1.4 (a) Series magnetic circuit (b) Equivalent magnetic circuit

A series magnetic circuit that has parts of different dimensions and materials is called a composite series magnetic circuit

1.6.2 Parallel magnetic circuit

A magnetic circuit that has more than one path for the magnetic flux is called a parallel magnetic circuit as shown in fig. 1.5.

Consider a coil of N turns wound on limb ABCDEF carries an electric current of I amperes.

The magnetic flux ϕ_1 set up by the coil divides at B into two paths viz. –

The magnetic ϕ_2 passes along the path BE.

The magnetic flux ϕ_3 passes along the path BCDE.

Therefore, the total flux is,

$$\phi_1 = \phi_2 + \phi_3 \tag{1.15}$$

The path BE and BCDE is in parallel and hence form a parallel magnetic circuit. In a parallel magnetic circuit, the mmf required for the whole parallel magnetic circuit is equal to the mmf required for any one of the parallel paths.

Let, $\mathcal{R}_1$=Reluctance of magnetic path BAFE

$\mathcal{R}_2$=Reluctance of magnetic path BE

$\mathcal{R}_3$=Reluctance of magnetic path BCDE

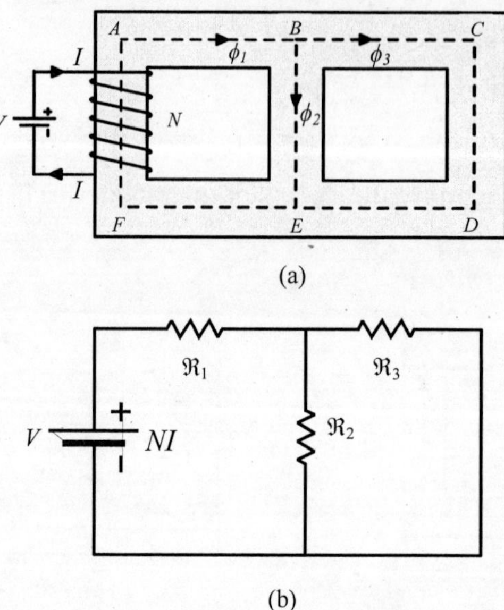

Fig.1.5 (a) Parallel magnetic circuit (b) Equivalent circuit

$$Total\ mmf = mmf\ of\ path\ \text{BAFE} + mmf\ of\ path\ BE + mmf\ of\ path\ BCDE$$

$$Total\ mmf = \phi_1\mathcal{R}_1 + \phi_2\mathcal{R}_2 + \phi_3\mathcal{R}_3 \tag{1.16}$$

1.7 Magnetomotive Force (mmf)

A current-carrying conductor produces a magnetic field around it. Magneto-motive force ($\mathcal{F}$) is defined as the product of current and the number of turns of the coil.

Let I= Current through the coil (A)

N= Number of turns in the coil

$$Magneto\ motive\ force = Current \times Turns$$

$$\mathcal{F} = I \times N \tag{1.17}$$

The Unit of mmf is ampere-turn (AT). Since N is dimensionless, its unit is written as ampere (A)

1.8 Magnetic Field Intensity (H)

The magnetic force acting on a unit North pole (1 Wb) when placed at a point in the magnetic field is called the magnetic intensity of the field or magnetic field intensity at that point. It is denoted by **H**.

In magnetic circuits, the ratio of magneto motive force per unit length of magnetic flux path is called magnetic field intensity. It is denoted by **H**, mathematically,

$$H = \frac{mmf}{length\ of\ magnetic\ path} \tag{1.18}$$

$$H = \frac{\mathcal{F}}{\ell} = \frac{\text{NI}}{\ell} \quad AT/m \tag{1.19}$$

$$\mathcal{F} = \text{H}\ell \tag{1.20}$$

1.9 Magnetic Flux (ϕ) and Magnetic Flux Density (B)

The region around a magnet where its poles exhibit a force of attraction or repulsion is called the magnetic field. The number of magnetic lines of force set up in a magnetic circuit is called magnetic flux. Its unit is Weber (Wb). It is analogous to electric current I in an electric circuit.

The equation introduced by Ohm's law for electric circuits is

$$Effect = \frac{Cause}{Opposition} \tag{1.21}$$

Similarly, equations can be applied to magnetic circuits. For a magnetic circuit, the effect desired is the flux ϕ. The cause is the magnetomotive force (mmf), which is an external force required to set up the magnetic flux lines within the magnetic material. The opposition to the setting up of the flux ϕ is the reluctance ($\mathcal{R}$), substituting we have

$$\phi = \frac{m.m.f.}{\mathcal{R}} \quad Wb \tag{1.22}$$

$$mmf \propto \phi \tag{1.23}$$

Equation 1.23 is known as Ohm's law of magnetic circuits.

> Ohm's law for magnetic circuit's states that the mmf is directly proportional to the magnetic flux hence as the magnetic flux decreases, the mmf also decreases.

1.9.1 Magnetic flux density

The magnetic flux density at a point is the flux per unit area at right angles to the flux at that point. It is, generally, represented by the letter '**B**'. it is measured in Testa (T) or in Wb/m^2.

$$\mathbf{B} = \frac{\phi}{A} \quad Wb/m^2 \tag{1.24}$$

(1 Wb/m^2 = 1 x $10^4 Wb/cm^2$=10^4 Gauss)

1.0 weber $\approx 10^8$ lines of force

Weber is commonly expressed in a multitude of other units.

$$Wb = \frac{kg.m^2}{s^2.A} = \text{V.s} = \text{H.A} = \text{T}.m^2 = \frac{\text{J}}{A} = 10^8 M.x \tag{1.25}$$

Where, Wb = Weber, T = Tesla, V = volt,

m = meter, J = joule, s = second,

H = Henry, A = ampere, $M.x$ = Maxwell

Both magnetic field intensity and magnetic flux density are vector quantities that flow in 3D. In most cases, only 2D models are used.

1.10 Magnetic Reluctance ($\mathcal{R}$)

Magnetic reluctance is defined as the opposition offered by a magnetic circuit to the production of magnetic flux. It is the property of the material that opposes the creation of magnetic flux in a magnetic material. It is also known as reluctance, magnetic resistance, or a magnetic insulator. fig.1.6 and fig. 1.7 represent the reluctance of a magnetic circuit and a rectangular bar through which flux flows.

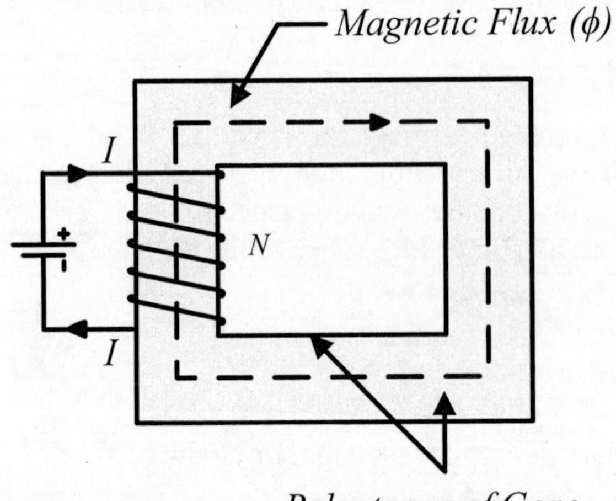

Fig. 1.6 Reluctance of magnetic core

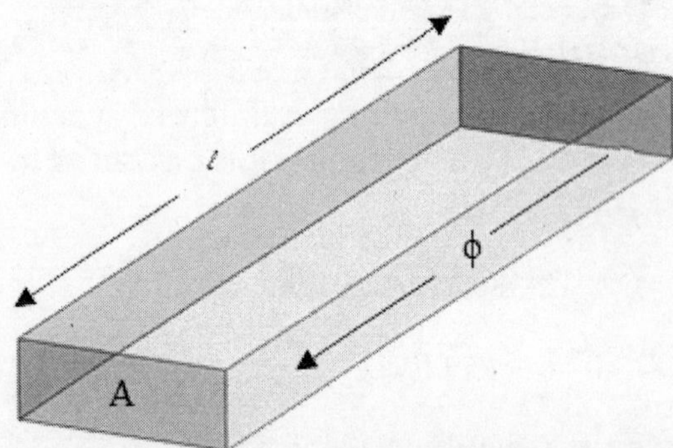

Fig.1.7 Reluctance of magnetic bar

In an electrical circuit, the resistance opposes the flow of current in the circuit and it dissipates the electric energy. The magnetic reluctance in a magnetic circuit is analogous to the resistance in an electric circuit as it opposes the production of magnetic flux in a magnetic circuit but it does not give rise to the dissipation of energy rather it stores magnetic energy. Further,

- Reluctance is directly proportional to the length of the magnetic circuit.
- It is inversely proportional to the area of the cross-section of the magnetic path.
- It is a scalar quantity and is denoted by $\mathcal{R}$.

Mathematically it can be expressed as

$$\mathcal{R} = \frac{\ell}{A\,\mu_0\mu_r} \quad (1.26)$$

The unit of reluctance is ampere-turns per Weber, AT/Wb or $1/\text{Henry}$ or H^{-1}

where,

ℓ = length of the magnetic path in meters

μ_0= permeability of free space (vacuum) = $4\pi \times 10^{-7}$ Henry/meter

μ_r= relative permeability of a magnetic material

A= Cross-sectional area in square meters (m^2)

1.10.1 Reluctance in a series magnetic circuit

In a series of magnetic circuits, the total reluctance equals the sum of the individual reluctances encountered around the closed flux path.

$$\mathcal{R} = \mathcal{R}_1 + \mathcal{R}_2 + \mathcal{R}_3 \ldots\ldots\ldots\ldots\ldots \mathcal{R}_n \quad (1.27)$$

$$\text{Where, } \mathcal{R} = \frac{\ell}{A\,\mu_0\mu_r}$$

Some of the applications of reluctance include:

- In a transformer, reluctance is mainly used to reduce the effect of magnetic saturation. Transformer cores are made such that the least air gap exists along the magnetic path. This ensures the lowest magneto motive force to set up a desired flux in the core. Any air gap in the magnetic circuit increases the reluctance of the magnetic circuit and hence decreases the flux.
- Synchronous reluctance motors, switched reluctance motors (SRM) and stepper motors work on the principle of variable reluctance. They are used for many constant speed applications such as constant speed drive, electric clock timers, signaling devices, recording instruments and drives etc.
- One of the main characteristics of magnetically hard materials is that it has a strong magnetic reluctance which is used to create permanent magnets. Examples are Tungsten steel, Cobalt steel, Chromium steel, Alnico etc.

- The speaker magnet is a soft magnetic material such as soft iron to minimize the effect of the stray magnetic field.
- Alternating magnetic fields emanating from simple devices like ON/OFF switch, Electric Arc welding, Television (TV), Cathode Ray Tube (CRT) etc. interferes with other devices such as Multimedia loudspeakers causing the audio output to distort. Such devices are magnetically shielded to reduce magnetic interference.

1.11 Magnetic Permeance ($\mathcal{P}$)

The permeability or magnetic permeability is defined as the ability of a material to allow magnetic lines of force to pass through it. It helps the development of the magnetic field in a magnetic circuit. Permeance is similar to Conductance in an electric circuit.

Magnetic permeance ($\mathcal{P}$) is the inverse of magnetic reluctance and is given as

$$\mathcal{P} = \frac{1}{\mathcal{R}} = \frac{A\,\mu_0\mu_r}{\ell} = \frac{A\,\mu}{\ell} \tag{1.28}$$

The unit of permeance is Henry.

1.12 Inductance (L)

Electrical energy is stored in inductance and capacitance. Energy storage is essentially required for the conversion of one form of energy to another form of energy. Electrical machines require the storage of electrical energy in coupling electromagnetic field where the conversion of one form of energy to another form takes place.

In the case of an electrical motor, electrical energy is converted into mechanical energy, and in an electrical generator mechanical energy is converted into electrical energy. In an electrical brake, a high-speed rotating shaft can be halted in lesser time by converting electrical energy to oppose rotation. In electrical energy conversion, it is the inductive parameter that is important because of its energy storage and hence conversion capabilities. It can also be linked to both electric and magnetic circuits. fig. 1.8 represents a simple inductive coil circuit.

The property of a coil due to which it opposes the change of current flowing through itself is called the inductance of the coil. The inductance of the coil is defined as the ratio of magnetic flux to current and describes how much magnetic energy can be stored in an inductor. The inductance (L)* of a conductor carrying the current I can be expressed as

---

* *The symbol L for inductance was chosen to honour Heinrich Lenz (1804–1865), whose pioneering work in electromagnetic induction was instrumental in the development of the field theory. The unit of inductance is the henry, named after Joseph Henry (1797–1878), the American scientist who discovered electromagnetic induction independently of and at about the same time as Michael Faraday (1791–1867) did in England.*

$$L = \frac{\phi}{I} \ Henry \tag{1.29}$$

Where the magnetic flux Φ is given by

$$\phi = \mu_0 \mu_r \int \boldsymbol{H}.\boldsymbol{ds} \tag{1.30}$$

$$\phi = \frac{NI}{\ell/_{A} \ \mu_0 \mu_r} \tag{1.31}$$

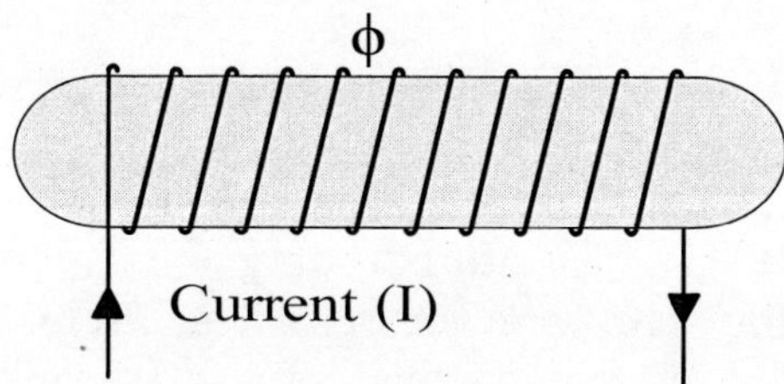

Fig.1.8 Inductive coil

An approximation of inductance for any coil of N number of turns can be calculated,

$$\text{L} = \frac{N^2 \mu \, A}{\ell} \tag{1.32}$$

$$\mu = \mu_0 \mu_r \tag{1.33}$$

It is seen that the inductance is proportional to the magnetic flux, which is proportional to the relative permeability μ_r of a magnetic core.

Where, L= Inductance of the coil in Henrys

N= Number of turns in coil (Straight wire=1)

μ= Absolute permeability of core material

μ_r= Relative permeability

μ_0= Permeability of free space, 1.26×10^{-6} Henry/m or Newton/A^2, (μ_r=1 for free space)

A = Area of coil in square meter

ℓ= Average length of coil in meters.

The inductance attained by a coil due to self-induced emf produced in the coil itself by the changing current flowing through it. If the current in the coil is increasing with respect to time, the self-induced emf is produced in the coil in a direction so as to oppose the rise of current i.e., the direction of self-induced emf is opposite to that of the applied voltage.

On the other hand, if the current in the coil is decreasing with respect to time, the self-induced emf is produced in the coil in such a direction to oppose the fall of current i.e., the direction of self-induced emf is in the same direction as that of the applied voltage. Self-inductance does not prevent the change of current but it delays the change of current flowing through a coil. The direction of induced emf thus produced follows Lenz's law. The coil only opposes the alternating current and time-varying current. It does not affect the steady-state DC current. In other words, the self-inductance of the coil will exhibit its presence to the alternating current and time-varying current, but it will not exhibit its presence to the steady state direct current.

> Although magnetic energy can be stored in a capacitive circuit also, but for increasing capacitive stored energy an extremely high voltage is required. Such high voltages are not deemed acceptable for safety reasons.

The devices based on the principle of electrostatics are electrostatic voltmeters and micro-electromechanical systems (MEMS). The stored energy in an inductive circuit can be increased by changing the number of turns and current. The heat loss thus generated by high current in the coil can be easily dissipated. Due to these reasons, electrical motors and generators are based on the principle of electromagnetic.

A solenoid without any iron core is regarded as an air-core coil, its inductance increases many folds when the same coil is wound on an iron core. The inductance of the magnetic core inductor can be expressed by the sum of the air-core inductance L_{AC} and an additional contribution ΔL due to the relative permeability of the core,

$$L_{MC} = L_{AC} + \Delta L \tag{1.34}$$

Where, L_{MC} = inductance of coil wound on a Magnetic core, L_{AC} = Air-core Inductance, ΔL= Additional contribution due to the presence of iron core. The inductance has been summarized in Table.1.1.

Table: 1.1 Inductance from different point of view

Circuit	Energy	Geometrical
$v_L = L\frac{di}{dt} = N\frac{d\phi}{dt}$ $L = \frac{v_L}{di/dt}$ $\int_{i(0)}^{i(t)} di = \frac{1}{L}\int_0^t v_L.dt$ $i(t) = \frac{1}{L}\int_0^t v_L.dt + i(0)$	$W = \int_0^t v_L i dt = \frac{1}{2}Li^2$ $L = \frac{2W}{i^2}$	$L = N\frac{d\phi}{di} = N\frac{\phi}{i}$ $L = \frac{N^2\mu A}{\ell} = \frac{N^2}{\mathcal{R}}$ Where $\mathcal{R} = \frac{\ell}{\mu A}$

1.13 Ampere's Law

According to Ampere's law, magnetic fields are related to the electric current. The law specifies the magnetic field that is associated with a given current or vice-versa, provided that the electric field doesn't change with time.

The magnetic field created by an electric current is proportional to the amplitude of the electric current with a constant of proportionality equal to the permeability of free space.

$$\oint \boldsymbol{B}.d\boldsymbol{S} = \mu_o I \tag{1.35}$$

Ampere's circuital law can be written as the line integral of the magnetic field surrounding the closed loop equals the number of times the algebraic sum of currents passing through the loop. The line integral of magnetic field intensity **H** about any closed path is exactly equal to the direct current enclosed by the path,

$$\oint \boldsymbol{H}.d\boldsymbol{L} = I_{enc} = \text{Current enclosed} \tag{1.36}$$

1.13.1 Applications of Ampere's Law: Applications of **Ampere's Law** are:

- Determine the magnetic induction due to a long current-carrying wire.
- Determine the magnetic field inside a toroid.
- Determine the magnetic field created by a long current-carrying conducting cylinder.
- Determine the magnetic field inside the conductor.
- Determining Boundary conditions between two surfaces carrying current.

1.14 Biot-Savart's Law

Definition: In magnetism, **Biot-Savart's law** defines the magnetic field produced at a point in space at some distance $'\boldsymbol{r}'$ from a current-carrying conductor and this magnetic field is produced due to current flowing in the conductor.

Statement: Biot Savart's law states that "magnetic field due to a current carrying conductor at a distance point $\boldsymbol{r}$ is inversely proportional to the square of the distance between the conductor and point, and the magnetic field is directly proportional to the length of the conductor, current flowing in the conductor". This is known as **Biot-Savart's law**. fig.1.9 shows the magnetic field due to a current-carrying conductor.

Mathematically, if $d\boldsymbol{\ell}$ represents the small section of a long current-carrying conductor having a current of I and $\boldsymbol{r}$ is the distance between the conductor and point let's say P and θ be the angle between $d\boldsymbol{\ell}$ and $\boldsymbol{r}$ so, magnetic field $d\boldsymbol{B}$ at point P due to Biot Savart's law as shown in the diagram is given as

$$d\boldsymbol{B} \propto \frac{I\, d\boldsymbol{\ell} \sin\theta}{r^2} \tag{1.37}$$

$$dB = K\frac{I\,d\ell \sin\theta}{r^2} \tag{1.38}$$

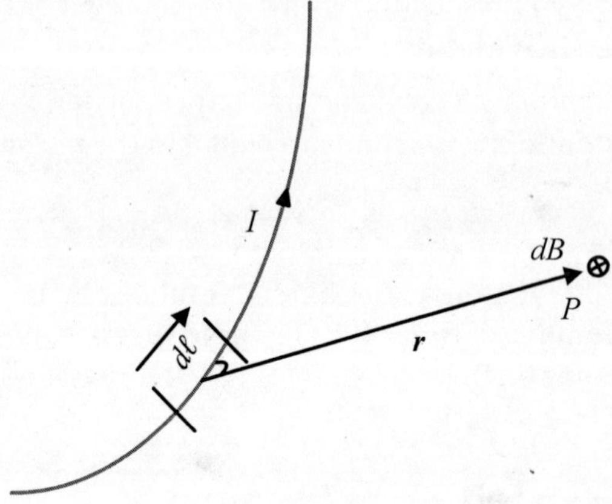

Fig. 1.9 Magnetic field due to current carrying conductor

Where K is the proportionality constant which depends upon the magnetic properties of the medium and the system of units employed. In SI system of unit.

$$K = \frac{\mu_0 \mu_r}{4\pi}$$

$$dB = \frac{\mu_0 \mu_r\, I\, d\ell \sin\theta}{4\pi r^2} \tag{1.39}$$

where μ_0 is known as the permeability of free space. This is the mathematical formula of the Biot Savart law.

1.14.1 Applications of Biot-Savart's Law

Biot -Savart Law is used in the calculation of

- Magnetic fields in space due to any current-carrying conductor.
- The force between two long and parallel current-carrying conductors.
- The Magnetic field on the axis of a circular current loop.
- The magnetic responses are even at the atomic or molecular level.

The importance of Biot-Savart's law is that

- Biot-Savart law is analogous to Coulomb's law in electrostatics.
- The law is applicable for very small current-carrying conductors too.
- The law is applicable also for symmetrical current distribution.

1.15 Effect of Magnetic Field

When flux produced by both the coils is additive, the total mmf for the magnetic circuit given in fig 1.10 is

$$\mathcal{F}_{mm} = NI = \phi\mathcal{R} \tag{1.40}$$

The reluctance of the magnetic circuit, for the length of the flux path ℓ_{iron} and cross-sectional area A_c, is

$$\mathcal{R} = \frac{\ell_{iron}}{\mu_o\mu_r A_c} \tag{1.41}$$

Neglecting the reluctance of iron, the core flux $\emptyset$ is

$$\phi = N_1I_1 + N_2I_2\frac{\mu_0\mu_r A_c}{\ell_{iron}} \tag{1.42}$$

In eqn. (1.42), the core flux produced is the result of the simultaneous action of both mmf. It is this result which determines the operating point of the core material. If equation (1.42) is given in terms of individual currents, the resultant flux linkage λ_1 of coil 1 is given as

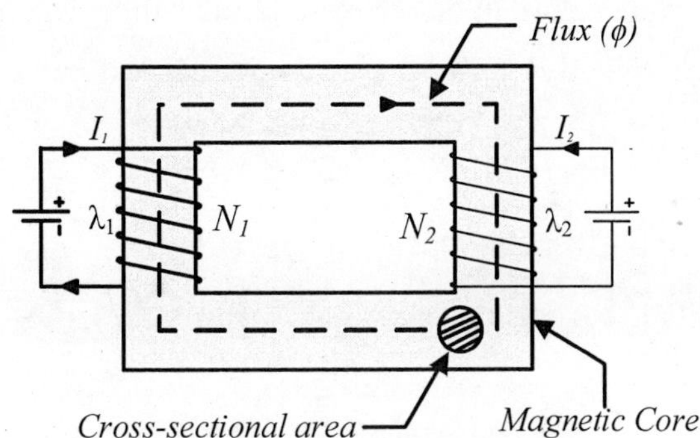

Fig. 1.10 Effect of magnetic field

$$\lambda_1 = N_1^2\frac{\mu_0\mu_r A_c}{\ell_{iron}}I_1 + N_1N_2\frac{\mu_0\mu_r A_c}{\ell_{iron}}I_2 \tag{1.43}$$

i.e., $\lambda_1 = L_{11}I_1 + L_{12}I_2$

Where self-inductance of coil 1 is

$$L_{11} = N_1^2\frac{\mu_0\mu_r A_c}{\ell_{iron}} \tag{1.44}$$

And mutual inductance of coil 1 with coil 2 is

$$L_{12} = N_1N_2\frac{\mu_0\mu_r A_c}{\ell_{iron}} \tag{1.45}$$

Similarly, the flux linkage λ_2 of coil 2, is given as

$$\lambda_2 = N_1 N_2 \frac{\mu_0 \mu_r A_c}{\ell_{iron}} I_1 + N_2^2 \frac{\mu_0 \mu_r A_c}{\ell_{iron}} I_2 \tag{1.46}$$

$$\lambda_2 = L_{21} I_1 + L_{22} I_2$$

$$\text{If } L_{12} = L_{21}, \qquad \text{then, } L_{12} = L_{21} = N_1 N_2 \frac{\mu_0 \mu_r A_c}{\ell_{iron}} \tag{1.47}$$

and mutual inductance

$$L_{22} = N_2^2 \frac{\mu_0 \mu_r A_c}{\ell_{iron}} \tag{1.48}$$

For a static magnetic circuit inductance L is constant. For static steady-state DC current, the induced emf in the coil is zero.

However, when the current changes due to either reason or for alternating current, the voltage drop across the coil is given as

$$e = L\frac{di}{dt} \tag{1.49}$$

For time-varying inductance, like in electromechanical energy conversion devices is given as differential of $\lambda = Li$

$$e = \frac{d}{dt}(Li) = L\frac{di}{dt} + i\frac{dL}{dt} \tag{1.50}$$

Power stored in the coil is given as

$$p = i\frac{d\lambda}{dt} \text{ watts or joules per second} \tag{1.51}$$

Thus, the change in magnetic stored energy dW in the magnetic circuit in a time interval t_1 to t_2 is given as

$$dW = \int_{t_1}^{t_2} p.dt = \int_{\lambda_1}^{\lambda_2} i.d\lambda \tag{1.52}$$

For a single winding system of constant inductance, the change in magnetic stored energy can be given as

$$dW = \int_{\lambda_1}^{\lambda_2} i\, d\lambda = \frac{i}{2L}(\lambda_2^2 - \lambda_1^2) \tag{1.53}$$

The total magnetic energy stored (W) at a given value of λ can be given for $\lambda_1 = 0$

$$\text{i.e.,} \qquad W = \frac{i}{2L}\lambda^2 = \frac{1}{2}Li^2 \ (1.54)$$

1.16 Magnetic Field Produced from Magnet

A bar magnet is having permanent magnetic properties. It is in the form of a bar of rectangular or circular cross-section. It is made up of iron, steel, or any other ferromagnetic substance or ferromagnetic composite. The magnetic force of it is the strongest at the pole. One of the oldest uses of the permanent magnet is in a Mariner's Compass, which was used to determine the direction of a moving ship.

1.16.1 MAGNETIC FIELD LINES

Magnetic field lines can be defined as imaginary lines that can be drawn on paper along the magnetic field acting around any magnetic substance using an elementary technique of compass. These lines of force can be numerically computed using electromagnetic simulation software based on FEM. The magnetic field lines possess certain properties. If two bar magnets are placed nearby then magnetic field lines are shown as in fig. 1.11.

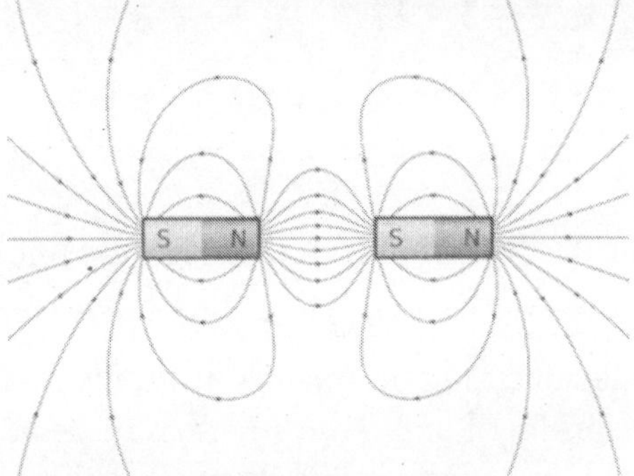

Fig. 1.11 Magnetic field lines

Let us understand the concept of magnetic field lines using the activity described below. Let us sprinkle iron filings on a sheet of paper and a bar magnet in between. When we tap the paper, we notice that the filings get aligned in the form of many lines. The patterns of the filings show us the magnetic field lines that surround this bar magnet as shown in fig. 1.12.

Scan to know more on magnetic circuit and its parameters.

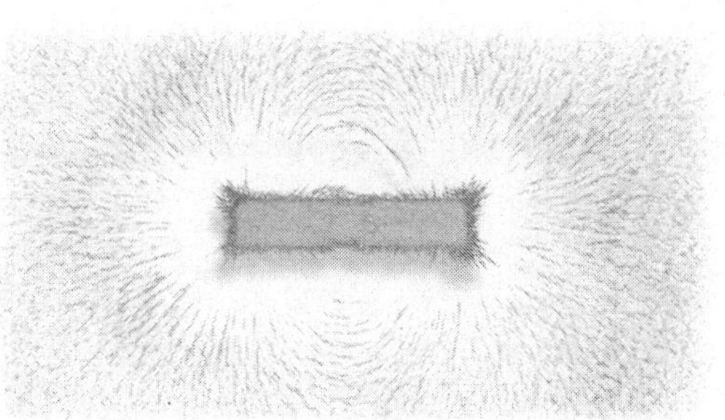

Fig.1.12 Iron filings near the bar magnet align themselves along the field lines.

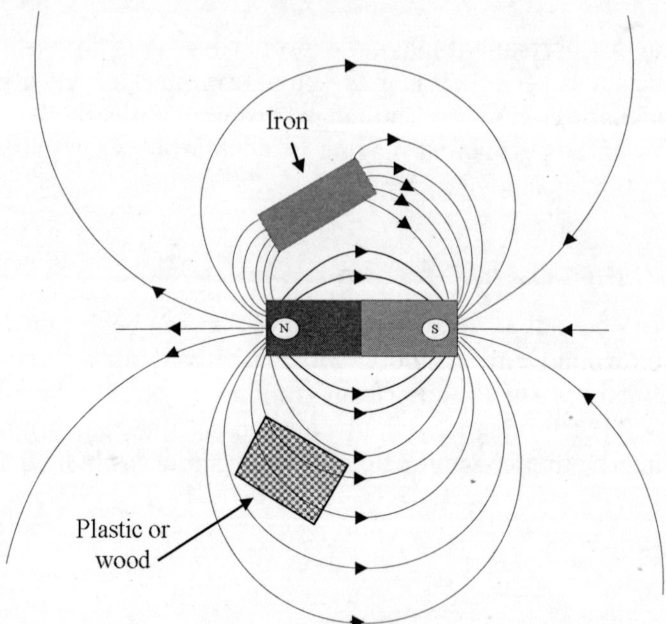

Fig.1.13 Effect of Iron on magnetic lines

We have observed the followings properties of magnetic lines of force

- The magnetic field lines of a magnet form continuous closed loops in 3D.
- They originate from the North or N Pole and terminate at the South or S Pole.
- The tangent to the field line at any point represents the direction of the net magnetic field **H** at that point.
- They are crowded where the magnetic field is strong, i.e. near the poles.
- The larger the number of field lines crossing per unit area, the stronger the magnitude of the magnetic field **H**.
- The magnetic field lines do not intersect.

The magnetic field lines bent away from their original positions to pass through the iron piece field. It happens by inserting a piece of iron inside a magnetic field as shown in fig. 1.13. The field will take the opportunity to pass through a lower reluctance material, the field is much larger in the iron.

When we place a ferromagnetic object into a static magnetic field, the magnetic field will penetrate the object's material, align its magnetic domains and that object itself will become magnetic with its own magnetic field. The alignment of the domains of the object will be identical to the magnet's and the two magnetic fields are going to amplify each other. Especially if they physically contact each other, they will act as one magnet with a common magnetic field.

1.17 Magnetic Field Due to Current Carrying Conductor

In 1820, it was discovered that an electrical current is always accomplished by a certain magnetic effect. Danish physicist, Hans Christian Oersted found that when current is passed through a conductor placed above the magnetic needle, the needle gets turned in a certain direction as shown in fig. 1.14. He also found that when the direction of the flow of current is reversed the magnetic needle is also deflected in the opposite direction. Through these observations, Oersted showed that electricity and magnetism were related phenomena.

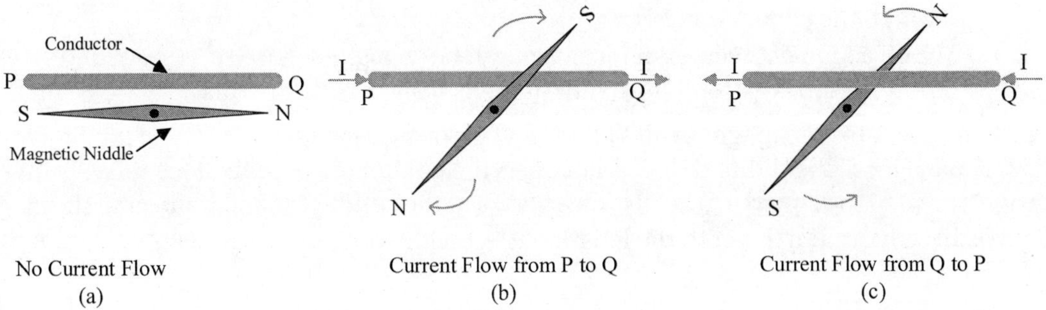

Figure. 1.14 Magnetic field effect

The field around the current-carrying conductor consists of lines of force, which encircles the conductor. It can be proved experimentally by passing a current-carrying conductor PQ in the board and plotting the field with the help of a magnetic needle on it as shown in fig. 1.15(a) and 1.15(b). It can be observed that when current is passed through the conductor from P to Q, the direction of force is in the clockwise direction and when current is passed from Q to P, the direction of the line of force is counterclockwise direction.

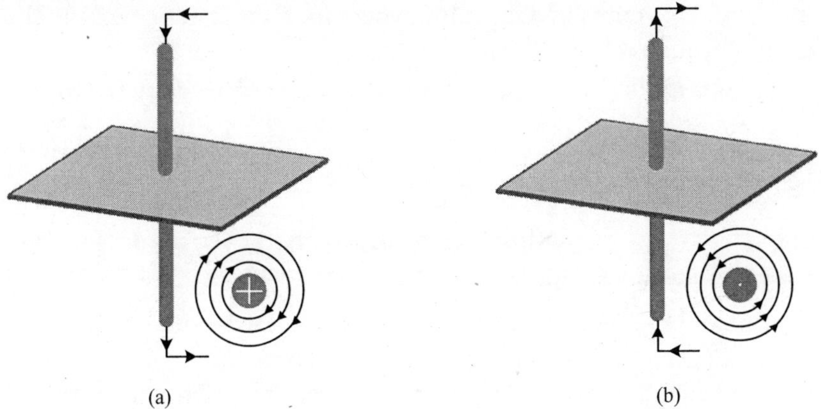

Fig.1.15 Properties of the lines of magnetic induction around a current-carrying conductor

From the above, it concludes that the properties of the lines of magnetic induction around a current-carrying conductor are

- Lines of magnetic induction are circles, symmetrical, and concentric with the axis of the conductor.
- The spacing between the lines of induction decreased as we move closer to the conductor.
- The direction of lines of magnetic induction depends on the direction of the flow of current through the conductor.
- Magnetic induction or flux density depends upon the amplitude of the current flowing through the conductor.
- The direction of the lines of force around the straight current-carrying conductor may be determined by the right-hand thumb rule or corkscrew rule.

Section 1.14 states the magnetic field due to a current-carrying conductor using Biot Savart's law. The magnetic field due to a current-carrying conductor at a distance point is inversely proportional to the square of the distance between the conductor and point, and the magnetic field is directly proportional to the length of the conductor, current flowing in the conductor.

$$d\boldsymbol{B} \propto \frac{I\,dl\sin\theta}{r^2} \tag{1.55}$$

1.18 Magnetic Permeability and Its Effect

Magnetic permeability also referred to as permeability in electromagnetism, is a property of a magnetic material that supports the formation of a magnetic field.

The term was coined by Oliver Heaviside in the year 1885. Magnetic permeability is a property that allows magnetic lines of force to pass through a material.

In other words, the magnetic permeability of a material can also be said to be its magnetization capability. This helps in determining how much magnetic flux can the material support when passing through it.

Magnetic permeability is defined as the ratio of the magnetic flux density (B) to the magnetic field intensity (H).

$$Magnetic\ permeability\ (\mu) = \frac{B}{H} \tag{1.56}$$

The SI unit of magnetic permeability is Henries per meter (H/m) or Newton·A^{-2}.

- It is a scalar quantity and is denoted by the symbol μ.
- Magnetic permeability helps us measure a material's resistance to the magnetic field or measure the degree to which a magnetic field can penetrate through a material.
- The magnetic permeability of a material is not constant. For a given temperature, it changes based on the intensity of the applied external magnetic field (H).
- Permeability also depends on several factors such as the nature of the material, humidity, position in the medium, temperature, and frequency of the applied force.

- Magnetic permeability is always positive and can vary with magnetic material. Meanwhile, the opposite of magnetic permeability is magnetic reluctivity.
- A material's permeability is important, which justifies the use of ferromagnetic material in electrical machines.

The different types of permeability include

- **Permeability of free space**

The permeability of free space is also known as the permeability of air or vacuum. It is represented by the ratio of magnetic flux density in vacuum and magnetizing field intensity

$$\mu_0 = \frac{B_o}{H} \tag{1.57}$$

- **Permeability of medium**

The ratio of magnetic flux density in the medium and magnetic field intensity. It is expressed as;

$$\mu = \frac{B}{H} \tag{1.58}$$

- **Relative permeability**

Relative permeability is a dimensionless quantity. It is the ratio of two quantities with the same units, so relative permeability has no unit.

$$\mu_r = \frac{\mu}{\mu_0} \tag{1.59}$$

The relative permeability of free space is 1.0

1.19 Analogy Between Electric and Magnetic Circuit

The analogy between electric and magnetic circuits is compared in the following table.

S.No.	Electric Circuit	Magnetic Circuit
1.	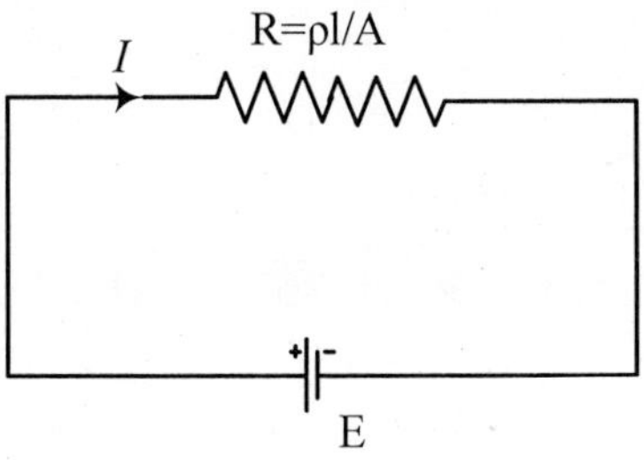 R=ρl/A I E	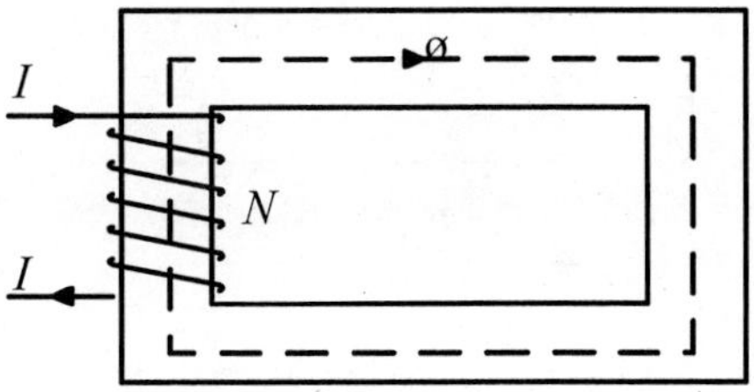 I ø N I

2.	The closed path for electric current is called an Electric Circuit	The closed path for magnetic flux is called Magnetic Circuit.
3.	$Current(I) = emf/resistance,$ $emf = voltage$	$flux = mmf/reluctance$ $mmf = NI$
4.	Unit of current (I) is measured in Amperes	Unit of flux (ϕ) is measured in Weber (Wb)
5.	Electromotive force is the driving force and the unit is Volts (V). $emf = \int \mathbf{E}.d\boldsymbol{\ell}$	Magneto motive force is the driving force and the unit is Ampere Turns (AT). $mmf = \int \mathbf{H}.d\boldsymbol{\ell} = NI$
6.	Resistance opposes the flow of the current $R = \frac{\rho\ell}{A}$ Measured in Ω	Reluctance opposes the flow of magnetic flux $\mathcal{R} = \frac{\ell}{A\,\mu_0\mu_r}$ Measured in (AT/Wb)
7.	Conductance = 1/Resistance	Permeance= 1/Reluctance
8.	Conductivity (σ): The property of a material that describes how the electric current in the component is related to the electrical potential difference (voltage) across it. Its unit is Siemens per meter, S/m .	Permeability (μ): Ratio of the magnetic flux density to the magnetic intensity. It is used to measure a material's resistance to the magnetic field or measure the degree to which a magnetic field can penetrate through a material. Its unit is H/m.
9.	Current density (δ) $\delta = \frac{I}{A}\ A/m^2$	Flux density(**B**) $\mathbf{B} = \frac{\phi}{A}\ Wb/m^2$ or Tesla
10.	Electrical Field intensity (E) $\mathbf{E} = \frac{V}{d} Volt/m$	Magnetic Field Intensity (H) $\mathbf{H} = \frac{mmf}{\ell} = \frac{NI}{\ell}\ AT/m$
11.	Voltage drop = IR	mmf drop = $\phi\,\mathcal{R}$

12.	In an electric circuit, electric current flows in the form of electrons. The direction of current is taken as opposite to the direction of electrons.	In a magnetic circuit, molecular poles are aligned; the flux doesn't flow but sets up in the magnetic circuit.
13.	In a series circuit, the current in all elements is the same, but voltage or emf is different across each element in the circuit.	In a series circuit, the same flux passes through all elements in series, and the sum of mmf's across the elements is equal to the applied mmf.
14.	In a parallel circuit voltage across all branches is the same and equal to the applied voltage whereas the current in the branches is different.	In a parallel circuit, the mmf of each branch is the same and equal to the applied mmf. The flux in each branch is different and their sum equals the resultant flux.
15.	There are many electric insulators. Air is a very good insulator and current can't pass through it.	There is no magnetic insulator; Fringe Flux can be set up even in the air.
16.	Kirchhoff current law and voltage law is applicable to the electric circuit	Kirchhoff mmf law and flux law are applicable to the magnetic flux.

1.20 Application of Magnets

Magnets are used in many applications including large industries to common human activity. Some common applications are given below:

1. Machines and devices

(a). Electric Motors: Types-DC (commutator and brushless), synchronous, induction start/synchronous run, hysteresis; rotary and linear; servo. Geometries of permanent magnet stator (conventional and iron fewer armatures), - permanent magnet rotor; inner or outer rotor; radial or axial field (disc) motors.

(b). Generators: Types- Magnetos, ignition or other pulse generators, tachometers, auxiliary exciters, alternators, multiphase synchronous machines, and homopolar DC machines. Geometries- permanent magnet rotor; radial or axial field; stator winding with or without iron.

(c). Electro-Mechanical Actuators: Linear-Force motors for valves, etc.; printer hammer mechanism; computer disc-drive head actuators; laser focusing and tracking (optic/magneto-optic recording: audio CDs, video, data); recorder pen positioners. Rotary-Disc drives; aircraft control surface actuators; materials handling robots.

(d). Measuring Instruments: Moving-coil (d'Arsonval and long scale geometries) and moving-magnet meters for many functions.

(e). Electric Current Control Circuit breakers, reed switches, miniature biased relays, thermostats, and eddy controls. Current motor over speed switch, arc blow out magnets.

2. Acoustic transducers

(a). Sound Generators: Loudspeakers, earphones, telephone receivers, ringers, buzzers, and ultrasonic generators.

(b). Sound Receivers: Dynamic microphones, ultrasound pickups.

(c). Other Audio Frequency Transducers, Phonograph Pickups.

3. Mechanical force and torque applications

(a). Contact Holding and Lifting: Machine-tool chucks, grippers, load-lifting magnets:(electrically switchable), tool holders, door catches, refrigerator seals, advertising signs, toys, and many more.

(b). Traction Devices: Conveyers, separators for ores and other materials, field gradient water purifiers, photocopier rollers.

(c). Couplings and Brakes: Synchronous torque couplings, linear followers, eddy current and hysteresis couplers and brakes, rotary-to-linear motion converter.

(d). Magnetic Bearings and Suspensions: Passive-watt-hour meters, ultra-centrifuges record player tone-arm support, textile spinning turbines. Partly active served systems-gyros, satellite momentum and energy wheels, laser beam scanner, turbo molecular pumps, electro-magnetic tracked vehicle levitation, hyperloop systems.

(e). Electro-Balances: Modern weighing devices from analytical balance to supermarket scales and truck weight stations.

4. Microwave/MM-wave devices, electro-ion beam control

(a). Power Tubes: Magnetrons (radar, kitchen ovens); PPM focusing for TWTS and klystrons; crossed-field amplifiers, gyrotron, etc.

(b). Waveguide Devices: Biasing ferrite or YIG elements in resonance filters, switches, and isolators.

(c). Laser Lenses, Particle Accelerators, Synchrotron Radiation Sources. Fee Electron deflecting magnets, wigglers, undulates.

(d). Mass Spectrometers: Deflecting magnets

5. Sensors, electric signal transducers

(a). Transducers Using Permanent Magnets: Inductive, Hall effect, magnetoresistive, temperature-sensitive elements.

(b). Quantities Measured: Position, velocity, acceleration, fluid and heat flow, pressure, vibration, temperature etc.

(c). Use Areas: Automotive, industrial, aerospace, computer peripherals (keyboards, read/write head sensors), office equipment.

6. Medical electronics and bioengineering

(a). NMR Imaging Devices: DC field source for MR: tomography.

(b). Mechanical Prostheses: Eyelid muscle assist, dental prostheses, stoma seals, valves, heart-assist pumps, artificial limbs.

(c). Surgical Clamps: For incisions and severed blood vessels.

(d). Diagnostic Aids: Catheters; sensors/transducers.

(e). Miniature Hearing Aids: External devices and implants.

7. Miscellaneous applications

(a). Magnetic Locks: Key and cylinder with encoded magnets

(b). Magnetic Jewellery: Necklaces, clasps, earrings

(c). Electronic Choke: Steady bias field for core

(d). Magnetic Bubble Memory: Bias field for bubble element

(e). Vacuum Technology: ton-getter pumps, vacuum gauges.

(f). Magnetic Card, ATM

Solved numerical examples

Q1. In the magnetic circuit of Fig. 1.16, the relative permeability of the ferromagnetic material is 1200. Neglect magnetic leakage and firing. All dimension is in centimeter, and the magnetic material has a square cross-sectional area. Determine the air gap flux, the air gap flux density, and the magnetic field intensity in the air gap.

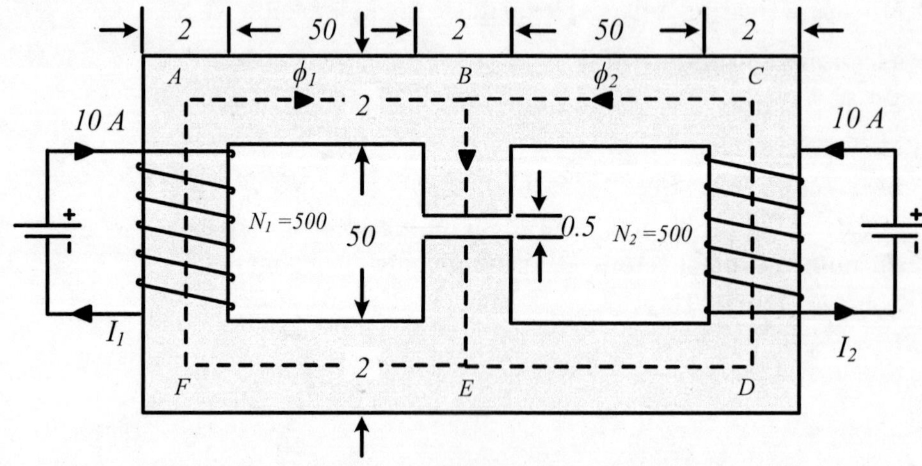

Fig. 1.16

Solutions:

Here mean magnetic paths of the fluxes are shown by the dashed line in fig.1.15, the equivalent

magnetic circuit

$$F_1 = N_1 I_1 = 500 \times 10 = 5000\ AT$$

$$F_2 = N_2 I_2 = 500 \times 10 = 5000\ AT$$

$$\mu_c = 1200\ \mu_o = 1200 \times 4\pi \times 10^{-7}$$

$$\mathcal{R}_{BAEF} = \frac{\ell_{BAEF}}{a_C\ \mu_c}$$

$$\mathcal{R} = \frac{3 \times 52 \times 10^{-2}}{1200 \times 4\pi \times 10^{-7} \times 4 \times 10^{-4}}$$

$$\mathcal{R} = 2.58 \times 10^6\ AT.Wb$$

For Symmetry

$$\mathcal{R}_{BCDE} = \mathcal{R}_{BAFE}$$

$$\mathcal{R}_g = \frac{\ell_g}{a_g\ \mu_a}$$

$$\mathcal{R}_g = \frac{5 \times 10^{-3}}{4\pi \times 10^{-7} \times 2 \times 2 \times 10^{-4}}$$

$$\mathcal{R}_g = 9.94 \times 10^6\ AT.Wb$$

$$\mathcal{R}_{BE(core)} = \frac{\ell_{BE(core)}}{a_c\ \mu_c}$$

$$\mathcal{R}_{BE(core)} = \frac{51.5 \times 10^{-2}}{1200 \times 4\pi \times 10^{-7} \times 4 \times 10^{-4}}$$

$$\mathcal{R}_{BE(core)} = 0.82 \times 10^6\ AT.Wb$$

The loop equation are

$$\phi_1(\mathcal{R}_{BAEF} + \mathcal{R}_{BE} + \mathcal{R}_G) + \phi_2(\mathcal{R}_{BE} + \mathcal{R}_G) = F_1$$

$$\phi_1(\mathcal{R}_{BE} + \mathcal{R}_G) + \phi_2(\mathcal{R}_{BCDE} + \mathcal{R}_{BE} + \mathcal{R}_G) = F_2$$

$$\phi_1(13.34 \times 10^6) + \phi_2(10.76 \times 10^6) = 5000$$

$$\phi_1(10.76 \times 10^6) + \phi_2(13.34 \times 10^6) = 5000$$

The air gap flux density is

$$\boldsymbol{B}_g = \frac{\phi_g}{a_g}$$

$$\boldsymbol{B}_g = \frac{4.134 \times 10^{-4}}{4 \times 10^{-4}}$$

$$\boldsymbol{B}_g = 1.034\ T$$

The magnetic field intensity in the air gap is

$$\boldsymbol{H}_g = \frac{\boldsymbol{B}_g}{\mu_0}$$

$$\boldsymbol{H}_g = \frac{1.034}{4\pi \times 10^{-7}}$$

$$\boldsymbol{H}_g = 0.822 \times 10^6\ AT/m$$

Q2. A coil of insulated wire of 300 turns and of resistance 2Ω is closely wound on iron ring as shown in fig. 1.17. The ring has a mean diameter of 0.25 m and a uniform cross-sectional area of 700 mm². Calculate the total flux in the ring when a DC supply of 6V is applied to the ends of the winding. Assume a relative permeability of 550.

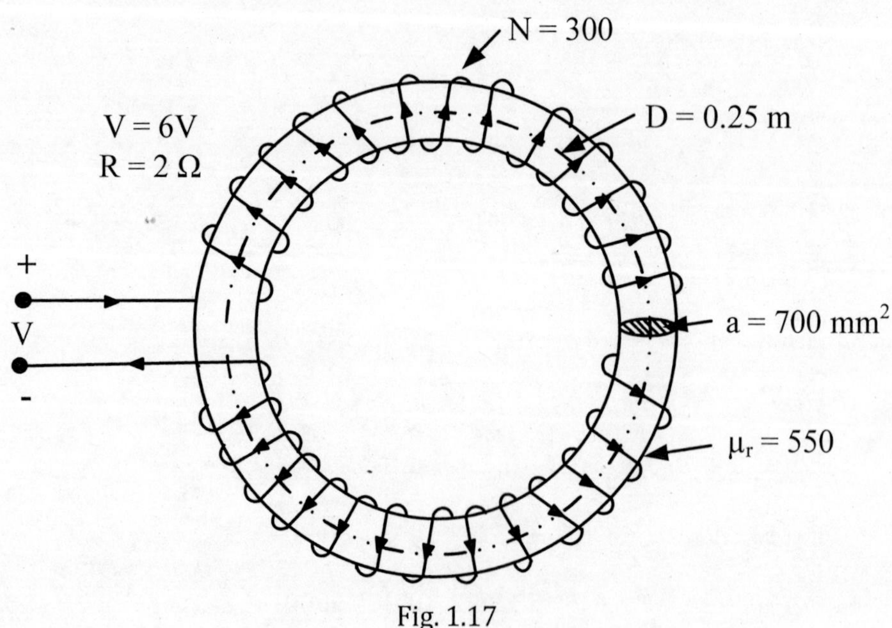

Fig. 1.17

Solution:

Mean length of iron ring, $l = \pi D = \pi \times 0.25 = 0.25\pi$ m

Area of cross-section, a_c = 700 mm² = 700×10^{-6} m²

Current flowing through the coil,

$$I = \frac{\text{Voltage applied across coil}}{\text{Resistance of coil}} = \frac{6}{4} = 1.5\,A$$

Total flux in the ring,

$$\phi = \frac{N\,I}{l\,/\,a\,\mu_o\,\mu_r}$$

$$= \frac{300 \times 1.5 \times 700 \times 10^{-6} \times 4\pi \times 10^{-7} \times 550}{0.25\pi}$$

= 0.2772 mWb

Q3. A coil is wound uniformly with 100 turns over a steel ring of relative permeability 900 having a mean diameter of 20 cm as shown in fig. 1.18. the steel ring is made of bar having circular cross-section of diameter 2 cm. if the coil has a resistance of 50 ohms and is connected to 250 V DC supply, calculate (i) the mmf of the coil, (ii) the field intensity in the ring, (iii) reluctance of the magnetic path.

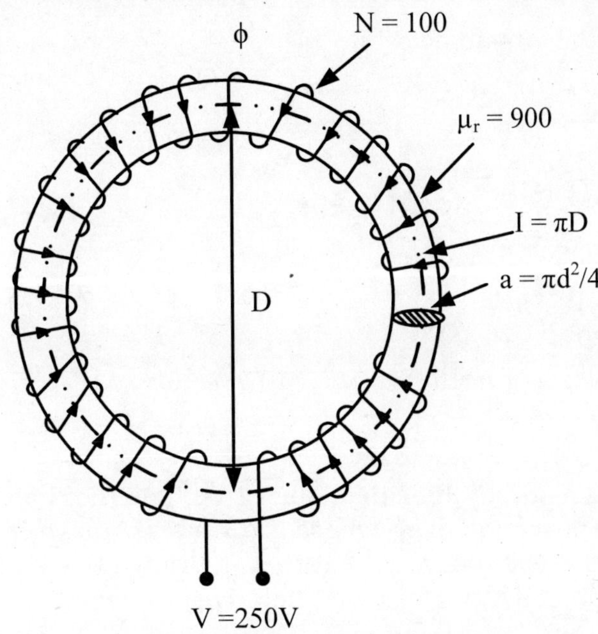

Fig. 1.18

Solution:

The magnetic circuit is shown in Figure

Current through the coil,

$$I = \frac{V}{R}$$

$$= \frac{250}{50}$$

$$= 5\text{ A}$$

mmf of the coil

$$NI = 100 \times 5$$

$$= 500\ AT$$

Field intensity

$$H = \frac{Nl}{\ell}$$

Where, l = πD = 0.2 π metre

$$H = \frac{500}{0.2\,\pi} = 796.17 \text{ AT/m}$$

Reluctance of the magnitude path,

$$\mathcal{R} = \frac{\ell}{a\,\mu_0\mu_r}$$

Where,

$$a = \frac{\pi}{4}d^2$$

$$= \frac{\pi}{4}0.02^2$$

$$= \pi \times 10^{-4} m^2$$

$$\mu_r = 900$$

$$\mathcal{R} = 0.2\,\pi/\pi \times 10^{-4} \times 4\pi \times 10^{-7} \times 900$$

$$= 17.684 \times 10^5 \text{ AT/Wb}$$

Q4. A magnetic core made of annealed sheet steel has the dimensions as shown in Figure. The cross-section everywhere is 25 cm² as shown in fig.1.18. The flux in branches A and B is 3500 mWb, but that in branch C is zero. Find the required ampere-turns for coil A and for coil C. Relative permeability of sheet steel is 1000.

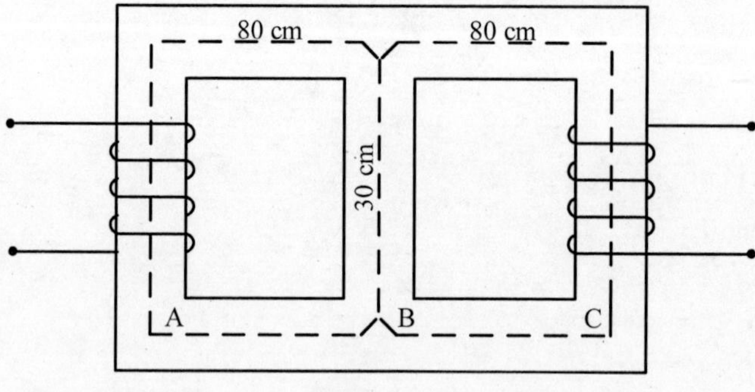

Fig. 1.18

Solution:

The given magnetic circuit is a parallel circuit. To determine the ATs for coil 'A', the flux distribution is shown in Figure

Since path 'B' and 'C' are in parallel with each other w.r.t. path 'A',

mmf for path 'B' = mmf for path 'C',

$$\phi_1 \mathcal{R}_1 = \phi_2 \mathcal{R}_2$$

$$\frac{3500 \times 10^{-6} \times 80 \times 10^{-2}}{a\,\mu_0\mu_r} = \phi_2 \times \frac{30 \times 10^{-2}}{a\,\mu_0\mu_r}$$

$$\phi_2 = 1312.5 \times 10^{-6}\text{Wb}$$

Total flux in the path 'A',

$$\phi = \phi_1 + \phi_2$$

$$= (3500 + 1312.5) \times 10^{-6}\text{Wb}$$

$$= 4812.5 \times 10^{-6}\text{Wb}$$

Actual (resultant) flux in path 'A'

$$= \phi - \phi_2 = 3500 \times 10^{-6}\ \text{Wb}$$

ATs required for coil 'A'

= ATs for path 'A' + ATs for path 'B' or 'C'

$$= \frac{3500 \times 10^{-6}}{4\pi \times 10^{-7} \times 1000 \times 25 \times 10^{-4}}(0.8 + 0.3)$$

$$= 1225.5$$

To neutralise the flux in section 'C', the coil produces flux of 1312.5 μ Wb in opposite direction.

ATs required for coil 'C'

= ATs for path 'C' only

$$= \frac{1312.5 \times 10^{-6} \times 0.8}{4\pi \times 10^{-7} \times 1000 \times 25 \times 10^{-4}} = 334.33$$

Q5. A wrought iron bar 20cm long and 2cm in diameter is bent into a circular shape as given in Fig. 1.19. It is then wound with 300 turns of wire. Calculate the current required to produce a flux of 0.5mWb in a magnetic circuit with an air gap of 1mm; μ_r(iron)=4000 (assume constant).

Solution:

$$\ell_i = 30cm = 0.3m;$$

Diameter, $d = 2cm$

So, Area,

$$a = \frac{\pi}{4}d^2 = \frac{\pi(2^2)}{4} \times (10)^{-4}\,m^2$$

$$a = \pi \times (10)^{-4}\,m^2$$

N = 300 turns

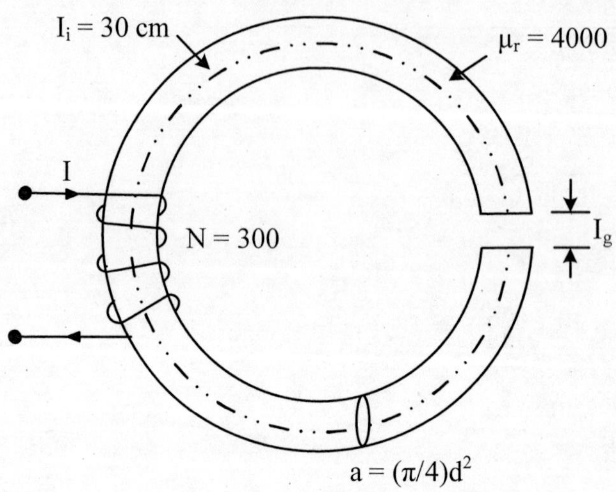

Fig. 1.19

$$N \times I = \frac{\varphi}{a}\left[\frac{l_i}{\mu_0\mu_r} + \frac{l_g}{\mu_0}\right]$$

$$I = \frac{0.5 \times 10^{-3}}{300 \times \pi \times 10^{-4}}\left[\frac{0.3}{4\pi \times 4000 \times 10^{-7}} + \frac{0.001}{4\pi \times 10^{-7}}\right] = 4.21\ A$$

Q6. Calculate the relative permeability of an iron ring when the exciting current taken by the 400-turn coil is 2A and the total flux produced is 1m Wb. The mean circumference of the ring is 0.5m and the area and cross-section is 10 cm².

Solution:

$$NI = \frac{\phi \times l}{a\mu_0\mu_r}$$

$$\mu_r = \frac{\phi \times l}{a\mu_0 NI}$$

Where, I = 2A; N = 400 turns; $\phi = 1m\ Wb = 1 \times 10^{-3} Wb$; $l = 0.5m$; $a = 10cm^2$ $= 10 \times 10^{-3} m^2$

$$\mu_r = \frac{1 \times 10^{-3} \times 0.5}{10 \times 10^{-4} \times 4\pi \times 10^{-7} \times 400 \times 2} = 497.61$$

Q7. A rectangular magnetic core is shown in fig. 1.20 has a square cross-section of area $16cm^2$. An air gap of 2mm is cut across one of its limbs. Find the exciting current needed in the coil having 800 turns wound on the core to create an air-gap flux of 4m Wb. The relative permeability of the core is 1500.

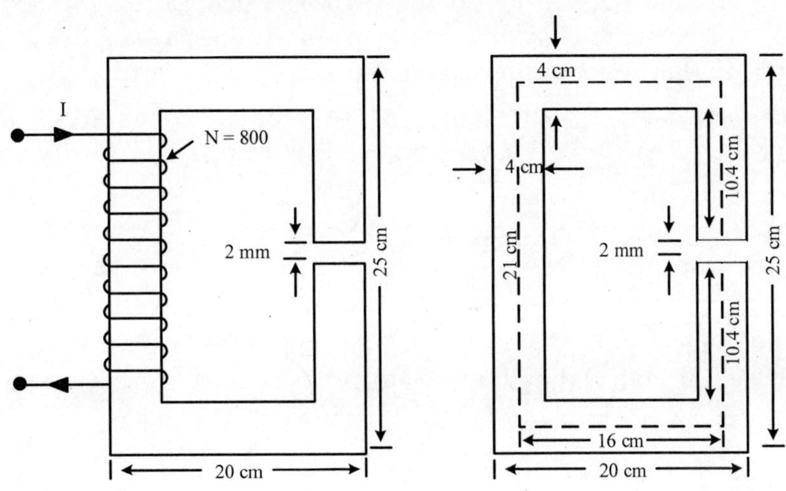

Fig. 1.20

Solution:

Here, Area of cross section, a $= 16cm^2 = 16 \times 10^{-4} m^2$;

Where, $l_g = 2mm = 2 \times 10^{-3} m$,

No. of turns, N = 800;

Flux, $\phi = 4m\ Wb = 4 \times 10^{-3} Wb$;

$\mu_r = 1500$

Flux density required,

$$B = \frac{\phi}{a} = \frac{4 \times 10^{-3}}{16 \times 10^{-4}} = 2.5T$$

Each side of the cross-section $= \sqrt{16} = 4\ cm$

Length of iron-path, $l_i = \left(25 - 2 \times \frac{4}{2} + 20 - 2 \times \frac{4}{2}\right) \times 2 - 0.2 = 73.8cm = 0.738m$

$$\text{Total ampere} - \text{turns required} = \frac{B}{\mu_0} l_g + \frac{B}{\mu_0 \mu_0} l_i$$

$$= \frac{2.5 \times 2 \times 10^{-3}}{4\pi \times 10^{-7}} + \frac{2.5 \times 0.738}{4\pi \times 10^{-7} \times 1500} = 3979 + 979 = 4958$$

$$\text{Exciting current required, I} = \frac{\text{Total ampere} - \text{turns}}{\text{N}} = \frac{4958}{800} = 6.19\ \text{A}$$

Multiple choice questions

Q1. Fringing at short air gap in a magnetic circuit is empirically accounted for by:

(a). Increasing the linear dimensions of the gap area by twice the gap length.
(b). Increasing the linear dimensions of the gap area by half gap length.
(c). Increasing the linear dimensions of the gap area by one gap length.
(d). Increasing the linear dimensions of the gap area by one-fourth gap length.

Q2. A circular iron core has an air-gap cut in it and is excited by passing direct current through a coil wound on it. The magnetic energy stored in the air gap and the iron core is:

(a). In direct ratio of their reluctances.
(b). In inverse ratio of their reluctances.
(c). Equally divided among them.
(d). Energy resides wholly in the iron core.

Q3. In a given magnetic circuit, a current of 1A flowing in the exciting winding produces a constant flux of 1.0 Wb. If the circuit reluctance is doubled, the exciting current should be:

(a). 1 A
(b). 0.5 A
(c). 2 A
(d). 1.5 A

Q4. The mutual inductance between two closely coupled coils is 1 H. The turns of one of the two coils are halved whereas those of the other are doubled. The value of mutual inductance now becomes:

(a). 2 H
(b). 0.5 H
(c). 1 H
(d). 1.5 H

Q5. Calculate the magnetic flux produced when the magnetic field is parallel to the surface area.

(a). Minimum
(b). Maximum
(c). Zero
(d). Dpend on the surface area

Q6. What is the unit of inductance is?

(a). V s/A
(b). H-turns^2
(c). Wb T/A
(d). All three are equivalent.

Q7. Match the following electric and magnetic quantities on the left with SI units given on the right:

1. Flux	A) AT/Wb
2. Magneto motive force	B) Wb
3. Reluctance	C) Wb/AT
4. Permeance	D) AT

Mark the correct answer below:

(a). 1B, 2D, 3A, 4C
(b). 1A, 2D, 3B, 4C
(c). 1D, 2C, 3B, 4A
(d). 1C, 2A, 3D, 4B

Q8. For the magnetic circuit shown in Figure 1.21 the mmf required to establish one-unit flux in the central limb is (permeability of iron core is infinite):

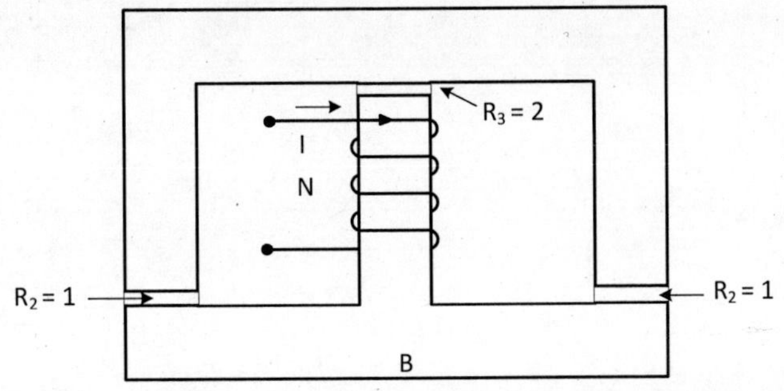

Fig. 1.21

(a). 5 AT
(b). 4 AT
(c). 2 AT
(d). 2.5 AT

Q9. The self-inductance of the coil wound on the toroid with air gap shown in Figure 1.22.

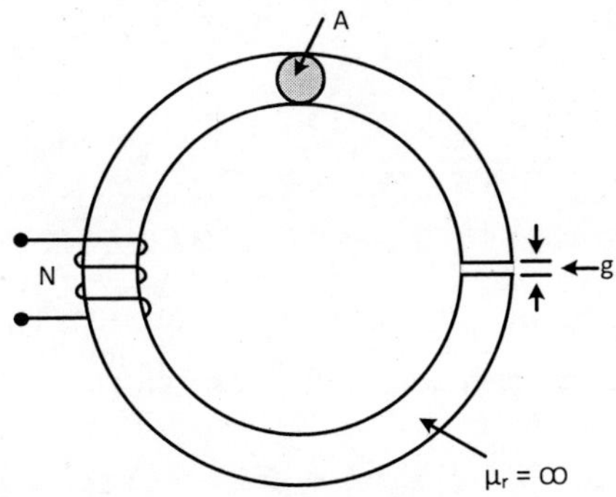

Fig. 1.22

(a). $AN/\mu_0 g$
(b). $\mu_0 AN/g$.
(c). $AN^2/\mu_0 g$
(d). $\mu_0 gN^2/g$

Q10. A coil of 1000 turns in wound on a core. A current of 2A flowing through the coil creates a core flux of 1 mWb. The energy stored in the magnetic field is:

(a). $\frac{1}{4}$ J
(b). $\frac{1}{16}$ J
(c). 1 J
(d). 2 J

Answer to Multiple Choice Questions

1	c	2	a	3	c	4	c	5	c	6	d	7	a	8	d
9	d	10	b												

Short answer type questions

Q1. What is a magnet?

Q2. What are permanent and temporary magnets?

Q3. What are magnetic poles?

Q4. What do you understand by a magnetic field?

Q5. What do you understand by axis?

Q6. Define and explain a magnetic circuit.

Q7. Mention at least four properties of magnetic lines of force.

Q8. Explain the term MMF.

Q9. Define relative permeability.

Q10. Define permeance of magnetic circuit.

Q11. What is a composite magnetic circuit?

Q12. State 'Ohms law' of a magnetic circuit.

Q13. Define leakage factor.

Q14. Why is it necessary to keep air gaps in magnetic circuits as small as possible?

Q15. Steel is alloyed with what material to reduce core loss?

EXERCISES

Q1. An iron ring has a cross-sectional area of 400 mm^2 and a mean diameter of 25 cm. It is wound with 500 turns. If the value of relative permeability is 500, find the total flux set up in the ring. The coil resistance is 400 Ω and the supply voltage is 200 V. **(Ans. 0.08 m Wb)**

Q2. An iron ring of mean diameter 22 cm and cross-section 10 cm^2 has an air gap 1 mm wide. The ring is wound uniformly with 200 turns of wire. The permeability of ring material is 1000. A flux of 0.16 mWb is required in the gap. What current should be passed through the wire? **(Ans. 1.076)**

Q3. An iron ring has a mean circumferential length of 60 cm with an air gap of 1 mm and a uniform winding of 300 turns. When a current of 1 A flows through the coil, find the flux density. The relative permeability of iron is 300. Assume $\mu_0 = 4\pi \times 10^{-7}$ H/m. **(Ans. 0.1256T)**

Q4. An iron ring has cross-section of 3 cm^2 and a mean diagram of 25 cm. An air gap of 0.4mm has been made by saw cut across the section. The ring is wound with 200 turns through which a current of 2 A is passed. If the total flux is 21×10^{-5} Weber, find μ for iron assuming no leakage. **(Ans. 2470)**

Q5. Determine magneto motive force, magnetic flux, reluctance, and flux density in case of a steel ring 30 cm mean diameter and a circular cross-section 2 cm in diameter has an air gap 1 mm long. It is wound uniformly with 600 turns of wire carrying a current of 2.5 A. Neglect magnetic leakages. The iron path takes 40% of the total magneto motive force. **(Ans. Total reluctance 4.221×10^6 ATs/Wb, Magnetic flux = 0.3554 mWb, Flux density = 1.131 Wb/m^2)**

Q6. A steel ring 10 cm mean radius and of circular cross-section of 1 cm in radius has an air gap of 1mm in length. It is wound uniformly with 500 turns of wire carrying a current of 3A. Neglect magnetic leakage. The air gap takes 60% of the total mmf. Find the total reluctance. **(Total reluctance = 4.221×10^6 ATs/Wb,)**

Books for further reading

1 Electric Machinery: Fitzgerald, Kingslay's, S. D. Umans, 7/e, McGraw-Hill, 2020.

2. Electric Machinery Fundamentals, S. Chapman, 4/e, McGraw Hill Education, 2017.

3. Electric Machines, Nagrath and Kothari, 5/e, McGraw Hill Education, 2017

4. Electric Machinery, P. S. Bimbhra, 7/e, Khanna Publishers. 2015

5. Electrical Machines, R. K. Srivastava, 2/e, Cengage Learning Pvt. Ltd.,2011

6. Electrical Machines, Smarajit Ghosh, 2/e, Pearson edu, 2012

7. Electrical Machines-I, D. K. Palwalia, N K Garg, P Kumar, G Jain. Ashirwad Publishers, 2020

REFERENCE BOOKS:

1. Electric Machinery and Transformer, Guru, Hiziroglu, 3/e, Oxford University Press, 2012

2. Basic Electric Machines, Vincent Del Toro, 1/e, Pearson Education India, 2016.

3. Performance and Design of A.C. machines, M. G. Say, CBS Publishers, 2002

For further reading scan to:-

Further Reading on Magnetic Circuits

Video lecture on Magnetic Circuits

Video lecture on Magnetic Circuits

Video lecture on Magnetic Circuits

2 Electromagnetic Force and Torque

UNIT SPECIFICS

Through this unit, the following aspects have been discussed:

- *Electromagnetic induction and characteristics of magnetic circuits;*
- *Dynamically and statically inducted emf;*
- *Magnetization;*
- *Electro-mechanical energy conversion and energy balance.*
- *Energy stored in the magnetic circuit;*
- *Determination of Magnetic force or torque*

The unit contains a detailed description of the topics. It gives solved and unsolved problems to illustrate further the theoretical study. A number of multiple-choice questions as well as questions of short and long answer types are also given. A list of references and suggested readings are given in the unit so that one can go through them for practice. QR codes have been provided in different sections and at the end of the unit, which can be scanned for relevant supportive knowledge.

RATIONALE

This unit on Electromagnetic Force and Torque will help students to acquire knowledge of electromagnetic induction, fundamental laws of electromagnetic induction, induced emf, magnetization or B-H curves, magnetic hysteresis, the principle of electromagnetic energy conversion, energy balance, and energy stored in the magnetic field. It gives an understanding of coupling field reactions and phenomena involved in singly & doubly excited magnetic systems. Further, the production of force or torque with graphic diagrams and examples is discussed.

PRE-REQUISITES

Mathematics: Vectors, Integrals, Differentiation, Algebra (Class XII)
Physics: Electricity and Magnetism (Class XII)

UNIT OUTCOMES

The list of outcomes of this unit is as follows:

U2-O1: Understand the concept and laws of electromagnetic induction.

U2-O2: Understand the principle of electromagnetic energy conversion and energy balance.

U2-O3: Analyse the energy balance and energy stored in the magnetic field.

U2-O4: Evaluate the coupling field reaction and phenomena involved in singly & doubly excited magnetic systems.

U2-O5: Analyse the production of force or torque.

Unit-2 Outcomes	**EXPECTEDMAPPINGWITHCOURSEOUTCOMES** (1-Weak Correlation; 2-Medium Correlation; 3-Strong Correlation)			
	CO-1	**CO-2**	**CO-3**	**CO-4**
U2-O1	3	-	1	1
U2-O2	3	1	1	1
U2-O3	2	1	-	1
U2-O4	1	-	1	1
U2-O5	1	-	2	2

2.1 Introduction

It is always advantageous to utilize electrical energy since it is cheaper, can be easily transmitted, is easy to control, and efficient. Electrical energy is normally generated from natural resources such as water, coal, diesel, wind, atomic energy, etc. From these resources, the mechanical energy is converted into electrical energy using suitable rotating machines acting as a generator. Solar power generation using photovoltaic (PV) cells and fuel cells are exceptions to it. For the utilization of electrical energy, it is converted into other forms of energy such as mechanical, heat, light etc. Large electrical power utilization is essentially required for rail transportation, industries, metal production, metal refinement etc. It is a well-known fact that electric drives have been universally adopted by the industry due to their inherent advantages. Energy conversion devices are always required at both ends of a typical electrical system. The devices or machines which convert mechanical energy into electrical energy and vice-versa are called electromechanical energy conversion devices.

Scan to know more on Electricity and Magnetism

The operation of all electrical machines such as DC machines, transformers, synchronous machines, induction motors, etc., rely upon their magnetic circuits. The closed path followed by the magnetic lines of force is called a magnetic circuit. The operation of all electrical devices (e.g., transformers, generators, motors, and electrical brakes etc.) depends upon the magnetism produced by their magnetic circuits. Therefore, to obtain the required characteristics of these devices, their magnetic circuits have to be designed carefully. In this chapter, we shall focus our attention on the fundamentals of magnetic circuits, electromagnetic force and torque, and their applications as electromechanical energy conversion devices.

2.2 Electro Magnetic Induction

The phenomenon by which an emf is induced in a circuit (and hence current flows when the circuit is closed) when changing magnetic flux linkage is called electro-magnetic induction. For illustration, consider a coil having a large number of turns to which the galvanometer is connected. When a permanent bar magnet is taken nearer to the coil or away from the coil, as shown in fig 2.1 (a), a deflection occurs in the needle of the galvanometer. Although, the deflections in the needle are opposite in two cases. On the other hand, if the bar magnet is kept stationary and the coil is brought nearer to the magnet or away from the magnet, as shown in fig. 2.1(b), again a deflection occurs in the needle of the galvanometer. The deflections in the needle are opposite in the two cases. However, if the magnet and the coil both are kept stationary, no matter how much flux is linking with the coil, there is no deflection in the galvanometer needle.

The following points are worth noting:

1) The deflection in the galvanometer needle shows that emf is induced in the coil. This condition occurs only when flux linking with the coil changes i.e., either the magnet or coil is in motion.

2) The direction of induced emf in the coil depends upon the direction of the magnetic field and the direction of motion of the coil.

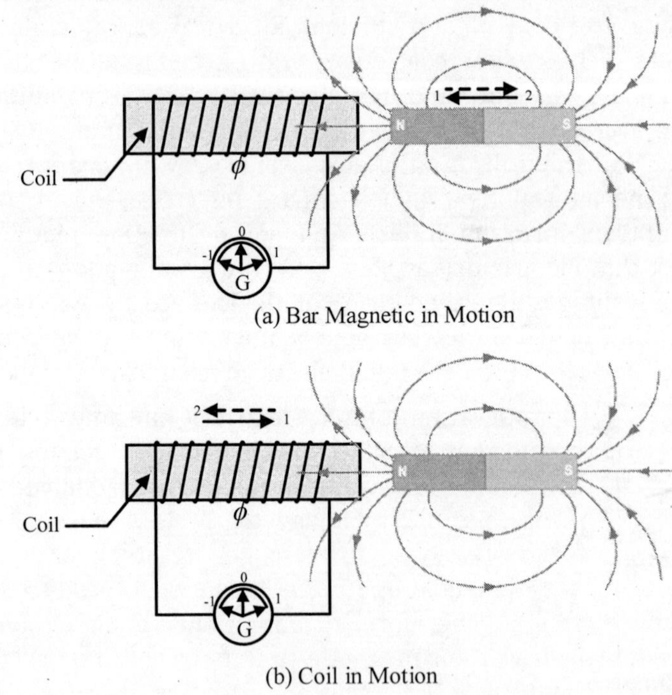

(a) Bar Magnetic in Motion

(b) Coil in Motion

Fig. 2.1 Electromagnetic induction

2.3 Faraday's Laws of Electromagnetic Induction

Michael Faraday summed up the conclusions of his experiments regarding electro-magnetic induction into two laws, known as Faraday's laws of electro-magnetic induction.

Scan to know more on Faraday's laws

First Law: This law states that "Whenever a conductor cuts the magnetic field, an emf is induced in the conductor".

Or

"Whenever the magnetic flux linking with any circuit (or coil) changes, an emf is induced in the circuit".

fig 2.2 shows a conductor connected to a galvanometer placed in the magnetic field of a permanent magnet. Whenever, the conductor is moved upward or downward i.e., across the field, there is deflection in the galvanometer needle which indicates that an emf is induced in the conductor. If the conductor has moved along (parallel) the field, there is no deflection in the needle which indicates that no emf is induced in the conductor.

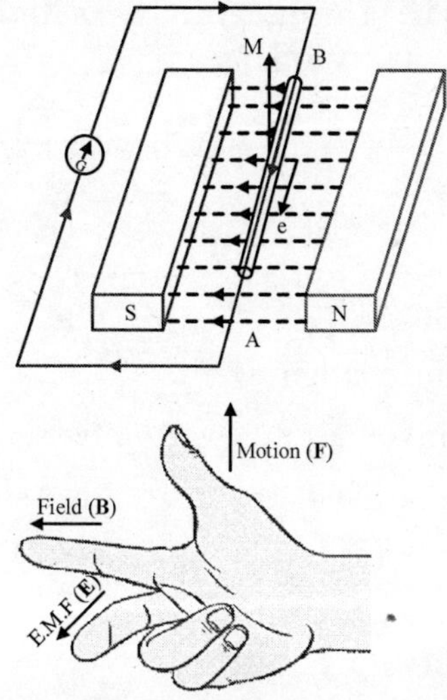

Fig. 2.2 Conductor moving in the field

For the second statement, consider a coil connected to a galvanometer is placed near a bar magnet, as shown in fig. 2.3. When the bar magnet (N-pole) is taken nearer to the coil [see fig. 2.3(a)]. there is deflection in the needle of the galvanometer. If now the bar magnet (N-pole) is taken away from the coil [see fig. 2.3(b)], again there is deflection in the needle of the galvanometer but in the opposite direction. The deflection in the needle of the galvanometer indicates that emf is induced in the coil.

Scan to know more on Moving magnetic near coil

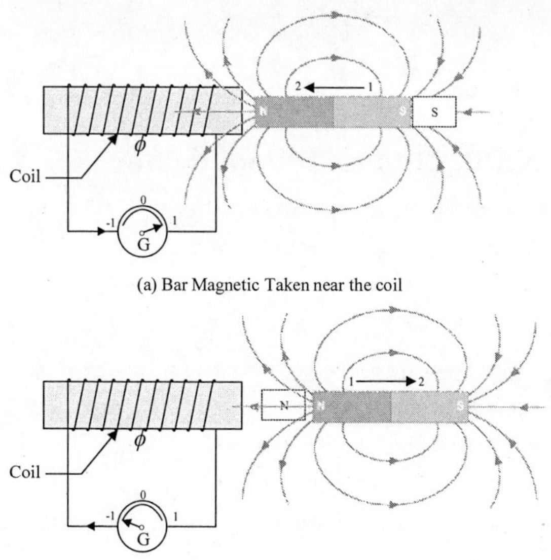

(a) Bar Magnetic Taken near the coil

(b) Bar Magnetic Taken away from the Coil

Fig. 2.3 Coil is stationary but the bar magnet (field) is moving

Second Law: This law states that "The magnitude of induced emf in a coil is directly proportional to the rate of change of flux linkages."

$$\text{Rate of changes of flux linkages} = \frac{N\,(\phi_2 - \phi_1)}{t}\ \text{Wb} - \text{turns/s} \tag{2.1}$$

Where,

N = Number of turns of the coils;

t = Time in seconds for the change

$(\phi_2 - \phi_1)$ = Change of flux in a single turn in Wb

∴ According to Faraday's second law of electro-magnetic induction;

$$\text{Induced emf } (e) \propto \frac{N\,(\phi_2 - \phi_1)}{t} \tag{2.2}$$

(taking proportionality constant, as unity)

$$\text{e} = \frac{N\,(\phi_2 - \phi_1)}{t} \tag{2.3}$$

In differential form,

$$e = N\,\frac{d\phi}{dt} = \frac{d\lambda}{dt}\ \text{Volt} \tag{2.4}$$

Usually, a minus sign is given to the right-hand side expressions which indicate that emf is induced in such a direction that opposes the cause (i.e. change in flux) which produces it (as per Lenz's law).

$$e = N\,\frac{d\phi}{dt}\ volt \tag{2.5}$$

2.4 The Direction of Induced emf

The direction of induced emf and hence current in a conductor or coil can be determined by either of the following two methods:

Fleming's Right-Hand Rule: This rule is applied to determine the direction of induced emf in a conductor moving across the field and is stated as under;

"Stretch, the first finger, second finger, and thumb of your right hand mutually perpendicular to each other. If the first finger indicates the direction of magnetic field, the thumb indicates the direction of motion of conductor then the second finger will indicate the direction of induced emf in the conductor."

Its illustration is shown in fig. 2.2. In vector notation in electromagnetic field theory, the electric field intensity (or emf induced in a coil) in a conducting material moving with speed $\boldsymbol{v}$ in a magnetic field $\boldsymbol{B}$ is given as

$$\boldsymbol{E} = \boldsymbol{v} \times \boldsymbol{B} \tag{2.6}$$

The direction of induced emf is in the direction of the cross product of speed $\boldsymbol{v}$ and magnetic flux density$\boldsymbol{B}$.

Lenz's Law: This law is more suitably applied to determine the direction of induced emf in a coil or circuit when flux linking with it changes. It is stated as under:

"In effect, electro-magnetically induced emf and hence current flows in a coil or circuit in such a direction that the magnetic field set up by it, always opposes the very cause which produces it."

Explanation: When the N-pole of a bar magnet is taken nearer to the coil as shown in fig. 2.3, an emf is induced in the coil and hence current flows through it in such a direction that side 'B' of the coil attains North polarity which opposes the movement of the bar magnet. Whereas, when the N-pole of the bar magnet is taken away from the coil as shown in fig. 2.3, the direction of emf induced in the coil is reversed and side 'B' of the coil attains south polarity in coil side close to N pole of bar magnet which again opposes the movement of the bar magnet. The force of attraction between the S-pole of the coil and the N-pole of the bar magnet opposes the force which takes the bar magnet away from it.

Scan to know more on Moving

2.5 Induced emf

When flux linking with a conductor (or coil) changes, an emf is induced in it. This change in flux linkages can be obtained in the following two ways:

- By either moving the conductor in the stationary magnetic field or moving the magnetic field system, keeping the conductor stationary. In both cases, the conductor cuts the magnetic field (as in the case of DC and AC generators). The emf induced is called dynamically induced emf or speed induced emf or motional emf.
- By changing the flux linking with the coil (or conductor) without moving either the coil or field system. If the bar magnet is replaced by a similar solenoid carrying alternating current, the magnetic flux coming from it will be alternating in nature experienced by the coil connected to the galvanometer. The *emf* induced in the coil will be statically induced emf, also referred to as transformer-induced emf. If such a coil system is suddenly excited by DC current, the change in flux linkage occurs momentarily giving a sudden deflection in the galvanometer.

2.5.1 Dynamically induced emf

The relative motion between the conductor and magnetic field system causes the magnetic flux to be cut by the conductor or change in flux linkage. The emf thus induced in the conductor is referred to as dynamically (speed or motional) induced emf. Considering a conductor of length l meter placed in the magnetic field of flux density B Wb/m^2 is moving at the right angle to the field at a velocity v meter/second as shown in fig. 2.4(a). Let the conductor be moved through a small distance dx meter in time dt second as shown in fig. 2.4(b). The active length of the conductor is l, perpendicular to the paper.

Area swept by the conductor,

$$A = l \times dx \tag{2.7}$$

Flux cut by the conductor,

$$\phi = B \times A = B.l.dx \tag{2.8}$$

According to Faraday's law of electromagnetic induction, the induced emf is

$$e = \frac{flux\ cut}{time} = \frac{\phi}{dt} = \frac{B.l.dx}{time} = B.l.v \tag{2.9}$$

Since $dx/_{dt} = v, velocity$

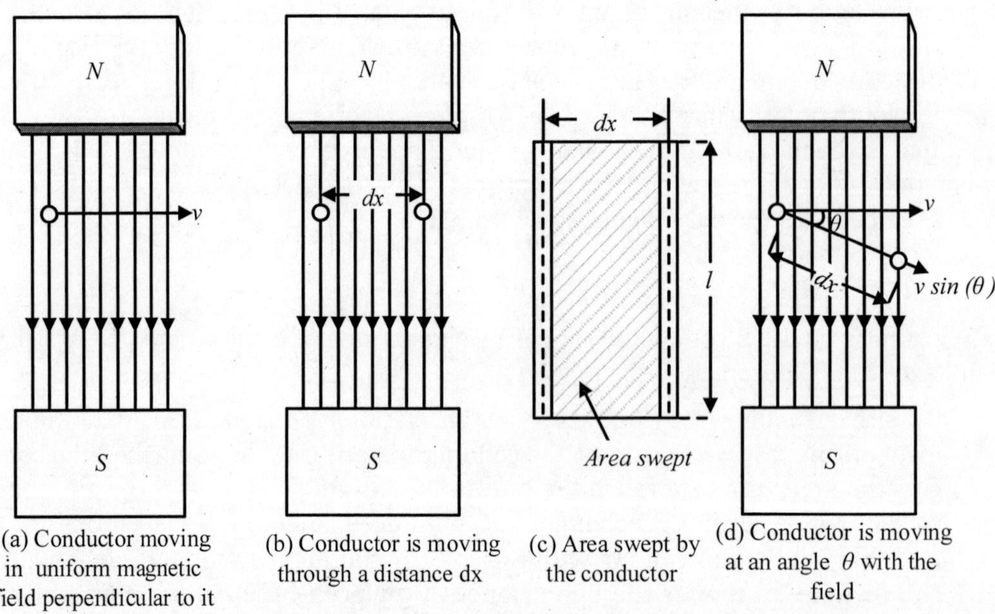

(a) Conductor moving in uniform magnetic field perpendicular to it

(b) Conductor is moving through a distance dx

(c) Area swept by the conductor

(d) Conductor is moving at an angle θ with the field

Fig. 2.4 Dynamically induced emf

Now, if the conductor is moved at an angle θ with the direction of the magnetic field at a velocity $v, meter/sec$ as shown in fig 2.4(d). A small distance covered by the conductor in that distance is dx in time dt second. Then the component of distance perpendicular to the magnetic field which produced emf is $dx \sin\theta$.

Area swept by the conductor,

$$A = l \times dx \sin\theta \tag{2.10}$$

Flux cut by the conductor in time dt,

$$\phi = B \times A = B.l.dx \sin\theta \tag{2.11}$$

Induced emf,

$$e = \frac{Bldx \sin\theta}{dt} = Blv \sin\theta \tag{2.12}$$

Since, $\frac{dx}{dt} = v, velocity$

According to Fleming's Right-Hand Rule, the direction of induced emf will be $\boldsymbol{v} \times \boldsymbol{B}$.

2.5.2 Statically induced emf

When the coil and magnetic field system both are stationary but the magnetic field linking with the coil changes (by changing the current producing the field), the emf thus induced in the coil is called statically induced emf.

The statically induced emf may be:

- Self-induced emf
- Mutually induced emf

Self-induced emf: The emf induced in a coil due to the change of flux produced by it linking with its own turns is called self-induced emf as shown in fig. 2.5. The direction of this induced emf is such that it opposes the cause which produces it (Lenz's law) i.e., change of current in the coil.

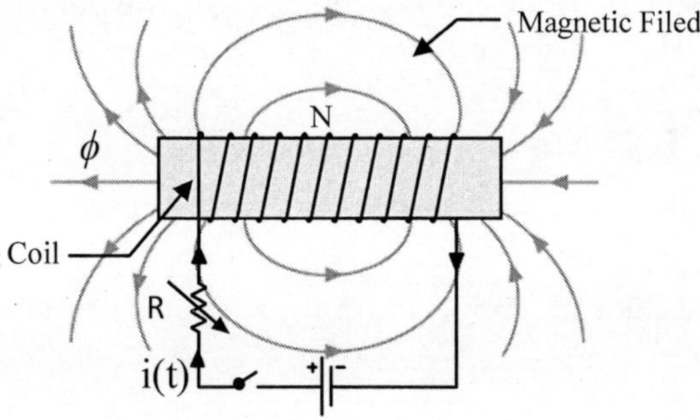

Fig. 2.5 Flux produced by coil linking with its own turns

Since the rate of change of flux linking with the coil depends upon the rate of change of current in the coil. The magnitude of self-induced emf will be directly proportional to the rate of change of current in the coil, i.e.,

$$e \propto \frac{d\,i}{d\,t} \quad or \quad e = L\frac{d\,i}{d\,t} \tag{2.13}$$

Here, '*i(t) or i*' is instantaneous current varying with respect to time. The direction of emf induced in coil opposes the applied voltage when the DC source is suddenly connected. The induced emf vanishes when steady state DC current flows in the circuit.

Mutually induced emf: The emf induced in a coil due to the change of flux produced by another (neighboring) coil, linking with it is called mutually induced emf, as shown in fig.2.6. Since the rate of change of flux linking with coil 'B' depends upon the rate of change of current in coil A, the magnitude of mutually induced emf in coil B, due to change in current in coil A will be directly proportional to the rate of change of current in coil 'A'.

$$e_m \propto \frac{d\,i_1}{d\,t} \quad or \quad e_m = M\frac{di_1}{d\,t} \tag{2.14}$$

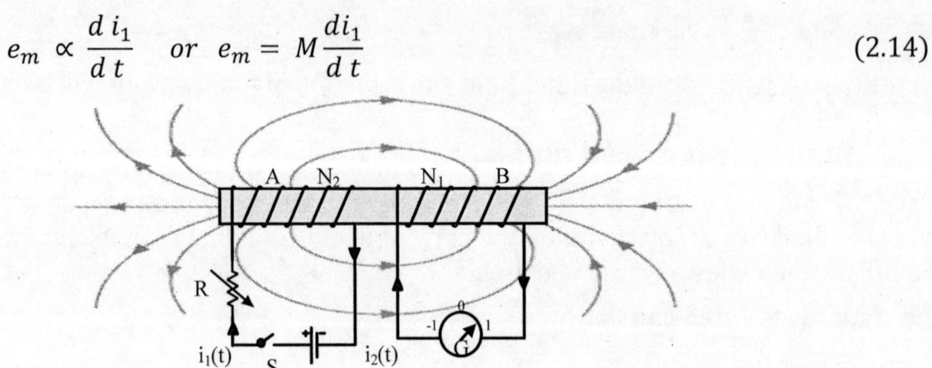

Fig. 2.6 Flux produced by coil-A linking with coil-B

Where M is a constant of proportionality and is called mutual inductance or co-efficient of mutual inductance. When a large number of current-carrying coils are present in an electromagnetic device, the mutually induced emf in a particular coil due to current changes in other coils will be obtained in a similar way considering the polarity (± sign) of the induced voltage.

2.6 Magnetization or B-H curve

The magnetization curve or B–H Curve indicates how the flux density (B) varies with the magnetizing force or magnetic field intensity (H).

For non-magnetic materials like air, copper, rubber, wood etc. the relationship between B and H is given by

$$B = \mu_0 H \tag{2.15}$$

Here, $\mu_0 = 4\pi \times 10^{-7} H/m$ is constant,

$$B \propto H \tag{2.16}$$

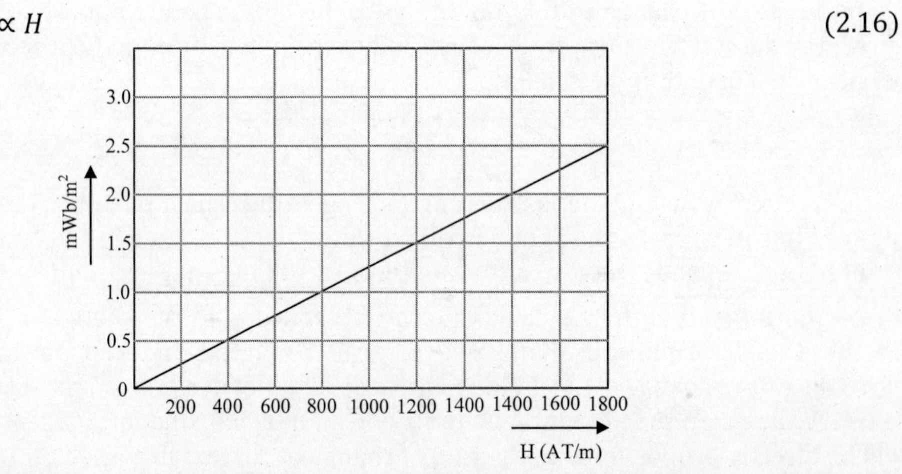

Fig. 2.7 B-H Curve for Non-Magnetic Material

Hence the B-H curve of a non-magnetic material is a straight line passing through the origin as shown in fig. 2.7. It can be observed that the curve never saturates no matter how great the flux density is; secondly, a large m.m.f is required to produce a given flux in the non-magnetic material e.g., air, glass, wood, plastic etc.

For magnetic materials such as iron, steel etc. the relation between B and H is given by

$$B = \mu_0 \mu_r H \tag{2.17}$$

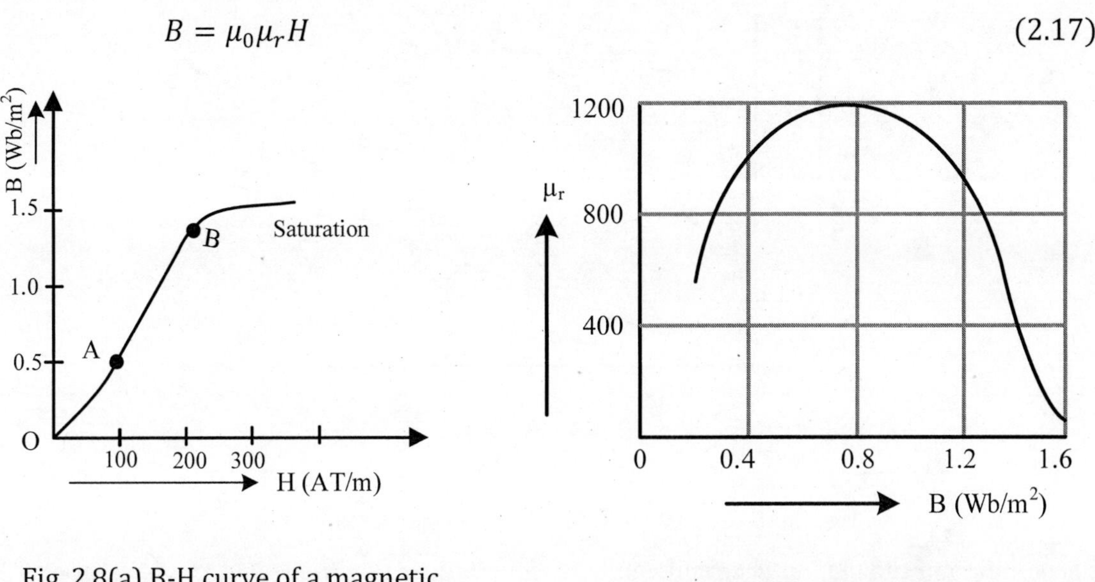

Fig. 2.8(a) B-H curve of a magnetic material

Fig. 2.8(b) μ_r-H curve

The μ_r is not constant but varies with flux density. Hence B-H curve of magnetic material is not constant. The general shape† of the B–H Curve of a magnetic material is shown in fig. 2.8(a). The shape of the curve is non-linear. This indicates that the relative permeability of a magnetic material is not constant but varies with respect to B or H. Lesser mmf (AT/m) results in comparatively large flux density as that of non-magnetic materials.

The value of relative permeability $\mu_r (= B/\mu_0 H)$ largely depends on the value of flux density at which the device operates at that instant of time. fig. 2.8(b) shows how relative permeability μ_r of magnetic material (for cast steel) varies with flux density.

---

†The shape of the curve is slightly concave up for low flux density (portion OA) and exhibits a straight-line character (Portion AB, μ_r of the material is almost constant) for medium flux density. For higher flux density the curve concave down and get saturated. Hence, any further increase in H does not increase B in same proportion.

The B–H curves of some of the common magnetic materials are shown in fig. 2.9. For electrical grade Silicon sheet steel, one requires lesser magnetic field intensity as that of cast –steel and cast iron for similar Flux density B.

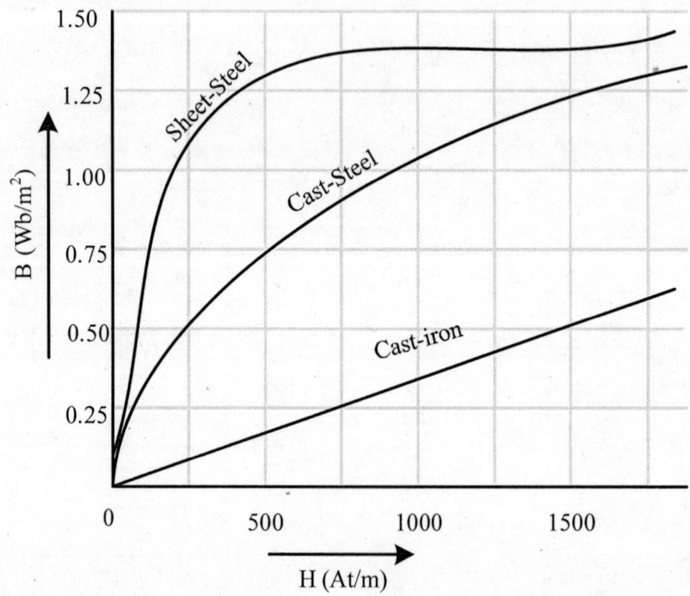

Fig. 2.9 B-H curve for different magnetic materials

The solution of the magnetic circuit can be easily obtained by the use of the B-H curve. The procedure is as under.

- Corresponding to the flux density B in the material, find the magnetizing force H from the B-H curve of the material.
- Compute the magnetic length l.
- m.m.f required $H \times l$

The use of the B-H curve for magnetic calculation saves a lot of time.

2.7 FLUX LINKAGE VS CURRENT CHARACTERISTICS OF MAGNETIC CIRCUIT

Figure 2.10 shows a typical variation of flux linkage (proportional to B or voltage) and Current (proportional to H). When the air gap decreases the flux linkage or flux density increases for the same current or H. The flux linkage current response is nonlinear and, in most cases, can be regarded as piece wise linear. The unsaturated part, the knee point where the machine tends to saturate and the saturated portion is three distinct regions of operations in the B-H curve. Most of the electrical machines are designed to operate at the knee point of the B-H curve under rated conditions. When the machine is slightly overloaded, it tends to operate in the saturation region, where flux density does not rise in the same proportion with respect to the increase in H as that of the unsaturated region and nonlinear region of the BH curve. The incremental rise of torque or power with an incremental H (current) is affected respectively for the motor and generator.

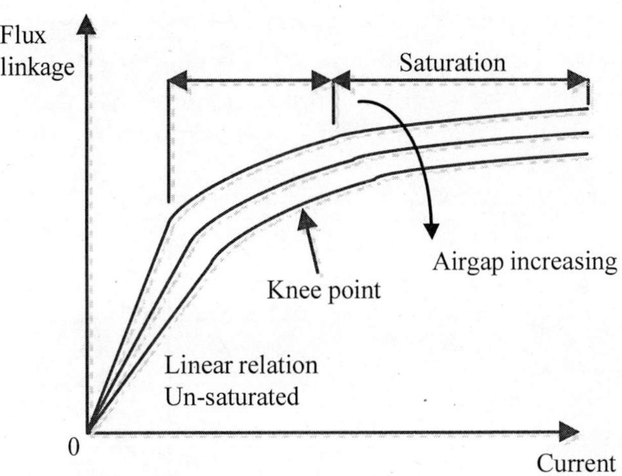

Fig. 2.10 The relationship between flux linkage and current while the air gap is changing

2.8 Magnetic Hysteresis

In India, the standard supply frequency is 50 Hz and the voltage is 415 Volt three phase and 230 Volt single phase for domestic use. Other countries use different standard voltages and supply frequencies for their industrial and domestic uses. The supply frequencies for aircraft (400 Hz) and spaceships (1000 Hz or more) are even higher than that of 50 Hz or 60 Hz. When the supply frequency is higher, the overall weight of the electromagnetic device for identical power output is highly reduced. This is required in airborne equipment like aircraft, drones, and spacecraft to have the lowest optimum weight.

A magnetic circuit may be subjected to alternating magnetization when the supply voltage is alternating current (AC) supply. In alternating magnetization, the magnetic core is magnetized first in one direction in the first half cycle of alternating current and then in the other direction in the next half cycle of AC supply. In this process of magnetization, the phenomenon of lagging of *magnetic flux density (**B**)* behind the *magnetizing force (**H**)* in a magnetic material is called ***Magnetic Hysteresis***. The term 'hysteresis' is derived from the **Greek** word ***hysterein*** meaning to **lag behind**.

In the case of Direct Current (DC) machines, the rotating armature core is subjected to rotational magnetization, i.e., the rotating armature always sees flux coming from the North pole and South poles in a cyclic manner, the currents flowing through armature conductors also change from positive DC to negative DC and negative DC to positive DC as the armature rotates. The commutator and brush arrangement act as a rotating rectifier.

2.8.1 Hysteresis loop

To understand the complete phenomenon of magnetic hysteresis, consider a coil of N turns wound on an un-magnetized iron bar AB [see fig 2.11(a)]. The magnetizing force or magnetic field intensity produced by the coil can be changed by varying the current through the coil. It can be seen that when the iron bar is subjected to one complete cycle of magnetization, the resultant *B-H* curve traces a loop *'abcdefa'* called as *hysteresis loop* [see fig. 2.11(b)].

Note that ***B*** always lags behind ***H***. Thus, at point 'b', ***H*** is zero but flux density ***B*** has a positive finite value ob. Similarly, at point 'e', ***H*** is zero, but flux density has a finite negative value 'oe'. This tendency of flux ***B*** to lag behind magnetizing force ***H*** in known as magnetic hysteresis.

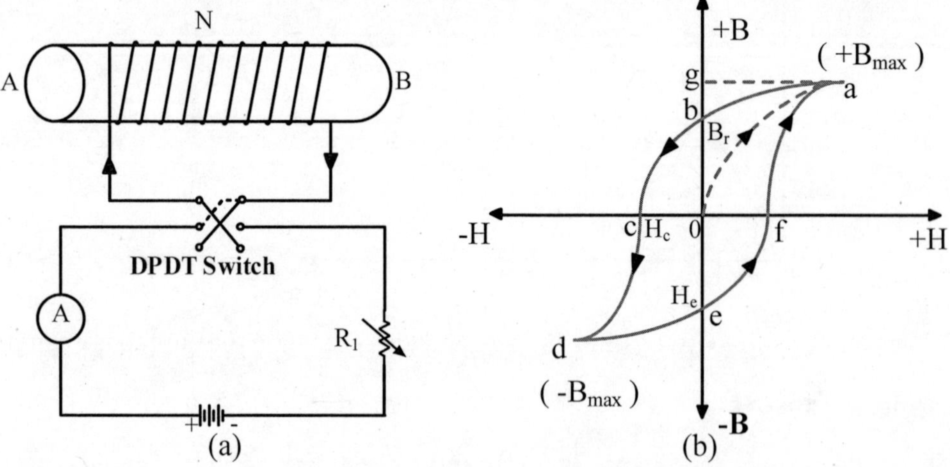

Fig. 2.11 Circuit and Hysteresis Loop

When the current in the coil is zero, the ***H*** is zero and hence ***B*** in the iron bar is zero. When ***H*** is increased by increasing the coil current, the magnetic flux density also increases until the point of maximum magnetic flux density ($+B_{max}$) at point 'a' is reached. If B_{max} is the maximum possible saturated flux density that can be obtained in the iron rod, the material is saturated and beyond this point, the magnetic flux density will normally not increase in the same proportion regardless of any increase in magnetizing force (***H***). The ideal ***B-H*** curve follows the path 'oa' (see the hysteresis loop), also referred to as the magnetization curve.

Now, if the ***H*** is gradually decreased by decreasing the coil current, it is found that the magnetic flux density does not decrease along the path 'oa' but instead follows the path **'**ab'. At point **'**b', the magnetizing force is zero but magnetic flux density in the material has a finite value (equal to 'ob') called residual flux density ($+B_r$). This value of flux density 'ob' retained by the magnetic material is called ***residual magnetism*** and the power of retaining this residual magnetism is called ***retentivity*** of the material.

To demagnetize the iron bar i.e., to remove the residual magnetism 'ob', the magnetizing force is reversed by reversing the coil current. When ***H*** is gradually increased in the reverse direction, the ***B-H*** curve follows the path **bc** so that when ***H*** = '-oc', the residual magnetism drops to zero. The values of ***H*** = '-oc' required to completely remove the residual magnetism is known as ***Coercive force*** *(H_c)*. Similarly, 'of' is the magnetizing force utilized to wipe off the residual magnetism 'oe', which is the reverse direction. Hence, 'cf' is the total coercive force required in one cycle of magnetization to wipe off the residual magnetism in either direction.

Now, if ***H*** is further increased in the reverse direction, the material again saturates in the reverse direction (point 'd'). Reducing ***H*** to zero and then increasing it in the positive direction traces the curve 'defa'. Therefore, when an iron bar is subjected to one complete cycle of magnetization, the ***B-H*** curve traces a closed loop 'abcdefa' called a *hysteresis loop*.

Area 'agbefa' shows energy stored in the magnetic field during the positive half cycle of magnetization, 'abefa' shows hysteresis loss per half cycle of H, and area 'agb' shows energy released by the magnetic field during the positive half cycle.

The area of hysteresis loop 'abcdefa' represents the hysteresis loss per cycle of magnetization.

2.8.2 HYSTERESIS LOSS

When a magnetizing force is applied, the magnetic material is magnetized and the molecular magnets are aligned in a particular direction. However, when the magnetizing force in a magnetic material is reversed, the internal friction of the molecular magnets opposes the reversal of magnetism, resulting in hysteresis and hence residual magnetism. To overcome this internal friction of the molecular magnets (or to wipe off the residual magnetism), a part of the magnetizing force is used. The work done by the magnetizing force against this internal friction of molecular magnets produces heat. This energy, which is wasted in the form of heat due to hysteresis, is called hysteresis loss.

Hysteresis loss occurs in all the magnetic parts of electrical machines where there is a reversal of magnetization. This loss results in the wastage of energy in the form of heat. Consequently, it increases the temperature of the machine which is undesirable, Therefore, a suitable magnetic material is selected for the construction of such parts, i.e., silicon steel is most suitable in which hysteresis loss is minimum. Cobalt steel operating at higher flux density though costly is also in use. Ferrite cores are used for high-frequency applications.

2.8.3 Importance of hysteresis loop

The shape and size of the hysteresis loop are material-dependent. The choice of magnetic material for a particular application depends upon the shape and size of the hysteresis loop. Consider the following cases of hysteresis loop to understand its importance.

Case:1

The smaller the area of the hysteresis loop of a magnetic material, the less is the hysteresis loss (e.g., silicon steel) as shown in fig.2.12. Therefore, the electrical grade silicon steel is widely used for making cores of transformers and rotating electric machines which are subjected to rapid cyclic reversals of flux density due to either alternating or rotating magnetizations.

Case:2

The larger the area of the hysteresis loop as shown in fig. 2.13 of hard steel indicates that this magnetic material has *high* retentivity and coercivity. The magnetism of hard steel does not destroy easily by an external magnetic field. Therefore, hard steel is quite suitable for making permanent magnets, but due to the large area of the hysteresis *loop*, there is *greater hysteresis loss for this reason hard steel is not suitable for* making cores of transformers and rotating electric machines.

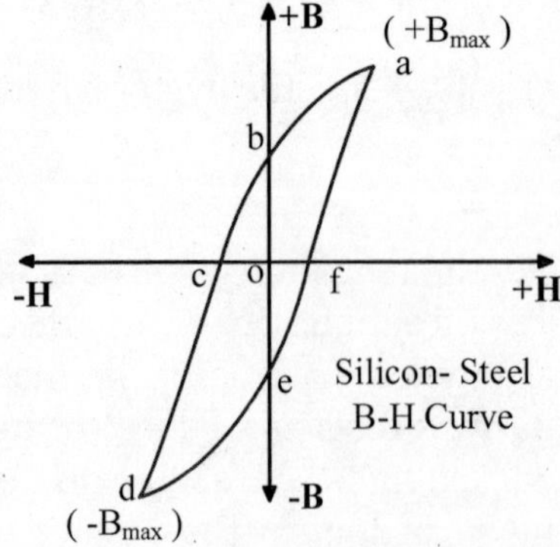

Fig.2.12 B-H Curve of Silicon Steel

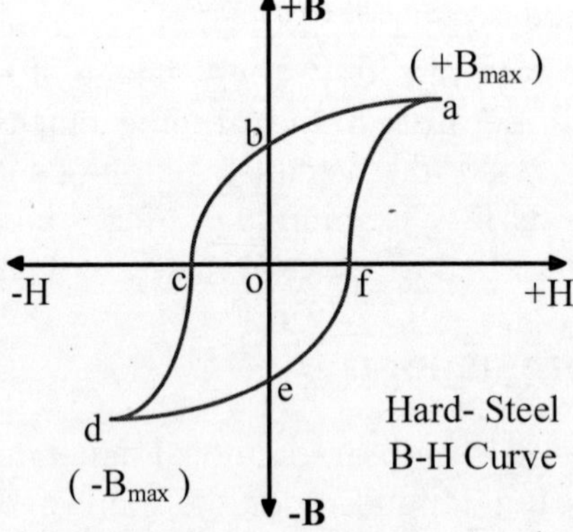

Fig. 2.13 B-H Curve of Hard Steel

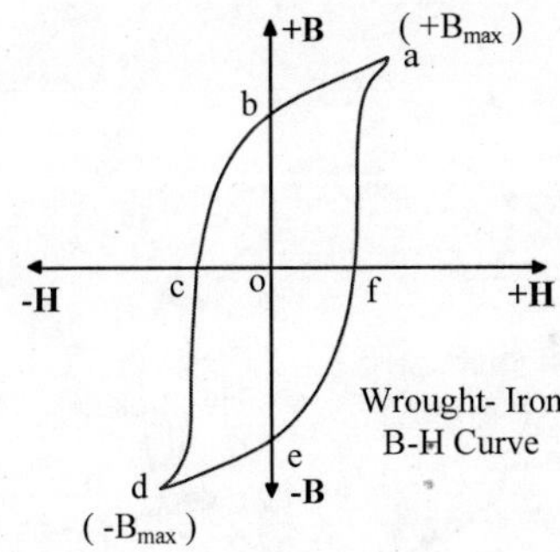

Fig. 2.14 B-H Curve for Wrought Iron

Case:3

The hysteresis loop for wrought iron has fairly good residual magnetism and coercivity. Therefore, it is used for making cores of electromagnets. The B-H curve for wrought iron shown in fig 2.14.

Magnetostriction *is a property of magnetic materials that causes the grains of ferromagnetic material to cyclic macroscopic change their shape or dimensions during the process of magnetization. The effect was first identified in 1842 by James Joule when observing a sample of iron*. This effect causes energy loss due to frictional heating in susceptible ferromagnetic cores. The effect is also responsible for mechanical vibration in the magnetic core causing low-pitched humming sound that can be heard coming from transformers, where oscillating AC currents produce a changing magnetic field.*

2.9 Magnetic Circuits

The closed path followed by magnetic flux is called a magnetic circuit. A magnetic circuit usually consists of magnetic materials (e.g., iron, soft steel, ferrite, etc.) having high permeability. In this circuit, magnetic flux starts from a point and finishes at the same point after completing its path.

Fig 2.15 shows a solenoid having N turns wound on an iron core (ring). When current I ampere is passed through the solenoid, magnetic flux ϕ Wb is set-up in the core.

Let,

ℓ = mean length of the magnetic circuit in m;

a = area of cross-section of core in m^2;

μ_r = relative permeability of core material.

Flux density in the core material,

$$B = \frac{\phi}{a} \quad Wb/\,m^2 \tag{2.18}$$

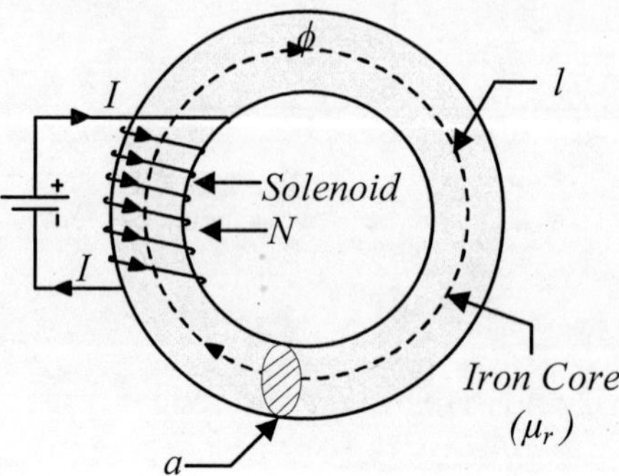

Fig.2.15 Magnetic Circuit

Magnetizing force in the core material.

$$H = \frac{B}{\mu_o \mu_r} = \frac{\phi}{a\mu_o \mu_r} \quad AT/m \tag{2.19}$$

According to work law, the work done in moving a unit pole once round the magnetic circuit (or path) is equal to the ampere-turns enclosed by the magnetic circuit. i.e.,

$$Hl = NI \tag{2.20}$$

$$\frac{\phi}{a\,\mu_0\,\mu_r} \times l = NI \tag{2.21}$$

$$Flux = \frac{m.m.f}{reluctance} = \frac{NI}{\mathcal{R}} \tag{2.22}$$

$$\phi = \frac{NI}{l/a\,\mu_0\,\mu_r} = \frac{NI}{\mathcal{R}} = \frac{NI\,a\mu_o\mu_r}{l} \quad Wb \tag{2.23}$$

The above expression reveals that the amount of flux set-up in the core is

- Directly proportional to the number of turns 'N' and the current flowing through it 'I' ampere i.e., NI, called magneto motive force (mmf). It shows that the flux increases if either of the two quantities increases and vice-versa.
- Inversely proportional to the reluctance of the magnetic path. The reluctance is the opposition offered to the magnetic flux by the magnetic path. The lower is the reluctance, the higher will be the flux, and vice-versa.

It may be noted that the above expression ($\phi = mmf/reluctance$) has a strong resemblance to Ohm's law for electric current (I = emf/resistance). Because of this similarity, the above expression is sometimes referred to as Ohm's law of the magnetic circuit.

2.10 Electro-Mechanical Energy Conversion Device

A device (machine) that makes possible the conversion of energy from electrical to mechanical form or from mechanical to electrical form is called an Electromechanical energy conversion device or Electromechanical transducer.

Depending upon the conversion of energy from one form to the other, the electro-mechanical device can be named a motor or generator.

Motor: An electro-mechanical device (electrical machine) that converts electrical energy or power (EI) into mechanical energy or power (ωT) is called a motor.

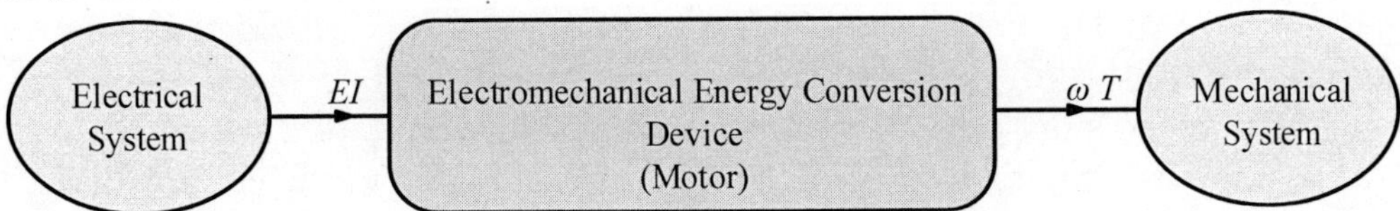

Fig. 2.16 Schematic flow of power in electrical Motor

Electric motors are used for driving industrial machines e.g., hammer presses, drilling machines lathes, shapers, blowers for furnaces, cranes, traction locomotives, electric vehicles (EV), electric hybrid vehicles (EHV), lifts, elevators, drones, etc., and domestic appliances e.g., refrigerators, fans, water pumps, toys, mixers, etc. The block diagram of energy conversion, when the electro-mechanical device works as a motor, is shown in fig. 2.16.

Generator: An electro-mechanical device (electrical machine) that converts mechanical energy or power (ωT) into electrical energy or power (EI) is called a generator.

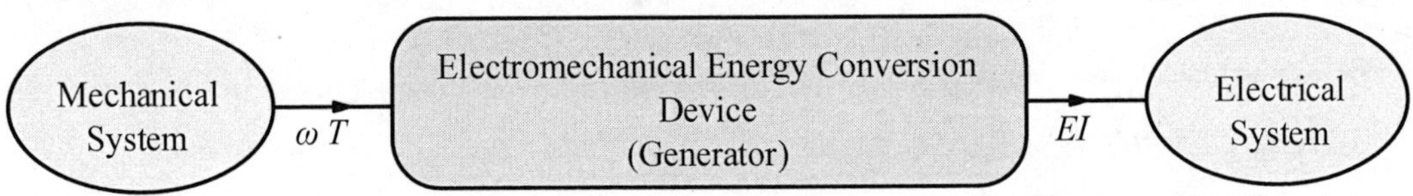

Fig. 2.17 Schematic flow of power in Generator

Generators are used in hydroelectric power plants, steam power plants, diesel power plants, nuclear power plants, wind turbines etc. In the above said power plants, various natural sources of energy are first converted into mechanical energy, and then it is converted into electrical energy with the help of generators. The block diagram of energy conversion, when the electro-mechanical device works as a generator, is shown in fig. 2.17. The same electro-mechanical device is capable of operating either as a motor, generator, or brake depending upon whether the input power is electrical or mechanical [see fig. 2.18]. Thus, the motoring and generating action is reversible. The generating action results in counter torque to the system opposing the driving torque. In an electric brake, the generated power is consumed within the system to cause opposing torque to halt or lower the speed.

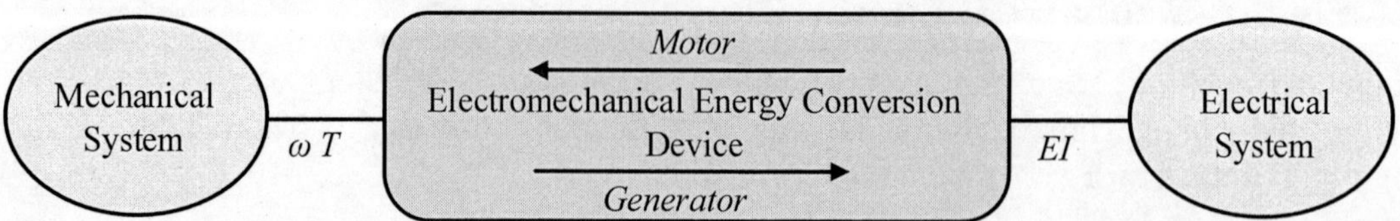

Fig. 2.18 Same machine can work as a generator or motor

The conversion of energy either from electrical to mechanical or from mechanical to electrical takes place through the magnetic field. During conversion, the whole of the energy in one form is not converted in the other useful form. The input power is divided into the following three parts:

- Most of the input power is converted into useful output power.
- Some of the input power is converted into heat losses (i^2R) which are due to the flow of current in the conductors, magnetic losses (hysteresis and eddy current losses), and friction losses.
- A small portion of input power is stored in the magnetic field of the electro-mechanical device.

2.11 BASIC PRINCIPLE OF ENERGY CONVERSION AND ENERGY BALANCE

According to it, energy can neither be created nor destroyed, it can merely be converted from one form into another. In an energy conversion device, out of the total input energy, some energy is converted into the required form, some energy is stored and the rest is dissipated for a motor, it can be written as

$$\begin{matrix} Total\ Electrical \\ Energy\ Input \end{matrix} = \begin{matrix} Mechanical\ Energy \\ output \end{matrix} + \begin{matrix} Total\ Energy \\ Stored \end{matrix} + \begin{matrix} Total\ Energy \\ dissipated \end{matrix} \quad (2.24)$$

For generator action, it can be written as

$$\begin{matrix} Mechenical\ Energy \\ Input \end{matrix} = \begin{matrix} Total\ Electrical \\ Energy\ output \end{matrix} + \begin{matrix} Total\ Energy \\ Stored \end{matrix} + \begin{matrix} Total\ Energy \\ dissipated \end{matrix} \quad (2.25)$$

Now the various forms of energy can be represented as

W_{ei} = Electrical energy input from the supply means

$W_{mo} =$ Mechanical energy output

The total energy stored in any device = Energy stored in the magnetic field, W_{es} + Energy stored in the mechanical system, W_{ms}

Total energy dissipated= Energy dissipated in an electric circuit as ohmic losses+ Energy dissipated as magnetic core loss (hysteresis and eddy-current losses) +Energy dissipated in the mechanical system (friction and windage losses)

Thus, the energy balance equation (2.24) can be written in more simplified terms as

$$W_{ei} = W_{mo} + (W_{es} + W_{ms}) + (Ohmic\ energy\ losses + Coupling\ field\ energylosses) + (Energy\ losses\ in\ mechanical\ system) \quad (2.26)$$

The subscripts *e, m, i, s,* and *o* stand for electrical, mechanical, input, stored, and output respectively. For example, subscript *ei* denotes electrical energy input, subscript *ms* denotes mechanical energy stored.

If the appropriate terms are grouped together, then the energy balance equation becomes,

$$(W_{ei} - Ohmic\ energy\ losses) = W_{mo} + W_{ms} + Energy\ losses\ in\ mechenical\ system + (W_{es} + Coupling\ field\ energylosses) \quad (2.27)$$

$$W_{elec} = W_{mech} + W_{fld} \quad (2.28)$$

Equation (2.26) leads to the electromechanical energy conversion model.

For the losses conversion system of fig. 2.19 equation (2.27) can be written in differential form as

$$dW_{elec} = dW_{mech} + dW_{fld} \quad (2.29)$$

Where,

dW_{elec}= Differential electrical energy input to coupling field

dW_{mech} = Differential mechanical energy output

dW_{fld}=Differential change in energy stored in the coupling field

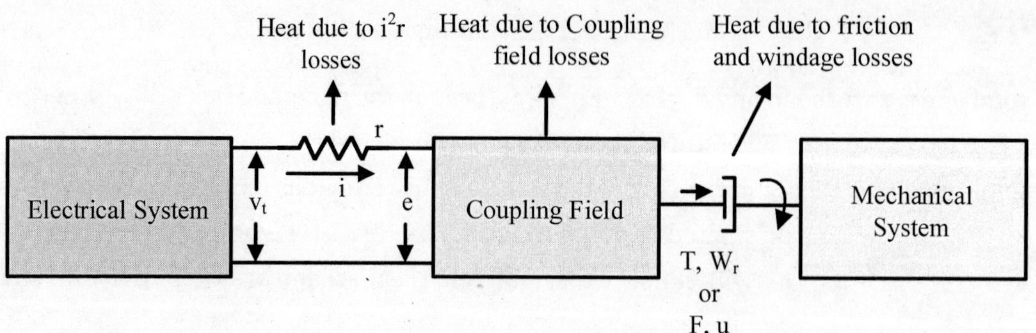

Fig. 2.19 Representation of electromechanical energy conversion system

From fig. 2.20, the differential electrical energy input in time dt is

$$dW_{ei} = v_t i \, dt \tag{2.30}$$

Ohmic loss in resistance r in time dt is $i^2 r \, dt$.

Differential electrical energy input to the coupling field,

$$dW_{elec} = dW_{ei} - ohmic\ loss$$

$$= (v_t - ir) i \, dt = ei \, dt \tag{2.31}$$

Equation (2.31) now becomes

$$eidt = dW_{mech} + dW_{fld} \tag{2.32}$$

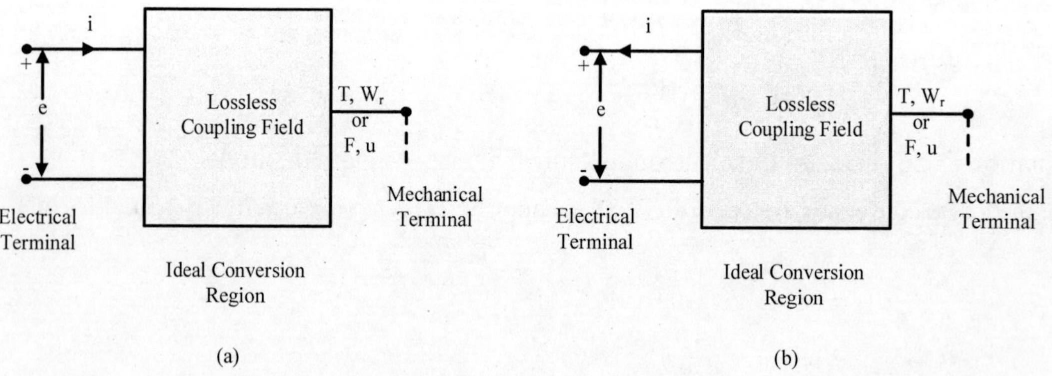

Fig. 2.20 Representation of lossless electrotechnical energy conversion system (a)Motoring mode (b) Generating mode

The energy balance equation (2.32) is obtained by applying the principle of conversion of energy to motoring mode. This equation (2.32) along with *Faraday's law* of induced *emf* form the fundamental basis for the analysis of energy conversion devices.

2.12 COUPLING FIELD REACTION

The energy stored in the coupling field must produce action and reaction on the electrical and mechanical systems for the conversion of energy.

- In a motor, this reaction is the counter *emf* 'e', the coupling field extracts energy proportional to e.i from the electrical system, converts and delivers energy proportional to $T.\omega_r$ (or F.u) to the mechanical system for rotary (or linear) motion.
- In the generator, this reaction is the counter torque, opposite to the applied mechanical torque of the prime mover, thus the coupling field extracts mechanical energy proportional to $T.\omega_r$ (torque. speed) from the mechanical system, converts and delivers it as electrical energy proportional to (e.i) to the electrical system.

Thus, it may be seen that the coupling field serves as the energy conversion region.

2.13 ENERGY STORED IN A MAGNETIC FIELD

Energy can be stored or retrieved from a magnetic system by means of an exciting coil connected to an electric source. Consider, for example, the magnetic system of an attracted armature relay as shown in fig. 2.21. The resistance of the coil is shown by a series of lumping outside the coil which then is regarded as an ideal lossless coil. The coil current causes magnetic flux to be established in the magnetic circuit. It is assumed that all the flux ϕ is confined to the iron core and therefore links all the N number of turns creating the coil flux linkages of

$$\lambda = N\,\phi \tag{2.33}$$

The flux linkage causes a reaction emf of

$$e = \frac{d\lambda}{dt} \tag{2.34}$$

to appear at the coil terminals with polarity (as per Lenz's law) shown in Fig. 2.21. The associated circuit equation is

$$v = iR + e \tag{2.35}$$

$$= iR + \frac{d\lambda}{dt} \tag{2.36}$$

The electric energy input into the ideal coil due to the flow of current i in timedt is

$$dW_e = ei\;dt \tag{2.37}$$

Assuming for the time being that the armature is held fixed at position x, all the input energy is stored in the magnetic field. Thus

$$dW_e = ei\;dt = dW_f \tag{2.38}$$

where dW_f, is the change in field energy in time dt. The expression for e in equation (2.34) is substituted in equation (2.38) as

$$dW_e = i\,d\lambda = F\,dx = dW_f \tag{2.39}$$

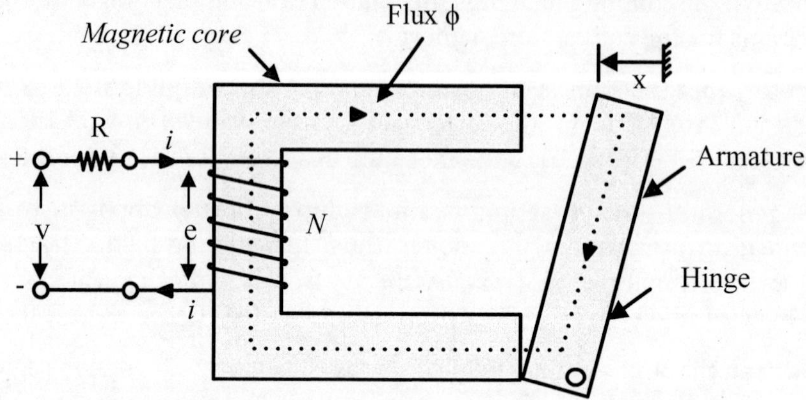

Fig. 2.21 Armature relay acting on the force of attraction

where $\mathcal{F} = Ni$, the magnetomotive force (mmf).

The relationship $i - \lambda$ or $\mathcal{F} - \lambda$ is a functional one corresponding to which the magnetic circuit in general is nonlinear (and is also history-dependent, i.e., it exhibits hysteresis). The energy absorbed by the field for finite change in flux linkages for flux is obtained from equation (2.39) as

$$\Delta W_f = \int_{\lambda_1}^{\lambda_2} i(\lambda)d\lambda = \int_{\phi_1}^{\phi_2} \lambda(\phi)d\phi \tag{2.40}$$

As the flux in the magnetic circuit undergoes a cycle $\phi_1 \rightarrow \phi_2 \rightarrow \phi_1$, an irrecoverable loss in energy takes place due to hysteresis and eddy-currents in the iron, assuming here that these losses are separated out and are supplied directly by the electric source. This assumption renders the ideal coil and the magnetic circuit as a conservative system with energy interchange between themselves so that the net energy is conserved.

The energy absorbed by the magnetic system to establish flux ϕ (or flux linkages λ) from initial zero flux is

$$W_f = \int_0^{\lambda} i(\lambda)d\lambda = \int_0^{\phi} \lambda(\phi)d\phi \tag{2.41}$$

This then is the energy of the magnetic field with a given mechanical configuration when its state corresponds to flux ϕ (or flux linkages λ).

The i-λ relationship is indeed the magnetization curve which varies with the configuration variable x (fig. 2.22): the air gap between the armature and core varies with position x of the armature. The total reluctance of the magnetic path decreases as x increases. The i-λ relationship for different values of $x$is shown in fig. 2.22. The relationship function can be expressed as

$$i = i\,(\lambda, x);\ \text{if}\ \lambda\ \text{is the independent variable} \tag{2.42}$$

$$\lambda = \lambda(i, x);\ \text{if}\ i\ \text{the independent variable} \tag{2.43}$$

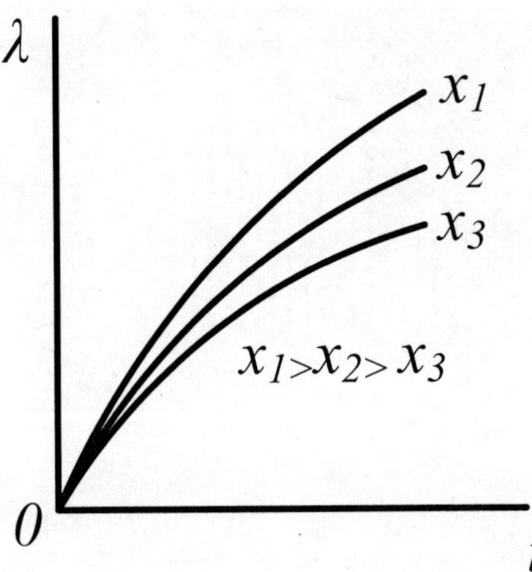

Fig. 2.22 i-λ relationship with variable x

Therefore, the field energy (equation (2.43) is in general a function of two variables,

$$W_f = W_f(\lambda, x) \tag{2.44}$$

$$W_f = W_f(i, x) \tag{2.45}$$

According to equations (2.44) and (2.45) field energy is determined by the instantaneous values of the system states (λ, x) or (i, x) and is independent of the path followed by these states to reach the present values. This means that the field energy at any instant is history-independent.

A change in λ with fixed x causes electric-magnetic energy interchange governed by the circuit equation (2.41) and the energy equation (2.44). Similarly, if x is allowed to change with fixed λ , energy with interchange between the magnetic circuit and the mechanical system.

As per equation (2.41) the field energy is the area between the λ-axis and i-λ curve as shown in fig. 2.23. A new term, co-energy is now defined as

$$W_f'(i, x) = i\lambda - W_f(\lambda, x) \tag{2.46}$$

wherein by expression λas $\lambda(i, x)$, the independent variables of W_f, become i and x. The co-energy on fig. 2.23 is shown to be the complementary area of the i-λ curve.

$$W_f' = \int_0^i \lambda di$$

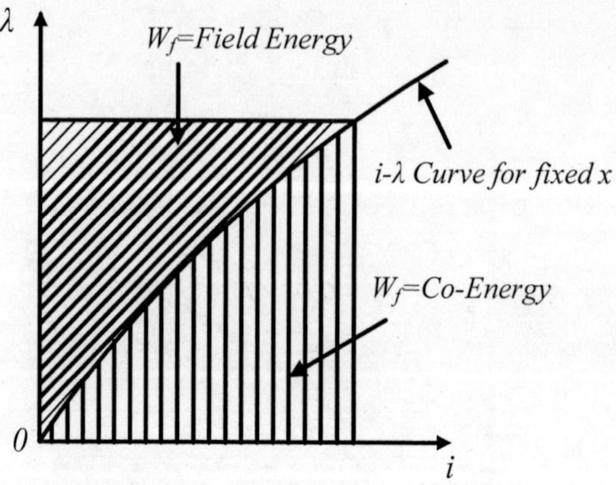

Fig. 2.23 Field energy and Co-energy

Linear Case

Electromechanical energy conversion devices are built with air gaps in the magnetic circuit which serve to separate the stationary and moving members. As a result, the i-λ relationship of the magnetic circuit is almost linear; also, the losses of magnetic origin are separately accounted for by semi-empirical methods. With the linearity assumption, the analysis is greatly simplified. Losses and certain nonlinear effects may then be incorporated at a later stage.

Assuming linearity,

$$W_f = \frac{1}{2}i\lambda = \frac{1}{2}\mathcal{F}_{mm}\phi = \frac{1}{2}\Re\phi^2 \tag{2.47}$$

where it is known, $\Re = \mathcal{F}_{mm}/\phi$ = reluctance of the magnetic circuit. Since the coil inductance

$$L = \lambda/i \tag{2.48}$$

the field energy can be expressed as

$$W_f = \frac{1}{2}\frac{\lambda^2}{L} = \frac{1}{2}Li^2 \tag{2.49}$$

In the linear case, the inductance L is independent of i but is a function of variable x. Thus the field energy is a special function of two independent variables λ and x, i.e.,

$$W_f(\lambda, x) = \frac{1}{2}\frac{\lambda^2}{L(x)} \tag{2.50}$$

The field energy is distributed throughout the space occupied by the field. Assuming no losses and constant permeability, the energy density of the field is

$$W_f = \int_0^B HdB = \frac{1}{2}HB = \frac{1}{2}\frac{B^2}{\mu} = \frac{1}{2}\mu H^2\ J/m^3 \tag{2.51}$$

Where H =magnetic field intensity (AT/m)

B=magnetic flux density (T)

The energy density expression of equation (2.51) is important from the point of view of design wherein the capability of the material is to be fully utilized in arriving at the gross dimensions of the device.

For the linear case, it easily follows from equation (2.47) that co-energy is numerically equal to energy, i.e.,

$$W_f' = W_f = \frac{1}{2}\lambda i = \frac{1}{2}\mathcal{F}_{mm}\phi \tag{2.52}$$

Also, in terms of the coil inductance

$$W_f' = \int_0^i \lambda . di = \frac{1}{2}Li^2 \tag{2.53}$$

Or in general

$$W_f'(i,x) = \frac{1}{2}L(\dot{x})i^2 \tag{2.54}$$

If A (m^2) and l (m) are the area and length dimensions of the field, then from equation (2.41)

$$W_f' = \frac{W_f'}{Al} = \int_0^{\lambda} \frac{iN}{l} d\left(\frac{\lambda}{NA}\right) = \int_0^B H\, dB \tag{2.55}$$

The expression for Co-energy density is

$$W_f' = \int_0^H B\, dH \tag{2.56}$$

which for the linear case becomes

$$W_f' = \frac{1}{2}\mu H^2 = \frac{1}{2}\frac{B^2}{\mu} \tag{2.57}$$

2.14 Basic Aspects and Physical Phenomena Involved in a Singly-Excited Magnetic System

A singly-excited system is the type of excitation system used in electromechanical energy conversion which requires only one coil to produce the magnetic field. In the singly-excited system, there is only one set of electrical input terminals and one set of mechanical output terminals. Examples of the singly-excited system are an electromagnetic relay, hysteresis motor, solenoid valve, etc.

In a singly excited system, a coil is wound around a magnetic core and is connected to a voltage source so that it produces a magnetic field. Due to this magnetic field, the rotor which is made up of ferromagnetic material experiences a torque urging it towards a region where the magnetic field is stronger, i.e., the torque exerted on the rotor tries to position it such that it gives minimum reluctance for the magnetic field. The reluctance depends upon the rotor angle.

2.14.1 Singly excited magnetic system

I. ***Electric energy input:*** Consider a simple magnetic system of a toroid, excited by a single coil as shown in fig. 2.24. The instantaneous voltage equation for the electric circuit is given as

$$As\ per\ kirchhoff's\ voltage\ law, v = ir \tag{2.58}$$

$$v_t = ir + e \tag{2.59}$$

$$e = \frac{d\lambda}{dt} \tag{2.60}$$

and

$$v_t = ir + \frac{d\lambda}{di}\, i\, dt \tag{2.61}$$

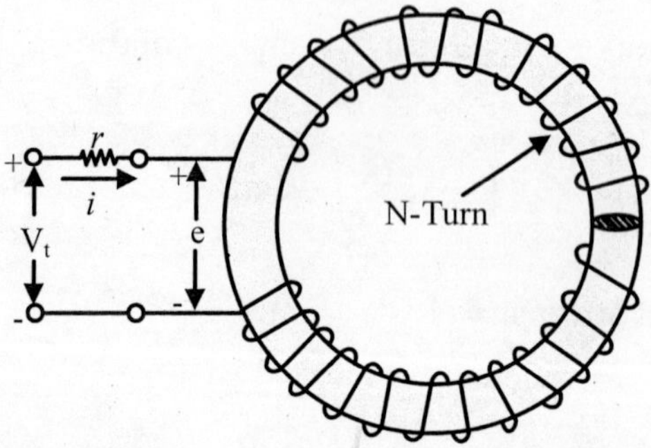

Fig. 2.24 Toroidal core excited from a single source

Here λ is the instantaneous flux linkages with the circuit. Multiplying both sides of equation 2.61 by it, we get

$$v_t idt = ri^2\ dt + i\ d\lambda \tag{2.62}$$

$$(v_t - ir)idt = i\ d\lambda \tag{2.63}$$

$$e\ idt = i\ d\lambda \tag{2.64}$$

$$\text{Since}\quad dW_{elec} = e\ idt = id\lambda \tag{2.65}$$

The flux linkages λ are equal to $NØ$ Wb-turns. Therefore, from the energy balance equation 2.32

$$dW_{elec} = id\lambda = iN\ d\phi = \mathcal{F}d\phi \tag{2.66}$$

In equation (2.66), ϕ is the instantaneous value of the coil flux and $\mathcal{F} = i\,N$ is the instantaneous coil m.m.f.

NOTE: The flow of charges or current against the reaction emf (e) causes the extraction of energy from electrical system.

II. ***Magnetic field energy stored:*** Consider a simple magnetic relay of fig. 2.25. Initially, the armature is in the open position. When switch S is closed, current ***i*** is established in the N-turn coil. Iron yoke is the South Pole. The flux setup depends on m.m.f. Ni and the reluctance of the magnetic path. The magnetic field thus produced, creates North and South poles as shown in fig. 2.25, and as a result of it, there is established a magnetic force tending to shorten the air-gap / strengthen the magnetic field B. If the armature is not allowed to move, the mechanical work done, dW_{mech} will be zero. According to equation (2.30) dW_{elec} is,

$$dW_{elec} = 0 + dW_{fld} \tag{2.67}$$

$$dW_{fld} = dW_{elec} \tag{2.68}$$

NOTE: This shows that when the movable part of any physical system is kept fixed, the entire electrical energy input is stored in the magnetic field.

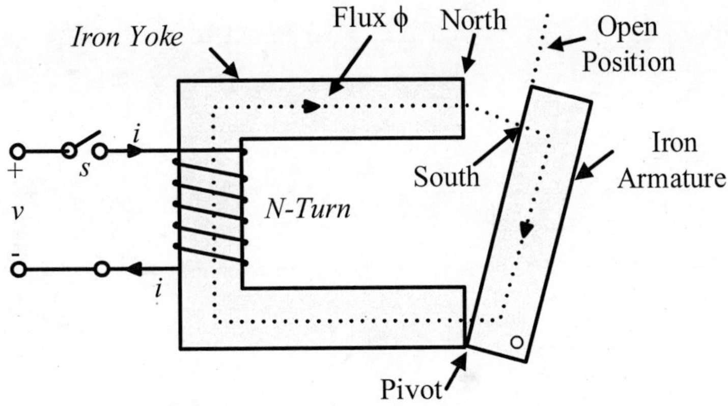

Fig. 2.25 Simple magnetic relay

fig. 2.26 (a) and (b): Pertaining to field energy and co-energy for a linear magnetic circuit, from Equation 2.66

$$dW_{fld} = dW_{elec} = i\,.\,d\lambda = \mathcal{F}.\,d\phi \tag{2.69}$$

If the initial flux is zero, then the magnetic field energy stored W_{fld} in establishing a ϕ flux or flux linkage λ_1 is given by

$$W_{fld} = \int_0^{\lambda_1} i \, . d\lambda = \int_0^{\phi_1} \mathcal{F} \, . d\phi \tag{2.70}$$

In equation 2.70, i and $\mathcal{F}$ must be expressed in terms of λ and ϕ respectively.

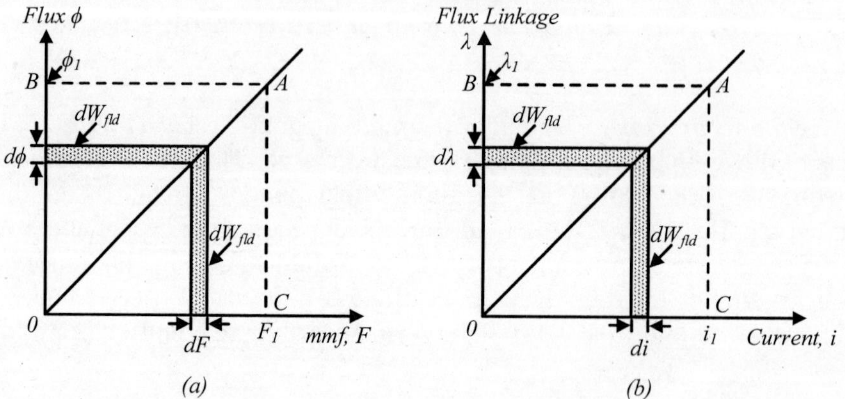

Fig. 2.26 (a) and (b): Pertaining to field energy and co-energy for a linear magnetic circuit

For fig. 2.26 (a),

$$W_{fld} = \int_0^{\phi_1} dW_{fld} = \int_0^{\phi_1} \mathcal{F} \, . d\phi = Area\ OABO \tag{2.71}$$

For fig. 2.26 (b),

$$W_{fld} = \int_0^{\lambda_1} dW_{fld} = \int_0^{\lambda_1} \mathcal{F} \, . d\lambda = Area\ OABO \tag{2.72}$$

In fig. 2.26(a) and (b)

$$\text{Area}OACO = \int dW_{fld} = \int_0^{\mathcal{F}_1} \phi \, d\mathcal{F} = \int_0^{i_1} \lambda \, di \tag{2.73}$$

This area OACO is called the co-energy W_{fed}

$$W_{fed} = \int_0^{\mathcal{F}_1} \phi \, dF = \int_0^{i_1} \lambda \, di \tag{2.74}$$

With no magnetic saturation.

$$Area\ OABO = Area\ OACO$$

$$W_{fld} = W_{fld}'$$

And $W_{fld} + W_{fld}' = \text{Area OCABO} = \phi_1 \mathcal{F}_{mm1} = \lambda_1 i_1$

In general, for a linear magnetic circuit,

$$W_{fld} = W_{fld}' = \frac{1}{2} \lambda i = \frac{1}{2} \mathcal{F}_{mm} \, \phi \tag{2.75}$$

NOTE: In equation 2.71, ϕ and λ must be expressed in terms of $\mathcal{F}$ and i respectively. Co-energy has no physical significance; it is however useful in calculating the magnetic forces.

The self-inductance L is defined as the magnetic flux linkages per ampere, i.e.,

$$L = \frac{\lambda}{i} \tag{2.76}$$

Therefore, from equation 2.75

$$W_{fld} = W'_{fld} = \frac{1}{2} Li^2 = \frac{1}{2}\frac{\lambda^2}{L} \tag{2.77}$$

This fact that field energy can be expressed in terms of circuit parameter L, clears the way for electric circuit approach to the analysis of electrical machines. i.e., the generalized theory of electrical machines. Thus, the field-energy approach serves as the physical basis for the generalized theory of electrical machines.

III. Mechanical Work Done

Consider the simple magnetic relay of fig. 2.27 again.

Case 1: *The armature is assumed to be held in the open position.*

Now when the switch is closed, the current area even from zero to $i_i = \frac{v_f}{r}$ and the flux linkage increases from zero to λ_1

So, $W_{fld} = Area\ OABO$

Case 2: *The armature is assumed to be closely held in the closed position* (fig. 2.28)

Now when the switch is closed, the current rises from zero to $i_i = \frac{v_f}{r}$, and the flux linkage rises from zero to λ_2. Remember that λ_2 is greater than λ_1 because during the armature movement reluctance decreases.

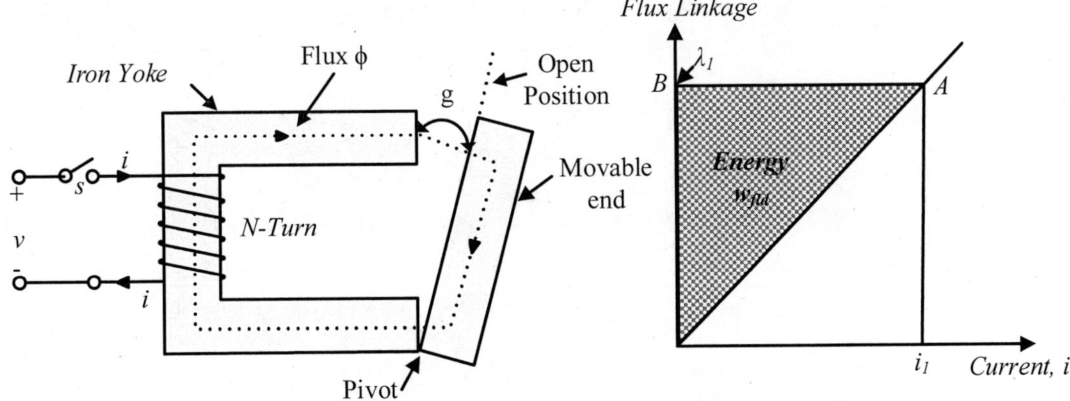

Fig. 2.27 Armature field in the open position

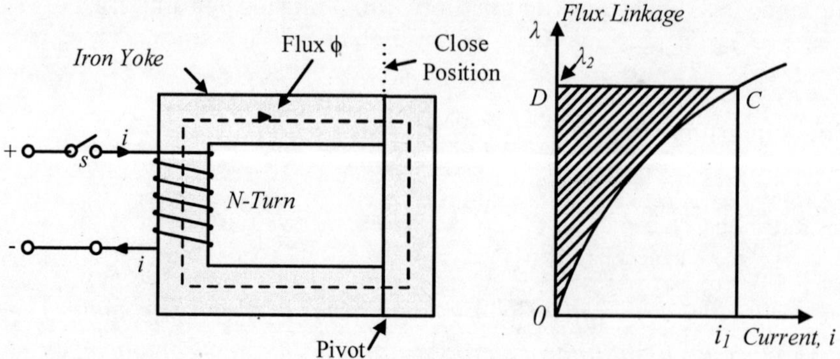

Fig. 2.28 Armature field in the closed position

NOTE: These increment in flux linkages induces a counter emf in the coil, which opposes the flow of exciting current i.

$$i = \frac{v - emf}{r} \tag{2.78}$$

The magnitude of counter emf induced in the exciting coil depends on how fast the armature moves, three cases are possible.

- ***Slow Movement:***

With the armature in the open position, the exciting current is i_i, the flux linkage is $\lambda_1$and the operating point is A (fig. 2.29). In the closed armature position, the flux linkages are λ_2 , current is i_i, and the operating point is C.Now change in the stored energy of the magnetic field W_{fld}, during the time the armature moves from the open point (point A) to the closed position (point C) is given by

W_{fld} = (Magnetic energy stored in the closed position)

− (Magnetic energy stored in the open position)

$$W_{fld} = Area\ OA'CDFO - Area\ OAA'FO$$

Electric energy input during this change is

$$W_{elec} = \int_{\lambda_1}^{\lambda_2} i_1 d\lambda = i_1(\lambda_2 - \lambda_1) = Area\ ACDFA'A \tag{2.79}$$

$$W_{elec} = W_{fld} + W_{mech} \tag{2.80}$$

$$Area\ ACDFA'A = Area\ OA'CDFO - Area\ OAA'FO + W_{mech} \tag{2.81}$$

$$W_{mech} = (Area\ ACDFA'A - Area\ OAA'FO) - Area\ OA'CDFO \tag{2.82}$$

$$W_{mech} = Area\ OACDFO - Area\ OA'CDFO \tag{2.83}$$

$$W_{mech} = Area\ OACA'O \tag{2.84}$$

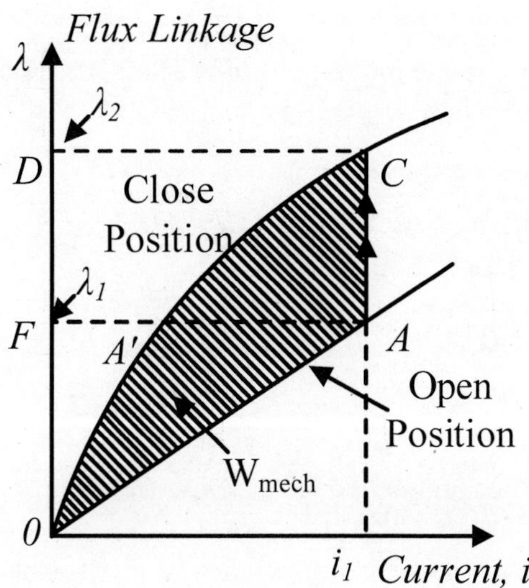

Fig. 2.29: Mechanical work done with slow armature movement

Equation 2.84 shows that the mechanical work done is equal to the area enclosed between the two magnetization curves at open and closed positions and the vertical $\lambda - i$ locus during the slow armature movement. This is shown by the cross-hatched area in fig. 2.30.

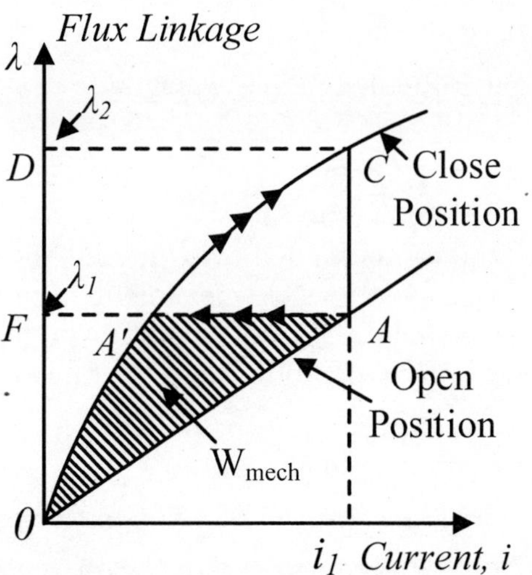

Fig. 2.30: With instantaneous armature movement

- ***Instantaneous movement***

Here the armature is assumed to move from the open to the closed position instantaneously. According to the constant flux linkage theorem, the flux linkages with an inductive circuit can't change suddenly. So here also, during the fast movement of the armature, the flux linkages don't change and remain constant at λ_1.

The operating point, therefore, travels horizontally from A to A'. After the armature has closed, the operating point travels from A' to C along the closed-position magnetization curve fig. 2.30, since the final operating point has to be C.

During this time instantaneous movement of the armature occurs from open (point A) to closed position (point A'), we have

Change in the magnetic stored energy,

$$W_{fld} = Area\ OA'FO - \text{Area OAA'FO} \quad (2.85)$$

$$W_{elec} = \int_{\lambda_2}^{\lambda_1} i d\lambda = 0\,(Constant\ flux\ linkage) \quad (2.86)$$

But $$W_{elec} = W_{fld} + W_{mech}$$

$$0 = Area\ OA'FO - Area\ OAA'FO + W_{mech}$$

Or $$W_{mech} = Area\ OAA'O \quad (2.87)$$

Equation 2.87 shows that the mechanical work done is equal to the area enclosed between the two magnetization curves at open and closed positions and the horizontal $\lambda - i$ locus during the instantaneous movement of the armature. This is indicated by the cross-hatched area in fig. 2.30.

During fast armature movement.

(a) There is no electrical energy input.

(b) Mechanical energy output = Reduction in the magnetic stored energy.

- ***Transient Movement***

The armature movement will neither be too slow nor too fast but will lie somewhere in between the two extreme limits discussed above. Initially, the armature movement is slow and as it is nearing the closed position, its movement becomes fast. The $\lambda - i$locus will therefore be $A\ C'C$ as illustrated in fig. 2.31. The operating point A reaches C' during this time armature moves from open to closed position. Since the final operating point has to be C.

During the time the armature moves from the open (point A) to the closed position (point C), we have a change in the magnetic stored energy,

$$W_{fld} = Energy\ stored\ in\ the\ closed\ position$$
$$- Energy\ stored\ in\ the\ open\ postion$$
$$= Area\ OA'C'D'FO - Area\ OAA'FO$$

$$W_{elec} = \int_{\lambda_2}^{\lambda_1} i d\lambda = Area\ AC'D'FA'A \quad (2.88)$$

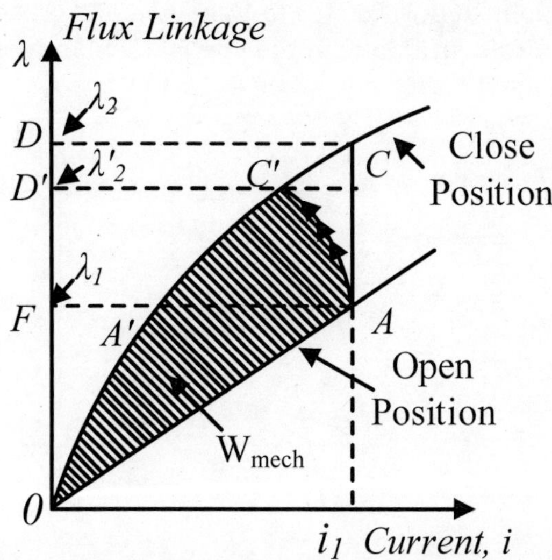

Fig. 2.31: Flux Linkage-current locus during transient movement of the armature

But $\quad W_{elec} = W_{fld} + W_{mech}$

$$Area\ AC'D'FA'A = Area\ OA'C'D'FO - Area\ OAA'FO + W_{mech}$$

$$W_{mech} = (Area\ OAA'FO + Area\ AC'D'FA'A) - Area\ OA'C'D'FO$$

$$= Area\ OAC'A'O \tag{2.89}$$

Equation 2.89 again shows that the energy converted to mechanical (or mechanical work done) is equal to the area enclosed between the two magnetization curves at open and closed positions and the $\lambda - i$locus during the transient movement of the armature.

2.14.2 DETERMINATION OF MAGNETIC FORCE (F_e) OR TORQUE (T_e)

The magnetic force tending to shorten the air gap increases as the gap length decreases

$$F_e(average) = \frac{Mechanical\ work\ done}{Distance\ travelled} \tag{2.90}$$

To obtain a suitable expression for it, the movable part is allowed a virtual displacement dx (or ds) in the direction of magnetic force F_e (or torque T_e), then its effect on the energy balance equation is investigated to obtain the magnitude and direction of magnetic force F_e or magnetic torque T_e.

Assume the armature to be at a distance g_1 from the open position, then a virtual displacement is in the direction of magnetic force F_e is considered.

a) 'a' is the operating point at position g_1 and λ_1 i_1 and are the corresponding values.

b) 'c' is the operating point at position (g+ dx), and $(\lambda_1 + d\lambda)$ and i_1, are the corresponding values. The mechanical work done in the virtual displacement dx is 'oabho'-$F_e dx$ over the virtual displacement dx may be taken as instantaneous.

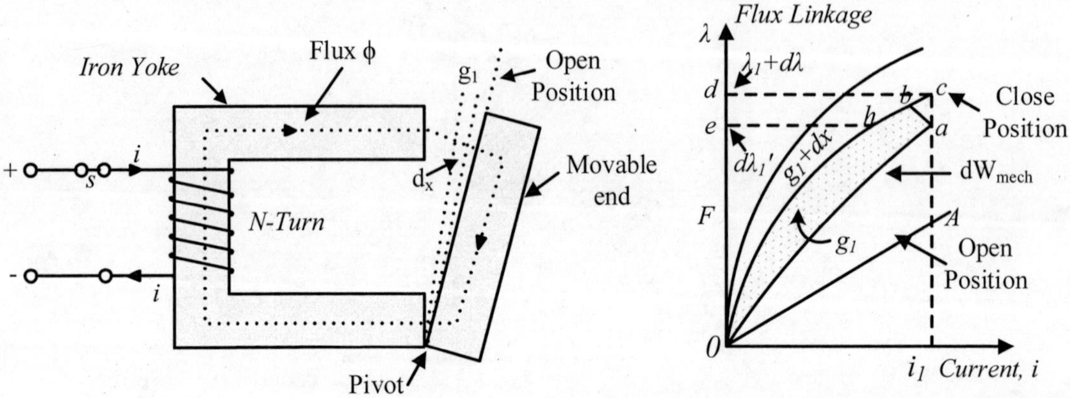

Fig. 2.32: Mechanical work done for differential movement of the armature

Now

$$dW_{ele} = \int_{\lambda_1}^{\lambda_2} id\,\lambda = 0, \;\; dW_{mech} = F_e\,dx \tag{2.91}$$

Putting these values in the energy balance equation.

$$0 = f_e dx + dW_{fld} \; of \; constant \; \lambda$$
$$f_e dx = -dW_{fld} \; at \; constant \; \lambda \tag{2.92}$$

Imp: Electrical energy flow during virtual displacement dx is zero.

The mechanical work $f_e dx$ is done at the expense of field energy stored, so it is indicated by the negative sign.

$$f_e = -\left(\frac{dW_{fld}}{dx}\right) \; for \; \lambda = constant \tag{2.93}$$

Also

$$f_e = -\left(\frac{dW_{fld}}{dx}\right) \; for \;\; \phi = constant \tag{2.94}$$

Note that W_{fld} must be expressed in terms of λ and x or ϕ and x. Given this equation 2.94 leads to the parametric equations for magnetic force as

$$f_e = -\frac{dW_{fld}(\lambda, \, x)}{dx} = -\frac{dW_{fld}(\phi, \, x)}{dx} \tag{2.95}$$

In the above expression for magnetic force, λ or ϕ are independent variables. As voltage is equal to the derivative of λ, this expression gives f_e for a voltage-controlled system.

Equations (2.95) give the magnitude of electromagnetic force f_e because the armature movement is linear. For angular movements of the armature, the electromagnetic torque T_e can be obtained from the parametric equation as

$$T_e = -\frac{dW_{fld}(\lambda,\ \theta)}{d\theta} = -\frac{dW_{fld}(\phi,\ \theta)}{d\theta} \tag{2.96}$$

Also

$$W_{fld} = \frac{1}{2}\frac{\lambda^2}{L} \tag{2.97}$$

Note

$$f_e = \frac{1}{2}\lambda^2\frac{d}{dx}\left[\frac{1}{L}\right] \tag{2.98}$$

Similarly For electromagnetic torque T_e

$$T_e = \frac{1}{2}i^2\frac{dL}{d\theta} = -\frac{1}{2}\lambda\frac{di(\lambda,\theta)}{d\theta} = \frac{1}{2}\ i\ \frac{d\lambda(i,\theta)}{d\theta} \tag{2.99}$$

2.15 Basic Aspects and Physical Phenomena Involved in Doubly Excited Magnetic System

Fig. 2.33 illustrates a simple model of a doubly excited magnetic system. This model consists of stator iron and rotor iron and both are of the salient pole type. The stator with N_s turns is energized from source 1 and the rotor with N_r turns is excited from source 2. The m.m.fs. produced by both the stator and rotor windings are in the same direction and magnetic torque T, is in the anticlockwise direction as shown in fig. 2.33. Therefore, the differential electrical energy input dW_{elec} from two energy sources 1 and 2, in fig. 2.33, is

$$dW_{elec} = i_s d\lambda_s + i_r, d\lambda_r \tag{2.100}$$

Here λ_s and $\lambda_r$are the instantaneous total flux linkages of stator and rotor windings respectively. Since the magnetic saturation is neglected, λ_s and λ_r can be expressed in terms of self and mutual inductances.

$$\lambda_s = L_s i_s + M_{sr} i_r \tag{2.101}$$

$$\lambda_r = L_r i_r + M_{sr} i_s \tag{2.102}$$

Where,

L_s = Self-inductance of the stator winding

L_r = Self-inductance of the rotor winding

M_{sr} = Mutual induction between stator & rotor windings

In fig. 2.32, initially, the space angle between the rotor and stator axis is θ, and both the currents i_s and i_r are assumed zeros. Now the stator and rotor coils are switched on to their respective energy sources so that the currents rise from zero to i_s and i_r, respectively. If the rotor is not allowed to move, the dW_{mech} is zero, and the energy balance equation is

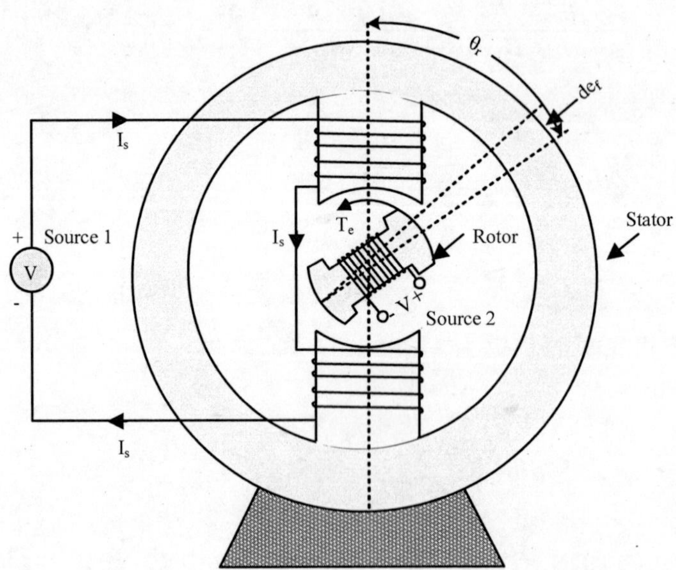

Fig. 2.33 Doubly excited magnetic system

$$dW_{elec} = 0 + dW_{fld} \tag{2.103}$$

Thus, with the rotor held fixed, all the electric energy supplied by the two supply sources is stored in the magnetic field. From equation 2.102

$$dW_{fld} = dW_{elec} = i_s d\lambda_s + i_r d\lambda_r \tag{2.104}$$

$$= i_s d\,(L_s i_s + M_{sr} i_r) + i_r d\,(L_r i_r + M_{sr} i_s) \tag{2.105}$$

Since the rotor is not allowed to move, the reluctances and therefore the inductances are constant. Because of this, the differential changes in inductances, i.e., dL_s, dL_s, and dM_{sr}, in equation 2.105 are all zeros.

Therefore, from equation 2.105

$$dW_{fld} = i_s L_s di_s + i_s M_{sr} di_r + i_r L_r di_r + i_r M_{sr} di_s \tag{2.106}$$

$$= i_s L_s di_s + i_r L_r di_r + M_{sr} d\,(i_s i_r) \tag{2.107}$$

The magnetic field energy stored in establishing the currents from zero to i_s *and* i_r is given by

$$W_{fld} = L_s \int_0^{i_s} i_s \, di_s + L_r \int_0^{i} i_r \, di_r + M_{sr} \int_0^{i_s i_r} d\,(i_s \, i_r) \tag{2.108}$$

$$= \frac{1}{2} i_s^2 L_s + \frac{1}{2} i_r^2 L_r + M_{sr} i_s \, i_r \tag{2.109}$$

For obtaining the magnetic torque T_e, assume the rotor to move through virtual displacement $d\theta_r$ in the direction of T_e as shown in fig. 2.32 now as the rotor moves, the reluctance and inductance vary.

So, the parameter dW_{elec} and $\theta \, dW_{fld}$ changes and

$$dW_{mech} = T_e d\theta_r \tag{2.110}$$

Substituting the values of dW_{elec} , dW_{mech} and dW_{fld} , the energy balance equation gives

$$\frac{1}{2} i_s^2 dL_s + \frac{1}{2} i_r^2 \, dL_r + i_s \, i_r dM_{sr} = T_e d\theta_r \tag{2.111}$$

$$T_e = \frac{1}{2} i_s^2 \frac{dL_s}{d\theta_r} + \frac{1}{2} i_r^2 \frac{dL_r}{d\theta_r} + i_s \, i_r \frac{dM_{sr}}{d\theta_r} \tag{2.112}$$

It is important to observe that the torque T_e depends on:

(a) The instantaneous values of current i_s *and* i_r

(b) The angular rate of change of inductance.

This equation above reveals some important results.

The North, and South Poles produced on stator by i_{s1} and South and North Poles produced on the rotor by i_{r1} attract each other tending to align their fields, the torque to be developed by the stator and rotor magnetic fields is the electromagnetic or interaction torque it should be noted that the reluctance torque ($\frac{1}{2} i_s^2 \frac{dL_s}{d\theta_r}$ *or* $\frac{1}{2} i_r^2 \frac{dL_r}{d\theta_r}$) does not depend on the direction of current in the stator or rotor winding but the interaction torque ($i_s \, i_r \frac{dM_{sr}}{d\theta_r}$) does depend on the direction of current i_s *and* i_r. The rotor tends to align itself along the minimum reluctance position.

2.16 Force as a Partial Derivative of Stored Energy to Position of a Moving Element

Consider a singly excited liner actuator as shown below.

R= Winding resistance, Ohm

V= Terminal voltage applied to the excitation winding, Volt

i= Instantaneous excitation winding current, Amp

The position of the movable plunger x, and the force acting on the plunger F with the reference direction chosen in the positive direction of the x-axis, as shown in the diagram.

After a time, interval dt the plunger is moved for a distance dx under the action of the force F. The mechanical work done by the force acting on the plunger during this time interval is

$$dW_m = F.dx \tag{2.113}$$

The amount of electrical energy that has been transferred into the magnetic field and converted into the mechanical work during this time interval can be calculated by subtracting the power loss dissipated in the winding resistance from the total power fed into the excitation winding as

$$dW_e = dW_f + dW_m \tag{2.114}$$

$$dW_e = vidt - Ri^2dt \tag{2.115}$$

$$dW_f = dW_e - dW_m \tag{2.116}$$

$$dW_f = eidt - Fdx \tag{2.117}$$

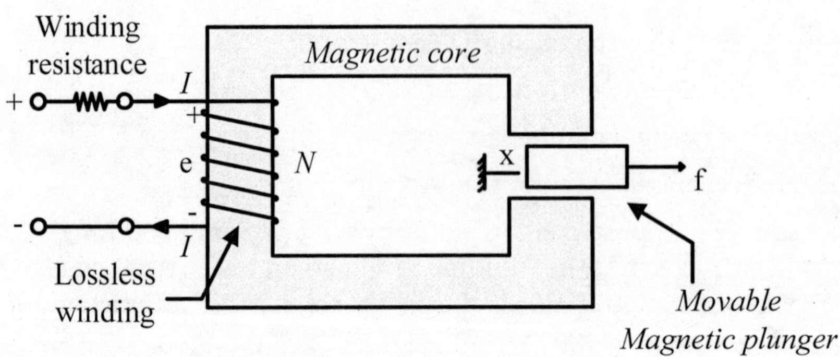

Fig.2.34 Excited static magnetic circuit

where dW_e, dW_f, *and* dW_f are incremental electrical energy input, stored field energy, and mechanical energy output, respectively.

From the above equation, we know that the energy stored in the magnetic field is a function of the flux linkage of the excitation winding and the position of the plunger. Mathematically, we can also write

$$dW_f(\lambda, x) = \frac{\partial W_f(\lambda,\ x)}{\partial \lambda}\partial\lambda + \frac{\partial W_f(\lambda,\ x)}{\partial x}\partial x \tag{2.118}$$

From the above two equations

$$i = \frac{\partial W_f(\lambda, \mathrm{x})}{\partial \lambda} \tag{2.119}$$

$$\mathrm{F} = -\frac{\partial W_f(\lambda, \mathrm{x})}{\partial \mathrm{x}} \tag{2.120}$$

The energy stored in the magnetic field is given as

$$W_f(\lambda, x) = \int_0^{\lambda} i(\lambda, x)d\lambda \tag{2.121}$$

For a magnetically linear system

$$w_f(\lambda, x) = \frac{\lambda^2}{2L(x)} \tag{2.122}$$

The force acting on the plunger will be

$$F = -\frac{\partial w_f(\lambda, x)}{\partial x}\left[\frac{\lambda}{2L(x)}\right]^2 \frac{dL(x)}{dx} \tag{2.123}$$

$$F = \frac{1}{2} i^2 \frac{dL(x)}{dx} \tag{2.124}$$

Hence above equation shows the value of force as a partial derivative of stored energy to the position of the moving element.

2.17 Production of Torque

The torque produced by the alignment of two fields (i.e., rotor field and stationery main field) varies in magnitude and direction depending upon the torque angle θ. Let us see, the effect of torque and θ on the torque produced in the following cases:

- permanent magnet
- electromagnet.

2.17.1 Permanent magnet

Consider a permanent magnet P, which is free to rotate about its axis. Let it be placed in the magnetic field of another permanent magnet Q as shown in fig. 2.35. Let,

θ =angle between the axis of two fields $\mathcal{F}_m$ and $\mathcal{F}_r$

l = length of magnet A

r = radius of the circle in which rotation takes places

F = Force acting on the North and South Pole of magnet A

$Torque$ = Force × Perpendicular distance

In the right-angled triangle "OAB", $AB = OA \sin\theta$

A distance perpendicular to force, AB= $r \sin\theta$

$$Torque = 2F \times r \sin\theta \tag{2.125}$$

$$T = 2F \times \frac{l}{2} \sin\theta \; (\_r = l/2) \tag{2.126}$$

$$T = Fl \sin\theta \tag{2.127}$$

Where F is the force of attraction,

$$F = \frac{m_1 m_2}{4\pi\mu_0\mu_r d^2} \tag{2.128}$$

$$T = K \sin\theta \tag{2.129}$$

Where $K = F \times l$ is constant

$$T \propto \sin\theta \tag{2.130}$$

The maximum torque will be produced when θ= 90° as said earlier.

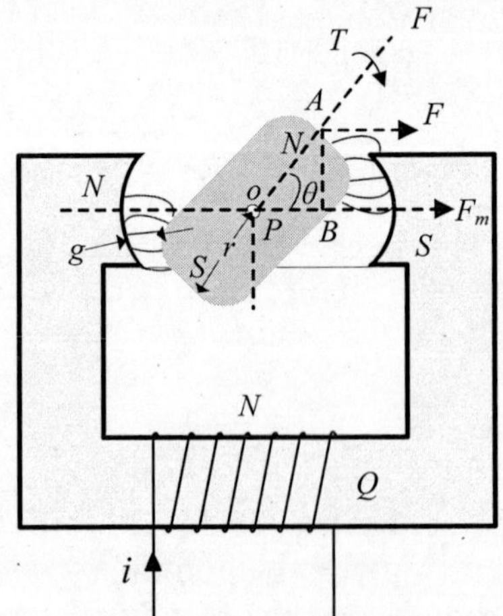

Fig.2.35 Torque developed at an instant when a permanent magnet is placed in the uniform magnetic field

2.17.2 Electromagnet

Consider an electromagnet, having only one coil carrying current. The axis of the field produced by the electromagnet $\mathcal{F}_r$ and axis of the main field produced by the permanent stationary magnet $\mathcal{F}_m$ are shown in fig.2.36. The angle between the two fields is θ. Due to the alignment of the two fields torque is developed.

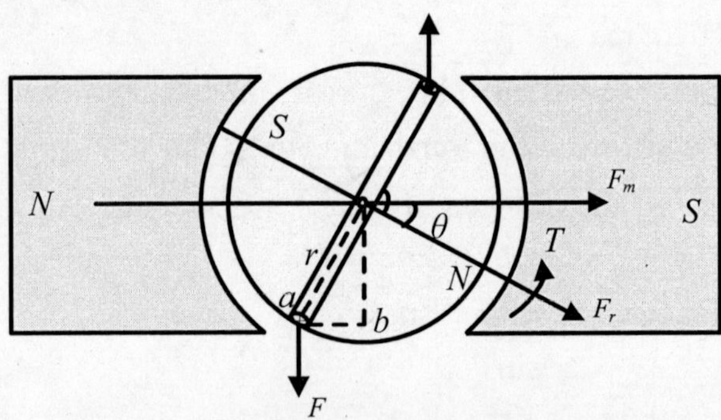

Fig. 2.36 Torque produced at an instant when an electromagnet is placed in a uniform magnetic field

The production of torque can also be explained by the concept of electromagnetic force acting on the current-carrying conductor placed in the magnetic field,

Let,

F =Force acting on the two conductors

r =radious of the circle in which the conductor rotates

θ = angle between the axis of two fields $\mathcal{F}_m$and $\mathcal{F}_r$

$$Torque = \text{Force} \times \text{Perpendicular distance} \tag{2.131}$$

In the right-angled triangle, angle $aob = \theta$

A distance perpendicular to force, $ab = oa \sin\theta$

$$= r\sin\theta \tag{2.132}$$

Total torque acting on the two conductors,

$$T = 2F_r \sin\theta \tag{2.133}$$

Where,

B =Flux density of the main field

i = Current flowing through the conductor

l = Effective length of conductor

$$T = 2Bilr \sin\theta \tag{2.134}$$

$$T = K_i \sin\theta \tag{2.135}$$

Where $K = 2Bilr$ is constant

$$T \propto \sin\theta \tag{2.136}$$

The magnitude of torque depends upon angle θ, it will be maximum when θ= 90°, when θ is positive torque is produced in one direction (say anti-clockwise), but when it is negative, the torque is produced in the other direction (say, clockwise). The direction of the torque is depending upon B and i. when either of the two is reversed the direction of the torque is reversed but if both are reversed the direction of the torque remains the same.

The direction of force acting on a current-carrying conductor when placed in the magnetic field can be determined by applying Fleming's left-hand rule and also $\mathbf{J} \times \boldsymbol{B}$. Here $\mathbf{J}$ is the current density vector and **B** is the flux density vector.

Fleming's Left Hand Rule states that if we arrange our thumb, forefinger, and middle finger of the left-hand perpendicular to each other, then the thumb points toward the direction of the force experienced by the conductor, and the forefinger points toward the direction of the magnetic field and the middle finger points towards the direction of the electric current.

2.18 Torque as a partial derivative of stored energy with respect to the angular position of a rotating element

For torque equation deriving with reference to the above section we can write

$$dW_e = id\lambda - iT\theta \tag{2.137}$$

The energy stored in the magnetic field is given as

$$W_f(\lambda, \theta) = \int_0^{\lambda} i(\lambda, \theta)d\lambda \tag{2.138}$$

$$i = \frac{\partial W_f(\lambda, \theta)}{\partial \lambda} \tag{2.139}$$

$$\mathrm{T} = -\frac{\partial W_f(\lambda, \theta)}{\partial \mathrm{x}} \tag{2.140}$$

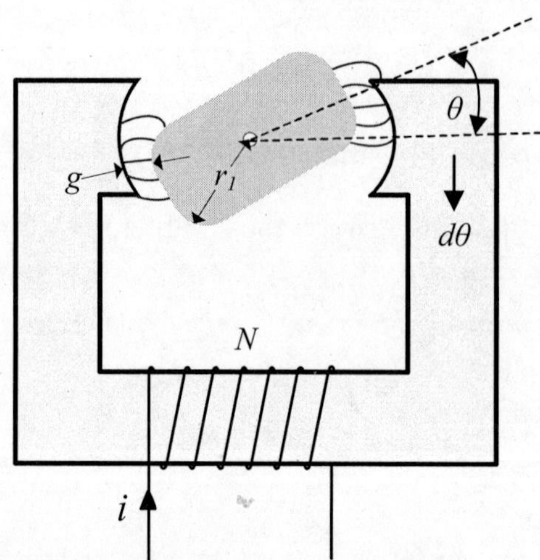

Fig. 2.37 Excited rotating actuator

For a magnetically linear system

$$W_f(\lambda, \theta) = \frac{\lambda^2}{2\mathrm{L}(\theta)} \tag{2.141}$$

The Torque acting will be

$$\mathrm{T} = -\frac{\partial w_f(\lambda, \theta)}{\partial \mathrm{x}} \left[\frac{\lambda}{2\mathrm{L}(\theta)}\right]^2 \frac{\mathrm{dL}(\theta)}{\mathrm{d}\theta} \tag{2.142}$$

$$T = \frac{1}{2} i^2 \frac{\mathrm{dL}(\theta)}{\mathrm{d}\theta} \tag{2.143}$$

Hence above equation shows the value of torque as a partial derivative of stored energy with respect to the angular position of the rotating element.

2.19 Examples:

A few examples are given below:

2.19.1 Galvanometer

A moving coil galvanometer is an electromagnetic device that can measure small values of current of the order of a few microamperes. It consists of permanent horseshoe magnets, coil, soft iron core, pivoted spring, non-metallic frame, scale, and pointer as shown in fig 2.38.

Principle

A current-carrying coil when placed in an external magnetic field experiences magnetic torque. The angle through which the coil is deflected due to the effect of the magnetic torque is proportional to the magnitude of current in the coil.

Construction and Diagram

The moving coil galvanometer is made up of a rectangular coil that has many turns and it is usually made of thinly insulated or fine copper wire that is wounded on a metallic frame. The coil is free to rotate about a fixed axis. A phosphor-bronze strip that is connected to a movable torsion head is used to suspend the coil in a uniform radial magnetic field. Essential properties of the material used for the suspension of the coil are conductivity and a low value of the torsional constant. A cylindrical soft iron core is symmetrically positioned inside the coil to improve the strength of the magnetic field and to make the field radial. The lower part of the coil is attached to a phosphor-bronze spring having a small number of turns. The other end of the spring is connected to binding screws.

The spring is used to produce a counter torque which balances the magnetic torque and hence helps in producing a steady angular deflection. A plane mirror which is attached to the suspension wire, along with a lamp and scale arrangement, is used to measure the deflection of the coil. The Zero-point of the scale is at the center.

Working of Moving Coil Galvanometer

Let a current I flow through the rectangular coil of N number of turns and a cross-sectional area A. When this coil is placed in a uniform radial magnetic field B, the coil experiences a torque T.

Let us first consider a single turn PQRS of the rectangular coil having a length l and breadth b as shown in fig.2.39. This is suspended in a magnetic field of strength B such that the plane of the coil is parallel to the magnetic field. Since the sides PQ and SR are parallel to the direction of the magnetic field, they do not experience any effective force due to the magnetic field. The sides PS and QR being perpendicular to the direction of the field experience an effective force F given by

$$F = Bil \tag{2.144}$$

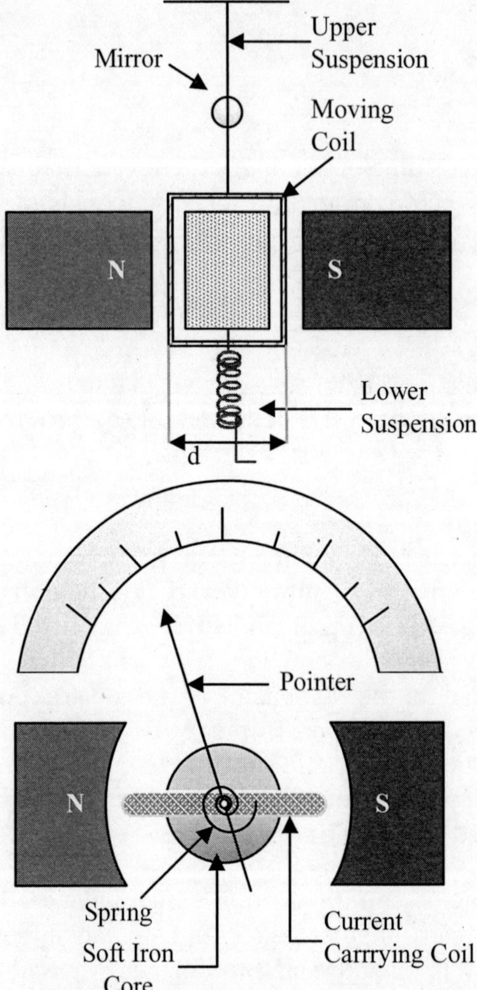

Fig. 2.38 Moving Coil Galvanometer

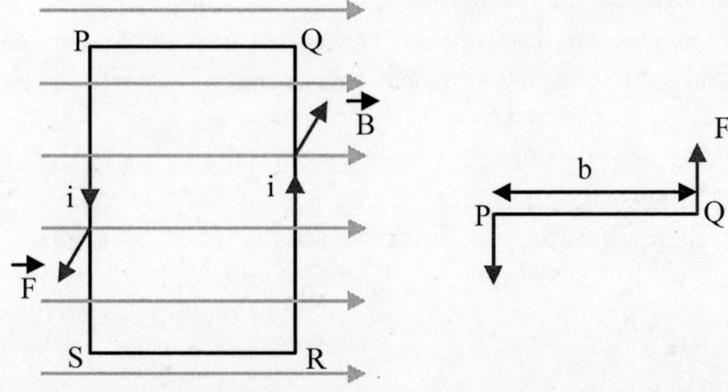

Fig. 2.39 Equivalent circuit of moving coil galvanometer

Using Fleming's left-hand rule, we can determine that the forces on PS and QR are in opposite directions to each other. When equal and opposite forces F called couple act on the coil, it produces a torque. This torque causes the coil to deflect.

We know that torque is equal to the product of force and the perpendicular distance between the forces that is

$$T = F \times b \tag{2.145}$$

Substituting the value of *F*, torque *T* acting on single-loop PQRS of the coil

$$T = Bil \times b \tag{2.146}$$

Where $l \times b$ is the area A of the coil,

Hence the torque acting on N turns of the coil is given by

$$T = BiAN \tag{2.147}$$

The magnetic torque thus produced causes the coil to rotate, and the phosphor bronze strip twists. In turn, the spring S attached to the coil produces a counter torque or restoring torque $k\theta$ which results in a steady angular deflection.

Under equilibrium condition:

$$k\theta = BiAN \tag{2.148}$$

Here k is called the torsional constant of the spring (restoring a couple per unit twist). The deflection or twist θ is measured as the value indicated on a scale by a pointer that is connected to the suspension wire.

$$\theta = \left(\frac{BAN}{k}\right) \times i \tag{2.149}$$

Therefore

$$\theta \propto i \tag{2.150}$$

The quantity BAN/k is a constant for a given galvanometer. Hence it is understood that the deflection that occurs in the galvanometer is directly proportional to the current that flows through it.

Sensitivity of Moving Coil Galvanometer

The general definition of the sensitivity experienced by a moving coil galvanometer is given as the ratio of change in deflection of the galvanometer to the change in current in the coil.

$$S = \frac{d\theta}{dI} \tag{2.151}$$

The sensitivity of a galvanometer is higher if the instrument shows a larger deflection for a small value of current. Sensitivity is of two types, namely current sensitivity and voltage sensitivity.

Current Sensitivity

The deflection θ per unit current I is known as current sensitivity θ/I.

$$\frac{\theta}{I} = \frac{NAB}{k} \tag{2.152}$$

Voltage Sensitivity

The deflection θ per unit voltage is known as Voltage sensitivity θ/V. Dividing both sides by V in equation 2.149.

$$\frac{\theta}{V} = \left(\frac{BAN}{Vk}\right) \times i \tag{2.153}$$

$$\frac{\theta}{V} = \left(\frac{BAN}{k}\right) \times \frac{i}{V} \tag{2.154}$$

$$\frac{\theta}{V} = \left(\frac{BAN}{k}\right) \times \frac{1}{R} \tag{2.155}$$

R stands for the effective resistance in the circuit.

It is worth noting that voltage sensitivity = Current sensitivity/ Resistance of the coil. Therefore, under the condition that R remains constant then voltage sensitivity is proportional to Current sensitivity.

Applications of Galvanometer

The moving coil galvanometer is a highly sensitive instrument that can be used to detect the presence of current in any given circuit. If a galvanometer is connected in a Wheatstone's bridge circuit, the pointer in the galvanometer shows null deflection, i.e., no current flows through the device. The pointer deflects to the left or right depending on the direction of the current.

The galvanometer can be used to measure:

- The value of current in the circuit by connecting it in parallel to low resistance.
- The voltage by connecting it in series with high resistance.

Advantages And Disadvantages of a Moving Coil Galvanometer

Advantages

- High sensitivity.
- Not easily affected by stray magnetic fields.
- The torque to weight ratio is high.
- High accuracy and reliability.

Disadvantages

- It can be used only to measure direct currents.
- Develops errors due to factors like aging of the instrument, permanent magnets, and damage of spring due to mechanical stress.

2.19.2 Relay Contact

The Relay contacts are opened or closed by a magnetic force. Relays are switching that open and close circuits electromechanically or electronically. An electromagnetic relay works on the force of attraction utilized either close or open single/multiple electrical points. Relays control one electrical circuit by opening and closing contacts in another circuit. An electromechanical relay diagram is shown in fig 2.40. When a relay contact is Normally Open (NO), there is an open contact when the relay is not energized. When a relay contact is Normally Closed (NC), there is a closed contact when the relay is not energized. In the absence of controlling voltage AC/DC, due to spring action, the electrical contacts are either NO or NC type depending upon the type of relay. Upon energization of the relay, the contacts are changed, and the electromagnetic force of the plunger acts against the restraining force of the holding spring. In either case, applying control electrical current to the relay, the contacts will change their state. These relays are commercially available with the single pole as well as with multi-pole ON/OFF with state feedback normally required in (PLC) Programming Logic Controllers. In industries, electro-pneumatic relays are also in use, in which electrical contacts are opened or closed using pneumatic control.

The basic parts and functions of electromechanical relays include:

I. **Frame:** Heavy-duty frame that contains and supports the parts of the relay.

II. **Coil:** Wire is wound around a metal core. The coil of the wire causes an electromagnetic field when the control voltage is there.

III. **Armature:** Relay's moving part. The armature opens or closes the contacts. An attached spring returns the armature to its original position.

IV. **Contacts:** The conducting part of the switch that makes (closes) or breaks (opens) a circuit.

An electromechanical relay transfers signals between its contacts through mechanical movement. It has three sections viz. *input section, control section* , and *output section*.

The *input section consists* of input terminals where a small control signal is to be applied. The *control section* has an electromagnetic coil that gets energized when the control input signal is applied to the input terminals and the *output section* consists of a movable armature and mechanical contacts – movable and stationary, the movement of the armature makes or breaks the electrical circuit.

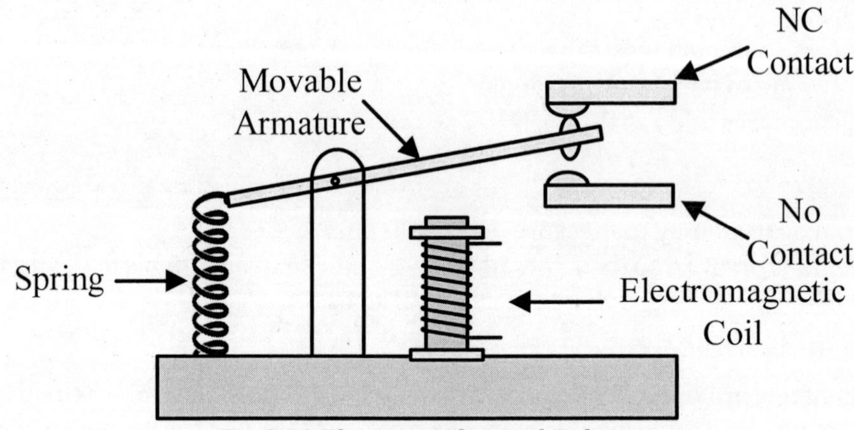

Fig.2.40 Electromechanical Relay

When an input control voltage is applied to the electromagnetic coil, it gets magnetized and the armature is attracted by the magnetic field produced by the coil. The movable mechanical contacts are attached to the armature, thus when the armature moves towards the electromagnet, the contacts close, making the output circuit switch on. When the control signal is removed, the armature comes back to its original position by the force of the spring, making the output circuit off.

2.19.3 Lifting Magnet

A lifting magnet is a type of permanent magnet capable of producing a stronger magnetic field than a normal magnet. This is because they are made of rare-earth metals and alloys such as NdFeB and ferrite etc. These lifting magnets find applications in various industries to lift metallic items or heavy objects. There are three main types of magnets used for lifting applications, in terms of magnet characteristics: permanent lifting magnets, electro-lifting magnets, and electro-permanent lifting magnets. There is no doubt that lifting magnets need to be powerful with a good weight-bearing capacity. They are made of specific magnetic materials such as iron alloys with precise configurations to achieve all the required properties. They also require ON and OFF features. A permanent lifting magnet has a block with the main body and a rotor. It has two magnets in the main body and a rotor each. When these two magnets are positioned in the same direction it produces a magnetic flux that lifts the metallic objects. When we change the direction of the magnet in the rotor, the load is released as there is no magnetic pull. An important point is, that the adhesion of objects to the magnet should be close and tight, without any air trapped in between.

Permanent lifting magnets are different from electro-permanent lifting magnets. In the latter, an electric coil is wound around the magnet, which passes an electric current to switch on the magnet. They also use magnetic energy to lift objects. The electric current is only used to switch the magnet on and is disconnected later.

Lifting magnets find their uses in simple door latch systems, levitated transport systems, and Hyperloop systems.

Benefits of Lifting Magnets

Apart from their huge application across industries, these magnets offer the following benefits:

- These magnets are durable and low on maintenance as they use only magnetic energy.
- Overall, they are cost-effective as they do not require any expensive infrastructure.
- They are easy to use and operated manually, hence there is no need to extensively train the workforce on their operation.
- As long as the operator follows all the proper steps, there are no safety issues.

Solved Numerical Examples

Q1. Calculate the relative permeability of an iron ring when the exciting current taken by the 600-turn coil is 1.2 A and the total flux produced is 1 m Wb. The mean circumference of the ring is 0·5 m and the area of cross-section is 10 cm².

$$NI = \frac{\phi \times l}{a\mu_0\mu_r}$$

$$\mu_r = \frac{\phi \times l}{a\mu_0 NI}$$

Where N = 600 turns; I = 1.2; ϕ = 1 m Wb = 1 × 10^{-3} Wb; l = 0.5 m; a = 10 cm^2 = 10 × 10^{-4} m^2

$$\mu_r = \frac{1 \times 10^{-3} \times 0.5}{10 \times 10^{-4} \times 4\pi \times 10^{-7} \times 600 \times 1.2} = 552.6\ Ans.$$

Q2. The ring-shaped core shown in fig. 2.41 below is made of a material having a relative permeability of 1000. The flux density in the smallest area of the cross-section is 2 T. If the current through the coil is not to exceed 1·5 A, compute the number of turns of the coil.

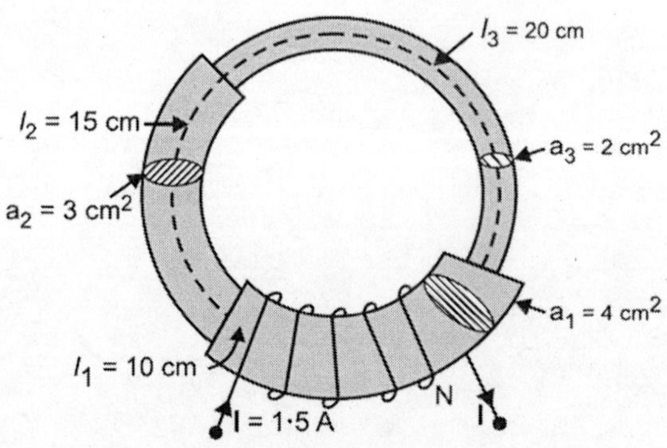

Fig. 2.41

Solution:

Flux in the core, $\phi = B \times A = 2 \times 2 \times 10^{-4} = 4 \times 10^{-4}\ Wb$

Total reluctance of the magnetic path,

$$S = S_1 + S_2 + S_3$$

$$= \frac{l_1}{a_1\mu_0\mu_r} + \frac{l_2}{a_2\mu_0\mu_r} + \frac{l_3}{a_3\mu_0\mu_r}$$

$$= \frac{1}{\mu_0 \mu_r}\left(\frac{l_1}{a_1} + \frac{l_2}{a_2} + \frac{l_3}{a_3}\right)$$

$$= \frac{1}{4\pi \times 10^{-7} \times 1000}\left(\frac{0.1}{4 \times 10^{-4}} + \frac{0.15}{3 \times 10^{-4}} + \frac{0.2}{2 \times 10^{-4}}\right)$$

$$= 13.926 \times 10^5 \; AT\ per\ Wb$$

Total mmf required, $NI = \phi S$

Or $N \times 1.5 = 4 \times 10^{-4} \times 13.926 \times 10^5$

The number of turns, N = 371.36 Ans.

Q.3 An iron ring of a mean length 1 m has an air gap of 1 mm and a winding of 200 turns. If the relative permeability of iron is 500 when a current of 1 A flows through the coil, find the flux density.

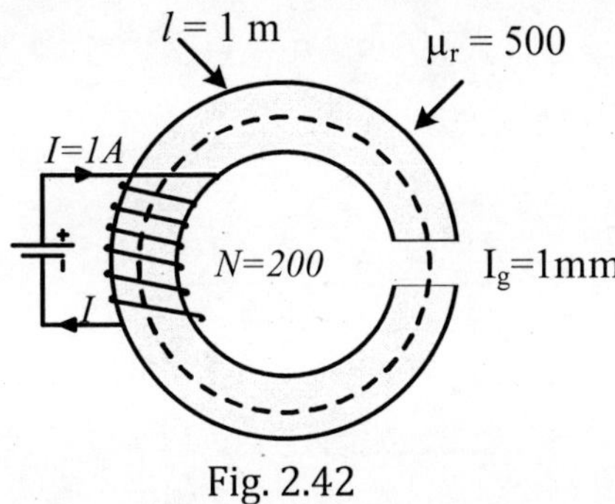

Fig. 2.42

Solution:

The magnetic circuit is shown in fig.

$$\text{Now, } mmf = flux \times reluctance$$

$$\text{i.e., } NI = \phi\left(\frac{l_i}{a\,\mu_0\mu_r} + \frac{l_g}{a\,\mu_0}\right)$$

$$\text{or} \quad NI = B\left(\frac{l_i}{\mu_0\mu_r} + \frac{l_g}{\mu_0}\right)$$

where N = 200 turns; I = 1A; $\mu_r = 500$

$$l_g = 1\ mm = 0.001\ m;$$

$$l_i = (1 - 0.001) = 0.999\ m$$

$$200 \times 1 = B\left(\frac{0.999}{4\pi \times 10^{-7} \times 500} + \frac{0.001}{4\pi \times 10^{-7}}\right)$$

$$B = \frac{200}{2385.73} = 0.0838\ \ Wb\ per\ square\ m$$

Q4. Determine magnetomotive force, magnetic flux, reluctance and flux density in case of a steel ring 30 cm mean diameter and a circular cross-section 2 cm in diameter has an air gap 1 mm long. It is wound uniformly with 600 turns of wire carrying a current of 2.5 A. Neglect magnetic leakage. The iron path takes 40% of the total magnetomotive force.

Solution:

Mmf of the magnetic circuit = $NI = 600 \times 2{\cdot}5 = 1500$ AT

As iron path takes 40% of the total mmf, the reluctance of iron in 40% and the rest of the reluctance (60%) is of air path.

$$\frac{S_a}{S_i} = \frac{60}{40} = \frac{3}{2} = 1.5$$

$$\text{Reluctance of air path, } S_a = \frac{l_a}{a\mu_0}$$

$$\text{Where, } l_a = 1 \times 10^{-3} m;\, a = \frac{\pi}{4}(2)^2 \times 10^{-4}\, m^2$$

$$S_a = \frac{1 \times 10^{-4}}{\pi \times 10^{-4} \times \pi \times 10^{-7}} = 2.533 \times 10^6\, AT\ per\ Wb$$

$$\text{Reluctance of iron path, } S_i = \frac{S_a}{1.5} = \frac{2.533 \times 10^6}{1.5} = 1.688 \times 10^6\ \ AT\ per\ Wb$$

$$\text{Total Reluctance, } S_a + S_i = (2.553 + 1.688) \times 10^6 AT\ per\ Wb$$

$$4.221 \times 10^6 AT\ per\ Wb$$

$$\text{Magnetic flux, } \phi = \frac{m.m.f}{reluctance} = \frac{1500}{4.221 \times 10^6} = 0.3554\, mWb$$

$$\text{Flux density, B} = \frac{\phi}{a} = \frac{0.3554 \times 10^{-3}}{\pi \times 10^{-4}} = 1.131\, Wb\ per\ square\ m$$

Q.5 The magnetic circuit shown in fig. 37 is made of cast steel. The rotor is free to turn about a vertical axis. The dimensions are shown in the figure.

A. **Derive an expression in mks rationalized units for the torque acting on the rotor in terms of the dimensions and the magnetic field in the two air gaps. Neglect the effects of fringing.**
B. **The maximum flux density in the overlapping portions of the air gaps is limited to approximately 130-kilo lines/in.^{2}, because of saturation in the steel. Compute the maximum torque in inch-pounds for the following dimensions: r_1 = 1.00 in.; h = 1.00 in.; g = 0.10 in**

Solution:

The torque can be derived from the derivative of air-gap reluctance, air-gap permeance, or of field energy.

(a) The field energy density is $\mu_0 {H_{ag}}^2/2$ and the volume of the two overlapping air gaps is $2gh\,(r_1 + 0.5g)\theta$. Consequently, the field energy is

$$W_{ag} = \mu_0 {H_{ag}}^2 gh\,(r_1 + 0.5g)\theta$$

At constant mmf H_{ag} is constant, and therefore differentiation of Eq. (1) with respect to θ at constant mmf in accordance with the following equation:

$$f = \left[\frac{(\partial W_{fld})}{\partial x}\right](F, x) = + \left[\frac{(\partial W_{fld})}{\partial x}\right](i, x), yields$$

$$\text{Torque, } T = \mu_0 {H_{ag}}^2 gh\,(r_1 + 0.5g) = \left[\left({B_{ag}}^2 gh(r_1 + 0.5g)\right)/\mu_0\right]$$

The torque acts in a direction to align the rotor with the stator pole faces.

(b) Convert the flux density and dimensions to mks units.

$$B_{ag} = [(130{,}000)/(6.45)] \times 10^{-4} \times 10^{-8} = 2.02\ Wb/m^2$$

$g = 0.1 \times 2.54\ \times 10^{-2} = 0.00254\ m$

$h = r_1 = 1.00 \times 2.54\ \times 10^{-2} = 0.0254\ m$

$\mu_0 = 4\pi \times 10^{-7}$

Using of these numerical values

$T = 5.56\ Nm$

$5.56 \times 0.738\ \times 12 = 49.3$ in. – lb.

Q.6 Find the self–inductance of a long solenoid shown in figure 2.5.

Solution: The magnetic induction inside a long solenoid, neglecting end effects, is constant,

$$B = \mu_0 N' I$$

where N' is the number of turns/meters. Thus

$$\phi = \left[\frac{\mu_0 I}{\ell}\right] \pi R^2,$$

where N is the total number of turns, ℓ is the length of the solenoid, and R is its radius. Then

$$L = (N\phi/\ell) = \left[\frac{\mu_0 N^2}{\ell}\right] \pi R^2$$

$$= \mu_0 N' \ell \pi R^2$$

Q.7 A coil with five series–connected turns rotate at a speed of 1200 rpm. The flux per pole is ϕ = 3 × 10⁶ maxwells; the number of poles is p = 6. What is the average emf induced in the coil? What are the amplitude and the effective value of the emf induced in the coil if the flux is sinusoidally distributed?

Solution: From the following equation:

$$E_{av} = 4N\left(\frac{p}{2}\right)\left(\frac{n}{60}\right)10^{-8}\ volt,$$

with $p = 6$ and $n = 1200$,

$$E_{av} = 4 \times 3 \times 10^6 \times 5 \times \left(\frac{6}{2}\right) \times \left[\frac{1200}{60}\right] \times 10^{-8}$$
$$= 3600\ volt$$
$$F = [(p \times n)/120]\ cps$$
$$= [(6 \times 1200)/120] = 60\ cps$$
$$E_m = 2\pi N\phi \times 10^{-8}\ volt$$
$$= 2\pi \times 60 \times 5 \times 3 \times 10^6 \times 10^{-8}$$
$$= 56.6\ volt$$

and

$$[(56.6)/\sqrt{2}] = 40\ volt$$

For a sinusoidally distributed flux, E_{av} must be $2/\pi\ E_m$

MULTIPLE CHOICE QUESTIONS

1. The perfect magnetic insulator is: -
   a. Copper
   b. Iron
   c. Rubber
   d. None of above
2. The ampere-turns are: -
   a. The product of the number of turns and current of the coil
   b. The number of turns of a coil through which current flowing
   c. The currents of all turns of the coil
   d. The turns of the transformer winding
3. What will be the current passing through the ring-shaped air-cored coil when the number of turns is 800 and ampere-turns are 3200: -
   a. 0.25
   b. 2.5
   c. 4.0
   d. 0.4
4. The magnetic reluctance of a material: -
   a. Increase with increasing cross-sectional area of material
   b. Does not vary with increasing the cross-sectional area of material
   c. Decreases with increasing the cross-sectional area of material
   d. Decrease with increasing the length of material
5. Mmf is analogous to: -
   a. The electric current in the electric circuit
   b. Current density in a conductor
   c. Electromotive force
   d. None of them

6. The magnetic strength of an electromagnet can be increased by: -
   a. Increasing current in the solenoid
   b. Increasing the number of turns of the solenoid
   c. Both (i) and (ii)
   d. None of above
7. Hysteresis loss in a magnetic material depends upon: -
   a. Area of the hysteresis loop
   b. Frequency of reversed field
   c. The volume of magnetic material
   d. All a, b and c
8. The direction of electro-magnetically induced emf is determined by: -
   a. Fleming right-hand rule
   b. Len's law
   c. Right-hand thumb rule
   d. Both (a) and (b)
9. The energy stored in the magnetic field is: -
   a. Given by the relation $LI^2/2$
   b. Directly proportional to the square of the current flowing through the coil
   c. Directly proportional to the inductance of the coil
   d. All a, b and c
10. The self-inductance of a solenoid of N-turns is proportional to: -
    a. N
    b. N^2
    c. $\frac{1}{N}$
    d. $\frac{1}{N^2}$
11. To reduce eddy current loss in the core of magnetic material: -
    a. The core is laminated
    b. The magnetic material used should have high resistivity
    c. Both (a) and (b)
    d. None of above

12. If two coils having self-inductance L_1 and L_2 and a mutual inductance M are connected in series with opposite polarity, then the total inductance of the combination will be: -
    a. $L_1 + L_2 + 2M$
    b. $L_1 - L_2 - 2M$
    c. $L_1 - L_2 + 2M$
    d. $L_1 + L_2 - 2M$

13. An inductive coil of 10 H develops a counter voltage of 50 V. What should be the rate of change of current in the coil: -
    a. $5\,{}^{A}/_{S}$
    b. $0.2\,{}^{A}/_{S}$
    c. $1\,{}^{A}/_{S}$
    d. $500\,{}^{A}/_{S}$
14. A machine that converts electrical energy into mechanical energy is called: -
    a. Electric generator
    b. Electric motor
    c. Both (a) and (b)
    d. None of these
15. In electromechanical energy conversion devices, the angle between the rotor field and the main magnetic field is called: -
    a. Mechanical angle
    b. Electrical angle
    c. Either (a) and (b)
    d. Torque angle
16. In an electromechanical device, when both the direction of rotation and direction of electromagnetic torque are in the same direction the machine works as a: -
    a. Generator
    b. Motor
    c. Both (a) and (b)
    d. None of these
17. In an electromechanical device, when both the direction of rotation and direction of mechanical torque are in the same direction, the machine works as a: -
    a. Generator
    b. Motor
    c. Both (a) and (b)
    d. None of these
18. In an electromechanical device, when induced emf (e) acts in the opposite direction to the direction of flow of current (i) in the armature conductor, the machine works as a: -
    a. Generator
    b. Motor
    c. Both (a) and (b)
    d. None of these

Keys to multiple choice questions

1.	D	2.	A	3.	C	4.	C	5.	C	6.	C
7.	D	8.	D	9.	D	10.	B	11.	C	12.	D
13.	A	14.	B	15.	D	16.	B	17.	A	18.	B

Short answer type questions

Q.1 What is a magnet?

Q.2 What are permanent and temporary magnets?

Q.3 What do you understand by a magnetic field?

Q.4 What are magnetic poles?

Q.5 What do you mean by magnetic axis?

Q.6 Define and explain a magnetic circuit.

Q.7 Mention at least four properties of magnetic lines of force.

Q.8 Explain the term mmf.

Q.9 Define relative permeability.

Q.10 Define the permeance of the magnetic circuit.

Q.11 Give similarities between electric and magnetic circuits.

Q.12 What is the composite magnetic circuit?

Q.13 State 'ohms law' of a magnetic circuit.

Q14. Define leakage factor.

Q.15 Why is it necessary to keep air gaps in magnetic circuits as small as possible?

Q16. Why does leakage occur in a magnetic circuit?

Q.17 What is magnetic firing?

Q18. What is hysteresis in a magnetic material?

Q.19 What do you understand by electromagnetic induction?

Q.20 Define Faraday's laws of electromagnetic induction.

Q.21 State Fleming's Right-hand rule as well as Fleming's Left-hand rule.

Q.22 State the right-hand cork screw rule.

Q.23 What do you mean by dynamically induced emf?

Q.24 Define self-inductance and give its unit.

Q.25 Distinguish between self-induced and mutually induced emf.

Q.26 Does inductance play any role in the DC circuit?

Q.27 What is a closed circuit?

Q.28 What is a short circuit?

Q.29 What is the type of energy being stored in a capacitor?

Q.30 What do you mean by electromechanical energy conversion devices?

Exercises:

1. A coil with an axial length of 25 cm and diameter of 20 cm has 200 turns. It is placed in a uniform radial flux of 0.002 Web/m2. If the coil is rotated at 25 revolutions per second, find the voltage induced in the coil.

   (Ans. 0.4 volts)

2. A circuit coil of 500 turns with a mean diameter of 50 cms is rotated about a vertical axis in the earth's field at 40 revolutions per second. Find the instantaneous value of emf induced in the coil when its plane is:
   (i) Parallel and
   (ii) Inclined at 30 degrees to the magnetic meridian. Take the value of H as 14.3 AT/m

   (Ans. (i) 0.443 V; (ii) 0.3836 V)

3. An iron ring has a cross-sectional area of 400 mm2 and a mean diameter of 25 cm. It is wound with 500 turns. If the value of relative permeability is 500, find the total flux set-up in the ring. The coil resistance is 400 ohms and the supply voltage is 200 V.

   (Ans. 0.08 m Wb)

4. An iron ring of a mean diameter 22 cm and cross-section 10 cm2 has an air gap of 1 mm wide. The ring is wound uniformly with 200 turns of wire. The permeability of ring material is 1000. A flux of 0.16 mWb is required in the gap. What current should be passed through the wire?

   (Ans. 1.076 A)

5. An iron ring has a cross-section 3 cm2 and a mean diameter of 25 cm. An air gap of 0.4 mm has been made by a saw cut across the section. The ring is wound with 200 turns through which a current of 2 A is passed. If the total flux is 21 X 10-5 Weber, find μ for iron assuming no leakage.

   (Ans. 2470)

6. A coil of 250 turns is wound on a magnetic circuit of reluctance 100000 AT/Wb. If a current of 2 A flowing in the coil is reversed in 5 ms, find the average emf induced in the coil.

   (Ans. 500 V)

Books for further reading

1 Electric Machinery, Fitzgerald, Kingslay, Umans, Tata McGraw-Hill.

2. Electric Machinery Fundamentals, Chapman, McGraw-Hill Higher Education.
3. Electric Machines, Nagrath and Kothari, Tata McGraw-Hill.
4. Electric Machinery, P.S.Bimbhra, Khanna Publishers.
5. Electrical Machines, R. K. Srivastava, 2/e, Cengage Learning Pvt. Ltd.-2011
6. Electrical Machines, Smarajit Ghosh, 2/e, Pearson edu, 2012
7. Electrical Machines-I, D. K. Palwalia, N K Garg, P Kumar, G Jain. Ashirwad Publishers-2020

REFERENCE BOOKS:

1. Electric Machinery and Transformer, Guru, Hiziroglu, Oxford University Press.

3. Basic Electric Machines, Vincent Deltoro, Prentice Hall.

3. Performance and Design of A.C. machines, M. G. Say

For further reading scan to:

Scan to know further video lecture on magnetic circuits

Scan to know further video lecture on magnetic circuits

Scan to know further video lecture on magnetic circuits

3 DC Machines

UNIT SPECIFICS

Through this unit, the following aspects have been discussed:

- *The basic construction of a DC machine;*
- *Visualization of the magnetic field produced;*
- *Air gap flux density distribution*
- *Induced EMF in DC Machine*
- *Armature winding and commutation;*
- *Derivation of back EMF equation,*
- *Armature MMF wave,*
- *Derivation of torque equation,*
- *Armature reaction and air gap flux density distribution with armature reaction.*

The unit describes the construction and operation of the DC machine. It contains a detailed description of the topics. It further gives solved and unsolved numerical problems to illustrate further to the theoretical study. A number of multiple-choice questions as well as questions of short- and long-answer types are also given. A list of references and suggested readings are given in the unit so that one can go through them for practice. Some QR codes have been provided in different sections which can be scanned for relevant supportive knowledge.

RATIONALE

This fundamental unit on DC machines will help students to get detailed knowledge of the construction of DC machines. It gives a visualization of the magnetic field produced and air gap flux density distribution. The concept of the type of armature winding and understanding of the magnetic circuits is discussed. It would enable us to obtain induced emf in the DC machine and derive a torque equation. The phenomenon of *armature reaction and airgap flux density distribution is reported in detail.*

PRE-REQUISITES

Mathematics: Vectors, Integrals, Differentiation, Algebra (Class XII)
Physics: Electricity and Magnetism (Class XII)

UNIT OUTCOMES

The list of outcomes of this unit is as follows:

U3-O1: *Understand the principal of DC Machine construction and operation.*

U3-O2: *Understand the armature winding design.*

U3-O3: *Visualization of flux density distribution, armature reaction*

U3-O4: *Determine the emf and torque of the DC Machine.*

Unit-3 Outcomes	***EXPECTED MAPPING WITH COURSE OUTCOMES*** *(1-Weak Correlation; 2-Medium Correlation; 3-Strong Correlation)*			
	CO-1	**CO-2**	**CO-3**	**CO-4**
U3-O1	-	3	3	-
U3-O2	1	3	3	-
U3-O3	2	3	3	-
U3-O4	1	2	3	-

3.1 Introduction

The first electrical machine based on the electromagnetic principle developed was a direct current (DC) machine. The first central power station developed by Thomas A. Edison to serve a part of New York City had DC generators in 1882. It operated at 110 V DC. In the US, Thomas Davenport is widely celebrated as the inventor of the first electric motor, and undoubtedly, he was the first to patent a useable electric motor in 1837. Davenport, however, was not the first person to build an electric motor, with various inventors in Europe having already developed more powerful versions by that time. In 1834, Moritz Jacobi presented a motor that was three times as powerful as the one Davenport would later patent, while Sibrandus Stratingh and Christopher Becker were the first to demonstrate a practical application for an electric motor, by running a small model car in 1835. The first practical DC motor was invented some years later in 1886 by Frank Julian Sprague, whose invention lead to the first motor-powered trolley system in 1887, and the first electric elevator in 1892. Sprague's DC motor was a hugely significant development, leading to a variety of applications that would reshape the face of industry and manufacturing. The word machine is commonly used to explain features that are common to both the motor and the generator. This is especially true for all dc machines. DC machines operate on DC voltages and currents. Direct currents do not vary with time. A DC machine is inherently ac but is made to appear DC at its terminals through a mechanical rectifier called a commutator. Similar to ac machines, DC machines operate based on the laws of induction. In a DC machine, the constant magnetic flux is established by fixed poles mounted on the stator (stationary part of the machine). Permanent magnets can be used as the poles or wind the field windings (excitation coils) around the poles. One of the major advantages of a wound machine is that one can control the flux in the machine by regulating the direct current in the field winding. The winding in which the electromotive force (emf) is induced is wound on the rotating part. The rotating part is called the armature and the winding is referred to as the armature winding. The Sectional view of a DC Machine is shown in fig. 3.1. DC machines are gradually giving way to AC machines driven by high-performance AC drive systems. One of the reasons for their declining popularity is the number of problems associated with the operation of commutator segments and brushes, including sparking, losses, wear and tear, and continuous and expensive maintenance requirements. Despite this background, a study of DC machines is still relevant. This is because DC machines still exist and are manufactured, and engineers will have to be familiar with their operation. DC machines have attractive operation and control properties, which are desirable to replicate when AC drives are designed. It is therefore important to the study of DC machines. The presentation in this chapter mostly considers the basic principles of DC machines to illustrate their operation. Further, developed equations for describing the performance of a DC machine.

Scan to know more on DC Machine

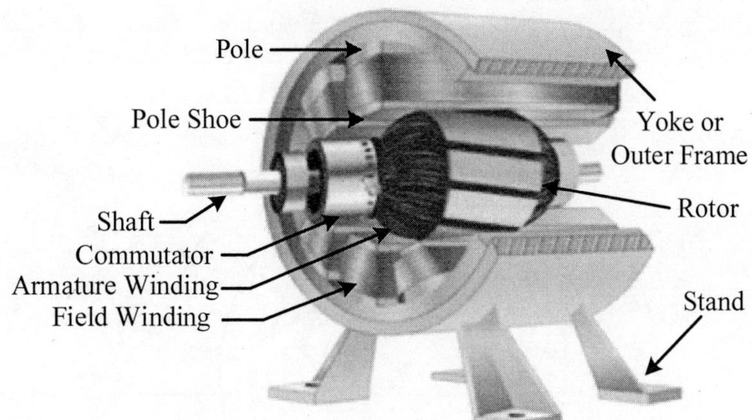

Fig. 3.1 Sectional view of DC machine

3.2 Construction of DC Machine

The cross-section of a DC machine is shown in fig. 3.1 and fig 3.2. A DC machine consists of two parts: the stator i.e., the stationary part, and the rotor i.e., the rotating part. The essential components like the Yoke, Pole core & pole shoes, Pole coil or field coil, Armature core, Armature winding or conductor, commutator, brushes & bearings of the machine have been discussed below.

3.2.1 Stator

The stator of a DC machine provides mechanical support for the machine and consists of the magnetic frame or yoke and the poles or field poles.

Magnetic Frame or Yoke

The outer cylindrical frame to which the main poles and inter poles are fixed is called a yoke or frame. It helps to fix the machine on the foundation. The main function of the yoke in the machine is to offer mechanical support intended for poles and protects the entire machine from moisture, dust, etc.

It serves two purposes:

- It provides mechanical support to the inner parts of the machine and acts as a protecting cover for the whole machine.
- It carries the magnetic flux produced by the poles and provides a low reluctance path for magnetic flux.

Scan to know more on DC Machine

The yoke is made of cast iron for smaller machines and for larger machines, it is made of cast steel or fabricated electrical grade rolled steel since electrical grade/rolled steel materials have better magnetic properties as compared to cast iron. The yoke serves the basic function of providing a highly permeable path for the magnetic flux. For small permanent-magnet (PM) machines, it can be a rolled-ring structure welded at yoke ends. For small wound machines, the field poles and the yoke are punched as one unit from thin laminated electrical grade sheet steel laminations. For large machines, the yoke is built using cast steel sections.

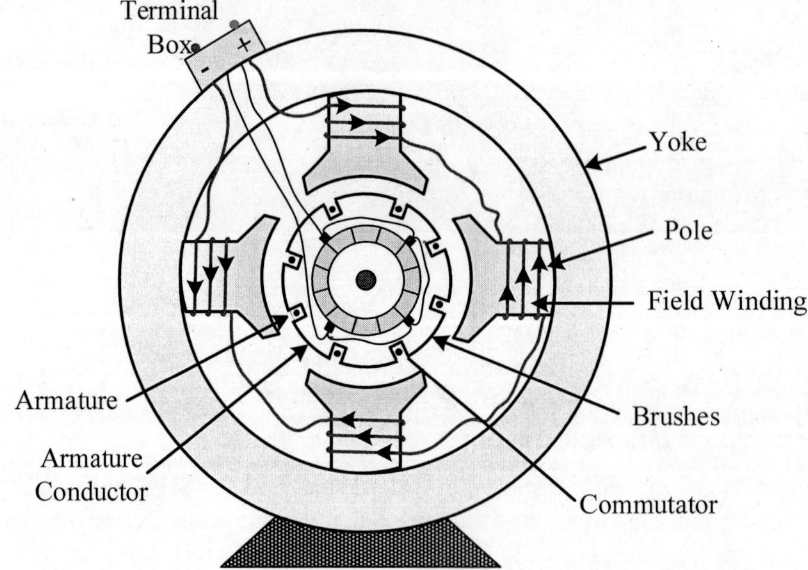

Fig. 3.2 The cross-sectional view of a 4-pole DC machine

Pole Core and Pole Shoes

The pole core and pole shoes are fixed to the magnetic frame or yoke by bolts. A pole of a DC machine is a supporting structure to the electromagnet. They serve the following purposes:

- They support the field or exciting coils.
- They spread out the magnetic flux over the armature periphery more uniformly.
- The pole shoes have a larger cross-section, and the reluctance of the magnetic path is reduced.

There are two main types of pole construction.

- The pole core itself may be a solid piece made out of either cast iron or cast steel but the pole shoe is laminated and is fastened to the pole face using countersunk screws
- In modern design, the complete pole cores and pole shoes are built of thin laminations of annealed steel which are riveted together under hydraulic pressure. The thickness of laminations varies from 0.25 mm to 1 mm.

For the construction of a pole or pole shoe low reluctance magnetic material such as cast steel or wrought iron laminations are riveted together under hydraulic pressure as shown in fig. 3.3.

Note that the cross-sectional area of the field pole is smaller than that of the pole shoe. This is done to provide sufficient room for the field winding and to decrease the mean turn length of the wire and thereby reduce its weight and cost.

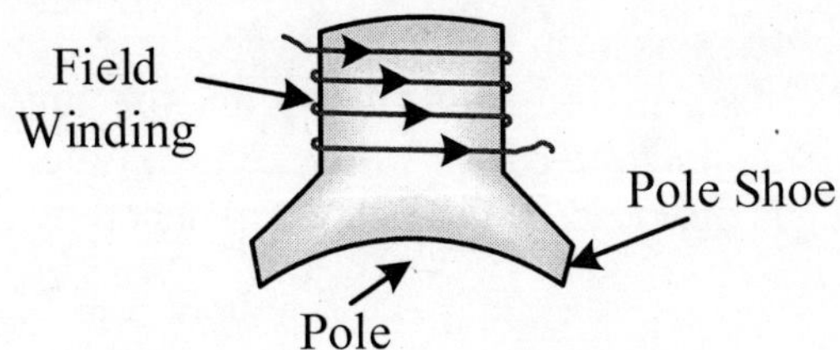

Fig. 3.3 Pole core and pole shoe

Field or Exciting Coils

The coils wound around the pole cores that carry the DC exciting current are called field coils as shown in Fig. 3.3. Enamelled standard wire gauge (SWG) copper wire is used for the construction of field or exciting coils. The function of the field system is to produce a uniform magnetic field when direct current is passed through it. The field coils of all the poles are connected in series in such a way that when current flows through them, the adjacent poles attain opposite polarity as shown in fig. 3.4

There are two types of field windings-a **shunt field winding** and a **series field winding.**

- The shunt field winding has a large number of turns of fine wire and derives its name from the fact that it is connected in parallel with the armature winding. A machine with a shunt field winding is called a **shunt machine.**

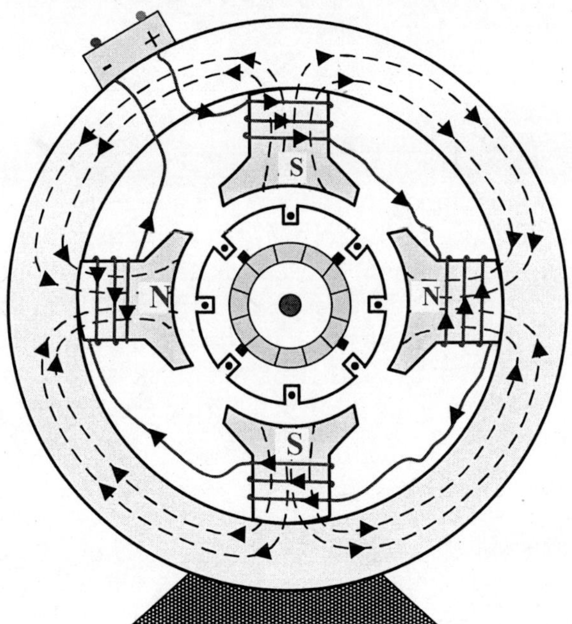

Fig. 3.4 Magnetic circuit of DC machine

- A series field winding contains almost full load current, as the name implies, is connected in series with the armature winding, and has comparatively fewer turns of thick wire. A **series machine** is wound only with series field winding.

- A **compound machine** has both windings. When both field windings in a compound machine produce fluxes in the same direction, the machine is said to be of the **cumulative** type. The machine is of the **differential** type when the field set up by the shunt field winding is opposed by the field established by the series field winding.

3.2.2 Armature core

The rotating part of a DC machine, which is surrounded by the fixed poles on the stator, is called the armature. It is cylindrical and keyed to the rotating shaft. The effective length of the armature is usually the same as that of the pole. Circular in cross-section, it is made of thin, highly permeable, and insulated electrical-grade sheet steel laminations. The laminations have axial slots (as shown in fig. 3.5) on their periphery to house the armature coils (armature winding).

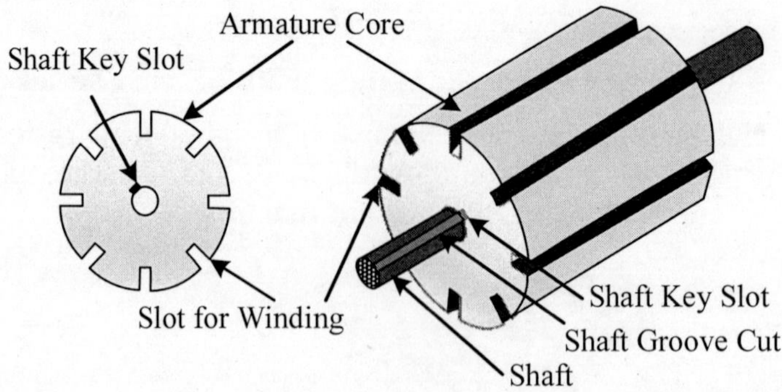

Fig. 3.5 Armature core

Armature is subjected to rotational magnetization Since armature is a rotating part of the machine, reversal of flux takes place in the core, hence hysteresis losses are produced. To minimize these losses silicon steel material is used for its construction. When it rotates, it cuts the magnetic field, and an emf is induced in it. This emf circulates eddy currents which result in eddy current loss in it. To reduce these losses, the armature core is laminated, in other words, we can say that about 0.3 to 0.5 mm thick stampings are used for its construction. Each lamination or stamping is insulated from the other by a varnish layer.

3.2.3 Armature winding

The insulated conductors housed in the armature slots are suitably connected. This is known as armature winding. The armature winding acts as the heart of a DC machine. It is a place where one form of power is converted to the other form i.e., in the case of a generator, mechanical power is converted into electrical power, and in the case of a motor, electrical power is converted into mechanical power. Based on connections, there are two types of armature windings named (*i*) Lap winding and (*ii*) Wave winding.

- ***Lap winding***: In lap winding the 'finish' of one coil is connected to the 'start' of the next coil. In this winding, the connections are such that the number of parallel paths is equal to the number of poles. Thus, if the machine has P poles and Z armature conductors, then there will be P parallel paths, and each path will have Z/P conductors in series. In this case, the number of brushes is equal to the number of parallel paths. Out of which half the brushes are positive and the remaining (half) are negative. fig. 3.8(a),(b) show the lap-connected armature winding.
- ***Wave winding***: In wave winding, the 'finish' of the one coil is connected to the 'start' of the coil one pole pitch apart. In this winding, the connections are such that the numbers of parallel paths are only two irrespective of the number of poles. Thus, if the machine has Z armature conductors, there will be only two parallel paths each having $Z/2$ conductors in series. In this case, the number of brushes is equal to two i.e., the number of parallel paths. fig. 3.8(c) shows the wave-connected armature winding.

3.2.4 Commutator

It is an important part of a DC machine. The commutator is of cylindrical shape and is made up of wedge-shaped hard-drawn copper segments. The segments are insulated from each other by a thin sheet of mica. The segments are held together by means of two V-shaped rings that fit into the V-grooves cut into the segments. Each armature coil is connected to the commutator segment through a riser. The sectional view of the commutator assembly is shown in fig. 3.6.

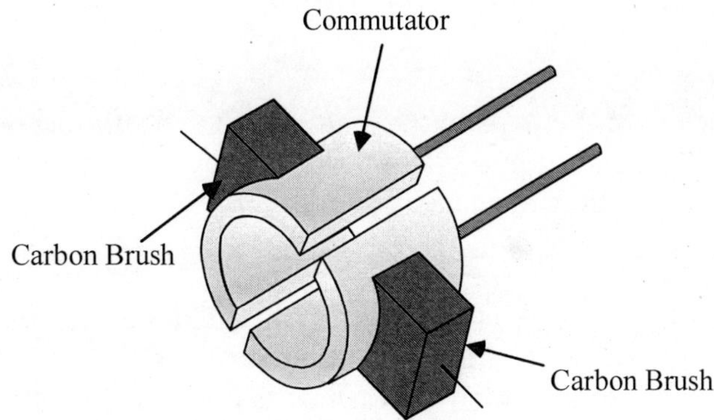

Fig. 3.6 Commutator

The commutator is a very well-conceived device that serves the function of a mechanical rectifier. It converts the alternating emf induced in the armature coils into a unidirectional emf.

It serves the following purposes:

- It connects the rotating armature conductors to the stationary external circuit through brushes.

- It converts the alternating current induced in the armature conductors into unidirectional current in the external load circuit in generator action, whereas, it converts the alternating torque into unidirectional (continuous) torque produced in the armature in motor action.

3.2.5 Brushes

Brushes are held in a fixed position on the commutator using brush holders and rockers. An adjustable spring inside the brush holder exerts constant pressure on the brush to maintain proper contact between the stationary brush and the rotating commutator. The poor contact results in excessive sparking and damage to the Commutator. On the other hand, too high a pressure results in excessive wear of the brush and overheating of the commutator through friction. There are many different brush grades, depending upon their composition. A brush may be made of carbon, carbon graphite, or a copper-filled carbon mixture. The graphite in a brush provides self-lubrication between the brush and the commutator. Although the brush holders are mounted on the brush gear, they are electrically insulated from it. A brush is electrically connected to the brush holder by a braided copper wire called the pigtail. Through these brush holders, we can establish an electrical connection between the external circuit and the armature coils.

3.2.6 Brush rocker

It holds the spindles of the brush holders. It is fitted onto the stationary frame of the machine with nuts and bolts. By adjusting its position, the position of the brushes over the commutator can be adjusted to minimize the sparking at the brushes.

3.2.7 End housings

End housings are attached to the ends of the main frame and support bearings. The bearing supports the shaft. The front housing supports or covers the bearing and the brush assemblies whereas the rear housing usually supports the bearing only.

3.2.8 Bearings

The bearings may be ball or roller bearings these are fitted in the end housings. Their function is to reduce friction between the rotating and stationary parts of the machine. Mostly high carbon steel is used for the construction of bearings as it is a very hard material.

3.2.9 Shaft

The shaft is made of mild steel with maximum breaking strength. The shaft is used to transfer mechanical power from or to the machine. The rotating parts like the armature core, commutator, cooling fan etc. are keyed to the shaft.

3.3 Simple Loop Generator

It is an important part of a DC machine. The commutator is of cylindrical shape and is made up of wedge-shaped hard-drawn copper segments. The segments are insulated from each other by a thin sheet of mica.

3.4. Armature Winding

Now in this section, we will discuss the winding of an actual armature. At the outer periphery of an armature core, slots are cut. In these slots, a number of conductors are placed which are connected in proper arrangement forming series–parallel paths depending upon the requirement. This arrangement of connections is known as *armature winding*. To understand the concept of armature winding, the meaning of the following term used to connect with armature winding should be desirable.

3.4.1 Conductor

The length of wire embedded in the armature core, lying in the magnetic field in which speed or motional e.m.f is induced, is called the conductor (or inductor) for example, length AB or CD in fig. 3.7 (a). It may be having one or more parallel strands. The total number of conductors in the armature winding is represented by the symbol Z. A single turn consists of two conductors.

3.4.2 Turn

Two conductors lying in a magnetic field connected in series at the back, as shown in fig. 3.7(a), so that emf induced in them is additive is known as a turn.

3.4.3 Coil and winding element

The two conductors AB and CD along with their end connections constitute one coil of the armature. The coil may be a single-turn coil having only two conductors, as shown in fig. 3.7(a), or it may be a multi-turn coil having more than two conductors as shown in fig. 3.7(b). In fig. 3.7(b), a three-turn coil is shown in which each coil side has three conductors.

The group of three conductors or wires constituting a coil side of a multi-turn coil is wrapped using cotton tape, as shown in fig. 3.7(c), and placed in the armature slot. A multi-turn coil can be represented by a single-line diagram as shown in fig. 3.7(d).

It may be noted that the beginning and the end of each coil must be connected to a commutator segment, there are as many commutators segment as coils for both lap and wave windings. The multi-turn coils are used to develop higher voltages. The total number of coils in the armature winding is represented by the symbol 'C'.

When the armature conductors are more, it is not feasible to use single turn coils because it will require large number of commutator segments and if used it will not give spark-less commutation. Moreover, it will not be economical due to use of more copper in the end connections.

3.4.4 Coil side

The single-turn or multi-turn has two sides called *coil sides.* Both the coil sides are embedded in two different slots as per the winding design as shown in fig. 3.7(b).

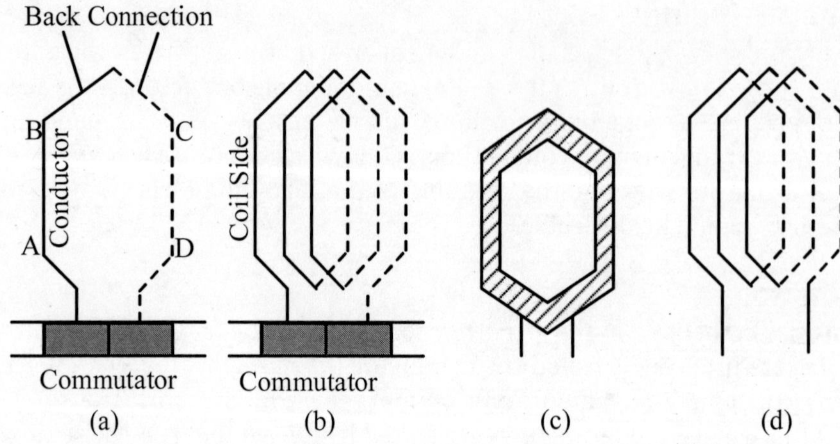

Fig. 3.7 Armature coil

3.4.5 Coil Group

A group of coils may have one or more coils. It is the product of the number of phases and poles in a rotating machine.

Coil group = the number of poles × the number of phases

3.4.6 Winding

When the number of coil groups is arranged on the armature in a particular fashion as per the design, it is called armature winding.

3.4.7 Inductance effect

All the coils have some inductance effect as current is changing in them. Due to the inductance effect the flow of current is opposed causing a reduction in resultant output voltage. Over-hanging end connections have some adverse effects due to inductance. In the DC machine, it causes the development of reactance voltage in the coil undergoing commutation and results in sparking at the trailing edge of the brushes'–poor commutation.

3.4.8 Front and back-end connector

A wire that is used to connect the end of a coil at the front to the commutator segment is called a front-end connector. Whereas, a wire that is used to connect one coil side to the other coil side at the back is called a back-end connector.

3.4.9 Pitch of the winding

It is defined as the distance around the armature between two successive conductors which are directly connected. Alternatively, it is the distance between the beginning of two consecutive turns.

$$Y = Y_B - Y_F \text{ ....for lap Winding}$$

$$= Y_B + Y_F \text{ ....for wave Winding}$$

In practical, coil pitch as low as eight-tenths of a pole pitch are employed without much serious reduction in the *emf.* Fractional-pitch winding are purposely used to effect substantial saving in the copper of the end connection and for improving commutation.

3.4.10 Pole pitch

It is defined as the number of armature slots per pole. It may also be defined as the number of armature conductors per pole. If there are 52 conductors for 4 poles then the pole pitch will be equal to 52/4 i.e., 13 conductors per pole. *Pole pitch is also defined as the armature periphery divided by the number of poles.*

3.4.11 Back pitch

The distance in terms of armature conductors, which a coil advances on the back of the armature between the first and last conductor or the distance between two coil sides of the same coil is called *back pitch.* It is also called the *coil span* or *coil spread* and is denoted by Y_B, as shown in fig. 3.8(a), (b), and (c)

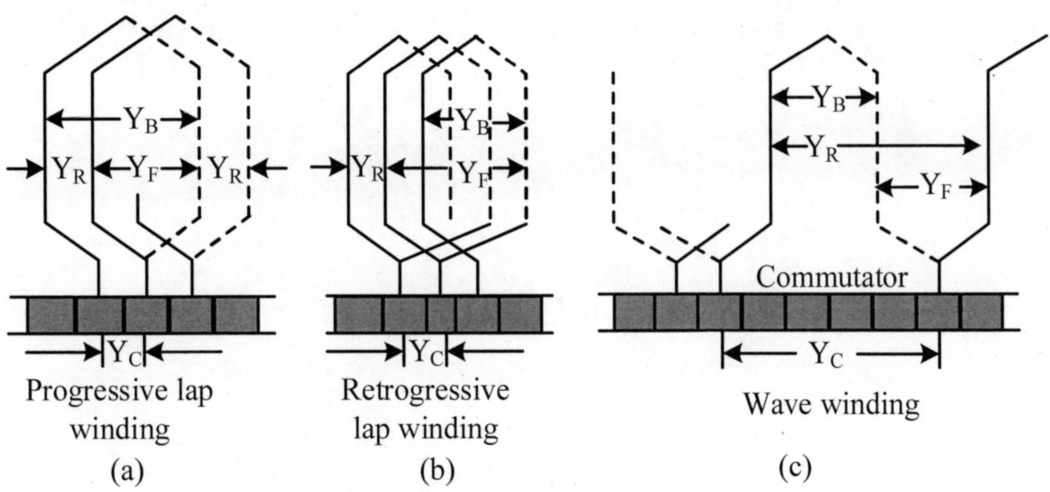

Fig. 3.8 Term used in coils

3.4.12 Front pitch

The number of armature conductors or elements spanned by a coil on the front is called front pitch. Alternatively, the distance in terms of the number of armature conductors or the number of slots between the second conductor of one coil and the first conductor of the next coil which are connected to the same commutator segment on the front is called *front pitch.* It is denoted by Y_F, as shown in fig. 3.8 (a), (b), and (c).

3.4.13 Resultant pitch

It is the distance between the beginning of one coil and the beginning of the next coil to which it is connected. It is denoted by Y_R as shown in fig. 3.8(a), (b), and (c).

3.4.14 Commutator Pitch

The distance measured in terms of commutator segments between the segments to which the two ends of a coil are connected is called *commutator pitch*. It is denoted by Y_C as shown in figs. 3.8(a), (b), and (c). the commutator pitch is the difference of Y_B and Y_F for lap winding whereas for wave winding it is the sum of Y_B and Y_F as shown in fig. 3.8(a) and (b).

3.4.15 Coil-Span and Coil-Pitch

The distance measured in terms of armature slots or armature conductors between two sides of a coil is called *coil span or coil pitch.* The coils are full-pitched, short-pitched (fractional Pitch), or over-pitched coils.

If the coil span or coil pitch is equal to the pole pitch then winding is called a full pitch-pitched coil. In this case, the induced emf in a coil is the arithmetic sum of the emf induced in two sides of the same coil since the coil sides are displaced by 180° electrical and the coil side lies under opposite poles (North and South). Therefore, maximum e.m.f. is induced in the coil. For example, in fig. 3.9. where the pole pitch is said 4. One side of coil-*A* is placed in slot No. 1 and the other side is placed in slot No. 5, then the coil span is 5-1 = 4 which is equal to pole pitch hence coil-*A* is called a full-pitched coil.

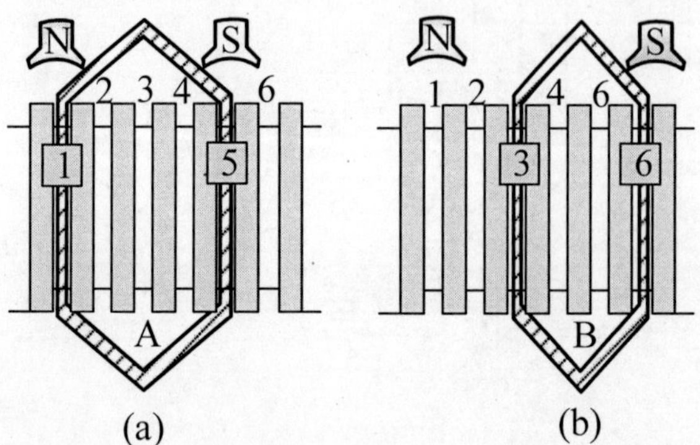

Fig. 3.9: Full-pitch and short-pitch winding

If the coil span is less than the pole pitch, then the winding is fractional or short-pitched. For example, in fig. 3.9(b). Whereas, in the case of coil-*B*, the coil span is 6–3 = 3 which is less than the pole pitch. In this case, there is a phase difference between the e.m.f hence, the resultant induced emf is reduced. because the two sides would fall under the influence of the same pole at some instant. The advantage of short-pitched winding is that in this case the copper used for end connections is reduced substantially which reduces the cost of the machine. It also reduced the sparking at brushes (Improve commutation). Moreover, it reduces the copper losses, mmf harmonics and improves efficiency to some extent.

3.4.16 Degree of re-entrant

If all armature conductors are included on returning to the starting point, the winding is called single re-entrant winding. It will be a double re-entrant if only half of the conductors are included in tracing through the winding once.

3.5 Types of Armature Winding

The continuous or closed armature windings are of two types:

1. Gramme-ring Winding 2. Drum Winding

3.5.1 Gramme-ring winding

The gramme-ring type of armature winding is an early form of armature winding. This is not used in modern DC machines because it is the earliest form of armature windings. This winding is very useful to understand the action of the commutator windings. In the ring winding DC machine, the core is usually made of steel laminations and insulated from each other. Fig.3.10 shows that the core has eight coils and each coil consists of two turns. S_i and F_i represent the start and end of the coil 'i'. The end of the coil 'i' is connected to the start of the coil '$i + 1$' and so on until the end of the last coil is connected to the start of the first coil.

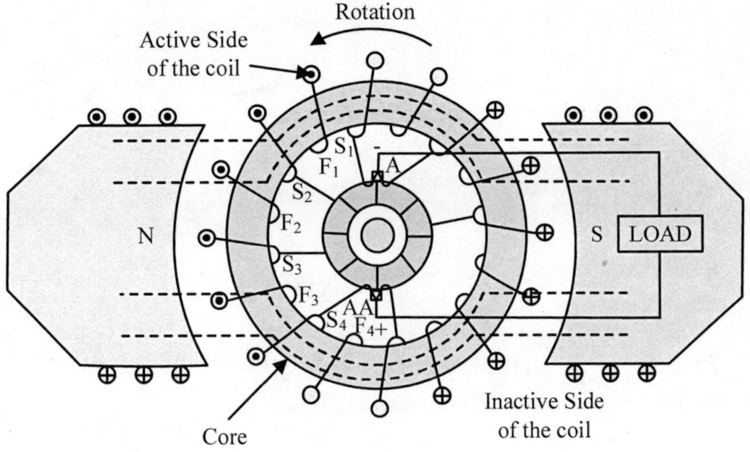

Fig.3.10 Ring winding

The commutator segment is connected to the junction between the neighbouring coils, that is, to the end of one coil and the start of another coil. The dotted lines show the flux lines. An electromagnetic field is set up by the electromagnet. For the anticlockwise direction of rotation of the armature, the emf induced in the coils under the north poles will be directed away from the papers and that of in the coils under the south poles will be directed into the papers. Hence, the emf of the coils under the north pole and the south pole is indicated by crosses and dots. The equivalent circuit in fig.3.10. The total emf induced around the closed loop will be zero.

For the brush position shown in fig. 3.11, there will be two circuits. These two circuits are parallel between the brushes where each circuit contains an equal number of coils generating emf acting from A to AA. Therefore, the current in each circuit will be half of the total current. The emf between the brushes A and AA depends on the emf/conductor as well as the number of conductors in each series circuit. If the brushes are placed such that one-half of the coil comes under the north pole and the remaining half comes under the south pole, the maximum emf will be across the brushes. To achieve this, the brushes are to be placed one pole pitch apart, that is, 180 electrical degrees. The number of conductors generating emf should not vary due to the rotation of the armature, and in this case, the voltage between the brushes A and AA will remain constant. The number of parallel paths in ring windings is equal to the number of poles, and hence two parallel paths exist in fig.3.10. For six poles, the number of parallel paths will be six. The drawback of gramme-ring winding is that emf is not induced in the inner portion of the coil.

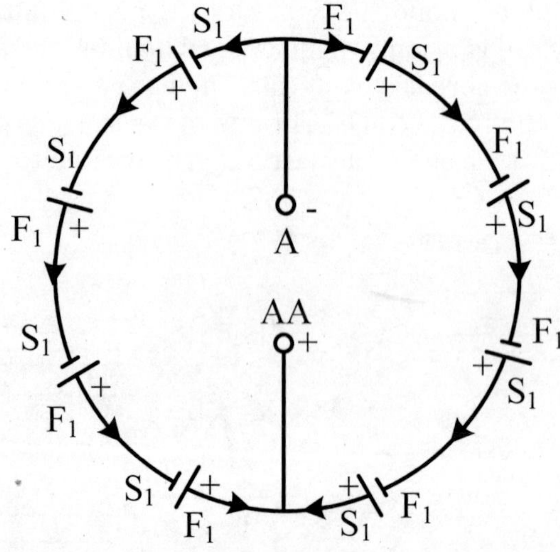

Fig.3.11 Equivalent circuit of ring winding

The following are the disadvantages of ring windings:

- Only half the coil is useful for generating the emf. The other half of the portion of winding was lying inside the core which was used only as connectors and so there was a waste of copper.
- As each turn was to pass through the center of the core, therefore, it was difficult to wind and require more labour, hence it is very expensive
- The maintenance and repairs were more costly.
- Insulation of winding was also difficult.
- Commutation conditions are not good.
- Construction was having a large air gap, so stronger field excitation was required to produce the required flux.

3.5.2 Drum winding

This type of winding has slots on the drum shape armature to house the armature conductors and connected to one another at the front and back through connectors. It has the following advantages.

- Each winding, placed on the armature slots, surrounds the core so that the entire length of the conductor, except the end connections, cut the main magnetic flux. Therefore, the voltage induced in this type of armature winding is higher than the Gramme-ring winding.
- The coils, before placing on the armature slots, can be pre-formed and insulated. Hence costs can be reduced.
- The two sides of the coil are placed under two different poles, one North Pole and another South Pole, hence the emf induced in them is always additive with the help of the end connection.
- Fractional pitch winding can be used in drum winding.
- The several conductors are placed in a single slot, the number of the slot gets reduced in the armature core, and the armature core teeth become mechanically stronger. The lamination and the protection of coils are also improved.
- The manufacturing cost will be reduced in the drum-type winding due to pre-formed coils.

The drum winding may be either single-layer or double-layer winding.

3.6 Single Layer Drum Winding

In this winding one conductor or one coil side is placed in each armature slot as shown in fig. 3.12(a) It is rarely used because of its cost factor.

3.7 Double Layer Drum Winding

The two-layer winding has two conductors or two coil sides per slot arranged in two layers. The conductors are arranged in such a way that one side of every coil is on the upper half of one slot. Another side of the same coil is placed on the lower half of another slot as shown in fig. 3.12(b). The windings in which two coil sides occupy each slot are commonly used in all medium-sized machines. In the case of large machines, more coil sides can be placed in a single slot as shown in fig. 3.13 (a) and 3.13(b). Placing several elements in a single slot gives fewer slots than segments that have the following advantages.

- Armature core teeth become stronger, providing less damage to the laminations and coils.
- Fewer slots reduce the number of coils to be wounded, which makes low manufacturing cost.
- For reduced slots, the number of commutation segments decreases. This decreases the number of coils connected to the adjacent segment. It results in less sparking in the commutator, which improves the commutation.

Two types of winding mostly employed for drum-type armature are known as lap winding and wave winding as mentioned in section 3.2.3.

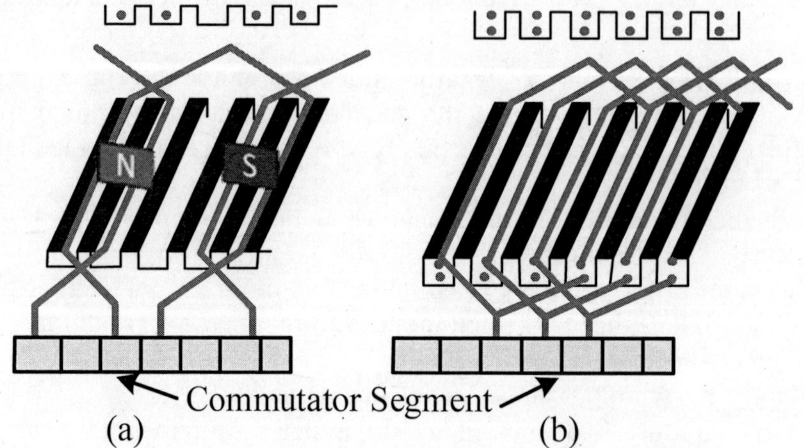

Fig. 3.12 Single-layer and double-layer drum winding

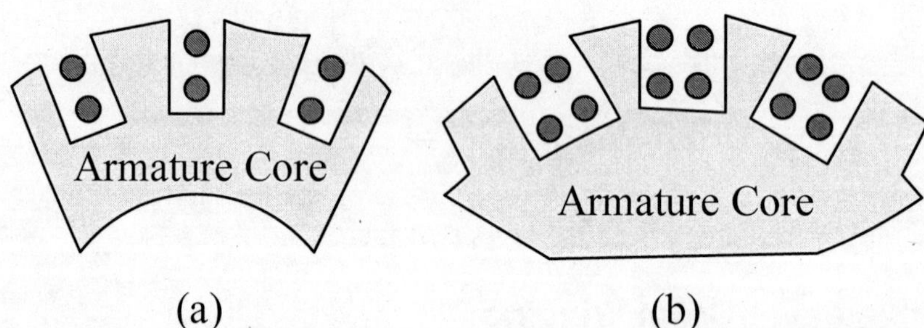

Fig. 3.13 Double-layer drum winding (a) Two coil sides per slot (b) Four coil sides per slot

3.8 Lap Winding

The winding is connected in such a manner that the end of the one coil is connected to the starting end of other coils (adjacent coil) of the same pole and so on then, the winding is known as *lap winding* as shown in fig. 3.14(a). In lap winding, the finishing end of one coil is connected to a commutator segment. The starting of an adjacent coil situated under the same pole is also connected to the same commutator segment. Since there is an overlapping of successive coils, and their parallel paths and poles are equal in number. In lap winding all the pole groups of coils generating emf in the same direction at any instant of time are connected in parallel by the brushes. Lap winding may be further classified as simplex (single) or multiplex (double or triple) windings.

- ***Simplex lap winding:*** In this winding, there are as many parallel paths as there are field poles on the machine.
- ***Double or duplex lap winding:*** In this case, two similar simplex windings are placed in alternate slots on the armature and connected to alternate commutator segments. Thus, each winding carries half of the armature current.
- ***Triple or triplex lap winding:*** In this case, three similar simplex windings are placed to occupy every third slot and connected to every third commutator segment. Thus, each winding carries one-third of the armature current. Similarly, there can be multiplex lap winding having even more than three simplex winding as per the requirement.

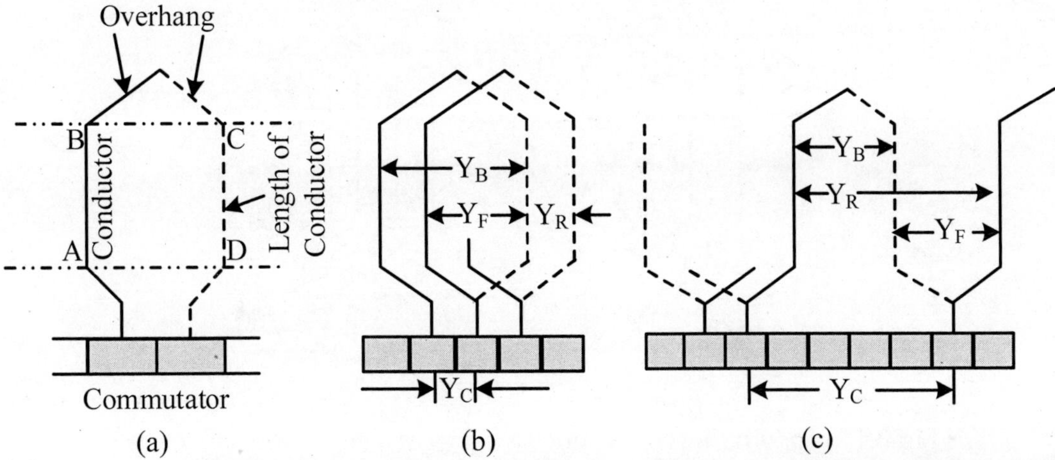

Fig. 3.14 (a) Single coil (b) Lap and (c) Wave winding

The purpose of employing multiplex lap winding is to increase the number of parallel paths enabling the armature to carry a large total current, at the same time reducing the conductor current to improve commutation conditions. The following points regarding lap winding should be carefully noted:

- The windings must be full pitched; that is, the front and back pitch must be approximately equal to the pole pitch.
- To place the coils properly on the armatures, the front pitch, as well as the back pitch, must be odd.
- The commutator segments are the images of the coil; that is, the number of commutator segments is equal to the number of the coils.
- The windings must close upon itself.
- The lap winding has been shown in fig.3.14(b).

The following points are important for the simplex lap winding:

- The back pitch (Y_B) and the front Pitch (Y_F) are add opposite signs. They cannot be equal. They differ by two or some multiple thereof.
- Y_B and Y_F must be nearly equal to pole pitch$\left(\frac{Z}{P}\right)$.

- The average pitch, $Y_A = \frac{Y_F+Y_B}{2}$ is equal to the pole pitch$\left(\frac{Z}{P}\right)$.
- The resultant pitch is the arithmetical difference between Y_B and Y_F. i.e., $Y_R = Y_B - Y_F$.
- The commutator pitch $Y_C = \pm 1$. In general, $Y_C = \pm m$. Here m is the multiplicity
- of winding.
- The number of slots for a two-layer winding as well as the commutator segment is equal to the number of coils.
- The number of parallel paths in the armature = mP. Here m is the multiplicity of the winding and P is the number of parallel paths.

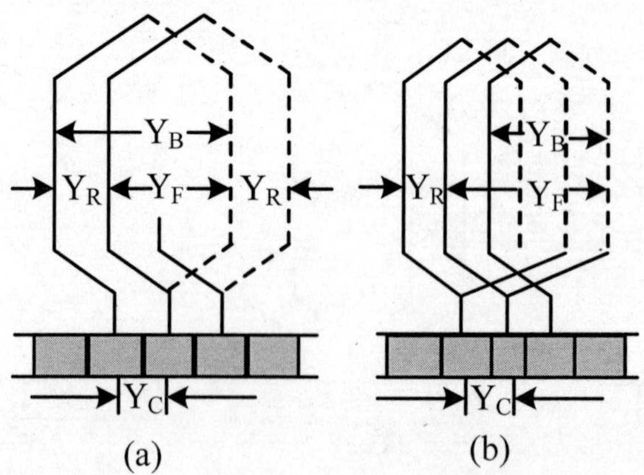

Fig. 3.15 Lap winding connection (a) Progressive (b) Retrogressive

A simple lap winding can be progressive or retrogressive, shown in fig, 3.15(a) and fig. 3.15(b), respectively. In progressive winding, the coil ends are connected to the commutator segments in ascending order, whereas in retrogressive winding, these are connected to the commutator segments in descending order. Therefore, for the simple lap winding, $Y_C = \pm 1$, Where '+' is for the progressive winding and '–' is for the retrogressive winding. For the progressive or right-handed winding, $Y_F = \frac{Z}{P} - 1$ and $Y_B = \frac{Z}{P} + 1$; and for the retrogressive or left-handed winding, $Y_F = \frac{Z}{P} + 1$ and $Y_B = \frac{Z}{P} - 1$.

Example. 3.1 Calculate the different pitches for 4 poles, lap wound DC machine having 24 slots each has 2 coils sides.

Solution: - Here, P = 4; No. of slots = 24; Number of coil side/ slot = 2

$$Z = 24 \times 2 = 48$$

$$\text{Pole pitch, } Y_P = \frac{Z}{P} = \frac{48}{4} = 12$$

$$\text{Back pitch, } Y_B = Y_P + 1 = 12 + 1 = 13$$

$$\text{Front pitch, } Y_F = Y_P - 1 = 12 - 1 = 11$$

Resultant pitch, $Y_R = Y_B - Y_F = 13 - 11 = 2$
Commutator pitch, $Y_C = 1$

3.9 Numbering of Coil and Commutator Segment

The winding diagram can be obtained by removing the armature periphery, cutting it along the slot, and laying it out flat so that the slots and conductors can be viewed to trace out the armature winding. To draw the winding diagrams, we number the coil only, not the individual turn. The upper side of the coil will be shown by a firm continuous line, whereas the lower side will be represented by a dotted line. The numbering of coil sides will be consecutive numbers i.e., 1, 2, 3,..... etc., and such that odd numbers are assigned to the top conductors and even numbers to the lower side for two-layer winding. The commutator segments will also be numbered consecutively, the number of the segments will be equal to the number of coils and the upper side will be connected to it.

Example.3.2 Draw the developed view of a simple lap winding of four-pole DC generators having 12 slots with two coil sides per slot.

Solution: - Pole pitch $= \frac{number\ of\ slots}{number\ of\ poles}\frac{12}{4} = 3$

The coil span taken here is equal to the full pole pitch of four slots/pole. It is assumed here that the winding is moving from left to right and the winding is behind the poles. Using a full-pitched coil span, the winding is drawn in a progressive way ($Yc = +1$). The upper sides of the coil are denoted as 1, 2,3, etc., and the lower sides of the same coil are denoted as 1′, 2′, 3′, etc. The table shows the sequence of the back connection and front connection.

Back Connection	Front Connection
1 is connected to 1' (4)	1' is connected to 2 (2)
2 is connected to 2' (5)	2' is connected to 3 (3)
3 is connected to 3' (6)	3' is connected to 4 (4)
4 is connected to 4' (7)	4' is connected to 5 (5)
5 is connected to 5' (8)	5' is connected to 6 (6)
6 is connected to 6' (9)	6' is connected to 7 (7)
7 is connected to 7' (10)	7' is connected to 8 (8)
8 is connected to 8' (11)	8' is connected to 9 (9)
9 is connected to 9' (12)	9' is connected to 10 (10)
10 is connected to 10' (13, i.e., 13-12 = 1)	10' is connected to 11 (11)

11 is connected to 11' (14, i.e., 14-12 = 2)	11' is connected to 12 (12)
10 is connected to 12' (15, i.e., 15-12 = 3)	12' is connected to 1 (1)

The numbers in brackets indicate the slots. When coil sides are placed in the first and fourth slot, there exist three slots of teeth.

The figure shows the sequence diagram

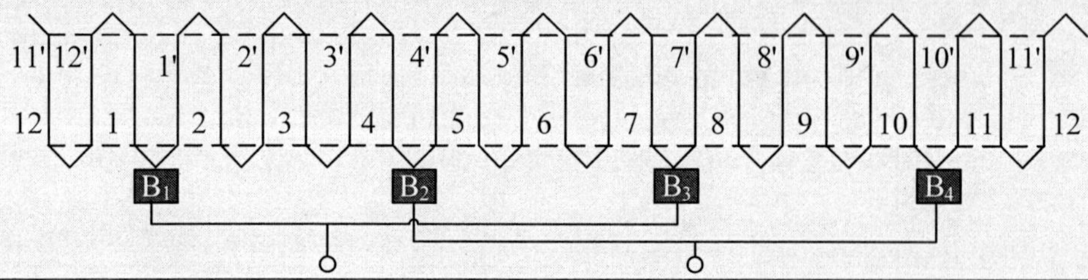

Example.3.3 Draw the developed winding diagram of a progressive lap winding for 4 poles, 16 slot single layer showing therein position of poles, the direction of motion, the direction of induced emf, and position of brushes.

Solution: - Here, the number of poles, P = 4

Number of coil sides, Z =16

Pole pitch, $Y_P = \frac{Z}{P} = \frac{16}{4} = 4$

For progressive winding, $Y_B = Y_P + 1 = 4 + 1 = 5$

$Y_F = Y_B - 2m = 5 - 2 \times 1 = 3 \; (here\; m = 1)$

The connections are to be made as per the following table:

Back End Connections	Front End Connections
1 to (1 + 5) = (6)	6 to (6 – 3) = (3)
3 to (3 + 5) = (8)	8 to (8 – 3) = (5)
5 to (5 + 5) = (10)	10 to (10 – 3) = (7)
7 to (7 + 5) = (12)	12 to (12 – 3) = (9)
9 to (9 + 5) = (14)	14 to (14 – 3) = (11)
11 to (11 + 5) = (16)	16 to (16 – 3) = (13)
13 to (13 + 5) = 18(18–16=2)	18 to (18 – 3) = (15)
15 to (15 + 5) = 20(20–16=4)	4 to (4 – 3) = (1)

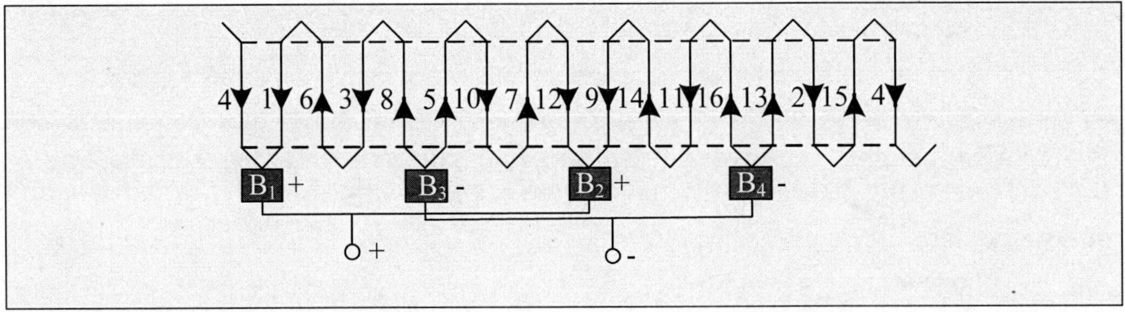

3.10 Wave Winding

In the wave winding the coil side is not connected back but rather progresses forward to another coil side placed under the next pole, as shown in fig. 3.14(c). In this way, the winding progresses, passing successively every North and South pole till it returns to the coil side from where it was started. It clearly shows that the connections of wave winding always progress in the same direction around the armature instead of moving forward and backward alternately like that of lap winding. As the shape of the winding is wavy, it is named as *wave winding.* In wave winding, all the coils carrying current in the same direction are connected in series. Therefore, the coils carrying current in a particular direction are connected in one series circuit, whereas the coils carrying current in the other particular direction are connected in another series circuit. Hence, there are two parallel circuits in wave windings.

The wave winding may be classified as progressive wave winding or retrogressive wave winding.

Progressive wave winding: If after completing one round of the armature, the winding falls in a slot to the right of its starting point the winding is known as *progressive wave winding.*

Retrogressive wave winding: If after completing one round of the armature. The winding falls in a slot to the left of its starting point, the winding is known as *retrogressive wave winding.*

The wave winding may also be a multiplex winding (double, triple, or more). A simplex wave winding has only two parallel paths irrespective of the number of pair of poles of the machine. However, a multiplex wave winding has 2 *m* parallel paths, where *m* is the multiplicity. For the wave winding the following points are important:

- Y_B and Y_F are odd and have the same signs.
- Y_B and Y_F are near equal to the pole pitch and differ by 2.
- The resultant pitch Y_R is the sum of Y_B and Y_F ($Y_R = Y_B + Y_F$).
- The commutator pitch $Y_C = \frac{Number\ of\ commutator\ Segment-1}{Number\ of\ Pairs\ of\ pole}$
- The average pitch (Y_A) is an integer and is given by

$$Y_A = \frac{Z \pm 2}{P} = \frac{\frac{Z}{2} \pm 1}{\frac{P}{2}} = \frac{Number\ of\ commutator\ Segment - 1}{Number\ of\ Pairs\ of\ pole}$$

- The Number of coils (N_C) can be obtained from $N_C = \frac{PY_A - 2}{2}$.
- The number of parallel paths =2m, where m is the multiplicity of the winding.

Example.3.4 A simple two-layer lap-winding four-pole generator has 16 coils. Draw the equivalent ring diagram with the position of the brushes.

Solution: - Here, the number of poles, P = 4

Number of coils, C =16

Number of coil sides, Z = 16 X 2 = 32

It is a two-layer winding.

∴ Number of coil sides in each slot = 2

Pole pitch, $Y_P = \frac{Z}{P} = \frac{32}{4} = 8$

For progressive winding, $Y_B = Y_P + 1 = 8 + 1 = 9$

$Y_F = Y_B - 2m = 9 - 2 \times 1 = 7$ $(here\ m = 1)$

The connections are to be made as per the following table:

Back End Connections	Front End Connections
1 to (1 + 9) = (10)	10 to (10 – 7) = (3)
3 to (3 + 9) = (12)	12 to (12 – 7) = (5)
5 to (5 + 9) = (14)	14 to (14 – 7) = (7)
7 to (7 + 9) = (16)	16 to (16 – 7) = (9)
9 to (9 + 9) = (18)	18 to (18 – 7) = (11)
11 to (11 + 9) = (20)	20 to (20 – 7) = (13)
13 to (13 + 9) = (22)	22 to (22 – 7) = (15)
15 to (15 + 9) = (24)	24 to (24 – 7) = (17)
17 to (17 + 9) = (26)	26 to (26 – 7) = (19)
19 to (19 + 9) = (28)	28 to (28 – 7) = (21)
21 to (21 + 9) = (30)	30 to (30 – 7) = (23)
23 to (23 + 9) = (32)	32 to (32 – 7) = (25)
25 to (25 + 9) = 34(34–32=2)	34 to (34 – 7) = (27)
27 to (27 + 9) = 36(36–32=4)	36 to (36 – 7) = (29)
29 to (29 + 9) = 38(38–32=6)	38 to (38 – 7) = (31)
31 to (31 + 9) = 40(40–32=8)	8 to (8– 7) = (1)

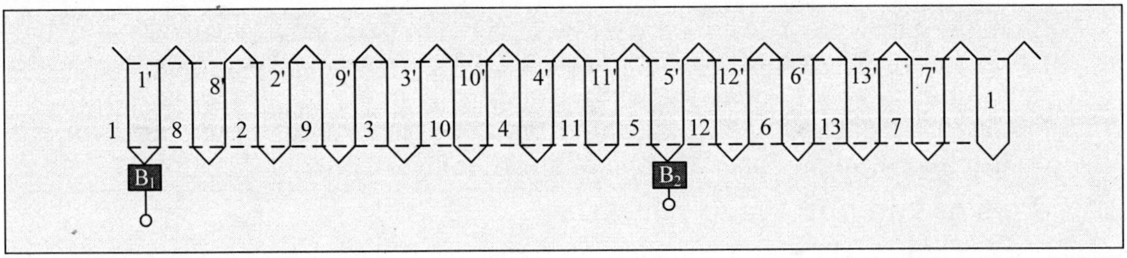

3.11 Dummy Coil

These coils are used in wave winding when the requirement of windings is not satisfied by armature punching. Since the dummy coils are not connected to the commutator, they do not affect the electrical characteristic of the windings. In other, it is defined as the coils placed on the armature which do not participate in the conversion of power but are employed only to make mechanical balancing of the armature known as *dummy coils*. In order to have symmetrical armature winding, a dummy coil should be avoided as far as possible. However, it may be used in low-rating commutator machines where an armature core with an existing number of slots has to use to avoid extra expenditure on new punching of armature stamping.

3.12 Equalising Connections and their Necessity

Equaliser Ring Equalizer rings are used in connection with the lap winding. It is the characteristic of lap winding that all conductors in any parallel path will be under one pair of poles. If the fluxes from all poles are exactly the same, then e.m.f induced in each parallel path is the same and carries the same current. If there is any inequality in the flux/pole due to slight variations in the air gap or in the magnetic properties of steel, there will be an imbalance of emf in various parallel paths resulting in unequal distribution of current at the brushes and circulating current flows within winding even at no load. This equalizer conductor which is in the form of copper rings is connected to equip-potential points on the backside of the armature. Such rings are called as 'Equalizer rings'. An equalizer ring has a low-resistance conductor wire, which connects the points in the armature winding at the same potentials. Hence, the function of equalizer rings is to avoid unequal distribution of current at the brushes thereby helping to get sparkless commutation.

Equalizer rings are not required for simplex wave winding even if there is magnetic unbalance. This is due to the fact that armature conductors are distributed under all poles. In view of this, both the parallel circuits are affected equally and there is therefore no need of equalizer rings in simplex wave windings. Circulating current results in overheating of the armature and poor commutation. Equalizing connection attempts to minimize the circulating current due to uneven emf present parallel path of lap winding. The existence of circulating current in lap-wound machines may due to any one or all of the following reasons:

- The air gap under different poles may not be the same, due to wear of the bearings or faulty assembly. As a result of it, some poles carry more flux than the other poles and, therefore, different e.m.fs are generated in the various parallel paths.

- The joints between pole cores and yokes or between pole- cores, and pole - faces may not be identical for all the poles. This results in different reluctance for the various pole and, therefore, different e.m.f are generated in the parallel circuits. (iii) The different reluctance for each magnetic path may also be due to the impurities or imperfections in the materials constituting the magnetic circuit.

3.13 Uses of Lap and Wave Winding

Wave windings give more emf for a given number of pole and armature conductors, whereas lap windings give the same emf for a greater number of conductors. Hence, the winding cost increases for the lap winding. Equalizing connections are not required in the wave winding, but this is mandatory for the lap winding. In lap winding, there are *P* parallel paths when the numbers of poles are *P*, whereas in wave winding, the numbers of parallel paths are two. Therefore, any inequality of pole fluxes in the wave winding cannot produce unequal voltages because this inequality of fluxes affects the two paths equally. In the lap winding, the inequality of fluxes produces unequal voltages and causes sparking at the brushes. Table 3.1 Show the comparisons between Lap and Wave windings.

Table 3.1 Comparison Between Lap Winding and Wave Winding

Lap Winding	Wave Winding
The lap winding can be defined as a coil that can be lapped back toward the succeeding coil.	The wave winding can be defined as the loop of the winding that can form the signal shape.
The connection of the lap winding is, the armature coil end is connected to the nearby commutator segment.	The connection of the wave winding is, the armature coil end is connected to the commutator segment at some distance apart.
The numbers of the parallel path are equal to the total of number poles.	The number of parallel paths is equal to two.
Another name for lap winding is multiple winding otherwise Parallel Winding.	Another name for wave winding is Series Winding otherwise Two-circuit.
The e.m.f of lap winding is Less.	The e.m.f of wave winding is More.
The number of brushes in lap winding is Equivalent to the number of parallel paths.	The number of brushes in wave winding is Equivalent to two.
The types of lap winding are Simplex lap winding & Duplex lap winding.	The types of wave winding are Progressive & Retrogressive.
The additional coil used in the lap winding is Equalizer Ring.	The additional coil used in the wave winding is the Dummy coil.
The winding cost of lap winding is High and less efficient.	The winding cost of wave winding is Low and more efficient.
The lap winding is used for high current, low voltage machines.	The applications of wave winding include low-current and high-voltage machines.

Example.3.5 A six-pole DC armature with lap-connected winding has 72 slots and 2 coil sides per slot. Determine the winding pitches and connections to the 6 equalizing rings.

Solution: - Number of poles, P = 6

Number of slots = 72

Number of coil sides, Z = 72 X 2 = 144

Number of parallel paths, A = P = 6 (for simplex lap winding)

Pole pitch,

$$Y_P = \frac{Z}{P} = \frac{144}{6} = 24$$

Back pitch,

$$Y_B = Y_P + 1 = 24 + 1 = 25$$

$$Y_F = Y_B - 2 = 25 - 2 = 23$$

Distance between the coils having the same potential i.e.,

$$Y_{eq} = \frac{Total\ number\ of\ coils}{Pair\ of\ poles}$$

$$= \frac{Z/2}{P/2} = \frac{144/2}{6/2} = 24$$

$$Total\ number\ of\ tapings = Number\ of\ equaliser\ rings\ \times pair\ of\ poles$$

$$n_r \times \frac{P}{2}$$

$$= 6 \times \frac{6}{2} = 18$$

$$Distance\ between\ adjacent\ tapings$$

$$= \frac{Total\ number\ of\ coils}{Total\ number\ of\ taps}$$

$$= \frac{144/2}{18} = 4$$

The equalizer rings are to be arranged as per the following table:

Ring No.	I	II	III	IV	V	VI
Coil No.	1	5	9	13	17	21
	25	29	33	37	41	45
	49	53	57	61	65	69

3.14 EMF Equation

As the armature rotates, a voltage is generated in the coil. In the case of a generator, the emf of rotation is called the generated emf or armature emf (E_g). In the case of a motor, the emf of rotation is known as back emf or counter emf (E_b). The expression is the same for both conditions of operation.

Consider a conductor rotating at N rpm in the field of P pole having a flux ϕ per pole and the number of parallel paths in the armature winding is A.

Thus, the total flux cut by one conductor in one revolution is $P\phi\ Wb$. The time to complete one revolution is t=60/N second.

The average induced emf in one conductor,

$$= \frac{flux\ cut\ per\ revolution\ in\ Wb}{time\ taken\ for\ one\ revolution\ in\ second}$$

$$e = \frac{P\phi}{t} = \frac{P\phi}{60/N} = \frac{P\phi N}{60} \tag{3.1}$$

The number of conductors connected in series in each parallel path is Z/A.

The average induced emf across each parallel path is

E=(average voltage per conductor) × (Number of conductors in series per path)

$$E = \frac{P\phi N}{60} \times \frac{Z}{A} = \frac{P\phi NZ}{60A}\ volt \tag{3.2}$$

$$\mathrm{E} = \frac{P\phi Zn}{A}\ volt \tag{3.3}$$

$$\mathrm{i.e.}\ n = \frac{N}{60};\ Where\ n\ in\ speed\ in\ r.p.s.$$

For a given machine, the number of poles and the number of conductors per path (Z/A) is constant.

$E = K\phi n$ where K $=\frac{PZ}{A}$ is constant or $E \propto \phi n$

$E = K_1\phi n$ where $K_1 = \frac{PZ}{60A}$ is constant or $E \propto \phi N$

$E \propto \phi\omega$ where $\omega=\frac{2\pi N}{60}$ is the angular velocity in radian/second

From the above expression, we conclude that emf is directly proportional to flux per pole and speed. The polarity of induced emf is dependent on the direction of the magnetic field and the direction of rotation.

When the machine is working as a generator the generated emf is $E_g = \frac{P\phi NZ}{60A}$; and when the machine is working as a motor the back emf is $E_b = \frac{P\phi NZ}{60A}$.

3.15 Torque Equation

The mechanical power developed by the armature is $T\omega$, Where T is the electromagnetic torque and ω_m is the armature angular velocity. If this torque is developed while the armature current is $I_a$at an armature-induced voltage E, then the armature power is EI_a. Thus, ignoring any losses in the armature the power developed is given as,

$$EI_a = T\omega \quad (3.4)$$

$$EI_a = \frac{2\pi N}{60} \times T \quad (3.5)$$

$$\frac{P\phi NZ}{60A} \times I_a = \frac{2\pi N}{60} \times T \quad (3.6)$$

$$T = \frac{P\phi ZI_a}{2\pi A}\ Nm \quad (3.7)$$

Alternately: We know that when the current carrying conductor is placed in the magnetic field a force is extracted from it which exerts torque $(F \times r)$.

This torque is produced due to the electromagnetic effect, hence is called *electromagnetic torque.*

Let r=Average radius of armature in meters.

l=Effective length of each conductor in meter

I_a=Total armature current

A = Number of parallel path

Average force on each conductor, $F = Bil$ Newton

The torque due to one conductor = $F \times r$ Nm

Total Torque developed in the armature,

$$T = ZFr \text{ Nm} \quad (3.8)$$

$$T = ZBilr \quad (3.9)$$

Now, the current in each conductor, $i = I_a/A$

Average flux density, $B = \phi la$

Where 'a' is the cross-sectional area of flux per pole at radius *r*.

$$\text{Now, } a = \frac{2\pi rl}{P}\ m^2,$$

$$B = \frac{P\phi}{2\pi rl}\ Tesla$$

By using this value, we get

$$T = Z \times \frac{P\phi}{2\pi rl} \times \frac{I_a}{A} \times l \times r \quad (3.10)$$

$$T = \frac{P\phi ZI_a}{2\pi A} \quad (3.11)$$

For a particular machine, the number of poles (P) and the number of conductors per parallel path (Z/A) are constant.

Hence, $T = K\phi I_a$

or

$T \propto \phi I_a$

Thus, we conclude that torque produced in the armature is directly proportional to flux per pole and armature current.

Example.3.6 A 4-pole lap wound DC generator has 780 conductors, a flux of 60 m Wb per pole, and is driven at 600 rpm. Find open circuit emf.

Solution: - Open circuit emf:

$$E_g = \frac{\phi ZNP}{60A}$$

Where, $\phi = 60m\ Wb = 60 \times 10^{-3}\ Wb$;

$Z = 780$;

$N = 600\ RPM$;

$P = 4$

A = P = 4 (lap winding)

$$E_g = \frac{60 \times 10^{-3} \times 780 \times 600 \times 4}{60 \times 4} = 468\ V\ (Ans.)$$

Example.3.7 A 4-pole, DC machine is having 460 wave wound conductors and running at 1000 rpm. The flux per pole is 20 mWb. What will be the voltage induced in the armature winding?

Solution: - Here, P = 4;

A = 2 (Wave wound);

Z = 460;

N = 1000 rpm;

$\emptyset = \mathbf{20}m\ Wb = 20 \times 10^{-3}\ Wb$

Generated voltage,

$$E_g = \frac{\emptyset ZNP}{60A} = \frac{20 \times 10^{-3} \times 460 \times 1000 \times 4}{60 \times 2} = 306.7\ V\ (Ans.)$$

Example.3.8 A 4-pole, DC machine has 100 slots in the armature with two coil-sides per slot, each coil has two turns. The flux per pole is 20 m Wb, the armature is lap wound and if rotates at 720 rpm, what is the induced emf (i) across the armature (ii) across each parallel path?

Solution: - Here, $P = 4$;

$$A = P = 4\ (Lap\ wound);$$

$$\emptyset = 20m\ Wb = 20\ \times\ 10^{-3}\ Wb;$$

$$N = 720rpm$$

No. of slots = 100 with 2 coil sides per slot and each coil has two turns

$$Z = 100 \times 2\ \times 2 = 400$$

Induced emf across the armature,

$$E_g = \frac{\emptyset ZNP}{60A} = \frac{20\ \times\ 10^{-3}\ \times 400\ \times 720\ \times 4}{60\ \times 4} = 96\ V\ (Ans.)$$

The voltage across each parallel path, $E_g =\ 96\ V\ (Ans.)$

Example.3.9 A 6-pole machine has an armature with 120 slots and 8 conductors per slot, the flux per pole is 0.03 Wb, and runs at 1200 rpm. Determine induced emf if winding is (i) lap connected and (ii) wave connected.

Solution: - Here, $P = 6$;

$$\emptyset =\ 0.03\ Wb;$$

$$N = 1200\ rpm$$

Number of slots = 120 each slot with 8 conductors

$$Z = 120 \times 8 = 960$$

(i) When lap connected: A = P = 6

Induced emf,

$$E_g = \frac{\emptyset ZNP}{60A} = \frac{0.03\ \times\ 960\ \times 1200\ \times 6}{60\ \times 6} = 576\ V\ (Ans.)$$

(ii) When wave connected: A = 2

Induced emf,

$$E_g = \frac{\emptyset ZNP}{60A} = \frac{0.03\ \times\ 960\ \times 1200\ \times 6}{60\ \times 2} = 1728\ V\ (Ans.)$$

Example.3.10 A DC generator carries 800 conductors on its armature with lap connections. The generator has 4 poles with 0.08 Wb useful flux. What will be the induced emf at its terminals if it is rotated at 1200 rpm? Also, determine the speed at which it should be driven to induce the same voltage with wave connections.

Solution: - Here, $P = 4;$

$Z = 800;$

$\phi = 0.08\,Wb;$

$N = 1200\,rpm$

A = P = 4 (When lap wound)

Induced emf,

$$E_g = \frac{\phi ZNP}{60A} = \frac{0.08 \times 800 \times 1200 \times 4}{60 \times 4} = 1280\,V\,(Ans.)$$

When the wave wound, let the speed be N' rpm but $E_g = 600\,V$

$$Now, N' = \frac{E_g \times 60A}{\emptyset ZP} = \frac{1280 \times 60 \times 2}{0.08 \times 800 \times 4} = 600\,rpm\,(Ans.)$$

Example.3.11 A wave wound armature of an eight-pole generator has 51 slots. Each slot contains 16 conductors. The voltage required to be generated is 300 V. What would be the speed of coupled prime mover if flux per pole is 0.05 Wb? If the armature is rewound as a lap wound machine and run by the same prime mover, what will be the generated voltage?

Solution: - Here, $P = 8;$

$\phi = 0.05\,Wb;$

Number of slots = 51;

Conductors per slot = 16

$Z = 51 \times 16 = 816$

When the machine is wave wound, A = 2 and $E_g = 300\,V$

Now,

$$E_g = \frac{\emptyset ZNP}{60A}\ or\ 300 = \frac{0.05 \times 816 \times N \times 8}{60 \times 2}$$

Speed,

$$N = \frac{300 \times 60 \times 2}{0.05 \times 816 \times 8} = 110.3\, rpm\ (ans)$$

When the machine is rewound as lap winding,

$A = P = 8$ and $N = 110.3$ rpm.

$$E_g = \frac{0.05 \times 816 \times 110.3 \times 8}{60 \times 8} = 75\, V\ (Ans.)$$

Example.3.12 A six-pole lap wound armature rotating at 350 rpm is required to generate 260 V. The effective flux per pole is about 0.05 Wb. If the armature has 120 slots, determine the suitable number of conductors per slot and hence determine the actual value of flux required to generate the same voltage.

Solution: - Here, $P = 6;$

$$A = P = 6;$$

$$N = 350\, rpm;$$

$$E_g = 260\, v;$$

$$\phi = 0.05\, Wb$$

Now,

$$E_g = \frac{\emptyset ZNP}{60A}$$

$$260 = \frac{0.05 \times Z \times 350 \times 6}{60 \times 6}$$

$$Z = \frac{260 \times 60 \times 6}{0.05 \times 350 \times 6} = \frac{260 \times 24}{7} = 891.4$$

No. of conductors/ Slot

$$= \frac{Z}{No. of\ slots} = \frac{260 \times 24}{7 \times 120} = 7.43 \cong 8\ (an\ integer)$$

For 8 conductors/ Slot, the Total number of conductors

$$Z = 120 \times 8 = 960$$

Actual value of flux required,

$$\phi = \frac{E_g \times 60A}{ZNP}$$

$$= \frac{260 \times 60 \times 6}{960 \times 350 \times 6} = 0.04640Wb\ (Ans)$$

Example.3.13 The emf generated by a 4-pole DC generation is 400 V when the armature is driven at 1200 rpm. Calculate the flux per pole if the wave wound generator has 39 slots having 16 conductors per slot.

Solution: - *Induced emf,*

$$E_g = \frac{\emptyset ZNP}{60A}$$

$Where, P = 4;$

$E_g = 400\ V;$

$N = 1200\ rpm;$

$Z = 39\ \times 16 = 624;$

$A = 2\ (Wave\ winding)$

∴ flux per pole,

$$\phi = \frac{E_g\ \times 60A}{ZNP}$$

$$= \frac{400\ \times 60\ \times 2}{624\ \times 1200\ \times 4}$$

$$= 0.016\ Wb$$

$$= 16\ mWb\ (ans)$$

Example.3.14 A 8-pole generator has an induced emf of 220 V when driven at 1500 rpm. The armature is lap wound and has 300 conductors. The radius of the pole shoe is 20 cm and it subtends an angle of 60°. Calculate the flux density in the air gap if the length of the pole shoe is 18 cm.

Solution: - Here, , $P = 8;$

$E_g = 220\ V;$

$Z = 300;$

$A = P = 8;$

$N = 1500\ rpm;$

$r = 0.2\ m;$

$l = 0.18\ m;$

Flux per pole

$$\phi = \frac{E_g\ \times 60A}{ZNP}$$

$$= \frac{220 \times 60 \times 8}{300 \times 1500 \times 8} = 0.029\ Wb$$

$$Pole\ shoe\ arc = 2\pi r \times \frac{\theta}{360}$$

$$= 2\pi \times 0.2 \times \frac{60}{360} = 0.21m$$

$$Area\ of\ pole\ shoe\ arc = pole\ shoe\ arc \times l$$

$$= 0.21 \times 0.0378\ m^2$$

$$Flux\ density\ in\ air\ gap = \frac{\phi}{area} = \frac{0.029}{0.0378} = 0.767\ tesla\ (Ans)$$

Example.3.15 A 8-pole DC generator has a rated armature current of 300 A. If the armature is lap-wound, determine the current flowing through each parallel path of the armature. What will be its value if the armature is wave wound?

Solution: - $\boldsymbol{Here}, P = 8;\ A = P = 8\ (lap\ winding);\ I_a = 300\ A$

For the lap-wound machine, the current in each parallel path

$$\frac{I_a}{A} = \frac{300}{8} = 37.5A\ (Ans)$$

For the wave-wound machine, the current in each parallel path

$$\frac{I_a}{A} = \frac{300}{2} = 150A\ (Ans)$$

Example.3.16 The induced emf in a DC machine is 250 volts at a speed of 1000 rpm. Calculate the electromagnetic torque developed at an armature current of 30 A.

Solution: - $Here, E = 250\ V; N = 1000\ rpm;\ I_a = 30\ A$

Power developed in the armature

$$\omega T_e = E_b I_a$$

$$T_e = \frac{E_b I_a}{\omega} = \frac{E_b I_a}{2\pi N/60}$$

$$= \frac{250 \times 30 \times 60}{2\pi \times 1000} = 71.61\ Nm\ (Ans)$$

3.16 Armature Reaction in DC Generator

When a DC generator is loaded, a current flows through the armature conductor in the same direction as that of the induced (or generated) emf, and the armature conductors carrying the current, produce their own magnetic field called armature field. *The effect of the armature field produced by the armature current-carrying conductors on the main magnetic field is known as the* **armature reaction.**

The following are the effects of armature magnetomotive force on the main field flux:

I. It partly weakens or demagnetizes the main field flux.
II. It cross-magnetizes or distorts the main field flux.

Reduction in the average main flux per pole reduced the generated voltage and torque., where the distortion of the main field flux results in an increase in maximum flux and influences the limits of successful commutation in the DC machine as well as increases the core loss. In this section, we investigate this effect and discuss the method of minimizing the problems arising from armature reaction.

Fig. 3.16 shows the 2-pole DC generator. When there is no load connected to the generator, the current in the armature conductor is zero. This main field flux, produced by field mmf $I_f N_f$ is shown by horizontal phasor $OA = \phi_f$ as shown in fig. 3.16(b).

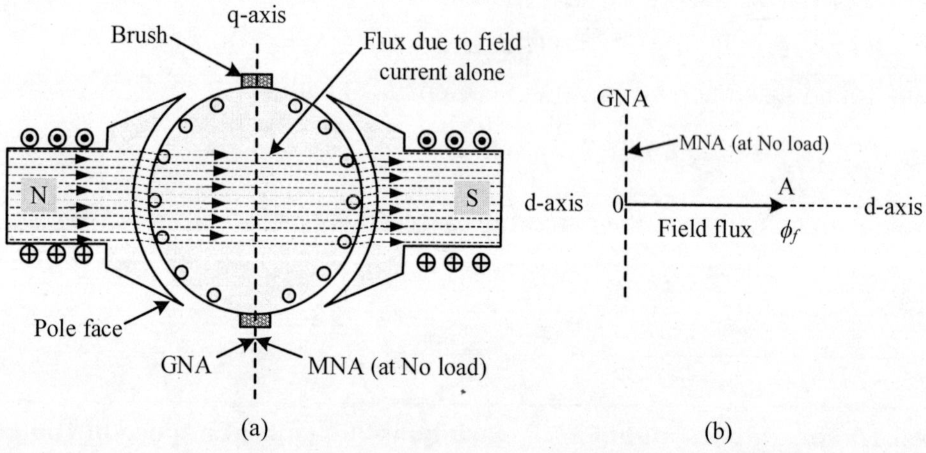

Fig. 3.16 Space distribution of main pole flux

When the DC machine is loaded, the current flows in the armature winding. These currents are shown in fig. 3.17(a), by dots under the main S-pole and by cross under the main N-pole. These armature currents set up armature flux shown by vertical flux lines in fig. 3.17(a), with field winding un-excited. This figure also shows armature flux ϕ_a by a vertical phasor OB. Flux ϕ_a is produced by armature mmf $I_a N_a$. If the DC machine is working as a motor, then its armature must rotate anti-clockwise, because of the fact that N, S pole of the main field attracts the armature produced S, N pole. In the case of a DC machine working as a generator, then its armature must be driven clockwise by the prime-mover.

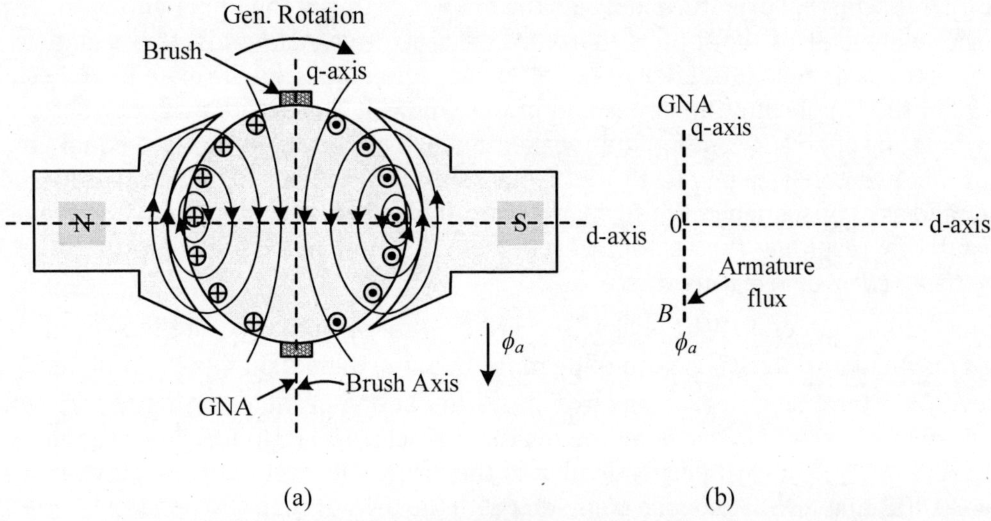

Fig. 3.17 Space distribution of armature flux

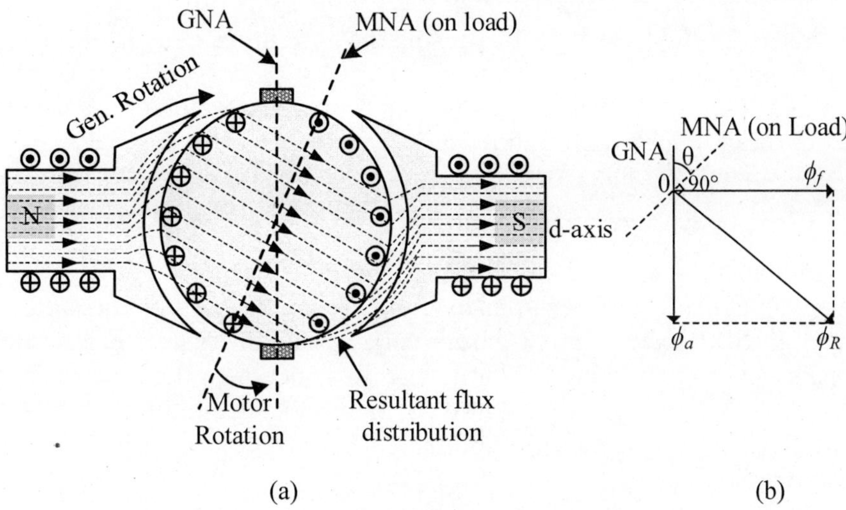

Fig. 3.18 Space distribution of resultant of both main field and armature flux

An examination of fig. 3.16 and fig. 3.17 reveals that the path of armature flux ϕ_a is perpendicular to the main flux path ϕ_f. In other words, the path of the armature flux crosses the path of the main-field flux. Thus, the effect of armature flux on the main field is entirely *cross magnetizing* and it is for this reason that the flux created by the armature mmf is called cross flux.

When the current flow in both the armature and field windings, the resultant flux distribution of obtained by superimposing the two fluxes of fig. 3.16 and 3.17. This is illustrated in fig.3.18(a). It is seen that armature flux aids the main field flux at the upper end of the N-pole and at the lower emf of the S-pole, therefore, at these two pole ends, the armature flux strengthens the main field flux. Likewise, the armature flux weakens the main field flux at the lower end of the N-pole and the upper end of the S-pole. If there is no magnetic saturation, then the amount of strengthening and weakening of the main field flux are equal and the resultant flux per pole remains unaltered from its no-load value. Magnetic saturation does not occur and as a consequence, the strengthening effect is less as compared to the weakening effect, and the resultant flux is decreased from its no-load value. This is called the *demagnetizing effect* of armature flux.

In the phasor diagram of 3.18(b) phasor sun of field flux ϕ_f and armature flux ϕ_a gives net flux ϕ_R. This resultant flux ϕ_R is seen to be more than the main field flux ϕ_f at no load. This is, however, not true because of magnetic saturation in one of the pole tips of each pole. A geometric neutral axis (GNA) is along the quadrature axis of the DC machine. The magnetic neutral axis (MNA) is always perpendicular to the axis of the resultant field flux. It is seen from fig. 3.18(a) that MNA at no load coincides with the GNA or q-axis. When the DC machine is loaded, fig. 3.18(a) show that MNA shifted from GNA. This shift depends upon the magnitude of the armature current. Thus, the greater the magnitude of armature (or load) current, the greater the shift of MNA from GNA. It may be therefore stated from above that the net effect of armature flux on the main-field flux is

(i) to distort the main-field flux thereby causing the non-uniform distribution of flux under the main poles,

(ii) to shift the MNA in the direction of rotation for a generator and against the direction of rotation for a motor and

(iii) to reduce the main-field flux from its no-load value due to magnetic saturation.

3.16.1 Graphical picture of armature reaction

For a better understanding of the interaction between main-field flux and armature flux, it is preferable and convenient to draw the developed diagram of armature conductors and poles. The developed view of fig.3.16(a), with no armature currents under each pole, is shown in fig. 3.19(a). The brushes are along the interpolar axis midway between the main-pole axes. The dotted lines in fig. 3.19(a) shows the distribution of main-field flux in the air gap with no armature currents. In fig. 3.19(b), the variation of main armature flux density along the air gap periphery is shown by solid lines.

In fig. 3.19(c), currents are indicated by dots under the S-pole and cross under N-pole. These armature currents, with field winding unexcited, produce magnetic flux as shown in fig. 3.19(d). as the armature flux produced by the armature conductor alone is normally normal to the main-field flux, as therefore, it is called *cross flux*. The mmf created by armature currents in a DC machine is triangular in nature and is depicted accordingly by the solid line in fig. 3.19(d).

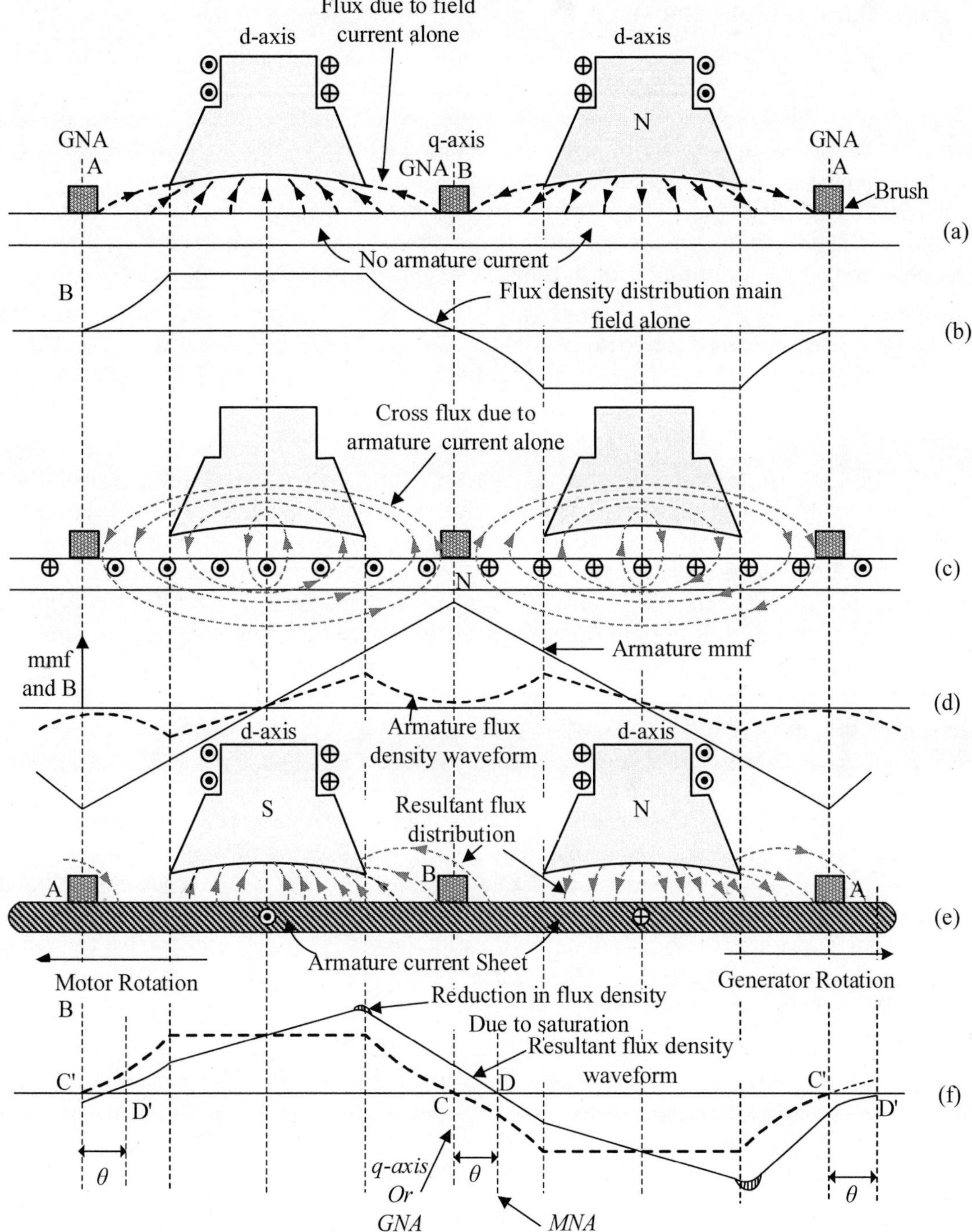

Fig.3.19 Flux distribution and flux density waveform respectively (a), (b) field current alone (c), (d) armature current alone, and (e), (f) both field and armature currents

The air-gap flux due to armature mmf is given by $\frac{armature\ mmf}{air-gap\ reluctance}$. Under the poles, the air gap is uniform, therefore, the air-gap flux variation under the poles is proportional to armature mmf and is shown in fig. 3.19(d). In between the poles i.e., in the interpolar region, the long air gap offers large reluctance, consequently, the armature flux is much smaller in this region, despite a large value of mmf. The armature flux density waveform created by armature currents is, therefore, saddle-shaped as depicted by the dotted curve in fig.3.19(d). It is observed from this figure that the armature flux-density waveform has *(i)* zero value at the center of the pole *(ii)* increases from zero to maximum value at the pole tips and then *(iii)* decreases rapidly to a minimum value at the middle of main poles.

When both armature and field windings carry currents, the resultant flux distribution is obtained by superimposing the two fluxes, field flux of fig. 3.19(a) and armature cross-flux of fig. 3.19(c). The resultant flux distribution, so obtained, is shown in fig. 3.19(e). This figure reveals the strengthening of the resultant flux at one pole tip and weakening at the other pole tip of each pole. As expected, this agrees with fig. 3.18(a).

For obtaining the resultant air-gap flux density waveform, when both field and armature windings carry currents add the corresponding flux-density ordinates of fig. 3.19(b) and fig. 3.19(d) at every point along the air-gap periphery. The resultant flux-density distribution in the air gaps is shown by the solid curve in fig. 3.19(f). It may be seen from the resultant flux-density distribution curve that the effect of cross-magnetizing armature mmf is to decrease the flux density under leading pole tips and to increase it under the trailing pole tips for generator operation. The magnetic saturation in the iron does occur and its effect is to increase the flux density under the trailing pole tips by a smaller amount than the decrease under the leading pole tips. The saturation effect is indicated by cross-hatched areas in fig. 3.19(f). Thus, under unsaturated conditions, the amount of flux increase under trailing pole tips is almost equal to the decrease under leading pole tips and the total flux per pole on load remains almost unchanged from its no-load value. Under saturated conditions, the amount of increase in flux is less than the decrease and therefore total flux per pole on load is less than its no-load value. Hence, under saturated conditions, the effect of cross-magnetizing armature mmf, i.e., cross flux, is to demagnetize the main field. But note that the demagnetizing effect of cross-flux is due to saturation only. fig. 3.19 (f) reveals that the point of zero flux density has shifted through an angle θ from C, C' to D, D' respectively. In other words, MNA has shifted from GNA by an angle θ.

For a motor, a dot under the south pole and a cross under the north pole result in an anti-clockwise rotation in fig. 3.16, fig. 3.17, and fig. 3.18. Therefore, the above results are also applicable to d.c. motor. Since the direction of rotation is reversed, the leading pole tips for a motor are the trailing pole tips for a generator. Hence for a motor, the effect of cross-magnetizing armature mmf is to decrease the flux under the trailing pole tips and to increase it under the leading pole tips.

Note from above that for a generator, the effect of the armature reaction is to distort the flux and shift the zero crossing of the flux density wave in the direction of rotation. In the case of the motor, the distortion of the flux and zero crossing of the flux-density wave is shifted against the direction of rotation. In constant flux d.c. machines, such as shunt machines, the flux distortion is much more prominent under heavy loads. In series and compound machines, the flux distortion is minimum, because with the increase of armature mmf., there is a corresponding increase in the field mmf.

The effects of armature m.m.f. described above, may be summarised as follows:

(i) The armature flux path is normal to the flux path of the main poles. That is why armature flux is called cross-flux or cross-magnetizing flux.

(ii) The armature section distorts the main-field flux distribution along the air-gap periphery. This distortion is in the direction of rotation for a DC generator and opposite to the direction of rotation for a DC motor. This also means that MNA is shifted in the direction of rotation for a generator and against the direction of rotation for a motor. This shift of MNA from GNA depends upon the magnitude of load (or armature) current.

(iii) The demagnetizing effect of armature mmf reduces the total flux per pole. This reduction is 1 to 5% from no-load to full-load.

Geometrical Neutral Axis: *The line passing through the geometrically central point between the two adjacent opposite magnetic poles is called* **geometrical neutral axis** (*GNA*).

Magnetic Neutral Axis: *The line passing through the magnetically neutral position between the two adjacent opposite magnetic poles is called* **magnetic neutral axis (MNA).** When a conductor (or coil) passes through these axes, no emf is induced in the conductor (or coil).

3.16.2 Effects of armature reaction

The effects of the armature reaction are summarised below:

1. Magnetic flux density is increased over one-half of the pole and decreased over the other half. But the total flux produced by each pole is slightly reduced and, therefore, the terminal voltage is slightly reduced. The effect of total flux reduction by armature reaction is known as the demagnetizing effect.
2. The flux wave is distorted and there is a shift in the position of the magnetic neutral axis (MNA) in the direction of rotation for the generator and against the direction of rotation for the motor.
3. Armature reaction establishes a flux in the neutral zone (or commutating zone). Armature reaction flux in the neutral zone will induced conductor voltage that aggravates the commutation problem unless brushes are placed in MNA.
4. Armature reaction lead to poor commutation (increases sparking at the brushes or at the commutator surface) and increases iron losses due to a rise in peak air gap flux density.
5. Armature is subjected to rotational magnetizing due to rotation in the bipolar magnetic field. The core loss increase due to a rise in peak flux density in the air gap.

3.17 Demagnetizing and Cross-Magnetizing Conductor

Fig. 3.20(a) and fig. 3.20(b) show the armature conductors that produce the demagnetizing and cross-magnetizing effects when the brush axis lies in the new position of the MNA, having a forward lead of θ. The direction of the currents through the conductors lying within angles *AOC* and *BOD*, shown in fig. 3.20, is such that the flux through the armature is from left to right, and hence produces demagnetizing effects, whereas the direction of the currents through the conductors lying within the angles *AOD* and *COB* is such that the combined flux is at right angles to the main flux, and hence causes cross-magnetizing effects, as shown in fig. 3.20(b).

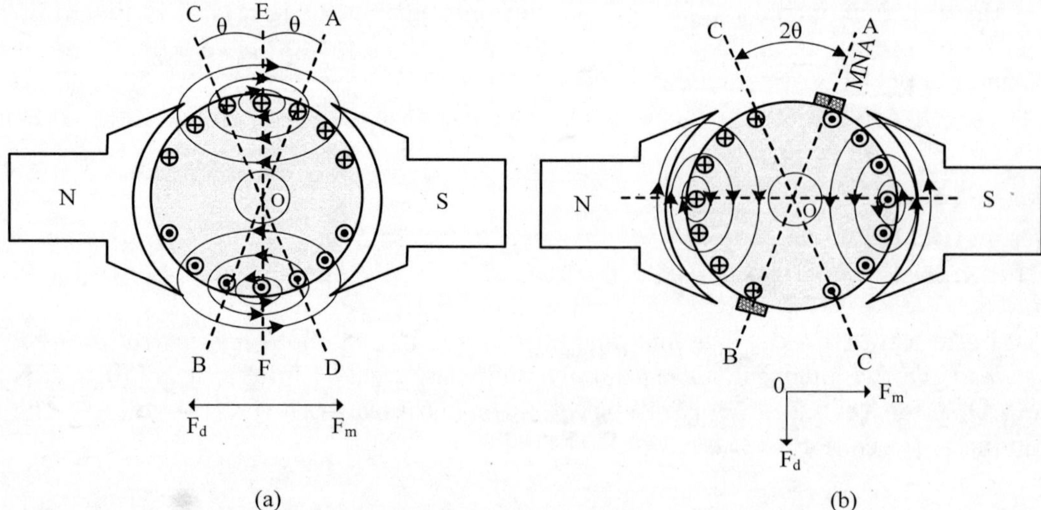

Fig. 3.20(a) Demagnetizing and (b) Cross-magnetizing effect

3.17.1 Demagnetizing ampere-turn per pole

To neutralize the demagnetizing effects, extra ampere-turns (AT) are added to the main field winding, and the calculation of the extra AT is required.

Let Z = Number of armature conductor

θ_m= Forward lead to a mechanical or geometrical or angular degree,

I_a = Armature current

Current per parallel path or current in each conductor, $I_c = \frac{I_a}{A}$

Total armature ampere-turn $= \frac{I_c Z}{2}$, since turn is $= \frac{Z}{2}$

Total ampere-turn per pole $= \frac{I_c Z}{2P}$

The Demagnetizing ampere-turn within the angle AOC and BOD

$$\frac{I_c Z}{2}\left(\frac{2\theta_m + 2\theta_m}{360}\right) = \frac{I_c Z}{2} \times \frac{4\theta_m}{360} = I_c Z \times \frac{2\theta_m}{360}$$

Therefore, the demagnetizing AT per pair of poles $(AT_d) = \frac{2\theta_m}{360} \times I_c Z$

Hence, the demagnetizing AT per pole

$$\frac{2\theta_m}{360} \times \frac{I_c Z}{2} = \frac{\theta_m}{360} \times I_c Z \tag{3.12}$$

3.17.2 Cross-magnetizing ampere-turn per pole

The cross-magnetizing conductor lies within the angle AOD and BOC, as shown in fig. 3.20(b). The total number of conductors including cross-magnetizing and demagnetizing is given by $\frac{Z}{P}$.

As derived earlier, demagnetizing conductors per pole pair $= \frac{2\theta_m}{360} \times Z$

Therefore, cross-magnetizing conductor per pole, $(AT_c) = \frac{I_c Z}{2P} - \frac{I_c Z \theta_m}{360}$

Hence, the cross-magnetizing AT_c per pole

$$= I_c Z \left(\frac{1}{2P} - \frac{\theta_m}{360}\right) \tag{3.13}$$

Example.3.17 A 8-pole generator has 700 wave wound conductors on its armature. It delivers 150 amperes at full load. If the brush lead is 6° mechanical, calculate the armature demagnetizing and cross-magnetizing ampere-turns per pole.

Solution: - Here, $P = 8$;

$A = 2\ (Wave\ wound)$;

$Z = 700$;

$I_a = 150\ A$;

$\theta = 6°$

Current in each conductor,

$$I_c = \frac{I_a}{A} = \frac{150}{2} = 75\ A$$

Demagnetizing ampere-turns/pole,

$$AT_d = I_c Z.\frac{\theta_m}{360} = 75 \times 700 \times \frac{6}{360} = 875\ AT/P(Ans.)$$

Example.3.18 A 250 kW, 500 V, 4 pole lap wound armature has 720 conductors. It is given a brush lead of 3° mechanical from its geometrical neutral axis (GNA). Calculate demagnetizing and cross-magnetizing ampere-turns per pole. Neglect shunt field current.

Solution: - Here, Load = 250kW;

V = 500V;

P = 4;

Z = 720;

$\theta = 3°\ (mech.)$

Load Current,

$$I_L = \frac{250 \times 1000}{500} = 500\ A$$

Armature current $I_a = I_L = 500\ A\ (Shunt\ field\ current\ neglected)$

No. Parallel paths, A = P = 4 (lap wound)

$$I_c = \frac{I_a}{A}$$

$$= \frac{500}{4} = 125\ A$$

Demagnetizing ampere-turns/pole

$$AT_d = I_c Z.\frac{\theta_m}{360}$$

$$= 125 \times 720 \times \frac{3}{360} = 750\ AT/Pole\ (Ans.)$$

Cross–magnetizing ampere-turns/pole

$$AT_c = I_c Z\left(\frac{1}{2P} - \frac{\theta_m}{360}\right)$$

$$= 125 \times 750 \times \left(\frac{1}{2 \times 4} - \frac{3}{360}\right) = 10937.5\ AT/Poles\ (Ans.)$$

Example.3.19 A four-pole motor has a wave wound armature with 720 conductors. The brushes are displaced backward through 4 degrees mechanical from GNA. If the total armature current is 80A, calculate demagnetizing and cross-magnetizing ampere-turns per pole.

Solution: - Here, P = 4;

A = 2;

Z = 720;

θ = 4;

$I_a = 80\ A$

Current in each conductor,

$$I_c = \frac{I_a}{A} = \frac{80}{2} = 40\ A$$

Demagnetizing ampere-turns/poles,

$$AT_d = I_c Z.\frac{\theta_m}{360} = 40 \times 720 \times \frac{4}{360} = 320\ AT/Poles\ (Ans.)$$

Cross–magnetizing ampere-turns/pole

$$AT_c = I_c Z\left(\frac{1}{2P} - \frac{\theta_m}{360}\right)$$

$$= 40 \times 720 \times \left(\frac{1}{2 \times 4} - \frac{4}{360}\right) = 3280\ AT/Poles\ (Ans.)$$

Example.3.20 A 250 V, 10 kW, 8 pole DC generator has single-turn coils. The armature is wave-wound with 90 commutator segments. If the brushes are shifted by 2 commutator segments at full load, calculate (i) total armature reaction ampere-turns (ii) demagnetizing ampere-turns and (iii) cross-magnetizing ampere-turns.

Solution: -Here, Load = 10kW;

V = 250V;

P = 8; A = 2 (Wave wound)

No. of segment = 90

Brush Shift = 2 commutator segments

Load current;

$$I_L = \frac{10 \times 1000}{250} = 40\ A$$

Armature current, $I_a = I_L = 40\ A$

Current per conductor,

$$I_c = \frac{I_a}{2} = \frac{40}{2} = 20\ A$$

Total armature conductors, Z = No. of commutator segment X No. of parallel path = 90 X 2 = 180

Brush shift, $\theta = \frac{2}{90} \times 360 = 8°\ (mech.)$

(i) Total armature - reaction ampere-turns/poles

$$= \frac{\mathrm{I_c Z}}{\mathrm{2P}} = \frac{40 \times 180}{2 \times 8} = 450\ \mathrm{AT/Pole\ (ans.)}$$

(ii) Demagnetizing ampere-turns/pole

$$= I_c Z.\frac{\theta_m}{360} = 40 \times 180 \times \frac{8}{360} = 160\ \mathrm{AT/Pole}(Ans.)$$

(iii) Cross–magnetizing ampere-turns/pole

$$= \frac{\mathrm{I_c Z}}{\mathrm{2P}} - \frac{I_c Z.\theta_m}{360}$$

$$= 450 - 160 = 290\ \mathrm{AT/Pole(ans.)}$$

3.18 Interpole

Interpole are small auxiliary poles attached to the stator yoke and situated exactly between the main poles as shown in fig 3.21. These are also known as commutating poles or compoles. For a generator, an inter pole should have polarity which is the same as that of the next main pole. These commutating poles induce an emf in the coil undergoing commutation which opposes and hence cancels the reactance voltage.

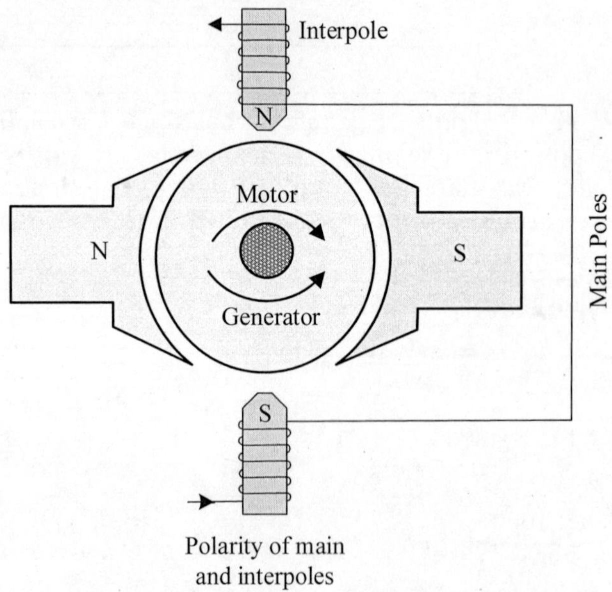

Fig. 3.21 Interpole and their polarity

Since the reactance voltage is proportional to the armature current, the inter poles are connected in series with the armature circuit. It is to be noted that the commutating field should be of proper magnitude. It should neither be too strong nor too weak. If it is too strong it will overcompensate the reactance voltage and the arc will be in the reverse direction and this is known as over commutation.

3.19 Compensating Winding

An armature reaction produces a demagnetizing and cross-magnetizing effect. The demagnetizing effect is compensated in high-power DC machines by incorporating a *few extra turns of thick insulated copper wire* to the main-field winding, whereas, to neutralize the cross-magnetizing effect a compensating winding is used. Large DC machines are subjected to the major fluctuation of loads. In the absence of compensating windings, the flux shifts backward and forward due to a change of load, which induces statically induced emf in armature conductors. The change of load determines the magnitude of this induced emf. The magnitude of this induced emf may be very high, resulting in a spark around the whole commutator circuit causing a ring of fire. In this case, a number of conductors or coils are embedded in the slots of the pole shoes and are connected in series with the armature winding in such a way that current flowing through these conductors or coils sets up a magnetic field which neutralizes the cross-magnetizing effect of armature field. This winding is known as *compensating winding,* as shown in fig. 3.22.

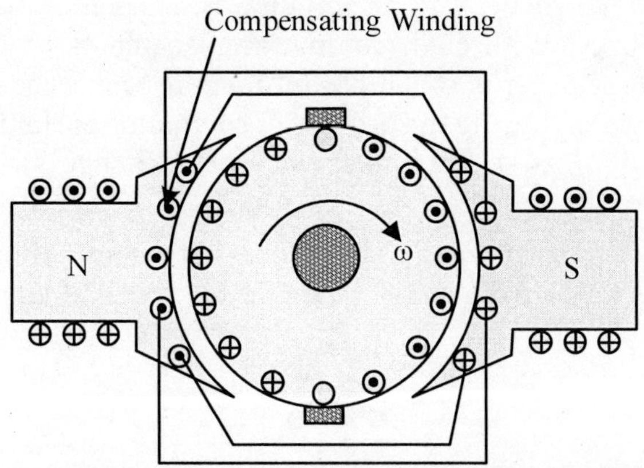

Fig. 3.22 Compensating winding

Connections are made in such a way that the direction of the current in these conductors is opposite to the direction of the current in the armature conductors. When current flows through this winding, it sets up a magnetic field that is equal and opposing to the cross-magnetizing effect of the armature field and neutralizes it. Thus, a sparkless or good commutation is obtained.

Let Z_a= Number of armature conductor/pole

A = Number of parallel paths

I_a= Current/armature conductors

Z_c = Number of compensating conductor/pole face

$$Z_c I_a = Z_a\left(\frac{I_a}{A}\right) \tag{3.14}$$

$$Z_c = \left(\frac{Z_a}{A}\right) \tag{3.15}$$

3.19.1 Number of compensating winding

The number of armature conductors/pole $= \frac{Z}{P}$

Therefore, the number of armature turns/pole$= \frac{Z}{2P}$

The number of armature-turn immediately under one pole $= \frac{Z}{2P} \times \frac{Pole\ arc}{Pole\ Pitch}$

The number of armature-turn immediately under one pole $= \frac{Z}{2P} \times 0.7$ (Approximately)

Therefore, the number of AT/pole for compensating winding

$$\frac{ZI}{2P} \times 0.7 = 0.7 \times Armature\ AT/Pole \tag{3.16}$$

3.20 Commutation

In a DC machine, one of the major functions is the delivery of current from the armature (rotating part) to the external circuit (stationary part) or vice-versa. This operation is conducted with the help of brushes and a commutator. Thus, the process in which a coil is short-circuited by the brushes through commutator segments while it passes from the influence of one pole to the other is called **commutation.** The carbon brush makes contact with the commutator segment during the process of commutation. In this process the current in a coil has to change from +I to -I. The time taken for this change is very short and is known as the **commutation period**.

Step-1: fig. 3.23(a) shows the beginning of the process of commutation in coil C. The current collected by the brush is I each from coils B and C i.e. a total of 2I amperes coil C is carrying current from right to left.

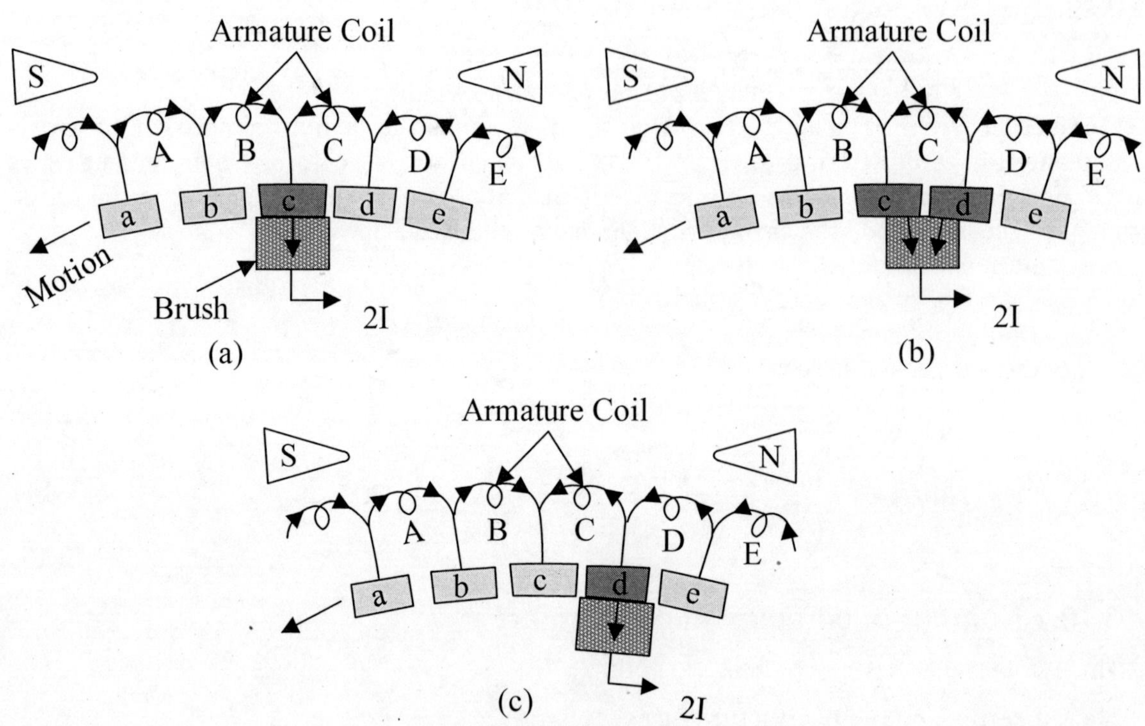

Fig. 3.23 Commutation

Step-2: In fig. 3.23(b) the brush is spanning the commutator segments c and d. The current to the brush is contributed equally by coils B and D. This current reaches the brush without passing through coil C, which is not carrying any current at this instant.

Step-3: fig. 3.23 (c) shows the operation at the next instant. Coil C is now moving under an S pole and the current is in the direction from left to right. Thus, the current in coil C must change from +I to -I during this period of short circuit which is about 2 msec. If the current does not reach the full value -I at the end of this period of short circuit, the difference of current would go from the commutator to the brush in the form of a spark. Such sparking causes heating, pitting, and roughening of the commutator surface.

Fig. 3.23 shows how the current in the short-circuited coil varies during the brief interval of the short circuit. Curve B Shows that the current changes from +I to -I linearly in the commutation period. Such a commutation is called an **ideal commutation or straight-line commutation**. Curve C shows **an accelerated or over-commutation** where the current attains its final value with a zero (quick) rate of change at the end of the commutation period. Usually, it provides a satisfactory commutation. Curve A shows a **retarded or under-commutation** where the final rate of change of current is very high. In such conditions sparking at the trailing edge of the brush is inevitable.

Suppose L is the inductance of the coil and the current changes from + I to - I in a time T_c (commutation time) a self-induced emf $L\frac{di}{dt}$ known as reactance voltage appears and is given as reactance voltage $\frac{2LI}{T_C}$ volts. The direction of this induced emf is such that it opposes the change in current. From the above discussion, it is concluded that for *satisfactory commutation,* the current in the coil which undergoes commutation must be reversed completely during its commutation period *Tc.*

3.20.1 Methods to improve commutation

Commutation may be improved by employing the following methods.

1. **By use of high-resistance brushes.** High-resistance carbon brushes help the current to be reversed in the coil undergoing commutation and reduce sparking at the brushes.
2. **By shifting of brushes.** In this method, brushes are shifted to the new position of *MNA* so that no *emf* be induced in the coil undergoing commutation. Thus, the sparking at the brushes is eliminated. But in this case, the position of *MNA* changes with the change in load on the machine, and simultaneously the position of brushes cannot be changed. Hence, this method is employed in the machine in which we do not have interpoles and the load on the machine remains almost constant.
3. **By use of interpoles or commutating poles.** In this method, narrow poles are placed in between the main poles of a DC machine which are re-energised to such an extent that they neutralize the field produced by the armature under load. Hence, no *emf* is induced in the coil which undergoes commutation.
4. **By use of compensating winding.** In this method, a number of conductors or coils are embedded in the slots provided at the pole shoes faces and carry current of such a magnitude and direction that the field produced by them neutralizes the armature field and improves commutation.

Solved Numerical Examples

Q1. A DC machine has a 4-pole, wave–wound armature with 46 slots and 16 conductors per slot. If the induced voltage in the armature is 480 V at 1200 rpm, determine the flux per pole.

Solution: - Here z = 16 x 46 = 736 and so, from the emf equation

$$\emptyset = \frac{60E}{nz}\left(\frac{a}{p}\right) = \frac{(60)(480)}{(1200)(736)}\left(\frac{2}{4}\right) = 16.3\ mWb$$

Q2. A 4–pole, lap–wound armature has 144 slots with two coil sides per slot, each coil having two turns. If the flux per pole is 20 mWb and the armature rotates at 720 rpm, what is the induced voltage?

Solution: - Here

$$\mathrm{p = a\ = 4, n = 720, \emptyset = 0.020, and\ Z = 144\ \times 2\ \times 2 = 576}$$

Substitute this value in the emf equation to obtain

$$E = \frac{(0.020)(720)(576)}{60}\left(\frac{4}{4}\right) = 138.24\ V$$

Q3. A 10–turn square coil of side 200 mm is mounted on a cylinder 200 mm in diameter. The cylinder rotates at 1800 rpm in a uniform 1.1 T field. Determine the maximum value of the voltage induced in the coil.

Solution: -

$$E_{max} = BNA\omega = \ (1.1)(10)(0.200)^2((2\pi\ \times 1800)|60) = 82.94\ V$$

Q4. The generator has 4 poles, is lap wound with 326 armature conductors, and runs at 650 rpm on full- load. If the bore of the machine is 42 cm (in diameter), its axial length is 28 cm, and each pole subtends an angle of 60°, determine the airgap flux density.

Solution: - $E = 252 = \frac{\emptyset nz}{60}\left(\frac{p}{a}\right)$ where $\phi = 71.35$ mWb

The pole surface area is

$$A = r\theta l = \ (0.21)(\pi|3)(0.28) = 0.0616\ m^2$$

Hence

$$B = \frac{\phi}{A} = \frac{71.35\ \times\ 10^{-3}}{0.0616} = 1.16\ T$$

Q5. Calculate the maximum value of EMF induced in the armature conductors of a machine driven at 600 RPM. The diameter of the machine is 200 cm and the length is 30 cm. The maximum value of flux density is 1 Wb/m².

Solution: - Given that

$$\mathrm{N} = 600\ \mathrm{RPM}, \mathrm{D} = 200\ \mathrm{cm} = 2\ \mathrm{m}, \mathrm{L} = 0.30\ \mathrm{m\ and}\ B_m = 1.0\ Wb/\ m^2$$

Revolution per second is

$$n = \frac{N}{60} = \frac{600}{60} = 10\ rps$$

Peripheral velocity is

$$V = \pi \times D \times n = 3.14 \times 0.3 \times 10 = 9.42\ m/s$$

The maximum EMF induced in a conductor is given as

$$E = B_m \times v \times L = 1.0 \times 9.42 \times 0.3 = 2.826\ V$$

Q6. In a DC machine, if the number of poles is eight, the average flux in the air gap is 0.095 Weber per pole, the total number of conductors in the armature is 1000, the number of parallel paths a is 8 and rotations per minute is 500. Calculate the EMF induced.

Solution: - Given that

$$p = 8, a = 8, \emptyset = 0.095\ Wb\ and\ N = 500\ RPM$$

Now, we have

$$E = E_C Z_C = p\emptyset n Z_C = Z_C B_{av} v L$$

$$E = \frac{p}{a}\emptyset\, n Z_a = \frac{8}{8} 0.095\ \frac{500}{60} 1000 = 791.7\ V$$

Q7. A two-pole DC motor is operating on an average flux density of 0.6 T. There are 24 slots in the machines and each slot has 10 conductors carrying 10 A current. The length of the machine is 0.3 m. Obtain torque if the machine diameter is 0.1 m.

Solution: - Given that $L = 0.3\ m, I = 10\ A, N = 10\ and\ B = 0.6\ T$

Force due to the single slot is

$$F = NIBL = 10 \times 10 \times 0.6 \times 0.1 \times 0.3 = 1.8\ N$$

Torque is

$$T = \frac{B_{av}.I_a.L.Z_a.D}{2}$$

$$= \frac{24 \times 10 \times 10 \times 0.6 \times 0.1 \times 0.3}{2}$$

$$= 21.7\ N\text{–}m$$

Q8. The armature of a four-pole DC machine has 100 turns and runs at 600 RPM. The EMF generated in an open circuit is 220 V. Find the useful flux per pole when the armature is: -

(a) Lap connected, and

(b) Wave connected

Solution: - The EMF induced is given as

$$E = \frac{p}{a} \emptyset Z \frac{N}{60}$$

(a) For lap winding, a = 4. Thus, flux is

$$\emptyset = \frac{60 \times E \times a}{p \times Z \times N}$$

$$\emptyset = \frac{60 \times 220 \times 4}{4 \times 200 \times 600} = 0.11\ Wb$$

(b) For wave windings, a = 2 flux is

$$\phi = \frac{60 \times E \times a}{p \times Z \times N}$$

$$\phi = \frac{60 \times 220 \times 2}{4 \times 200 \times 600} = 0.055\ Wb$$

Q9. A six-pole 100kw, 440-volt lap-connected DC machine has 500 conductors. Find the number of turns on each interpole, if interpolar air gap is 1.0 cm and interpolar density is 0.28 Wb. Neglect leakage of the iron parts of the magnetic circuit.

Solution: -

Given full load current,

$$I_a = \frac{100 \times 1000}{440} = 227.3\ A$$

Number of armature conductors $Z = 500$

Number of poles, $P = 6$

Number of parallel paths for lap winding, $a = 6$

Interpolar air gap, $l_g = 1.0\ cm$

Flux density, $B_i = 28\ Wb/m^2$

Ampere turn for each interpole is

$$AT_i = \frac{Z.I}{2.a.P} + \frac{B_c.l_g}{\mu}$$

$$= \frac{500 \times 227.3}{2 \times 6 \times 6} + \frac{0.28 \times 0.01}{4 \times 3.14 \times 10^{-7}} = 3807.58\ AT$$

$$I_a N_c = AT_i$$

$$N_c = \frac{3807.58}{227.3} = 16.75 \cong 17\ Turns$$

Q10. Calculate the torque developed in a six-pole DC motor. The armature is carrying a current of 20 A and has 1000 conductors connected in the lap. The flux per pole is 25 mWb.

Solution: - Given that

poles p = 6

number of parallel paths a = 6

flux per pole = 0.25 mWb

Total conductors Z = 1000

current I =20A

The developed torque is

$$T = \frac{p}{a}.\frac{Z.I_a.\phi}{2\pi} = \frac{6 \times 1000 \times 20 \times 25 \times 0.001}{6 \times 2 \times \pi} = 79.6\ N - m$$

Multiple Choice Questions

Q1. In the lap winding the number of brushes is equal to

e. 2
f. 4
g. Number of Poles
h. 8

Q2. Which winding on a DC generator is preferred for generating large currents

a. Lap Winding
b. Progressive wave winding
c. Retrogressive wave winding
d. None of above

Q3. For the wave winding, the average pitch must be

a. odd
b. even
c. odd or even
d. fractional

Q4. For the fixed number of poles and armature conductors, which winding will give higher emf?

a. Lap winding
b. Wave winding
c. None of the above
d. Both

Q5. As a result of the armature reaction, the total mutual air gap flux in the DC generator is approximately a percentage

a. 5
b. 25
c. 50
d. 60

Q6. In a duplex winding for a 4-pole machine, the number of parallel paths not will be

a. 2
b. 4
c. 8
d. 15

Q7. In the lap winding the number of brushes is equal to

a. 2
b. 4
c. Number of Poles
d. 8

Q8. The function of the commutator in the DC machine is

a. To collect current from conductors
b. To conduct the current to the brushes
c. To change AC to Pulsating DC
d. Both a and c are correct

Q9. A 4-pole lap wound armature has 480 conductors and a flux per poole of 25 mWb. The emf generated, when running at 600 rpm, will be

a. 240
b. 120
c. 60
d. 30

Q10. The armature of the DC machine in laminated

a. To reduce the hysteresis losses
b. To reduce eddy current losses
c. To reduce inductivity of armature
d. To reduce the mass of the armature

Q11. Why is the air gap between the stator and armature of an electric machine kept as small as possible?

a. To get the stronger magnetic field
b. To make the rotation easier
c. To reach a higher speed of rotation
d. To improve air circulation

Q12. Interpole are meant for

a. Increasing the speed of the motor
b. Decrease counter emf
c. Reduced sparking at the commutator
d. Convert armature current to DC

Q13. If the flux of a DC machine approximately zero, its speed will be approaches

a. Infinity
b. Zero
c. A stable value nearer to the rated speed
d. None of these

Q14. The function of equalizing connections in a lap wound DC generator is

a. To neutralize the armature reaction effect
b. To avoid unequal distribution of current at brushes
c. To avoid short circuit current
d. None of the above

Q15. The function of compensating winding placed in slots in the pole shoes is

a. To neutralize the cross-magnetizing effect
b. To neutralize the demagnetizing effect
c. To neutralize both the effect
d. To avoid flashover around the commutator

Q16. If the armature current of a DC machine is increased keeping the field flux constant, then the developed torque

a. Increase proportionally
b. Decrease in inverse proportion
c. Remain constant
d. Increase proportionally to the square of the current

Q17. In a DC machine, the armature mmf is always directed along the

a. Polar axis
b. Brushes axis
c. Interpole axis
d. None of the above

Q18. The wave form of the armature mmf in a DC machine is

a. Square
b. Rectangular
c. Triangular
d. Sinusoidal

Keys to multiple choice questions

1.	C	2.	A	3.	C	4.	B	5.	A	6.	A
7.	C	8.	D	9.	B	10.	B	11.	A	12.	C
13.	A	14.	B	15.	D	16.	A	17.	B	18.	C

Short answer type questions

Q1. Define statically and dynamically induced emf.
Q2. Why is a commutator employed in DC machines?
Q3. Define coil pitch and coil span.
Q4. Define resultant pitch.
Q5. How will you define a magnetic neutral plane (mnp)?
Q6. Why equalizer rings are used?
Q7. For what type of DC machine, wave winding is employed?
Q8. What is armature reaction?
Q9. What for the brushes are employed in the DC generator?
Q10. For what type of DC machine lap winding is employed?

Q11. What is the role of interpoles and compensating winding in DC machines?
Q12. Name the factor that opposes the reversal of current in a coil undergoing commutation.
Q13. Give the relation between electrical and mechanical angle in case of a rotating machine.
Q14. What are equalizer rings and why are they used?
Q15. What is the basic principle on which a generator operates?
Q16. In what way compensating winding is connected to the armature?
Q17. What is linear commutation?
Q18. Explain bad commutation.
Q19. In a DC machine without interpole to get improved commutation, whether the brush shift Should be varied with change in load or brush shift should be fixed?
Q20. In large DC machines electro-magnets are preferred over permanent magnets, why?
Q21. State the effects of armature reaction in DC machines.
Q22. Define the factors and state the benefits of short-pitched coils.
Q23. Why the armature of a DC machine is made of laminated silicon steel?
Q23. Which are the different types of armature windings commonly used in DC machines?
Q24. Explain the difference between integral slot and fractional slot winding.
Q25. Why fractional pitch winding is preferred over full pitch winding?
Q26. Why are the interpoles of a DC machine tapered?
Q26. Why is the pole shoe section of a DC machine made larger than the pole core?
Q27. In which type of armature winding equalizer connections are used?
Q28. What are the disadvantages of armature reaction?
Q29. What is the function of compensating winding in a DC machine?
Q30. In small DC machines, cast iron yokes are preferred, why?

Exercises

Q1. Design and draw a 2-layer progressive duplex lap winding for a 4-pole DC generator with 24 slots. Show the position of the brushes and their polarity. Also draw a diagram representing the number of parallel paths thus formed.

Q2. Prepare a winding diagram for a four-pole wave-connected armature of a DC generator having 16 coil sides.

Q3. Draw a developed winding diagram of progressive lap winding for 4 pole, 22 slot with one coil side per slot, single layer winding showing there in position of poles, direction of motion, the direction of induced emf, and position of brushes.

Q4. A 250 kW, 500 V, 4 pole lap wound armature has 720 conductors. It is given a brush lead of 3° mechanical from its geometrical neutral axis (GNA). Calculate demagnetizing and cross-magnetizing ampere-turns per pole. Neglect shunt field current. **10500** (Ans.)

Q5. A current of 72A is supplied by a four-pole lap wound DC generator with 480 conductors on its armature. The brushes are given an actual lead of 12° (mechanical). Calculate the cross-magnetizing ampere-turns per pole. (**Ans.** 792)

Q6. The armature of a four-pole 200 V, lap wound generator has 400 conductors and a speed of 300 rpm. Determine the useful flux per pole. If the number of turns in each field coil is 900, what is the average induced emf in each field coil on breaking its connection if the magnetic flux set-up by it dies away completely in 0.1 seconds? (**Ans.** 0.1 Wb; 900 V)

Q7. A 420 V, four-pole, 25 kW DC generator has a wave-connected armature winding with 846 conductors. The mean flux density in the air-gap under the interpoles is 0.5 Wb/m2 on full load and the radial gap length is 0.4 cm. Calculate the number of turns required on each interpole. (**Ans.** 79.62)

Q8. A four-pole, wave wound DC armature has a bore diameter of 84 cm. It has 600 conductors and the ratio of pole arc to pole pitch is 0.8. If the armature is running at 360 rpm and the flux density in the air gap is 1.2 tesla, determine the induced emf in the armature if the effective length of the armature is 20cm. **912.4 V** (Ans.)

Q9. A four-pole, lap wound DC armature has a bore diameter of 70 cm. It has 540 conductors and the ratio of pole arc to pole pitch is 0.72. If the armature is running at 500 rpm and the flux density in the air gap is 1.2 tesla, determine the induced emf in the armature if the effective length of the armature is 20 cm. (**Ans.** 427.68 V)

Q10. A 250 V, 10 kW, 8-pole DC generator has single-turn coils. The armature is wave-wound with 90 commutator segments. If the brushes are shifted by 2 commutator segments at full load, calculate (i) total armature reaction ampere-turns (ii) demagnetizing ampere-turns and (iii) cross-magnetizing ampere-turns.
450 (Ans.) **160** (Ans.) **290** (Ans.)

Q11. Estimate the number of turns needed on each commutating pole of a six-pole generator delivering 200 kW at 200V, given that the number of armature conductors is 540 and the winding is lap-connected interpole air gap is 1.0 cm and the flux density in the interpole air-gap is 0.3 Wb/m2. Neglect the effect of iron parts of the circuit and of leakage.

9.886 (Ans.)

Q12. A 4-pole machine has an armature with 90 slots and 8 conductors per slot, the flux per pole is 0.05 Wb, and runs at 1200 rpm. Determine induced emf if winding is (i) lap connected and (ii) wave connected. (**Ans.** 720 V; 1440 V)

Q13. The electromagnetic torque developed in a DC machine is 80 Nm for an armature current of 20 A. Find the torque for a current of 30 A. What is the induced emf for a speed of 900 rpm?
376.8 volt (Ans.)

Q14. A 4-pole, DC machine has 144 slots in the armature with two coil sides per slot, each coil has two turns. The flux per pole is 20 m Wb, the armature is lap wound and if rotates at 720 rpm, what is the induced emf (i) across the armature (ii) across each parallel path?
138.24 V (Ans.), **138.24 V** (Ans.)

Q15. A four-pole, lap wound DC machine is having 500 conductors on its armature and running at 1000 rpm. The flux per pole is 30 m Wb Calculate the voltage induced in the armature winding. What will be induced emf if the armature is rewound for wave winding?
(**Ans.** 250 V; 500 V)

Books for further reading

1 Electric Machinery, Fitzgerald, Kingslay, Umans, Tata McGraw-Hill.

2. Electric Machinery Fundamentals, Chapman, McGraw-Hill Higher Education.
3. Electric Machines, Nagrath and Kothari, Tata McGraw-Hill.
4. Electric Machinery, P.S. Bimbhra, Khanna Publishers.
5. Electrical Machines, R. K. Srivastava, 2/e, Cengage Learning Pvt. Ltd.-2011
6. Electrical Machines, Smarajit Ghosh, 2/e, Pearson edu, 2012
7. Electrical Machines-I, D. K. Palwalia, N K Garg, P Kumar, G Jain. Ashirwad Publishers-2020

REFERENCE BOOKS:

1. Electric Machinery and Transformer, Guru, Hiziroglu, Oxford University press.

3. Basic Electric Machines, Vincent Deltoro, Prentice Hall.

3. Performance and Design of A.C. machines, M. G. Say

For further reading scan to: -

Lecture Notes on DC Machines

Video animation on DC Machines operation

Video lecture on DC Machines

Video lecture on Principal of operation of DC Machines

4 DC Machine-Motoring and Generation

UNIT SPECIFICS

This unit discusses the following aspects of DC machines as motors and generators:

- *Armature circuit equation for DC machine,*
- *Types of excitations*
- *Characteristic and voltage build-up of DC generator,*
- *Back EMF with armature reaction,*
- *Characteristics of DC motors.*
- *Speed control of DC Motors.*
- *Testing of DC machines*

The unit on DC machines as motors and generators contain the fundamental aspects of its operation, control, and testing. The unit covers the mathematical analysis of the armature circuit equation. It covers the type of connection of the DC machine. The characteristics of different types of DC machines and their applications are discussed in detail. It further describes methods for testing DC machines. The unit gives solved and unsolved numerical problems to illustrate further the theoretical study. A number of multiple-choice questions as well as questions of short- and long-answer types are given. A list of references and suggested readings are given in the unit so that one can go through them for practice. Some QR codes have been provided in different sections which can be scanned for relevant supportive knowledge.

RATIONALE

This unit on DC machines is designed to provide students with a comprehensive understanding of the principles and operation of DC machines, both as motors and generators. The unit will cover various aspects of DC machine theory and practical applications, enabling students to develop a solid foundation in this field. The unit will then delve into the operation of DC machines as motors. Students will gain knowledge of the working principle of DC motors, including the interaction between the magnetic field produced by the field coils and the current flowing through the armature. Various motor characteristics, such as torque, speed, and efficiency, will be explained, along with the methods for controlling these parameters.

The unit will cover the operating characteristics and performance analysis of DC machines. Students will learn how to calculate and interpret key parameters, such as back EMF, armature current, terminal voltage, power output, and efficiency. They will gain insights into the relationships between these variables and how they can be optimized for specific applications. The students will explore the methods used to control and regulate the operation of DC machines. They will learn about various control techniques, including armature control, field control, and voltage control. The unit will also cover the importance of speed control and the different methods employed, such as armature resistance control, field flux control, and voltage control.

Overall, this unit on DC machines will provide students with a comprehensive understanding of the working principles and operational aspects of DC machines, both as motors and generators. By combining theoretical knowledge with practical experience, students will develop the necessary skills to analyze, design, and optimize DC machine systems for various applications.

PRE-REQUISITES

Mathematics: Vectors, Integrals, Differentiation, Algebra (Class XII)
Physics: Electricity and Magnetism (Class XII)

UNIT OUTCOMES

The list of outcomes of this unit is as follows:

U4-O1: Knowledge of DC Machine Components.

U4-O2: Understanding of DC Motor and DC Generator Operation

U4-O3: Analysis of DC Machine Operating Characteristics.

U4-O4: Analyse the torque, speed, and efficiency characteristics of DC motors.

Unit-4 Outcomes	EXPECTED MAPPING WITH COURSE OUTCOMES (1-Weak Correlation; 2-Medium correlation; 3-Strong Correlation)			
	CO-1	CO-2	CO-3	CO-4
U4-O1	-	1	1	-
U4-O2	1	3	2	-
U4-O3	1	3	3	-
U4-O4	1	3	3	-

4.1 INTRODUCTION

In Chapter 3, the basics of DC machines were introduced, focusing on the conversion of mechanical energy to electrical energy and vice versa. The principle underlying this energy conversion is the production of dynamically induced electromotive force (emf). In a DC generator, the field coils are located on the stator (stationary part), while the armature is situated on the rotor (rotating part). The field coils are excited by a direct current (DC) supply. The rotor is rotated by a prime mover, such as a turbine or an electric motor, causing it to cut the flux produced by the field coils. As the armature rotates and cuts through the flux, an alternating emf is induced in the armature windings. This induced emf constantly changes in a cyclic manner as the armature rotates. However, for many practical applications, a unidirectional emf is required. To achieve this, a commutator is used. The commutator consists of a set of copper segments attached to the armature shaft. The armature windings are connected to these segments. As the armature rotates, the brushes (made of carbon or graphite) remain in contact with the commutator segments. The brushes provide a means for connecting the armature windings to an external circuit. The commutator functions by reversing the connections of the armature windings at the precise moment when the induced emf changes its direction. This reversal of connections ensures that the output of the DC generator is a unidirectional emf. The commutator segments and brushes effectively convert the alternating emf induced in the armature into a direct current. This Chapter will provide a deeper understanding of the construction, operation, control, and applications of DC machines working as DC motors or as DC generators. It covers the construction and components, and detailed exploration of the construction of DC machines, including the stator, rotor, field coils, armature core, armature windings, commutator, and brushes. Students will gain an understanding of the materials used, the arrangement of components, and the role of each part in the operation of the machine.

Scan to know more on DC Machine

It further illustrates the concept of voltage regulation and detailed discussion of voltage regulation techniques for DC generators. Students will learn about the effects of armature reaction and voltage drop in the armature circuit. They will explore methods such as compounding, series field control, and parallel field control to regulate the generated voltage.

The analysis of various losses in DC machines, including copper losses, iron losses, and mechanical losses gives an understanding of how these losses affect the overall efficiency of the machine. They will learn methods to calculate and minimize losses, leading to improved efficiency.

The different methods used for starting DC motors, such as resistance starting, and regenerative and dynamic braking techniques employed to bring DC motors to a stop give an understanding of starting and braking in DC motors. Various speed control methods for DC motors, such as armature voltage control, field flux control, and armature resistance control give an understanding of the principles behind each method and their applications in different scenarios. The further exploration of the applications of DC machines in various industries, such as electric vehicles, robotics, power generation, and industrial processes provides insights into the specific requirements and challenges in these applications and understand how DC machines are employed to meet those needs effectively.

This Chapter introduces other aspects of the DC Machine.

4.2 Armature Circuit Equation for Motoring and Generation

Generating Action:

Armature is rotated at speed n rotation per second (rps) or N rotation per minute (rpm) using prime mover. The basic circuit model for a DC machine is given in fig. 4.1.

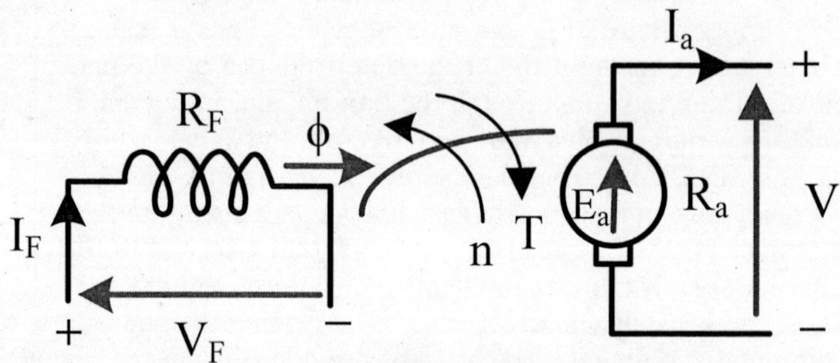

Fig. 4.1. Circuit model for DC machine

The machine operates in generating mode when the armature current is in the direction of induced emf E_a as shown in fig. 4.2.

$$V = E_a - I_a R_a, \qquad E_a > V \tag{4.1}$$

$$P_{mech} = E_a I_a = P_{elect} \tag{4.2}$$

Total output power,

$$P_o - VI_a \tag{4.3}$$

$$E_a I_a - VI_a = I_a^2 R_a \tag{4.4}$$

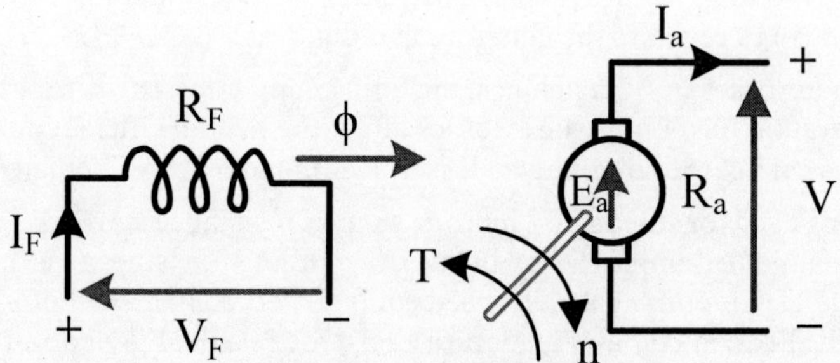

Fig. 4.2. Circuit model for DC machine generating mode

In this mode, the torque of electromagnetic origin is in the opposite direction of rotation of the armature.

$$[P_{mech}]_{gross} = Shaft\ Power = [P_{mech}]_{net} + Rotational\ loss$$

Motoring Action

In motoring mode, the armature current I_a flows opposite to the emf induced i.e., E_a. The basic circuit shows this mode in fig. 4.3.

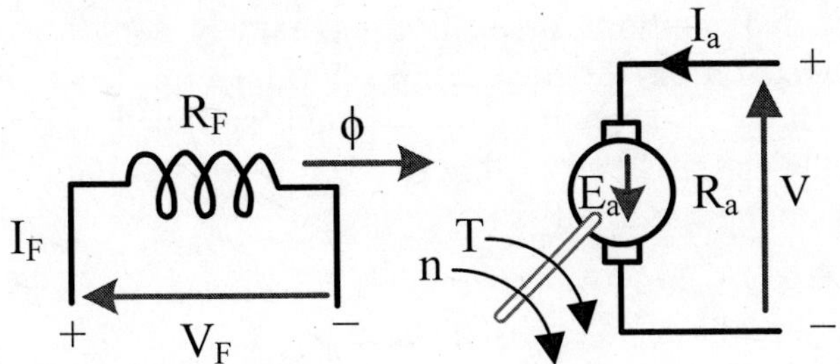

Fig. 4.3. Circuit model for DC machine motoring mode

E_a is called as back emf here as it opposes the armature current under dynamic conditions of motoring.

$$V = E_a + I_a R_a \tag{4.5}$$

Electrical power connected to mechanical form

$$P_{elec/net} = E_a I_a \tag{4.6}$$

Power input

$$P_i = V I_a \tag{4.7}$$

$$V I_a - E_a I_a = I_a^2 R_a = armature\ copper\ loss \tag{4.8}$$

The back emf is given by (refer to section 4.19)

$$E_b = P\phi NZ \tag{4.9}$$

The motor develops electromagnetic torque T in the direction of rotation. The load torque opposes the motor torque.

4.3 Types of DC machine

The classification of the DC machines based on the field configuration is as follows:

- Homopolar machines
- Heteropolar machines

Homopolar machines: These types of machines are used where low voltage and high currents are required. Faraday's disc dynamo is an example of this type of machine.

The homopolar generators are widely used in defense as well as in metal refining. These generators generate very large values of DC current at very low voltage.

Heteropolar machines: The DC machines that are commonly used fall under this category and operate on bipolar construction.

4.4 Type of DC Generator

The magnetic flux in a DC machine is produced by field coils carrying DC current. The production of magnetic flux in the machine by circulating current in the field winding is called excitation. There are two methods of excitation, namely separate excitation, and self-excitation. In separate excitation, the field coil is energized by a separate DC source. In self-excitation, the current flowing through filed winding is supplied by the machine itself. D.C. generators are generally classified according to the methods of their field excitation. Based on these criteria, they can be classified as:

i. Separately excited DC Generator
   a. Field winding connected to a DC source
   b. Permanent Magnet
ii. Self-excited DC Generator – these are further classified as:
   a. Shunt wound DC Generator
   b. Series wound DC Generator
   c. Compound wound DC Generator.
      - Long shunt compound wound Generator
      - Short shunt compound wound Generator

Video lecture on DC Generator

These types of machines could be a generator or a motor.

4.4.1 Separately excited DC generator

In this type of generator, the field coils are energized from an independent external source, as shown in fig. 4.4. The flux produced by the poles depends upon the field current within the unsaturated region of magnetic material of the poles (i.e., $\phi \propto I_f$), but in the saturated region, the flux almost remains constant. The field coil produces alternates static poles N-S in armature air gap.

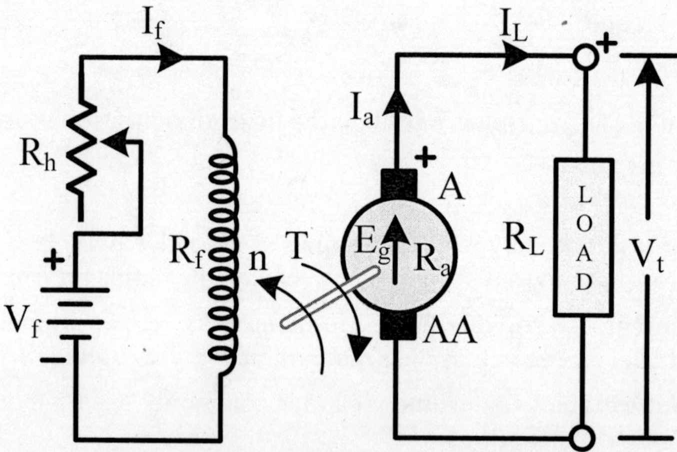

Fig. 4.4 Separately excited DC generator

In separately excited generator the load is connected across the armature winding. Hence the armature current is equal to the load current.

$$I_a = I_L \tag{4.10}$$

Where, I_a is the armature current and I_L is the load current.

- **Field current,**

  Due to Separate excitation, the field current I_f will remain constant.

$$I_f = \frac{V_f}{R_f} = Constant \tag{4.11}$$

Where, V_f = External DC excitation applied to field winding; R_f = Resistance of field winding

- **Terminal voltage,**

$$V_t = \ Generated\ voltage - Armature\ voltage\ drop(I_a R_a)$$

$$V_t = E_g - I_a R_a \tag{4.12}$$

If contact brush drop per brush is v_b then the terminal voltage

$$V_t = E_g - I_a R_a - 2v_b \tag{4.13}$$

- **Power developed,**

$$P_g = E_g I_a \tag{4.14}$$

- **Power delivered to the load or output power,**

$$P_L = V_t I_L = V_t I_a \tag{4.15}$$

- **Armature current and load current,**

$$I_a = I_L = \frac{V_t}{R_L} \tag{4.16}$$

$$I_a = I_L = \frac{E_g - I_a R_a}{R_L} \tag{4.17}$$

- **Internally induced voltage (E_g)**

$$E_g = \frac{P\phi NZ}{60A} \tag{4.18a}$$

- **Counter torque (T)**

$$T = \frac{P_g}{\omega}, \qquad Where\ \omega = \frac{2\pi N}{60} \tag{4.18b}$$

The internally induced emf is proportional to the speed (N) and flux ϕ. For a separately excited generator, if the field current I_f is constant, then the internally induced emf is proportional to the speed. The polarity of internally induced emf is decided by the direction of rotation. If we reverse the direction of rotation of armature then the polarity of "E_g" will reverse.

Similarly, if we change the direction of field current by changing terminals at DC source connected to field winding, it will also result in reversal of polarity of "E_g".

4.4.2 Self-Excited generator

A DC generator whose field winding is supplied current by the generator itself is called a self-excited DC generator. In such machine the field coil may be connected in parallel with the armature, in series with the armature, or partially in series parallel with the armature winding. The self-excited generator can be classified as:

a) Shunt wound generators
b) Series wound generators
c) Compound wound generators

a) Shunt wound generator

In shunt wound generator the field coil is in shunt or parallel with the armature winding. Its conventional diagram is shown in fig. 4.5. In shunt wound DC generator the full terminal voltage is applied across the field winding. A very small current I_{sh} flows through it because this winding has many turns of fine wire having very high resistance R_{sh}.

EMF E_g is dynamically induced EMF when the armature is being rotated by an external prime mover. This prime mover may be any electrical motor, diesel/petrol/ gas engine or water/wind turbine etc.

The field resistance is deliberately kept high because the changes in load current should not disturb the value of field current (I_{sh} should remain constant).

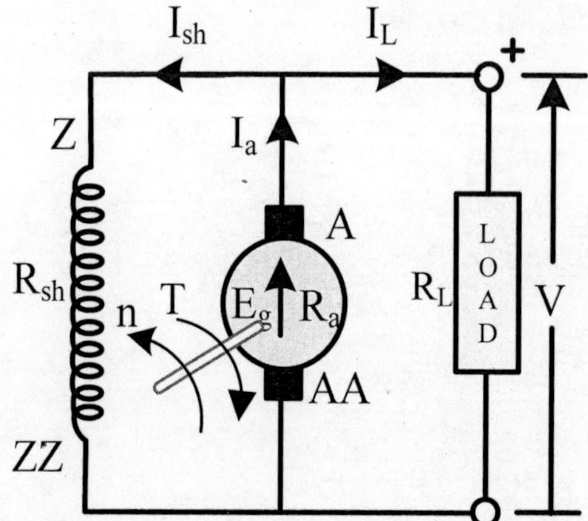

Fig. 4.5 Shunt wound DC generator

Here, the load is connected across the armature winding. Hence the armature has to supply the field current as well as load current.

$$I_a = I_L + I_{sh} \tag{4.19}$$

Where, I_a is armature current, I_{sh} is shunt field winding current and I_L is the line current.

- ***Shunt field current,***

The shunt field current I_{sh} is practically constant at all loads, therefore DC shunt machine is considered to be a constant flux machine. The load voltage is equal to the voltage across the shunt field winding. Hence field current is given by

$$I_{sh} = \frac{V_t}{R_{sh}} \tag{4.20}$$

Where, V_t = Terminal voltage; R_{sh} = Resistance of shunt field winding

It produces the necessary field flux (ϕ) required for generating operation. In most cases, It is assumed that field flux per pole (ϕ) is proportional to the field current I_{sh}. Under heavy load conditions, when the machine is saturated, field flux is almost constant.

- **Terminal voltage,**

$$V_t = \text{ Generated voltage} - \text{Armature voltage drop } (I_a R_a)$$

$$V_t = E_g - I_a R_a \tag{4.21}$$

If contact brush drop per brush is v_b then the terminal voltage

$$V_t = E_g - I_a R_a - 2v_b \tag{4.22}$$

- **Power developed,**

$$P_g = E_g I_a \tag{4.23}$$

- **Power delivered to the load or output power,**

$$P_L = V_t I_L \tag{4.24}$$

- **load current,**

$$I_L = \frac{V_t}{R_L} \tag{4.25}$$

$$I_L = \frac{E_g - I_a R_a}{R_L} \tag{4.26}$$

- **Internally induced voltage (E_g)**

$$E_g = \frac{P\phi NZ}{60A} \tag{4.27}$$

- **Counter torque (T)**

$$T = \frac{P_g}{\omega}, \quad Where\ \omega = \frac{2\pi N}{60} \tag{4.28}$$

b) Series wound generator

In a series-wound generator, the field winding is connected in series with the armature winding as is shown in fig. 4.6. In series wound generator the full line current I_L or armature current I_a flows through armature winding. Since the series field winding carries full load current, it has a few turns of thick wire having low resistance.

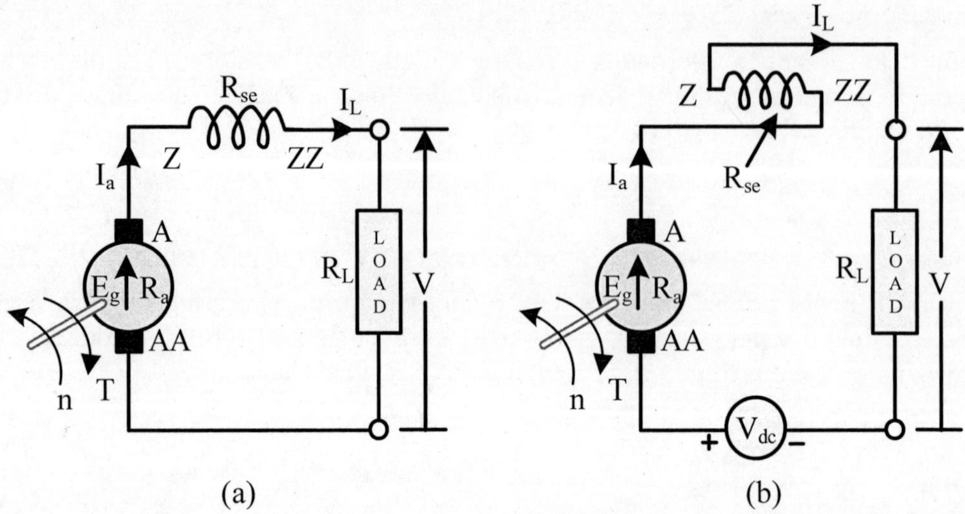

Fig. 4.6 Series Wound DC generator (a) self-excited (b) Buck/ Boost

Here, the armature and field winding are connected in series, the current flowing through them is the same and given by.

$$I_a = I_{se} = I_L \tag{4.29}$$

Where, I_a is armature current, I_{se} is series field winding current and I_L is the line current.

- **Terminal voltage,**

The armature is being rotated by a prime mover and the armature circuit is closed as shown in Fig 4.6(a), the terminal voltage due to self-excitation is,

$$V_t = \text{Generated voltage} - \text{Armature voltage drop } (I_a R_a) - \text{Series winding voltage drop } (I_a R_{se})$$

$$V_t = E_g - I_a R_a - I_{se} R_{se} \tag{4.30}$$

If contact brush drop per brush is v_b then the terminal voltage

$$V_t = E_g - I_a R_a - I_{se} R_{se} - 2v_b \tag{4.31}$$

Where, R_{se} = Resistance of series field winding and $I_a = I_{se} = I_L$

$$V_t = E_g - I_a (R_a + R_{se}) - 2v_b$$

If armature reaction voltage drop is considered

$$V_t = E_g - I_a (R_a + R_{se}) - 2v_b - \mathit{Armature\ reaction\ voltage\ drop} \tag{4.32}$$

- **Power developed,**

$$P_g = E_g I_a \tag{4.33}$$

- **Power delivered to the load or output power,**

$$P_L = V_t I_L = V_t I_a \tag{4.34}$$

The flux developed by the series field winding is directly proportional to the current flowing through it (i.e., $\phi \propto I_{se}$). But it is only true before magnetic saturation. Once DC series generator is saturated, the voltage drops to zero. This is the drawback of DC series generator. These generators can be considered for high power DC boost and buck required in DC Grid voltage of Renewable Energy resources and Electric Vehicle (EV). In such cases DC series generator will be connected in series with DC source and load as shown in Fig 4.6 (b).

c) *Compound wound generator*

In a compound wound generator, there are two sets of field windings on each pole. One of them is connected in series and the other is connected in parallel with armature. The series winding has few turns of thick wire and the shunt is having many turns of fine wire. Compound wound generators are of two types:-

i. *Long shunt compound generator*

Long shunt compound generator in which shunt field winding is connected in parallel with the combination of both armature and series field winding as shown in fig. 4.7.

- **Series field current,**

Here, the armature and field winding are connected in series, the current flowing through them is same and given by.

$$I_a = I_{se} = I_{sh} + I_L \tag{4.35}$$

- **Shunt field current,**

$$I_{sh} = \frac{V_t}{R_{sh}} \tag{4.36}$$

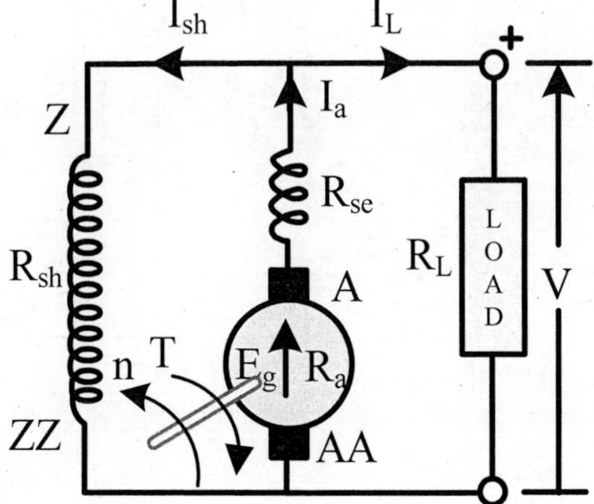

Fig. 4.7 Long shunt compound DC generator

- **Terminal voltage,**

$$V_t = \textit{Generated voltage} - \textit{Armature voltage drop}(I_a R_a) - \textit{series winding voltage drop}(I_a R_{se})$$

$$V_t = E_g - I_a R_a - I_{se} R_{se} \tag{4.37}$$

If contact brush drop per brush is v_b then the terminal voltage

$$V_t = E_g - I_a R_a - I_{se} R_{se} - 2v_b \tag{4.38}$$

Where, R_{se} = Resistance of series field winding and $I_a = I_{se}$

$$V_t = E_g - I_a(R_a + R_{se}) - 2v_b$$

If armature reaction voltage drop is also considered

$$V_t = E_g - I_a(R_a + R_{se}) - 2v_b - Armature\ reaction\ voltage\ drop \tag{4.39}$$

- **Power developed,**

$$P_g = E_g I_a = E_g I_{se} \tag{4.40}$$

- **Power delivered to the load or output power,**

$$P_L = V_t I_L \tag{4.41}$$

ii. Short shunt compound generator

Short shunt compound generator in which shunt field winding is connected in parallel only with the armature winding as shown in fig. 4.8.

- **Series field current,**

Here, the load and series field winding are connected in series, the current flowing through them is the same and given by.

$$I_{se} = I_L \tag{4.42}$$

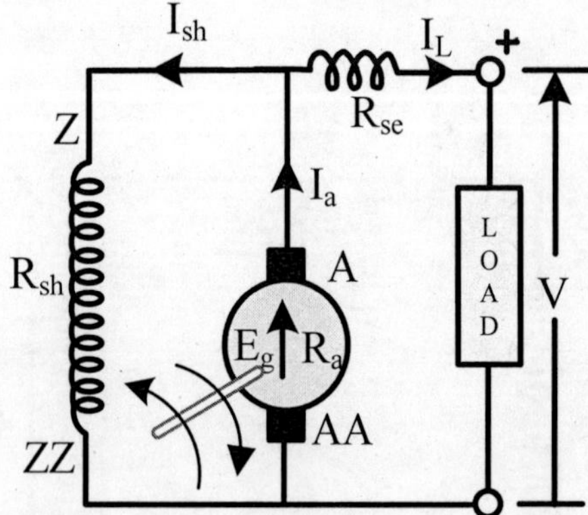

Fig. 4.8 Short shunt compound DC generator

- **Shunt field current,**

$$I_{sh} = \frac{V_t - I_L R_{se}}{R_{sh}} \tag{4.43}$$

$$= \frac{E_g - I_a R_a}{R_{sh}} \tag{4.44}$$

$$I_a = I_{sh} + I_L = I_{sh} + I_{se} \tag{4.45}$$

- **Terminal voltage,**

$$V_t = \text{ } Generated\ voltage - Armature\ voltage\ drop(I_a R_a) - series\ winding\ voltage\ drop(I_{se} R_{se})$$

$$V_t = E_g - I_a R_a - I_{se} R_{se} \quad (4.46)$$

If contact brush drop per brush is v_b then the terminal voltage

$$V_t = E_g - I_a R_a - I_{se} R_{se} - 2v_b \quad (4.47)$$

Where, R_{se} = Resistance of series field winding and $I_a = I_{se}$

$$V_t = E_g - I_a(R_a + R_{se}) - 2v_b$$

If armature reaction voltage drop is also considered

$$V_t = E_g - I_a(R_a + R_{se}) - 2v_b - Armature\ reaction\ voltage\ drop \quad (4.48)$$

- **Power developed,**

$$P_g = E_g I_a \quad (4.49)$$

- **Power delivered to the load or output power,**

$$P_L = V_t I_L \quad (4.50a)$$

- **Counter torque (T)**

$$T = \frac{P_g}{\omega}, \quad Where\ \omega = \frac{2\pi N}{60} \quad (4.50b)$$

Cumulatively and Differentially compound-wound generators

In compound wound DC generators, the field is produced by the shunt as well as series winding. Generally, the shunt field is stronger than the series field. When the series field assists or aids the shunt field, the generator is called a **cumulatively compound wound generator** as shown in fig. 4.9(a). However, when the series field *opposes* the shunt field, the generator is known as a **differentially compound wound generator** as shown in fig. 4.9(b).

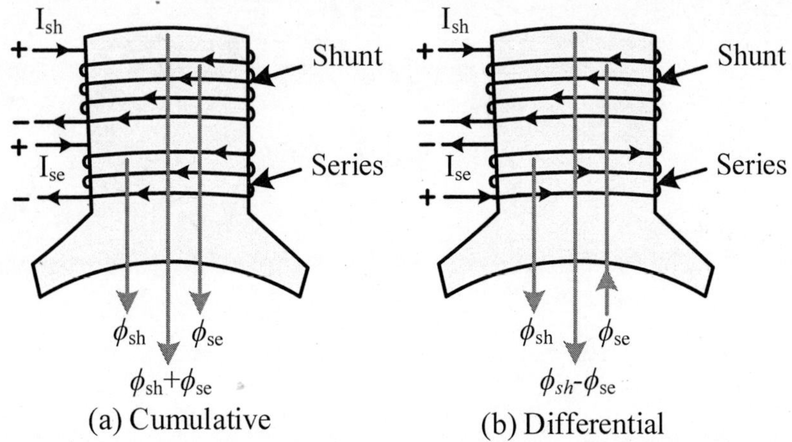

Fig. 4.9 Winding position and direct current in shunt and series winding

4.5 Voltage Regulation

The degree of change in armature terminal voltage due to the application of load is known as voltage regulation. When a generator delivers load current, the counter torque due to Lenz's law opposes the drive torque of the prime mover. This results in a drop in speed as well as the generated EMF. The terminal voltage further drops due to an increase in ohmic voltage drop in the armature circuit and armature reaction. If there is little change in voltage from no load to full load, the generator is said to have good voltage regulation. If there is an appreciable change in voltage from no load to full load, the generator is said to have poor voltage regulation. The change in load voltage from no load to full load expressed as a percentage of the rated terminal voltage is known as percent voltage regulation.

$$\%\, Voltage\ regulation = \frac{E_0 - V}{V} \times 100 \tag{4.51}$$

Example 4.1: A 8-pole DC shunt generator has 60 slots on its armature with 12 conductors per slot with wave winding. The armature and field winding resistance is 0.6 ohm and 50 ohms respectively. The generator is supplying a resistive load of 10 ohms at terminal voltage of 250 V when running at a speed of 800 rpm. Find the armature current, the generated emf, and the flux per pole.

Solution: -

$$I_L = \frac{V}{R_L} = \frac{250}{10} = 25A$$

$$I_{sh} = \frac{V}{R_{sh}} = \frac{250}{50} = 5A$$

Armature current,

$$I_a = I_L + I_{sh} = 25 + 5 = 30\ A\ (Ans)$$

Generated emf,

$$E_g = V + I_a R_a$$

$$= 250 + (30\ \times 0.6) = 268\ V\ (Ans)$$

Now,

$$Z = 60\ \times 12 = 720; N = 800\ RPM; P = 8; A = 2;$$

$$E_g = \frac{P\phi NZ}{60A}$$

$$268 = \frac{8 \times \phi\ \times 800\ \times 720}{60\ \times 2}$$

Or

$$\emptyset = 6.98\ m\ Wb\ (Ans)$$

Example 4.2: A six-pole shunt generator with lap-connected armature has field and armature resistance of 40 Ω and 0.2 Ω respectively. The generator is supplying a load of 5 kW at 200 V. Calculate the armature current, current in each conductor, and generated emf.

Solution: -

Load current,

$$I_L = \frac{5 \times 1000}{200} = 25\ A$$

Shunt field current;

$$I_{sh} = \frac{V}{R_{sh}} = \frac{200}{40} = 5\ A$$

Armature current;

$$I_a = I_L + I_{sh} = 25 + 5 = 30\ A\ (Ans)$$

Current in each conductor,

$$I_c = \frac{I_a}{A} = \frac{30}{6} = 5\ A\ (Ans)$$

Generated emf,

$$E_g = V + I_a R_a$$
$$= 200 + (30 \times 0.2) = 206\ V (Ans)$$

Example 4.3: A load of 30 kW at 230 V is supplied by a compound DC generator. If the series, shunt field, and armature resistances are 0·07, 120, and 0.3 ohms respectively. Calculate the generated emf when the generator is connected as long shunt.

Solution: -

$$Load = 30kW = 30 \times 10^3 W$$
$$V = 230\ V;\ R_a = 0.3\ \Omega;\ R_{se} = 0.07\Omega;\ R_{sh} = 120\Omega$$

Line Current,

$$I_L = \frac{30 \times 10^3}{230} = 130.43\ A$$

Shunt field current,

$$I_{sh} = \frac{V}{R_{sh}} = \frac{230}{120} = 1.91\ A$$

Armature current,

$$I_a = I_L + I_{sh} = 130.43 + 1.91 = 132.34\ A$$

Generated emf,

$$E_g = V + (I_a R_a) + (I_a R_{se})$$
$$= 230 + (132.34 \times 0.3) + (132.34 \times 0.07) = 278.96\ V\ (Ans)$$

Example 4.4: A load of 12.5 kW at 230V is supplied by a short-shunt cumulatively compound DC generator. If the armature, series and shunt field resistances are 0·5, 0·4 and 80 ohms respectively. Calculate the induced emf and the load resistance.

Solution:

$$I_L = \frac{12.5 \times 1000}{230} = 54.34\ A$$

$$I_{sh} = \frac{V + I_L R_{se}}{R_{sh}} = \frac{230 + (54.34 \times 0.4)}{80} = 3.14\ A$$

$$I_a = I_L + I_{sh} = 54.34 + 3.14 = 57.48\ A$$

Induced emf,

$$E_g = V + I_L R_{se} + I_a R_a$$

$$= 230 + (54.34 \times 0.4) + (57.48 \times 0.5) = 280.48\ V\ (Ans)$$

Load resistance,

$$R_L = \frac{V^2}{P} = \frac{(230)^2}{12.5\ \times 1000} = 4.232\ ohm\ (Ans)$$

Example 4.5: A 6-pole DC shunt generator with 860 wave-connected armature conductor running at 400 rpm supplies a load of 12.5-ohm resistance at terminal voltage of 230 V. The shunt field resistance is 200 ohms and the armature resistance is 0.24 ohms. Find the armature current, induced emf, and flux per pole.

Solution: -

$$Load\ current, I_L = \frac{V}{R} = \frac{230}{12.5} = 18.4\ Amp$$

$$Shunt\ field\ current, I_{sh} = \frac{V}{R_{sh}} = \frac{230}{200} = 1.15\ Amp$$

Armature current, $I_a = I_L + I_{sh} = 18.4 + 1.15 = 19.55\ Amp$

Induced emf=230+(19.55× 0.24)=234.69 V

$$E_g = \frac{PZ\emptyset N}{60A}$$

$$234.69 = \frac{6 \times 860 \times \emptyset \times 400}{60 \times 2}$$

$\emptyset = 13.64\ mWb$ (Ans)

Example 4.6: A The useful flux of a six-pole, lap wound DC generator is 0.05 Wb. Its armature carries 180 turns and each turn has 0.003-ohm resistance, if its armature current is 40 A, running at a speed of 800 rpm, (i) Determine its terminal voltage (ii) for the same machine if generated emf is 280 V. find the flux per pole

Solution: -

Here, $P = 6; A = 6(lap\ wound);$

$\emptyset = 0.05\ Wb; N = 800\ rpm; I_a = 40\ A$

No. of armature conductors,

$$Z = 2 \times No.of\ turns = 2 \times 180 = 360$$

Induced emf,

$$E = \frac{PZ\emptyset N}{60A} = \frac{6 \times 360 \times 0.05 \times 800}{60 \times 6} = 240\ V$$

No. of turns connected in each parallel path,

$$= \frac{180}{6} = 30\ turn$$

Resistance of each parallel path,

$$= 30\ \times 0.003 = 0.09\ ohm$$

Six resistances, each of 0.09 ohm, are connected in parallel

$\therefore$ Armature resistance,

$$R_a = \frac{0.09}{6} = 0.015\Omega$$

Terminal voltage,

$$V = E - I_a R_a\ \ (generator\ action)$$

$$= 240 - (40 \times 0.015) = 239.40\ V\ (Ans)$$

(ii) induced emf

$$E_g = \frac{PZ\emptyset N}{60A}$$

$$280 = \frac{6 \times 360 \times \emptyset \times 800}{60 \times 6}$$

$\emptyset = 0.058\ mWb$ (Ans)

Example 4.7: A two-pole DC shunt generator with a wave wound armature having 800 conductors has to supply a load of 300 lamps each of 100 W at 200 V. Allowing 5 V for the voltage drop in the connecting leads between the generator and the load and brush drop of 1.5 V. Calculate the speed at which the generator should be driven. The flux per pole is 20 m Wb and the value of $R_a = 0.06\Omega$ and $R_{sh} = 41\ \Omega$.

Solution: -

Refer conventional circuit diagram of the DC shunt generator

$$Total\ Load = 300 \times 100W = 30KW$$

$$I_L = \frac{30 \times 1000}{200} = 150\ A$$

Voltage drops in leads,

$$V_L = 5\ V$$

The voltage across shunt field winding,

$$V_{sh} = V + V_L = 200 + 5 = 205\ V$$

$$I_{sh} = V_{sh}/R_{sh} = 205/41 = 5\ A$$

$$I_a = I_L + I_{sh} = 150 + 5 = 155\ A$$

Armature drop,

$$= I_a R_a = 155 \times 0.06 = 9.3\ V$$

Total brush drop,

$$2v_b = 1.5\ V$$

Generated emf,

$$E_g = V + I_a R_a + V_L + 2v_b$$

$$= 200 + 9.3 + 5 + 1.5 = 215.8\ V$$

Now,

$$E_g = \frac{P\emptyset NZ}{60\ A}$$

Or

$$215.8 = \frac{2 \times 20 \times 10^{-3} \times N \times 800}{60 \times 2}$$

Or

$$N = \frac{215.8 \times 60 \times 2}{2 \times 20 \times 10^{-3} \times 800} = 809.25\ rpm\ (Ans)$$

4.6 Characteristics of DC Generator

Important characteristics to determine the relationship between different quantities of a DC generator are as follows:

a) ***No-load saturation characteristic (E_0/I_f):*** It gives the relation between the no-load generated emf in armature (E_0) and the field or exciting current (I_f) at rated speed. It is also called *open-circuit characteristic (O.C.C) or magnetization characteristic.* This characteristic is the same for the separately excited and the self-excited generators.

b) ***External characteristic (V_t/I_L):*** It shows the relation between the terminal voltage V_t and the load current I_L. It is also called *performance characteristic* or sometimes *voltage regulating curve.*

c) ***Internal characteristic (E_g/I_a):*** It practically gives the relation between the emf in armature conductor (E_g) (after considering the demagnetizing effect of armature reaction) and armature current I_a. It is also known as *total characteristic.*

4.7 No-load Saturation Characteristics or Magnetisation Curve of DC Generator

The no-load saturation characteristic shows the relation between the no-load generated emf in the armature (E_0) and the field current (I_f), at a specified speed. In Figure 4.10, the field circuit is provided with variable excitation. The armature terminals are connected to the load through a two-pole main switch. A voltmeter is connected across the armature and an ammeter is connected in series with the field coil. Reduce the field current to zero and run the armature at a specified speed. Get the reading of the voltmeter and mark the point '*a*' on the graph. To plot the characteristics, take field current (I_f) along X-axis and no-load generated emf (E_0) along Y-axis. Increase the field current (excitation) in steps and get the corresponding voltmeter readings. Plot these values on the graph. The curve thus obtained shows the no-load characteristics or open circuit characteristics (*O.C.C.*) of the generator as shown in fig. 4.11.

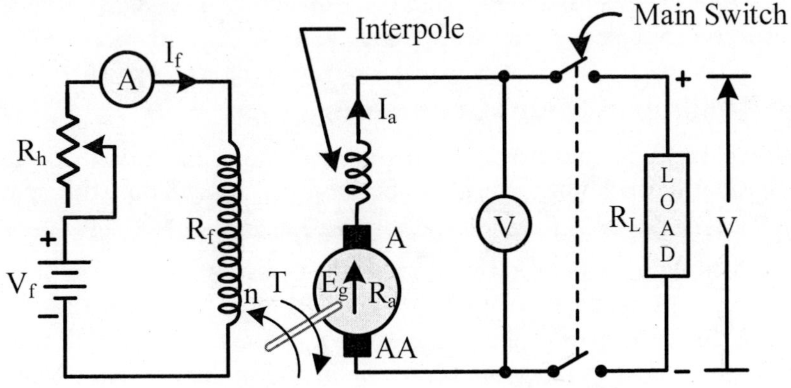

Fig. 4.10 Separately excited generator

While analyzing the curve, the following points are worth noting:

- During the reading of the ammeter and voltmeter, it is observed that the curve starts from point '*a*' instead of '*O*' when the field current is zero.
- It is observed that the voltmeter gives a reading at the zero value of field excitation current. This is due to a small amount of magnetism in field poles. This is called residual magnetism, which is usually sufficient to produce 2 to 3 percent of the normal terminal voltage. In some special cases, it is purposely increased to 10 percent or more.

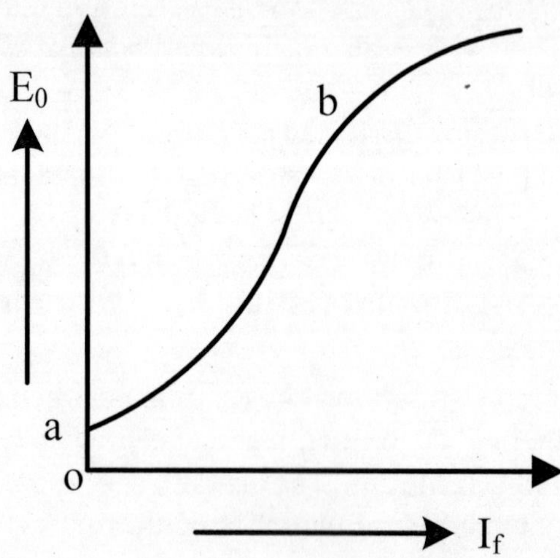

Fig. 4.11 No Load (OCC) characteristic

- The initial part of the curve (*ab*) is almost a straight line because at this stage the magnetic material is unsaturated and it has high permeability. In this zone, the flux produced is proportional to the current.
- After point '*b*' (also referred to as knee point) the curve bends and the generated emf (E_0) becomes almost constant. It is because after point '*b*', the poles (magnetic material) start getting saturated. The rise in voltage for a current rise ΔI is comparatively less than that of the unsaturated region.

4.8 Voltage Build-up of Shunt Generators

A shunt or self-excited dc generator supplies its own field excitation. Its field winding is connected in parallel with the armature as shown in fig. 4.12. Thus, the armature voltage supplies the field current.

This generator will build up a desired terminal voltage. Assume that the generator in fig. 4.12 has no load connected to its armature and is driven at a certain speed ω radian per second by a prime mover. In this section, we shall study the conditions under which the voltage build-up takes place. The voltage in a DC generator depends upon the presence of a residual flux in the field poles of the generator. A small voltage E_{ar} will be generated and given by

$$E_{ar} = k\phi_{res}\omega \tag{4.52}$$

This voltage is of the order of 1V or 2V. it causes an initial field current I_f to flow in the field winding and it is given by

$$I_f = \frac{E_{ar}}{R_f} \tag{4.53}$$

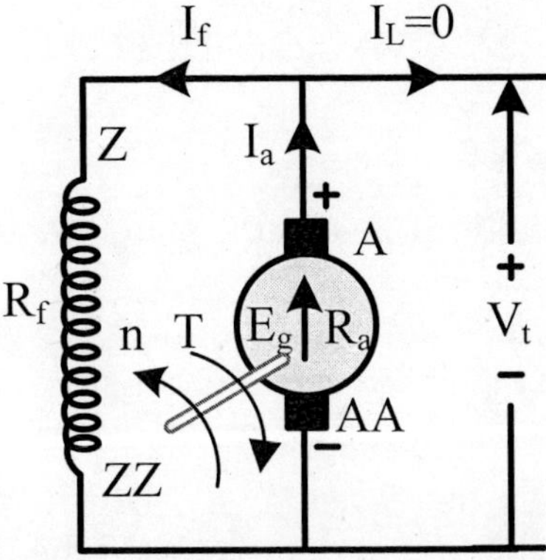

Fig. 4.12 Equivalent circuit of a shunt DC generator

The field current produces a magnetomotive force in the field winding, which increases the flux. The increase in flux gradually increases the generated voltage E_g and it increases the terminal voltage V_t. With the increased armature voltage V_t the field current I_f increases further. This in turn increase flux ϕ and consequently generated emf E_g increases further. The process of voltage build-up continues. Figure 4.13 show the voltage build-up of DC shunt generator. The generator will fail to self-excite when field current due to residual magnetism does not aid the original field flux. A reversal of either armature terminals or field terminals is required so that the developed field aids or supports the residual flux. Changing the direction of rotation of the prime mover is also helpful.

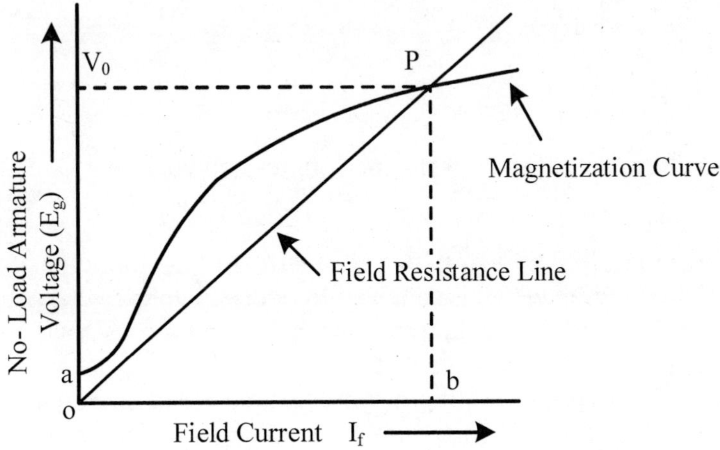

Fig. 4.13 Voltage build-up of a DC shunt generator

The effect of magnetic saturation in the pole faces limits the terminal voltage of the generator to a steady-state value. The E_g versus I_f the curve is the magnetization curve shown in fig. 4.13.

For the field circuit

$$V_t = I_f R_f \tag{4.54}$$

The straight line given by equation (4.54) is called the field resistance line.

4.9 Critical Field Resistance

Fig. 4.14 shows the voltage build-up in the DC shunt Generator for various field circuit resistance.

A decrease in the resistance of the field circuit reduces the slope of the field-resistance line resulting in a higher voltage. If the speed remains constant, an increase in the resistance of the field circuit increases the slope of the field resistance line, resulting in a lower voltage.

If the field circuit resistance is increased to R_c, which is termed as the ***critical field resistance***, the field resistance line becomes tangent to the initial part of the magnetization curve. When the field resistance is higher than this value, the generator fails to self-excite.

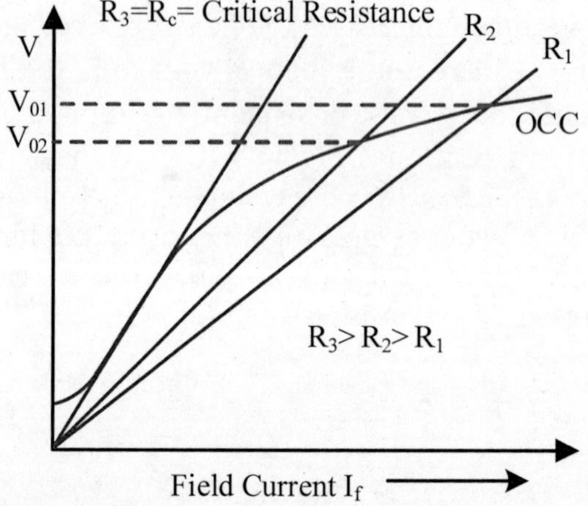

Fig. 4.14 Effect of resistance on no-load voltage

Critical resistance of a field winding for self-excitation: It is that maximum value resistance of a field winding which is required to build-up voltage in a generator. If the value of field resistance is more than this value, the generator would not build-up the voltage.

Critical load resistance: The minimum value of load resistance on a DC shunt generator with which it can be in position to build-up is called its critical load resistance.

4.10 Critical Speed

Fig. 4.15 shows the variation of no-load voltage with fixed R, and variable speed of the armature. The magnetization curve varies with the speed and its ordinate for any field current is proportional to the speed of the generator. If the field resistance is kept constant and the speed is reduced, all the points on the magnetization curve are lowered, and the point of intersection of the magnetization curve and the field resistance line moves downwards. At a particular speed, called the critical speed, the field-resistance line becomes tangential to the magnetization curve. Below the critical speed, the voltage will not build up.

Critical speed of a DC shunt generator for self-excitation: It is the speed of a DC shunt generator at which shunt field resistance will represent the critical field resistance. At critical speed, the DC shunt generator will just self-excite.

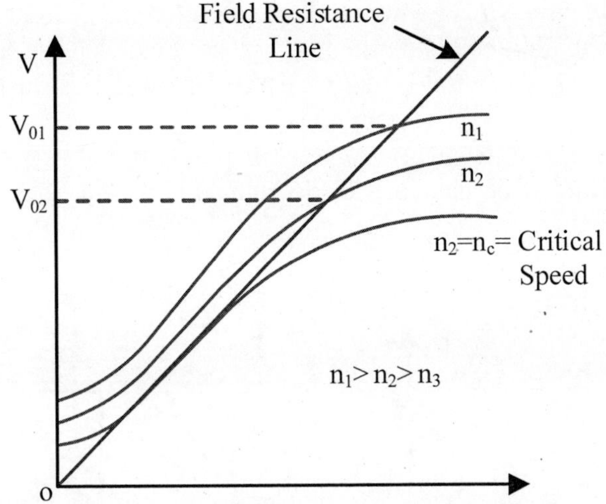

Fig. 4.15 Variation of no-load voltage with speed

In brief, the following conditions must be satisfied for voltage build-up in a self-excited DC generator.

a) There must be sufficient residual flux in the field poles.
b) The field terminals should be connected in such a way that the field current increases flux in the direction of residual flux.
c) The field circuit resistance should be less than the critical field circuit resistance.
d) The speed of the prime mover is greater than the critical speed

If in a DC machine's armature (with unexcited Field/open field circuit) being run at rated speed does not develop a voltage of the order of few volts, it means the machine has lost its residual magnetism. If there is no residual flux in the field poles, disconnect the field from the armature circuit and apply a DC voltage to the field winding. This process is called flashing the field. It will induce some residual flux in the field poles.

4.11 Load Characteristics of DC Shunt Generator

It is also called the external or performance characteristics of shunt generator. It shows the relation between the terminal voltage V_t on load and the load current I_L.In a shunt generator, the field circuit is directly connected across the armature. The terminal voltage decreases due to an increase in the generator current. The following three factors are responsible for the decrease in terminal voltage:

a) ***Armature circuit resistance:*** The generator has armature resistance, and hence voltage drop occurs. This resistance includes the resistance of (i) the copper conductors of the armature windings, (ii) contact resistance between the brushes and the commutator, and (iii) the brushes themselves.
b) ***Armature reaction:*** Due to the flow of currents through the armature conductors, a flux known as armature flux surrounds these conductors. The direction of this flux is such that it reduces the strength of the main flux, and hence reduces the generated voltage as well as the terminal voltage.
c) ***Reduction of field current:*** The decrease of the terminal voltage decreases the field current because the field coil is connected across the armature. This decrease in the field current decreases the field flux, which in turn further decreases the terminal voltage.
d) Counter torque acts against prime mover torque. If speed is not being regulated, a drop in speed result in decrease of speed-induced emf.

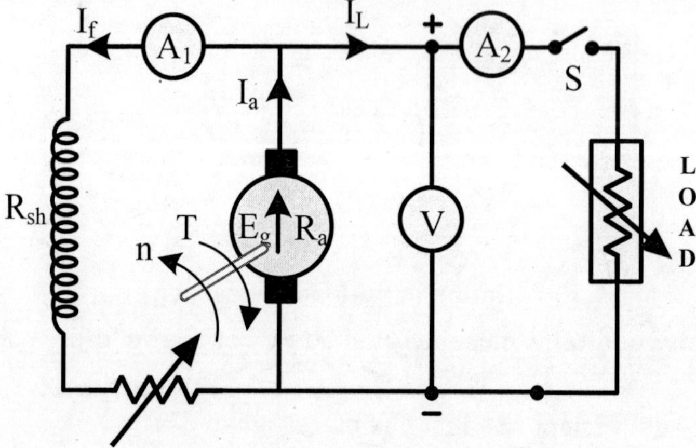

Fig. 4.16 Schematic circuit diagram for load test of DC shunt generator

To obtain external or load characteristics connect an ammeter A_1 and rheostat in the field circuit and an ammeter A_2 and voltmeter V on the load side as shown in fig. 4.16. Apply a variable load across the terminals. At start the load is disconnected and run the generator at the rated speed. No-load generated emf E_g will appear across the voltmeter. Then connect the load through switch S and increase the load gradually keeping the field current (ammeter reading A_1) constant with the help of rheostat R_h. Take the readings of voltmeter V and ammeter A_2 at various instants. The curve plotted by these readings is shown in fig. 4.17.

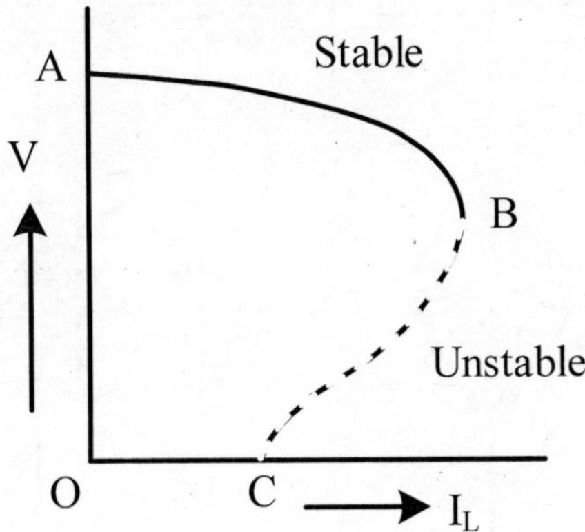

Fig. 4.17 Terminal characteristic of self-excited DC shunt generator

Here, we note that

a) At no-load, the voltage across the terminals is maximum and is considered to be equal to generated emf E_g.

b) As the load is increased gradually, the load current I_L increases but the terminal voltage decreases. The decrease in voltage is because of armature resistance, armature reaction, and decreases in field current as stated earlier.

c) During the initial stable portion of the curve AB, the tendency of the voltage drop due to armature resistance is more than the armature reaction.

d) At point B these two effects neutralize each other.

e) After point B, the armature reaction dominates and the curve turns back (BC portion of the curve). Since the BC part is unstable - with normal instruments, these measurements cannot be taken. This is an unstable portion of the curve.

f) The point C at which the external characteristic cuts the current axis corresponds to a gradual short circuit.

g) For obtaining point C, the prime mover is brought to rest and the load is removed with a short circuit. The prime mover is again started at the rated speed, the ammeter A_2 value is recorded.

4.12 Load Characteristics of DC Series Generators

In DC series generator the field winding is connected in series with the armature and load as shown in fig. 4.18. Therefore, full armature current I_a flows through the load. When load increases, I_a increases which increase flux ϕ and consequently generated emf $E_g$is also increased. This, correspondingly increases the terminal voltage V_t. Thus, a series generator has a rising characteristic (curve OA) as shown in fig. 4.19.

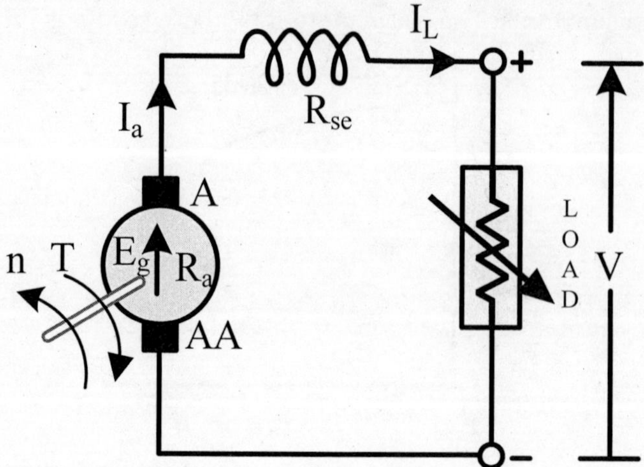

Fig. 4.18 Schematic circuit diagram of DC series generator

However, at higher loads, the terminal voltage begins to reduce because of the excessive demagnetizing effects of armature reaction. Ultimately, the terminal voltage V_t reduces to zero at load current $I_L = OB$ as shown in fig. 4.19.

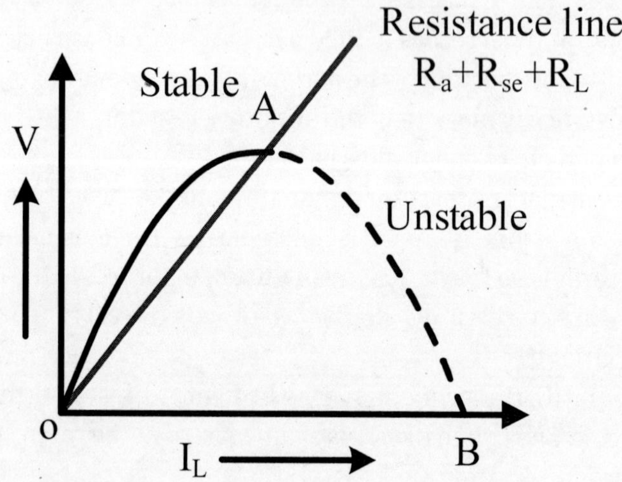

Fig. 4.19 Curve DC series generator

4.13 Load Characteristics of Compound Generator

In the application where constant terminal voltage is essential, the shunt generator is not suitable, because its terminal voltage decreases with the increase in load on it. However, it can be made suitable for such applications by connecting a few turns in series with the armature as shown in fig. 4.20. Such generators are known as compound generators.

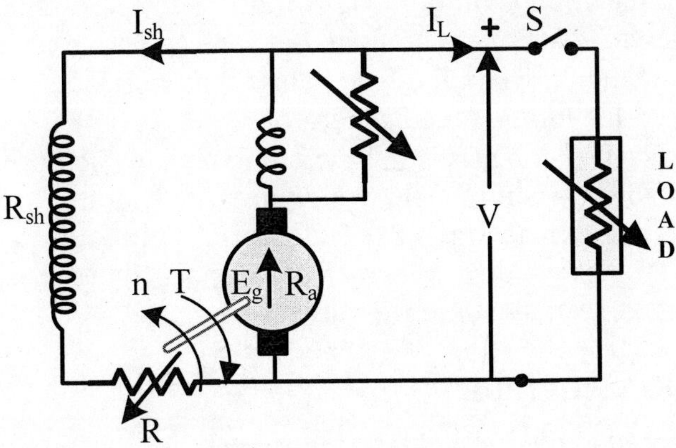

Fig. 4.20 Schematic circuit diagram of compound DC generator

The field produced by these series field winding turns assists the field produced by the shunt winding. In such generators when load current increases, the flux increases which increases the induced emf. This extra induced emf compensates for the voltage drop in the armature resistance and the demagnetizing effect due to armature reaction. Hence, the terminal voltage V_t remains substantially constant.

A cumulatively – compound wound generator is shown in fig. 4.20. In which shunt field flux and series field flux are in the same direction. Its level of compounding can be changed by varying the amount of current passing through the series field winding by connecting a by-pass rheostat R_h.

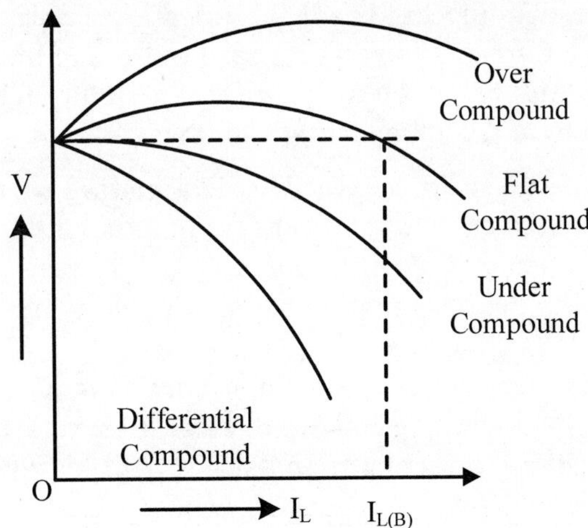

Fig. 4.21 Load Characteristic of compound generator

- The field current is adjusted such that the terminal voltage V on full load remains the same as that on no-load, the generator is called *level* or *flat compounded generator* as shown in fig. 4.21.

- When the terminal voltage on full-load is more than its terminal voltage at no-load, the generator is called an over-compounded generator.
- when the terminal voltage on full-load is less than no-load voltage, the generator is called to be as under compounded generator.
- The field produced by the series field winding acts in the opposite direction to the field produced by the shunt field winding, the generator is called differentially compounded as shown in fig. 4.21. Differential Compound generator finds their application in DC Arc welding. When electrodes are short-circuited for initiation of the arc, the voltage drops to limit the current.

4.14 Losses in DC Generator

A part of energy dissipated in the form of heat in the surrounding air while the conversion of mechanical energy into electrical energy is called losses in the generator. These losses affect the efficiency of the generator. A reduction in these losses leads to higher efficiency. Thus, the objective in the design of a DC machine is the reduction of losses. The losses that occur in DC machines can be divided into the following basic categories:

- Electrical or copper losses (I^2R Losses)
- Core losses or iron losses
- Brush losses
- Mechanical losses
- Stray-load losses
- Dielectric loss

a. Electrical or copper losses or winding losses

These variable losses are winding losses. The copper losses are present because of the resistance of the windings. Currents flowing through these windings produce ohmic losses (that is, I^2R Losses). The windings that may be present in addition to the armature winding are field windings, interpole, and compensating windings.

- *Armature copper losses:* $= I_a^2 R_a$, where I_a is armature current and R_a is armature resistance. There losses are about 30 per cent of total full-load losses.
- *Copper loss in shunt field of a shunt machine:* $= I_{sh}^2 R_{sh}$, where I_{sh} is the current in the shunt field and R_{sh} is the resistance of the shunt field winding. The shunt regulating resistance is included in R_{sh}.
- *Copper loss in the series field of a series machine:* $= I_{se}^2 R_{se}$ where I_{se} is the current through the series field winding and R_{se} is the resistance of the series field winding.
- *In a compound machine,* both shunt and series field losses occur. These losses are about 20% of full load losses.
- *Copper loss in interpole winding:* $= I_a^2 R_i$, where R_i is the resistance of interpole windings.
- *Copper loss in compensating winding:* $= I_a^2 R_c$, where R_c is the resistance of compensating winding.

- *Brush contact loss:* $I_a^2 R_b$ or $2I_a V_b$, these losses are generally included in armature copper losses. Here V_b is the voltage between the brush and commutator segment.

b. Magnetic Losses/Core Losses/Iron Losses

The core losses are the hysteresis loss and eddy-current loss. Since machines are usually operated at constant flux density and constant speed, these losses are almost constant. These losses are about 20 percent of full-load losses.

- **Hysteresis loss.** Whenever a magnetic material is subjected to cyclic reversal of magnetic flux, this loss occurs. It is due to retentivity property of the material. It is expressed with reasonable accuracy by the following expression

$$P_h = K_h V f B_{max}^{1.6} \tag{4.55}$$

where, K_h = Hysteresis constant in j/m^3

V = Volume of magnetic material in m^3

f = frequency of magnetic reversal in cycle/second and

B_{max} = maximum flux density in the magnetic material in Tesla.

It occurs in the rotating armature which is subjected to rotational magnetization. The flux density at any point of armature active surface in airgap changes from N-S in a cyclic manner as armature rotates. To minimize this loss, the armature core is made of silicon steel which has low hysteresis constant.

- **Eddy current loss.** When flux linking with the magnetic material changes (or flux is cut by the magnetic material) an emf is induced in it which circulates eddy currents through it. These eddy currents produce eddy current loss in the form of heat. It is expressed with reasonable accuracy by the expression

$$P_e = K_e B_{max}^2 f^2 t^2 V \tag{4.56}$$

where,

K_e = constant called coefficient of eddy current; its value depends upon the nature of magnetic material; t = thickness of lamination in m

The major part of this loss occurs in the armature core. To minimize this loss, the armature core is laminated with insulated thin sheets of Cold Rolled Grain-Oriented Silicon Sheet Steel (CRGOS) (0.3 to 0.5 mm) since this loss is directly proportional to the square of the thickness of the laminations.

c. Brush Losses

Brush losses are the losses taking place between the commutator and the carbon brushes. It is the power loss at the brush contact point. The brush loss depends upon the brush contact voltage drop and the armature current Ia. It is given by the equation shown below:

$$P_{BD} = V_{BD} I_a \tag{4.57}$$

The voltage drops occurring over a large range of armature currents, across a set of brushes is approximately constant. If the value of the brush voltage drop is not given then it is roughly taken as 2V.

d. *Mechanical Losses*

The losses that take place because of the mechanical rotation are known as mechanical losses. Mechanical losses are divided into bearing friction loss and windage loss. Bearing friction loss is referred to as friction loss. Rotating parts of the machine cuts the surrounding medium (air) which requires certain electrical power termed as windage loss. These losses are very small. To reduce these losses proper lubrication of bearings is essentially required.

e. *Stray load losses*

The sum of the iron losses and mechanical losses in a DC machine is at loaded condition known as stary losses i.e.

$$\text{Stray Losses} = \text{Iron Losses} + \text{Mechanical Losses}$$

These losses are the miscellaneous type of losses. The following factors are considered in stray load losses.

- The distortion of flux is because of the armature reaction.
- Short circuit currents in the coil, undergoing commutation.

These losses are very difficult to determine. Therefore, it is necessary to assign a reasonable value to the stray loss. For most machines, stray losses are taken by convention to be one percent of the full load output power.

f. *Constant Losses*

The losses in a DC machine which remain the same at all loads are called constant losses. The constant losses in a DC machine are:

i. Iron losses
ii. Mechanical losses
iii. Shunt field copper losses

g. *Variable Losses*

The losses in the DC machine mentioned earlier which vary with load are called variable losses. The variable losses in a DC machine are:

i. Armature copper losses
ii. Series field copper losses
iii. Interpole winding copper losses
iv. Compensating winding copper losses

4.15 Power flow Diagram

The mechanical power is supplied to the generator which is converted into electrical power. While conversion, various losses occur in the machine. The power flow diagram for a DC generator is shown in fig. 4.22.

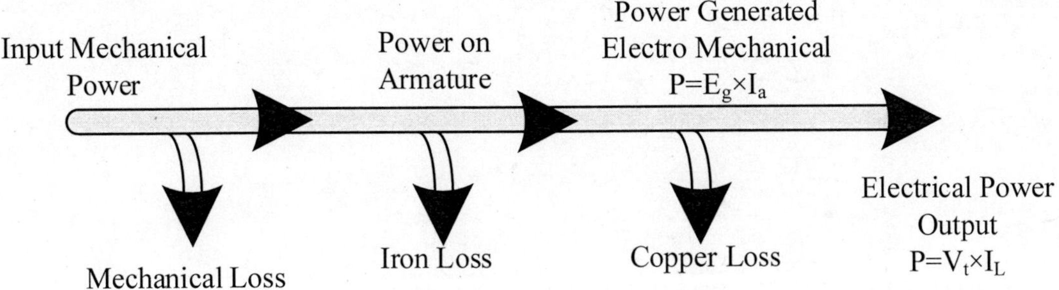

Fig. 4.22 Power flow diagram of DC generator

4.16 Efficiency of DC Generator

The *efficiency* of a *DC generator* is defined as the ratio of output electrical power to mechanical input power.

$$\text{Efficiency}, \eta = \frac{Electrical\ Power\ Output}{Mechanical\ Power\ Input} \tag{4.58}$$

Where,

Power output $= V_t I_L\ watt$

Power Input = Power Output + Variable Losses + Constant Losses

Since the shunt field current I_{sh} is very small as compared to line current, therefore,

$$I_L \cong I_a\ (neglecting\ I_{sh}) \tag{4.59}$$

$$Variable\ losses = I_L^2 R_a \tag{4.60}$$

$$Constant\ losses = P_c \tag{4.61}$$

$$Terminal\ Voltage = V_t \tag{4.62}$$

Then, the power input

$$= V_t I_L + I_L^2 R_a + P_c \tag{4.63}$$

$$\eta = \frac{V_t I_L}{V_t I_L + I_L^2 R_a + P_c} \tag{4.64}$$

Condition for maximum efficiency

From equation (4.64)

$$\eta = \frac{1}{1 + \frac{I_L R_a}{V_t} + \frac{P_c}{V_t I_L}} \tag{4.65}$$

Now, efficiency will be maximum when the denominator of eq. (4.65) in minimum i.e.,

$$\frac{d}{dI_L}\left(1 + \frac{I_L R_a}{V_t} + \frac{P_c}{V_t I_L}\right) = 0 \tag{4.66}$$

$$\frac{R_a}{V_t} - \frac{P_c}{V_t I_L^2} = 0 \tag{4.67}$$

$$\frac{R_a}{V_t} = \frac{P_c}{V_t I_L^2} \tag{4.68}$$

$$I_L^2 R_a = P_c \tag{4.69}$$

i.e., Variable losses = constant losses

Hence, the efficiency of a DC machine will be maximum when the line current is such that the constant loss is equal to the variable loss.

The line current at which the efficiency of DC machine is maximum.

$$I_L = \sqrt{\frac{P_c}{R_a}} \tag{4.70}$$

4.17 Application of DC Generator

The application of DC generator based on their characteristic are given below:

I. **Separately excited DC generators:** These generators are more costly than self-excited generators as they require a separate source for their field excitation. But their response to the change in field resistance is quicker and more precise. Therefore, these are employed where the quick and definite response to control is important such as the Ward–Leonard System of speed control. A separately excited DC generator connected to a resistive load has its counter torque proportional to speed. In lab electrics, it is frequently used for determining torque-speed characteristics, and power-speed characteristics of drive motors and engines.

II. **Shunt-wound DC generators:** As they provide constant terminal voltage, they are best suited for battery charging. Along with field regulators, they are also used for light and power supply purposes.

III. **Series-wound DC generators:** The developed voltage drops to zero when the machine is saturated. These generators have very few applications. They can be employed as series boosters for increasing DC voltage across the DC feeders.

IV. **Compound-wound DC generators**

- ***Over-compounded type:*** These are more suited for lighting and power services, as they compensate for the voltage drop in the lines, and voltage at the terminals of the load remains constant.
- ***Differential-compounded type:*** They are normally employed as arc welding sets. In such cases, the generator is practically short-circuited every time the electrode touches the metal plates to be welded for initiating the arc. The fall in voltage helps in limiting current when initially electrodes are short-circuited.

Example.4.8 A DC generator is connected to 230 V DC mains. The current delivered by the generator to the mains is 150 A. The armature resistance is 0.2 ohm. The generator is driven at a speed of 800 rpm Calculate (i) the induced emf (ii) the electromagnetic torque (iii) the mechanical power input to the armature neglecting iron, winding and friction losses, (iv) Electrical power output from the armature, (v) armature copper loss.

Solution: -

The induced emf,

$$E_g = V + I_a R_a = 230 + (150 \times 0.2) = 260\ V\ (Ans)$$

Using the relation, $\omega T = E_g I_a$

Electromagnetic torque,

$$T = \frac{E_g I_a}{\omega} = \frac{E_g I_a}{2\pi N} \times 60 \qquad \because \omega = \frac{2\pi N}{60}$$

$$= \frac{260 \times 150 \times 60}{2\pi \times 800} = 465.52\ N\text{-}m\ (Ans)$$

Neglecting iron, winding and friction losses,

Input to armature $= \omega T\ or\ (E_g I_a)$

$$= \frac{2\pi\ NT}{60} = \frac{2\pi \times 800 \times 465.52}{60} = 38999.31\ W\ (Ans)$$

Electrical power output $= VI_a = 230 \times 150 = 34500\ W\ (Ans)$

Armature copper losses $= I_a^2 R_a = (150)^2 \times 0.2 = 4500\ W\ (Ans)$

Example.4.9 A shunt generator supplies 150 A at 230 V. Armature resistance is 0.03-ohm, shunt field resistance is 56 ohms. If the iron and friction losses amount to 1400 watts, find (i) emf generated; (ii) copper losses; (iii) b.h.p. of the engine driving the generator.

Solution: -

Shunt field current,

$$I_{sh} = \frac{V}{R_{sh}} = \frac{230}{56} = 4.10\ A$$

Armature current,

$$I_a = I_L + I_{sh} = 150 + 4.10 = 154.10\ A$$

Generated or induced emf,

$$E_g = V + I_a R_a = 230 + (154.10 \times 0.03) = 234.62\ V\ (Ans)$$

Armature copper loss;

$$= I_a^2 R_a = (154.10)^2 \times 0.03 = 712.40\ W$$

Shunt field copper loss;

$$= I_{sh}^2 R_{sh} = (4.10)^2 \times 56 = 941.36\ W$$

Total copper losses;

$$= 712.40 + 941.36 = 1653.76\ W\ (Ans)$$

Output power;

$$= VI_L = 230 \times 150 = 34500\ W$$

Input power;

$$= 34500 + 1400 + 1653.76 = 37553.76\ W$$

B.H.P. of the engine driving the generation

$$= \frac{37553.76}{735.5} = 51.05\ H.P.\ (Ans)$$

Example.4.10 A long shunt dynamo running at 1500 rpm supplies 30 Kw at a terminal voltage of 230 V. The resistance of armature, shunt, and series field is 0.05, 130 and 0.03 ohms respectively. Overall efficiency at the above load is 80%. Find:

a. **Copper loss**
b. **Iron and friction loss**
c. **Torque developed by the prime mover**

Solution: -

$$I_{L,}\ Load\ current = \frac{30000}{230} = 130.43\ Amp$$

$$I_{f,}\ Shunt\ field\ current = \frac{230}{130} = 1.76\ Amp$$

$$I_{a,}\ Armature\ current = 130.43 + 1.76 = 132.19\ Amp$$

Input power = 30,000/ 0.80 = 37500 watts

Total losses in the machine = Input – Output
= 37500 – 30000 = 7500 watts

i. Copper losses:

Power loss in series field-winding + armature winding

$$= 132.19^2 \times 0.08\ Watts = 1397.93\ watts$$

Power loss in shunt field circuit

$$= 1.76^2 \times 130 = 402.68\ Watts$$

Total copper loss = 1397.93 + 402.68 = 1800.68 watts

ii. Iron and friction losses = Total losses – Copper losses

$$= 7500 - 1800.6 = 5699.39\ Watts$$

iii. Let T = Torque developed by the prime-mover

At 1500 r.p.m., angular speed,

$$\omega = 2\pi \times \frac{1500}{60} = 157.07\ rad./sec$$

$$T \times \omega = Input\ power$$

$$T = \frac{37500}{157.07} = 238.74\ N\text{-}m\ (Ans.)$$

Example.4.11 A 10 Kw, 250 V, d.c., 6-pole shunt generator runs at 1000 r.p.m. when delivering full-load. The armature has 534 lap connected conductor. Full load Cu loss is 0.64 Kw. The total brush drop is 1 volt. Determine the flux per pole. Neglect shunt current.

Solution: - Since shunt current is negligible, there is no shunt Cu loss. The copper loss occurs in armature only.

$$I = I_a = \frac{10{,}000}{250} = 40\,A\,;$$

$$I_a^2 R_a = Arm.\,Cu\,loss;\; 40^2 \times R_a = 0.64 \times 10^3 \;\therefore R_a = 0.4\,\Omega$$

$$I_a R_a\;drop = 0.4 \times 40 = 16\,V$$

$$Brush\;drop = 2 \times 1 = 2\,V$$

$$Generated\;emf, E_g = 250 + 16 + 1 = 267\,V$$

$$\text{Now}, E_g = \frac{\emptyset ZN}{60}\left(\frac{P}{A}\right) Volt\,;\; 267 = \frac{\emptyset \times 534 \times 1000}{60}\left(\frac{6}{6}\right)$$

$$\emptyset = 30 \times 10^{-3}\,Wb = 30\,mWb$$

Example.4.12 A long-shunt compound-wound generator gives 240 volts at F.L. output of 100 A. The resistance of various windings of the machine are: armature (including brush contact) 0.1 Ω series field 0.02 Ω interpole field 0.025 Ω shunt field (including regulating resistance) 100 Ω. The iron loss at F.L is 1000 W; windage and friction losses total 500 W. Calculate F.L efficiency of the machine.

Solution:

$$\text{Output} = 240 \times 100 = 24{,}000\,W$$

$$\text{Total armature circuit resistance} = 0.1 + 0.02 + 0.025 = 0.145\,\Omega$$

$$I_{sh} = \frac{240}{100} = 2.4\,A$$

$$I_a = 100 + 2.4 = 102.4\,A$$

$$\text{Armature circuit copper loss} = 102.4^2 \times 0.145 = 1521\,W$$

$$\text{Shunt field copper loss} = 2.4 \times 240 = 576\,W$$

$$\text{Iron loss} = 1000\,W;\; Friction\;loss = 500\,W$$

$$\text{Total loss} = 1521 + 1500 + 576 = 3597\,W$$

$$\eta = \frac{24000}{24000 + 3597} = 0.87 = 87\%$$

Example.4.13 A shunt generator has full load current of 196 A at 220 V. The stray losses are 720 W and the shunt field coil resistance is 55 Ω. If it has a full load efficiency of 88%, (i) find the armature resistance and load current corresponding to maximum efficiency. (ii) find the total loss and load current to maximum efficiency when full load efficiency of 80%

Solution:

$$\text{Output} = 220 \times 196 = 43120\ W$$

$$\eta = 88\%\ (Overall\ efficiency)$$

$$\text{Input} = \frac{43120}{0.88} = 49000\ W$$

$$\text{Total losses} = 49000 - 43120 = 5880\ W$$

$$\text{Shunt field current} = \frac{220}{55} = 4A$$

$$I_a = 196 + 4 = 200A$$

$$\text{Shunt Cu loss} = 220 \times 4 = 880\ W;\ Stray\ losses = 720\ W$$

$$\text{Constant losses} = 880 + 720 = 1600$$

$$\text{Armature Cu loss} = 5880 - 1600 = 4280\ W$$

$$I_a^2 R_a = 4280\ W;\ 200^2 R_a = 4280$$

$$R_a = \frac{4280}{200^2} = 0.107\ \Omega$$

For maximum efficiency,

$$I_a^2 R_a = constant\ losses = 1600\ W$$

$$I_L = \sqrt{1600/0.107} = 122.34\ Amp$$

(ii) When efficiency is 80%

$$\text{Input} = \frac{43120}{0.80} = 53900\ W$$

$$\text{Total losses} = 53900 - 43120 = 10780\ W$$

$$\text{Armature Cu loss} = 10780 - 1600 = 9180\ W$$

$$R_a = \frac{9180}{200^2} = 0.2295\ \Omega$$

For maximum efficiency,

$$I_L = \sqrt{1600/0.2295} = 83.49\ Amp$$

Example.4.14 The hysteresis and eddy current losses in a DC machine running at 1000 rpm. are 250 W and 100 W respectively. If the flux remains constant, at what speed will total iron losses be halved?

Solution:

$$\text{Total loss } W = W_h + W_e = AN + BN^2$$

Now,

$$W_h = 250\ W$$

$$A \times \left(\frac{1000}{60}\right) = 250$$

$$A = 15$$

$$W_e = 100\ W$$

$$B \times \left(\frac{1000}{60}\right)^2 = 100$$

$$B = 9/25$$

Let N be the new speed in rps at which total loss in one half of the loss at 1000 rpm.

Hence new loss=

$$= \frac{250 + 100}{2} = 175\ W$$

$$175 = 15\ N + \left(\frac{9}{25}\right) N^2 \ or\ 9N^2 + 375\ N - 4375 = 0$$

$$\text{N} = \frac{-375 \pm \sqrt{375^2 + 36 \times 4375}}{2 \times 9}$$

$$= \frac{-375 \pm 546}{18}$$

$$= 9.5\ rps$$

$$= 570\ rpm$$

Note: It may be noted that at the new speed

$$W_h = 250 \times \left(\frac{570}{100}\right)^2 = 142.5W$$

$$W_e = 100 \times \left(\frac{570}{1000}\right)^2 = 32.5\ W.$$

$$Total\ loss = 142.5 + 32.5 = 175\ W.$$

4.18 Types of DC Motor

Similar to DC generator, the DC motor can also be classified as

- Permanent Magnet DC motor
- Separately Excited DC motor
- Series Wound DC Motor
- Shunt Wound DC motor
- Compound Wound DC Motor

Scan to learn about DC Motor working

4.18.1 Permanent Magnet DC Motor

A Permanent magnet DC (PMDC) motor is a DC motor whose poles are made of permanent magnet i.e., field flux required by in the air gap is developed by set of permanent magnets fixed on the stator. The permanent magnets of a PMDC motor are radially magnetized and mounted on the inner periphery of the cylindrical steel stator. The stator also serves as a return path for the magnetic flux. Field coils are usually not required. However, some of these motors do have coils wound on the poles. If they exist, these coils are intended only for recharging the magnets in the event they lose their strength. The rotor of this motor is similar to the rotor of a conventional DC motor i.e., rotor of a PMDC motor consists of armature core, armature windings, and commutator. Stationary carbon brushes are kept pressed on the commutator surface as in a conventional DC motor. The material used for permanent magnets is ceramics and rare earth magnetic material which have high residual flux as well as high coercivity. A 2-pole PMDC motor is shown in fig. 4.23.

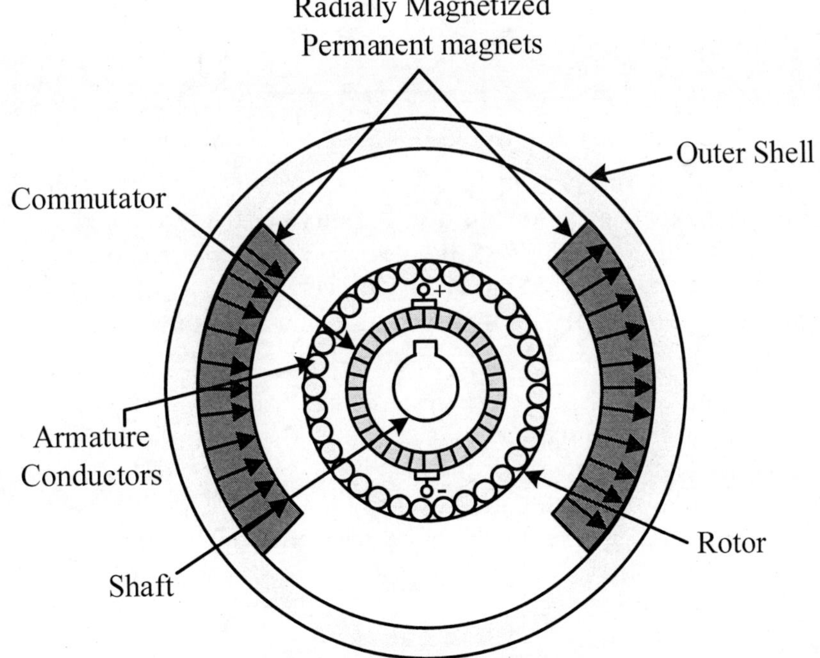

Fig. 4.23 2-pole PMDC motor

Small fractional and sub-fractional kW motors are now constructed with permanent magnet poles. Most of the PMDC motors operate on 6 V, 12 V or 24 V DC supply obtained from batteries or rectifiers. The torque is developed by interaction between current-carrying rotor conductors and the magnetic flux set up by the permanent magnets. Ferrite magnets have comparatively lesser remanent flux density as that of rare earth magnets, however, they have larger coercivity.

The equivalent circuit of a PMDC motor is shown in fig. 4.24. Since the field flux in a PMDC motor is developed by permanent magnets, the field winding is not shown in the equivalent circuit.

The major advantage of PMDC motors is that they do not have any field winding, so they require no excitation current. Thus, there is no continuous loss of energy in the field. Because of absence of field windings, there is a saving in space and PMDC motors are smaller in size and cheaper in cost in comparison to corresponding rated conventional DC motors.

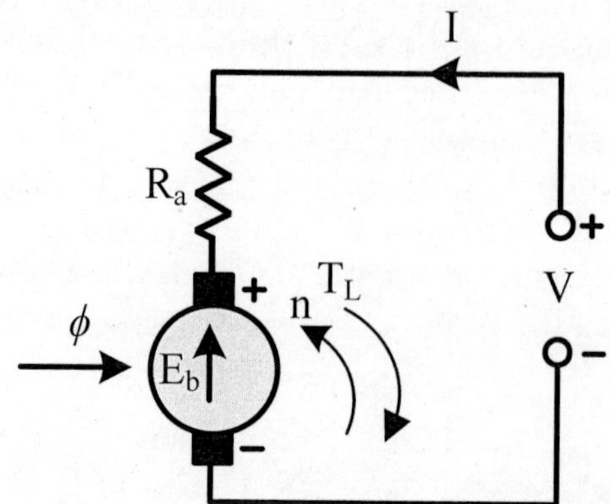

Fig. 4.24 Equivalent Circuit of PMDC Motor

The limitation of PMDC motors is that the excessive current in the armature winding may demagnetize the permanent magnets and also the flux density produced in the air gap by permanent magnets is limited. The risk of permanent magnetism getting destroyed by armature reaction (at starting/reversing or heavy overloads) has been greatly reduced by the new permanent magnetic materials like samarium-manganese-germanium and neodymium-iron-boron.

These motors offer shunt-type characteristics and can only be armature controlled i.e., in PMDC motors, speed and torque are controlled through the adjustment of voltage applied to the motor terminals, armature rheostat control and chopper control. PMDC motors are used extensively in automobiles as starter motors and for windshield wipers and washers, for blowers used in heaters and air-conditioners, to raise and lower windows, in slot cars and electric tooth brushes etc. In a Brushless PMDC motor, the commutator is replaced by electronic circuit (commutator) which Switches ON/OFF torque-producing coils. Switching is synchronized with rotor position. The rotor position is sensed using either Hall Effect sensor and Latch IC (Integrated Circuit) or resolver.

4.18.2 Separately Excited DC Motor

These motors have field coils similar to those of a shunt wound machine. But the armature and field coils are fed from different supply sources, as shown in fig. 4.25 and may have different voltage ratings.

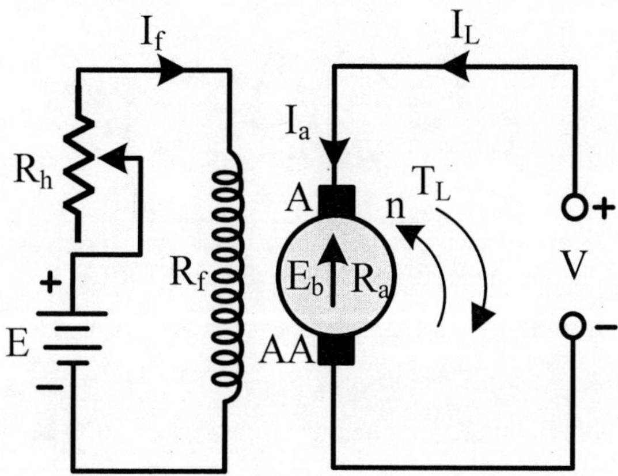

Fig. 4.25 Separately Excited DC Motor

In a separately excited DC motor,

Armature Current, I_a=Line Current, $I_L = I$ (4.71)

Back emf developed, $E_b = V - IR_a$ (4.72)

Power drawn from Mains supply, P=VI where V is supply voltage (4.73)

Mechanical Power developed,

P_m = Power input to armature−Power lost in armature

$= VI - I^2R_a$

$= I(V - IR_a)$

$P_m = E_bI$ (4.74a)

Torque, $T = P_m/\omega$ (4.74b)

4.18.3 Series Wound DC Motor

In series wound DC motor field winding is connected in series (R_{se}) with the armature as shown in fig. 4.26. The series winding consists of few turns of thick wire. The cross-sectional area of the wire used for field coil has to be fairly large to carry the armature current, but owing to the higher current, the number of turns of wire in them need not be large.

In a DC series motor Armature Current

Armature Current, I_a= Series field Current, $I_{sc} = I_a = I$ (4.75)

Back emf developed, $E_b = V - I(R_a + R_{se})$ (4.76)

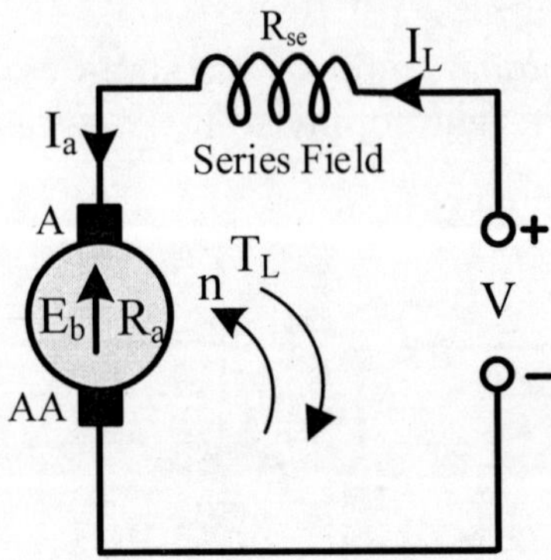

Fig. 4.26 Series Wound DC Motor

Power drawn from Mains supply, P=VI where V is supply voltage (4.77)

Mechanical Power developed,

P_m = Power input to armature−Power lost in armature

$= VI - I^2(R_a + R_{se})$

$= I(V - IR_a - IR_{se})$

$P_m = E_b I_a$ (4.78*a*)

Torque, $T = P_m/\omega$ (4.78*b*)

4.18.4 Shunt Wound DC Motor

In shunt wound DC motor the field windings are connected in parallel with the armature. The field winding consists of large number of turns of comparatively fine wire to provide large resistance. The field current much be less than the armature current. Sometimes as low as 5%. The connection diagram of shunt wound DC motor is shown in fig. 4.27, the current supplied through the motor is given according to KCL i.e.

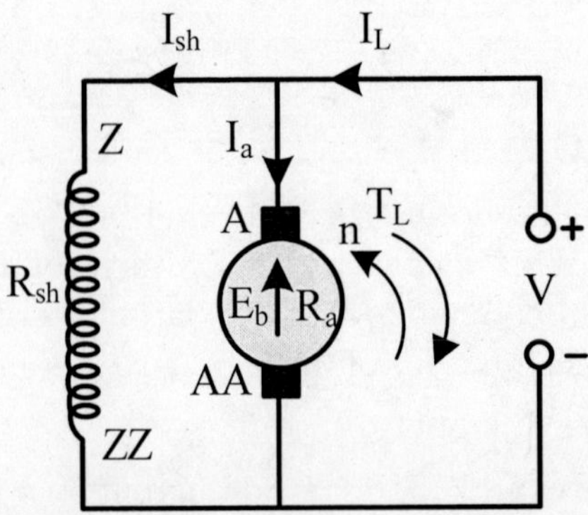

Fig. 4.27 Shunt Wound DC Motor

Input line current, $I_L = I_a + I_{sh}$ (4.79)

Where, I_a is armature current and I_{sh} is the shunt field winding current and is given by the expression

$$I_{sh} = \frac{V}{R_{sh}} \quad (4.80)$$

Where, V is the supply voltage and R_{sh} is the shunt field circuit resistance.

Back emf developed, $E_b = V - I_a R_a$ (4.81)

Power drawn from Mains supply, $P = VI_L$ where V is supply voltage (4.82)

Mechanical Power developed,

P_m = Power input to armature−Power lost in armature and shunt field

$$= VI_L - VI_{sh} - I_a^2 R_a$$

$$= V(I_L - I_{sh}) - I_a^2 R_a$$

$$= VI_a - I_a^2 R_a$$

$$= I_a(V - I_a R_a)$$

$$P_m = E_b I_a \quad (4.83a)$$

$$\text{Torque, T} = P_m/\omega \quad (4.83b)$$

4.18.5 Compound wound DC Motor

A compound wound DC motor has both shunt and series field windings. The shunt field is normally having more amp-turn. Compound wound motor are two types namely cumulative compound wound and differential compound wound motor.

- **Cumulative Compound Wound DC Motor**

Cumulative compound motor is one in which the field winding is connected in such a way that the direction of flow of current is the same in both the field windings, as shown in fig. 4.28.

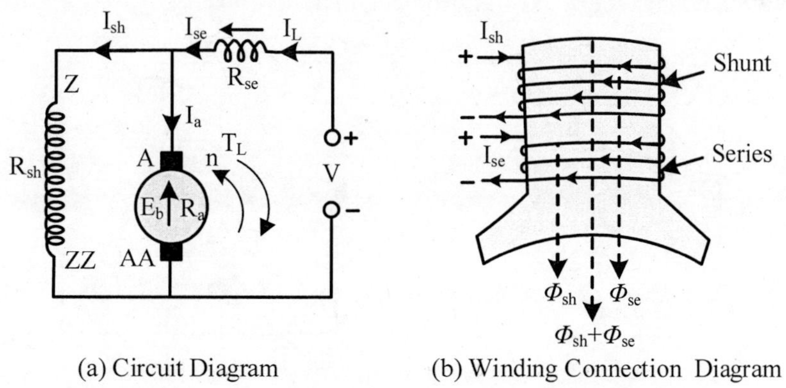

(a) Circuit Diagram (b) Winding Connection Diagram

Fig. 4.28 Cumulative compound wound DC Motor

In this type of motor flux due to series field winding strengthens the field due to the shunt field winding. In these motors, the flux produced by both the windings is in the same direction, i.e.,

$$\phi_r = \phi_{sh} + \phi_{se} \tag{4.84}$$

- **Differential Compound Wound DC Motor**

In differential compound wound DC Motor the field windings are connected in such a way that the direction of flow of current is opposite to each other in the two field windings, as shown in fig. 4.29.

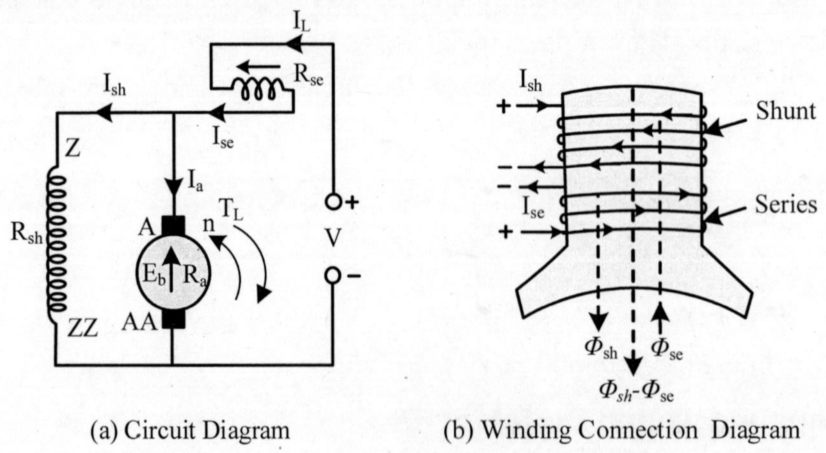

(a) Circuit Diagram (b) Winding Connection Diagram

Fig. 4.29 Differential compound wound DC Motor

Compound wound DC motors, like compound wound DC generators, may be either long-shunt connected or short-shunt connected as shown in fig. 4.30. In long shunt-connected compound wound motors, series field, and armature are connected in series with each other and in parallel with the shunt field. In short shunt compound wound motors, the armature and shunt field are in parallel with each other and the pair is in series with the series field. Long shunt connection sometimes results in simpler wiring. Changing from long to short shunt, or vice-versa, has little effect on motor performance.

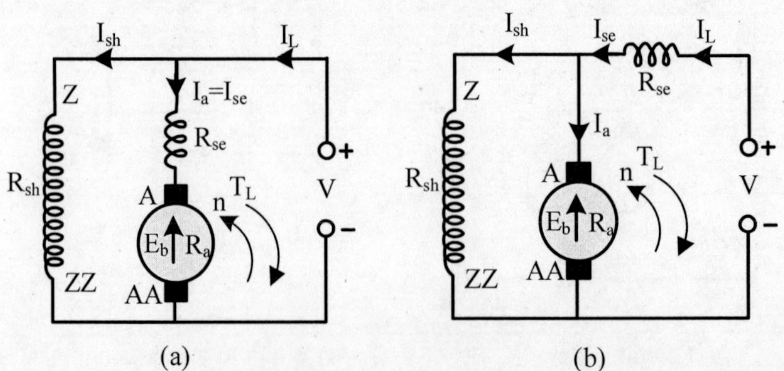

(a) (b)

Fig. 4.30 Compound Wound DC Motor (a) Long Shunt (b) Short Shunt

4.19 Armature Torque of a Motor

The electrical power which is supplied to a DC motor is converted into mechanical power. If T_a be the gross torque developed by the armature rotating at a speed N rpm, the mechanical power (P_m) developed is expressed as follows.

$$P_m = T_a \times \omega \tag{4.85}$$

$$P_m = E_b \times I_a \tag{4.86}$$

From equations 4.85 and 4.86, we have

$$E_b \times I_a = T_a \times \omega \tag{4.87}$$

Where, E_b back emf given by equation.

$$E_b = \frac{P\phi NZ}{60A} \tag{4.88}$$

$$and\ speed, \omega = \frac{2\pi N}{60} \tag{4.89}$$

From expressions (4.87, (4.88), and (4.89)

$$\frac{P\phi NZ}{60A} \times I_a = T_a \times \frac{2\pi N}{60} \tag{4.90}$$

$$T_a = \frac{P\phi ZI_a}{2\pi A} \tag{4.91a}$$

For a particular machine, the number of poles (P) and the number of conductors per parallel path (Z/A) are constant.

$$T_a = K\phi I_a, \qquad where\ K = PZ/2\pi A(Constatnt)$$

$$T_a = K\phi I_a \tag{4.91b}$$

From the above equation of torque, we find that $T_a \propto \phi I_a$

a. In case of series motor, ϕ is directly proportional to I_a (before saturation) because field winding carries full armature current. Hence,

$$T_a \propto I_a^2 \tag{4.92}$$

b. For shunt motor ϕ is practically constant, Hence

$$T_a \propto I_a \tag{4.93}$$

4.20 Shaft Torque and Brake Horse Power

The whole of the armature torque, as calculated above, is not available for doing useful work. because a certain percentage of it is required for supplying iron and friction losses in the motor. In DC motor the torque which is available for doing useful work is known as shaft torque T_{sh}. The shaft torque (T_{sh}) is somewhat less than the torque developed in the armature.

$$T_{sh} = \frac{Output\ in\ watt}{Speed\ in\ radians\ per\ second} \tag{4.94}$$

$$T_{sh} = \frac{E_b I_a}{2\pi N/60} \quad N\text{-}m \tag{4.95}$$

$$T_{sh} = \frac{60 \times E_b I_a}{2\pi N} \quad N\text{-}m \tag{4.96}$$

$$T_{sh} = 9.55 \times \frac{E_b I_a}{N} \tag{4.97}$$

The difference between the armature torque and the shaft torque $(T_a - T_{sh})$ is known as the lost torque and is due to the formation of the torque. The load torque T_L opposes the developed electromagnetic torque on the motor.

Brake Horse Power

In the case of the motor, the mechanical power available at the shaft is known as Brake Horse Power. If T_{sh} is the shaft torque in Newton Meter and N is the speed in rpm then,

$$\text{Useful output power} = \omega T_{sh} \tag{4.98}$$

$$= \frac{2\pi N \times T_{sh}}{60} \tag{4.99}$$

$$output\ in\ B.H.P = \frac{2\pi N \times T_{sh}}{60 \times 745.7} \tag{4.100}$$

Example.4.15 The armature resistance of DC shunt motor of 0.3 ohm, it draws 30 A from 230 V supply and is running at a speed of 100 radian per second. Determine induced emf, Electromagnetic torque and speed in rpm and rps.

Solution:

Here, V=230V, $I_a = 30\ A, R_a = 0.3\ ohm, \omega = 100\ rad/sec$

Induced emf,

$$E = V - IR_a = 230 - (30 \times 0.3) = 221\ V\ (Ans.)$$

Electromagnetic torque,

$$T = \frac{EI_a}{\omega} = \frac{221 \times 30}{100} = 66.3\ N\text{-}m\ (Ans.)$$

Speed in rpm,

$$N = \frac{60\omega}{2\pi} = \frac{60 \times 100}{2\pi} = 955.4\ rpm\ (Ans.)$$

Speed in rps,

$$N = \frac{\omega}{2\pi} = \frac{100}{2\pi} = 15.91\ rps\ (Ans.)$$

Example.4.16 A D.C. motor takes an armature current of 110 A at 480 V. The armature circuit resistance is 0.2 Ω. The machine has 6-poles and the armature is lap-connected with 864 conductors. The flux per pole is 0.05 Wb. Calculate the speed and the gross torque developed by the armature.

Solution: -

$$E_b = 480 - 110 \times 0.2 = 458\,V,$$

$$\emptyset = 0.05\,Wb, \qquad Z = 864$$

Now,

$$E_b = \frac{P\emptyset\,NZ}{60A}$$

$$458 = \frac{6 \times 0.05 \times 864 \times N}{60 \times 6}$$

$$N = 636\,rpm$$

$$T_a = \frac{P\emptyset\,ZI_a}{2\pi A} = \frac{6 \times 0.05 \times 864 \times 110}{2\pi \times 6} = 756.30\,N\text{-}m$$

Example.4.17 A 230 V, 4-pole, wave-wound DC series motor has 782 conductors on its armature. It has armature and series field resistance of 0.75 ohm. The motor takes a current of 40 A. Estimate its speed and gross torque developed if it has a flux per pole of 25 mWb.

Solution: -

$$E_b = V - I_a R_a$$

$$E_b = 230 - 40 \times 0.75 = 200\,V$$

$$E_b = \frac{P\emptyset\,NZ}{60A}$$

$$200 = \frac{4 \times 25 \times 10^{-3} \times N \times 782}{60 \times 2}$$

$$N = 306\,rpm$$

Gross torque or armature torque

$$T_a = \frac{P\phi Z I_a}{2\pi A} = \frac{4 \times 25 \times 10^{-3} \times 782 \times 40}{2\pi \times 2}$$

$$N = 248.91\,rpm$$

Example.4.18 A D.C shunt machine develops an emf of 250 V at 1500 rpm. Find its torque and mechanical power developed for an armature current of 50 A. State the simplifying assumptions.

Solution: - A given D.C. machine develops the same emf in its armature conductor whether running as a generator or as a motor. Only difference is that this armature emf is known as back emf. when the machine is running as a motor.

Mechanical power developed in the armature

$$= E_b I_a$$

$$= 250 \times 50 = 12{,}500\ W$$

$$T_{sh} = 9.55\ E_b I_a / N$$

$$= 9.55 \times 250 \times \frac{50}{1500}$$

$$= 79.6\ N - m$$

Example.4.19 A cutting tool exerts a tangential force of 400 N on a steel bar of diameter 10 cm which is being turned in a simple lathe. The lathe is driven by a chain at 840 r.p.m. from a 220 V d.c. Motor Which runs at 1800 r.p.m. Calculate the current taken by the motor if its efficiency is 80%. What size is the motor pulley if the lathe pulley has a diameter of 24 cm?

Solution:

$$T_{sh} = Tangential\ force \times radius$$

$$= 400 \times 0.05 = 20\ N\text{–}m$$

$$\text{Output power} = T_{sh} \times \frac{2\pi N}{60}\ watt$$

$$= 20 \times 2\pi \times \left(\frac{840}{60}\right)\ watt = 1.760\ kW$$

$$\text{Motor } \eta = 0.8$$

$$\text{Motor input} = 1{,}760/0.8 = 2{,}200\ W$$

$$\text{Current drawn by motor} = 2200/220 = 10\ A$$

Let N_1 *and* D_1 be the speed and diameter of the driver pulley respectively and N_2 *and* D_2 the respective speed and diameter of the lathe pulley.

$$\text{Then } N_1 \times D_1 = N_2 \times D_2\ or\ 1{,}800 \times D = 840 \times 0.24$$

$$D_1 = 840 \times 0.24/1{,}800 = 0.112\ m = 11.2\ cm$$

Example.4.20 A 220-V DC shunt motor runs at 500 rpm. when the armature current is 50 A. Calculate the speed if the torque is doubled. Given that $R_a = 0.2\ \Omega$.

Solution: - As seen from the torque equation, $T_a \propto \emptyset I_a\ , since\ \emptyset\ is\ constant, T_a \propto I_a$

$$T_{a1} \propto I_{a1}\ and\ T_{a2} \propto I_{a2}$$

$$T_{a2}/T_{a1} = I_{a2}/I_{a1}$$

$$2 = I_{a2}/50\ or\ I_{a2} = 100\ A$$

Now, $N_2/N_1 = E_{b2}/E_{b1}$

Since $\emptyset$ remains constant.

$$E_{b1} = 220 - (50 \times 0.2) = 210\ V$$

$$E_{b2} = 220 - (100 \times 0.2) = 200\ V$$

$$N_2/500 = 200/210$$

$$N_2 = 476\ r.p.m.$$

Example.4.21 The hysteresis and eddy current losses in a d.c. machine running at 1000 rpm are 250 W and 100 W respectively. If the flux remains constant, at what speed will be total iron losses be halved?

Solution:

$$\text{Total loss } W = W_h + W_e = AN + BN^2$$

$$W_h = 250\ W \therefore A \times \left(\frac{1000}{60}\right) = 250; A = 15$$

$$W_e = 100\ W \therefore B \times \left(\frac{1000}{60}\right)^2 = 100\ ; B = 9/25$$

Let N be the new speed in rps at which total loss in one half of the loss at 1000 r.p.m.

$$\text{New loss} = \frac{250 + 100}{2} = 175\ W$$

$$175 = 15\ N + \left(\frac{9}{25}\right) N^2;\ \ 9N^2 + 375\ N - 4375 = 0$$

$$\text{N} = 9.5\ rps = 570\ rpm$$

It may be noted that at the new speed

$$W_h = 250 \times \left(\frac{570}{1000}\right) = 142.5W\ and\ W_e = 100 \times \left(\frac{570}{1000}\right)^2 = 32.5\ W$$

$$Total\ loss = 142.5 + 32.5 = 175\ W$$

4.21 Characteristic of DC Motor

The performance of a DC motor can be easily referred from its *characteristic* curves. The characteristics of a motor show relation between the two quantities. The following characteristics can be obtained:

a. **Speed (N) and Armature current (I_a) Characteristics:** It is the curve drawn between speed N and armature current I_a. It is also known as *speed characteristics.*
b. **Torque (T) and Armature current (I_a) Characteristics:** It is the curve drawn between torque developed in the armature T and armature current I_a. It is also known as *electrical characteristic.*
c. **Speed (N) and Torque (T) characteristics:** It is the curve drawn between speed N and torque developed in the armature T. It is also known as *mechanical characteristics.*

The following relations are important while discussing motor characteristics:

Electromagnetic Torque $T \propto \phi \, I_a$. (4.101)

Back emf $E_b \propto N\phi$ (4.102)

Speed, $N \propto \frac{E_b}{\phi}$ (4.103)

4.22 Characteristic of DC Series Motor

In DC series motor the series field winding carries the armature current as shown in fig. 4.31 Therefore, the flux produced by the series field winding is proportional to the armature current before magnetic saturation. After magnetic saturation flux becomes constant.

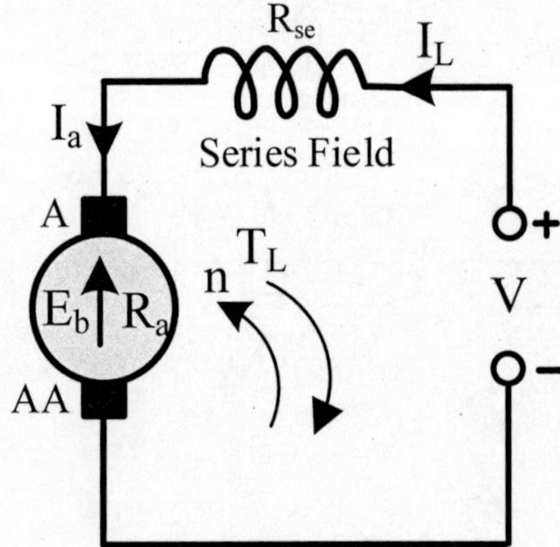

Fig. 4.31 Circuit diagram of DC series motor

a. Speed (N) and Armature current (I_a) Characteristics:

In case of DC series motor, the mmf due to the field coil increases in direct proportion to the line current or armature current. Hence the value of flux varies with load current according to the ordinary magnetization curve. From the speed equation, it is obvious that speed is proportional to back emf E_b and inversely proportional to flux ϕ. The motor speed N is given by

$$N \propto \frac{E_b}{\phi} \tag{4.104}$$

$$N \propto \frac{V - I_a(R_a + R_{se})}{\phi} \tag{4.105}$$

With the increase in armature current voltage drop in armature circuit and series field $I_a(R_a + R_{se})$ increases and, therefore, back emf E_b decreases. Under normal conditions at a low value of I_a, the voltage drops $I_a(R_a + R_{se})$ are negligibly small in comparison with V. Therefore

$$N \propto \frac{V}{\phi} \tag{4.106}$$

If applied voltage V is constant then speed is inversely proportional to the flux.

$$N \propto \frac{1}{\phi} \tag{4.107}$$

In a series motor, the flux $\emptyset$ is produced by the armature current flowing in the field winding so that $\phi \propto I_a$. Hence the series motor is variable flux machine. Equation 4.107 now becomes

$$N \propto \frac{1}{I_a} \tag{4.108}$$

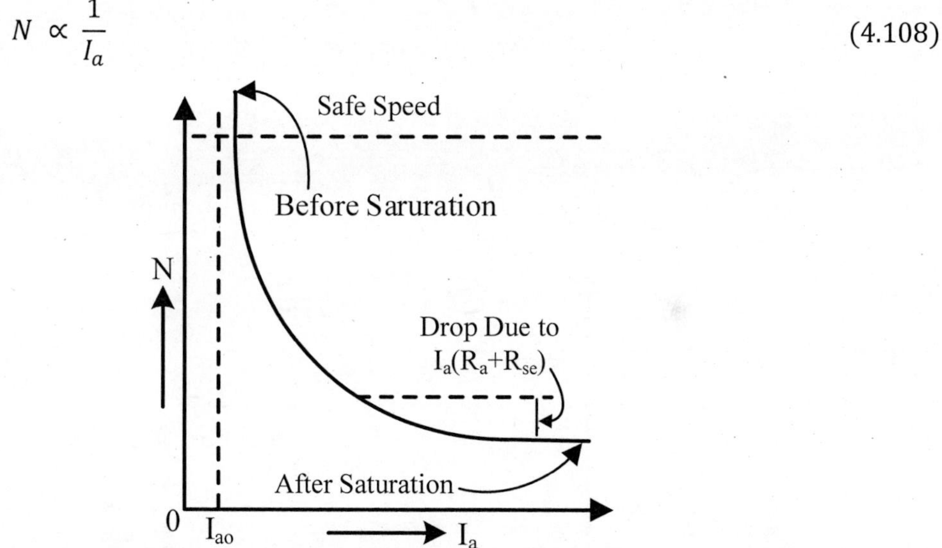

Fig. 4.32 Speed armature current characteristic of DC series Motor

Thus, for the series motor, the speed is inversely proportional to the armature current or load current. The speed-load current characteristic is a rectangular hyperbola before magnetic saturation as shown in fig. 4.32. After magnetic saturation, flux becomes constant, then

$$N \propto E_b \propto V - I_a(R_a + R_{se}) \tag{4.109}$$

Thus, after magnetic saturation, the speed armature-current curve follows a straight-line path and speed decreases slightly as shown in fig. 4.32

Equation (4.108) shows that when the load current is small, the speed will be very large. Therefore, at no load or at light loads there is a possibility of dangerously high speeds, which may damage the motor and its coupling due to large centrifugal forces.
Hence a series motor must never run unloaded. It should always be coupled to a mechanical load either directly or through gearing. It should never be coupled by belt, which may break at any time. With increases in load current, speed decreases.

b. Torque (T) and Armature current (I_a) Characteristics:

From the expression of mechanical torque T, it is obvious that torque is directly proportional to the product of flux per pole ϕ and armature current I_a. Up to saturation point flux is proportional to field current and hence to armature current, because $I_a = I_f$. Therefore, at light load mechanical torque is proportional to the square of armature current.

$$T \propto \phi\, I_a \tag{4.110}$$

In series motor, Before magnetic saturation

$$\phi \propto I_a \tag{4.111}$$

Hence,

$$T \propto I_a^{\,2} \tag{4.112}$$

Therefore, this portion of the curve (OA) is a parabola passing through the origin as shown in fig. 4.33.
However, after magnetic saturation, the flux ϕ becomes constant.

$$T \propto I_a \tag{4.113}$$

Hence, after magnetic saturation, the curve (AB) becomes a straight line.

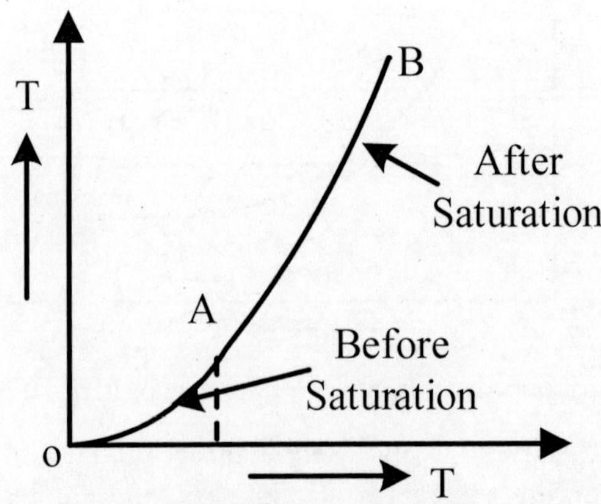

Fig. 4.33 Torque armature current characteristic of DC series motor

It is seen that before magnetic saturation $T \propto I_a^2$. When the load is applied to this motor at start, it takes a large current, and heavy torque is produced which is proportional to the square of this current. Thus, this motor is capable to pick up heavy loads at the start and therefore best suited for electric traction. Although these machines in traction are being replaced by induction and synchronous motor whose torque and speed characteristics are modified similar to that of DC series motor with the help of variable voltage variable frequency (VVVF) inverter.

c. Speed (N) and Torque (T) characteristics:

The speed-torque characteristic of DC series motor is derived from the first two characteristics. At low value of load, I_a is small, torque is small but the speed is very high. As load increases, I_a increases, torque increases but the speed decreases rapidly. Thus, for increased torque, speed decreases rapidly as shown in fig. 4.34.

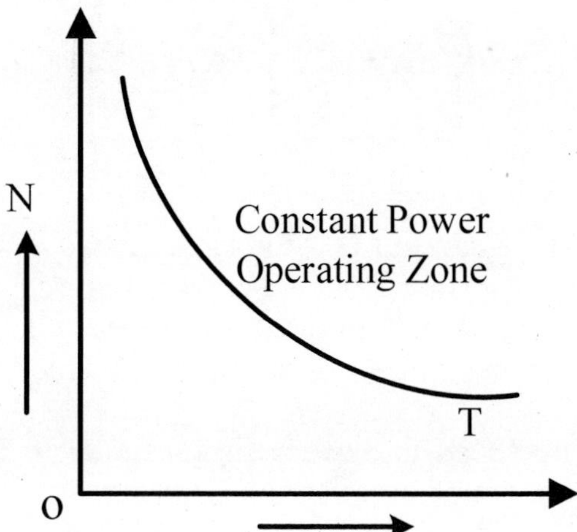

Fig. 4.34 Speed-torque characteristic of DC series motor

This characteristic shows that the D.C. series motor has a high torque at a low speed and a low torque at a high speed. Hence the speed of the D.C. series motor changes considerably with increasing load.

It is a very useful characteristic for traction purposes, hoists and lifts where at low speeds high starting torque is required to accelerate large masses. In the operating zone of speed torque Characteristics of DC series motor, the product of voltage and armature current, i.e., electrical input power remains almost constant. Due to this reason, DC series motors are also referred to as constant power drive.

- DC series motor on AC supply works satisfactorily. Such motors are referred to as universal motor.
- Household mixers, vacuum cleaners etc. are some of its applications. High-power AC series motors require compensating windings.

4.23 Characteristic of DC Shunt or Separately Excited Motor

The circuit diagram of DC shunt motor is shown in fig. 4.35. In these motors, the shunt field current that is given by equation (4.80) i.e., $I_{sh} = V/R_{sh}$ remains constant since the supply voltage V is constant. Hence, the flux in DC shunt motors is practically constant. Somewhat flux may decreases due to armature reaction at heavy load.

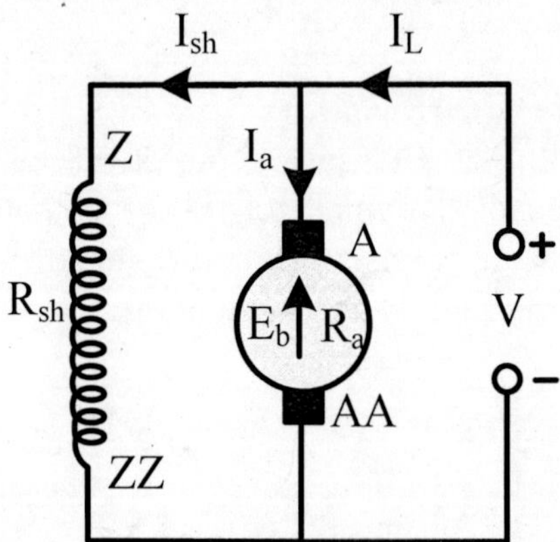

Fig. 4.35 Circuit diagram of shunt motor

a. Speed (N) and Armature current (I_a) Characteristics:

If applied voltage V is kept constant, the field current will remain constant, hence flux will have maximum value on no load but will slightly decreases due to armature reaction, as the load increase. However, for most purposes the flux is considered to be constant, neglecting armature reaction effect.

From speed equation, the speed is directly proportional to back emf E_b and inversely proportional to the flux ϕ as stated in equation (4.88). Since flux is considered to be constant as mentioned above.

$$N \propto E_b \tag{4.114}$$

$$N \propto (V - I_a R_a) \tag{4.115}$$

Equation (4.86) is the equation of a straight line with a negative slope. That is, the speed N of the motor decreases linearly with the increase in armature current as shown in fig. 4.36(a). If the armature voltage drops ($I_a R_a$) is negligible, the speed of the motor will remain constant for all values of load as shown by the dotted line AB.

Since $I_a R_a$ at full load is very small compared to V, the drop in speed from no load to full load is very small. The decrease in N is partially neutralized by a reduction in Ø due to armature reaction. Hence for all practical purposes, the shunt motor may be taken as a constant-speed motor.

Moreover, the characteristic curve does not start from a point of zero armature current because a small current (as shown in Fig. 4.36(b)), no-load armature current I_{a0}, is necessary to maintain rotation of the motor at no-load.

Since there is no appreciable change in the speed of a DC shunt motor from no-load to full load that is why it is considered to be a constant speed motor. This motor is best suited where almost constant speed is required and the load may be thrown off totally and suddenly.

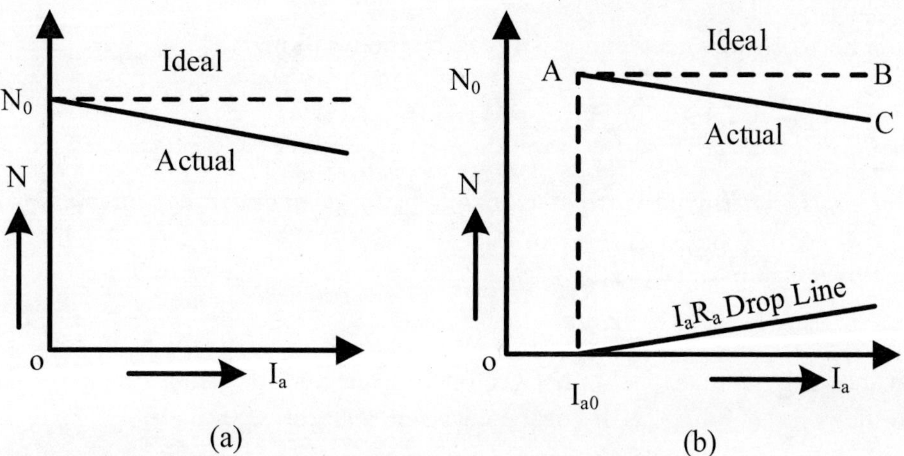

Fig. 4.36 Speed armature current characteristic of (a)shunt (b) separately excited DC motor

b. Torque (T) and Armature current (I_a) Characteristics:

From the equation of torque of DC motor, torque is directly proportional to the product of flux and armature current as stated in equation (4.110). Since in case of DC shunt motor flux is considered to be constant, torque increase with increase in load current (or armature current). The torque-armature current characteristic is a straight line passing through origin as shown in fig. 4.37.

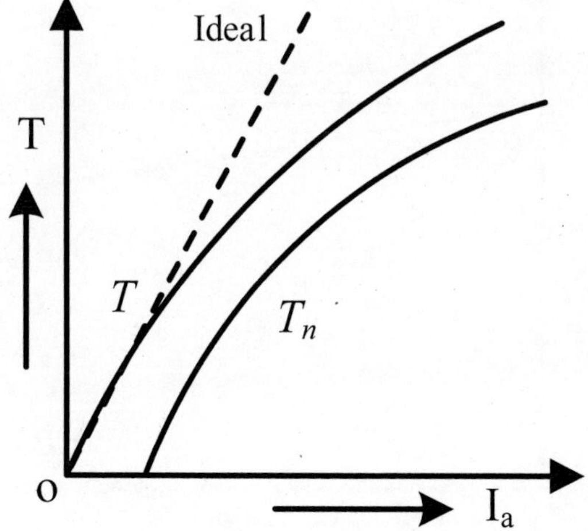

Fig. 4.37 Torque armature current characteristic of a shunt or separately excited DC motor

If the effect of armature reaction is taken into account, the value of ϕ decreases slightly with the increase in armature current. Hence at higher value I_a the gross or total torque T decrease slightly.

The relation between various torques is given by the relation

$$T_n = T - (T_f + T_w) \tag{4.116}$$

$T_n = net\ torque\ or\ useful\ torque\ or\ load\ torque\ at\ the\ output\ shaft$

$T = gross\ or\ total\ torque$

$T_f = frictional\ torque$

$T_w = windage\ torque$

The graph showing the relationship between net torque and the armature current is a curve parallel to the corresponding gross torque curve. It is slightly below it.

It is clear from the characteristic curve that a large armature current is required at the start if machine is on heavy load. Thus, shunt motor should never be started on load.

c. Speed (N) and Torque (T) characteristics:

The speed torque characteristic is derived from the first two characteristics. When load torque increases, armature current I_a increases but speed decreases slightly. Thus, with the increase in load or torque, the speed decreases slightly as shown in fig. 4.38. It causes field failure, field flux drops to zero, shunt motor run at exorbitant high speed. A such situation is to be avoided.

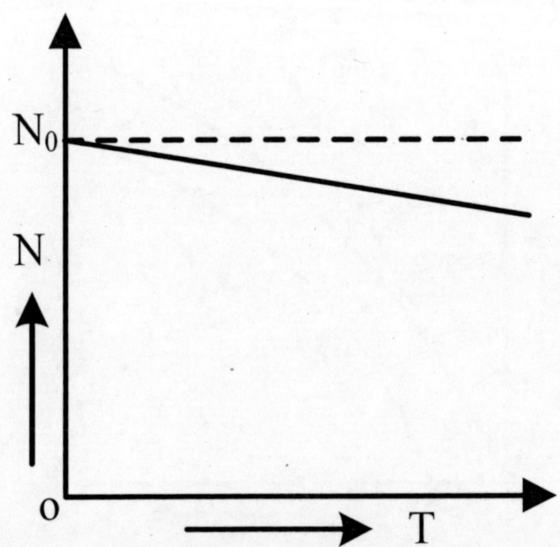

Fig. 4.38 Speed Torque characteristic of a shunt or separately excited DC motor

4.24 Characteristic of DC Compound Motor

A compound motor has both shunt and series field winding, so its characteristics are intermediate between the shunt and series motors. The cumulative compound motor is generally used in practice.

a. Operating characteristic of cumulative compound wound DC motor

The characteristic of cumulative compound wound motors are the combination of shunt and series characteristics and lie between those of shunt and series motor.

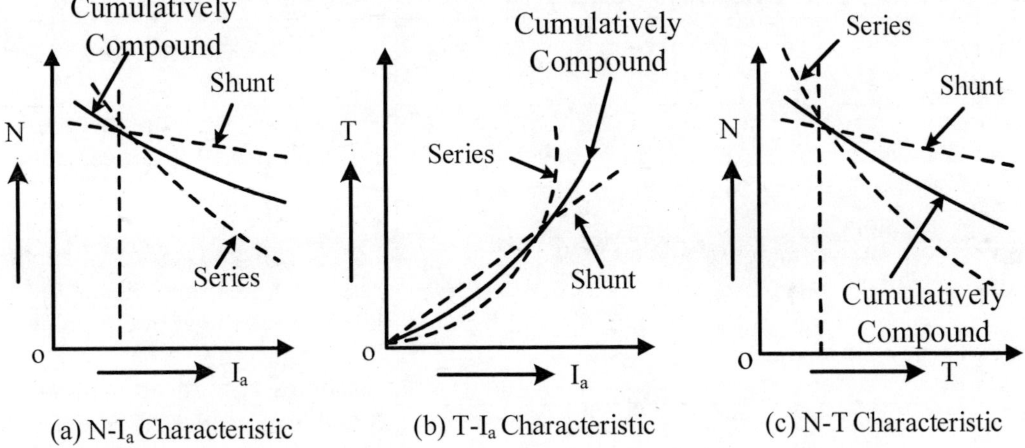

Fig. 4.39 Characteristics of cumulatively compound DC motor

As the load increase, the flux due to series field winding increase and cause the torque greater than it would have been with shunt field winding alone for a given machine and for a given current. The increase in flux due to series field winding on account of increase in load cause the speed to fall more rapidly than it would have done in shunt motor. The characteristic is shown in fig. 4.39.

b. Operating characteristic of differential compound wound DC motor

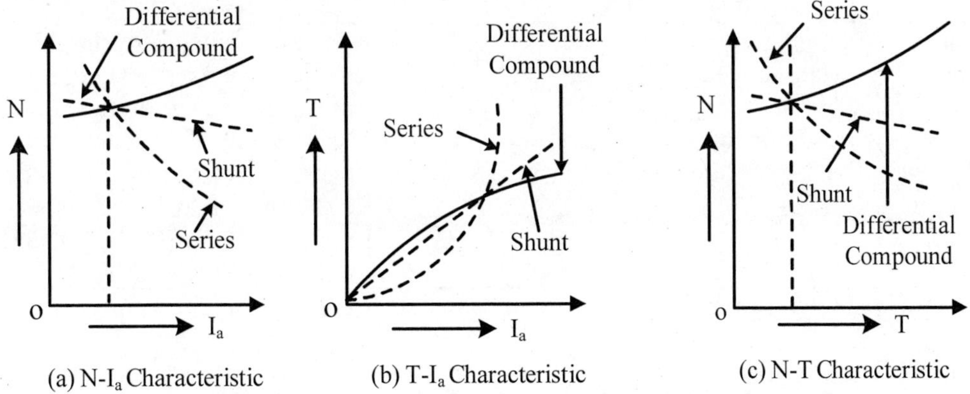

Fig. 4.40 Characteristics of differential compound DC motor

In differential compound wound motor, since the series field opposes the shunt field, the resultant flux decreases with the increase in load. Thus, the machine runs at a higher speed then it would do as a shunt motor. The decrease in flux with the increase in load causes the torque to be less than that of a shunt motor as shown in fig. 4.40.

Speed with load torque/armature current: Very large Speed is achieved when the motor is overloaded. The motor may even start in the opposite direction if started with a heavy load. Due to these faces, these differentially compound motors are rarely used.

4.25 Summary of Characteristic and Application of DC Motor

Type of Motor	Characteristic	Application
Separately excited DC motor	Very accurate speed can be obtained Best suited for applications requiring speed variation from very low value to high value	For paper machines, diesel-electric propulsion of ships, steel rolling mills etc.
Shunt motor	Approximately Constant Speed Adjustable Speed Medium starting torque, up to 1.5 times full load torque	For constant-speed applications requiring medium starting torque. For lathes, centrifugal pumps, blowers and fans, machine tools, woodworking machines, reciprocating pumps etc.
Series motor	Variable speed Adjustable varying speed High starting torque The speed regulation is widely variable, Very high speed at no-load Speed control by series resistance When series motor is used in traction locomotive, series-parallel control is used for starting and speed control.	Suitable for drives requiring high starting torque and where adjustable, varying speed is required. For cranes, hoists, trolley cars, conveyors, electric locomotives, electric vehicles (EV) etc. The load must be positively connected and not belted To prevent overspeed, the lightest load should not be less than 15 to 20% of full load torque.
Cumulative compound DC motor	Starting torque is high. Variable speed Adjustable varying speed Speed control is not used but may be up to 125% by field control	Used for drive requiring high starting torque and only fairly constant speed, pulsating load with flywheel action. For shears, conveyors, crushers, bending rolls, punch presses, hoists, elevators, heavy planers, ice-making machines, Air compressors, rolling mills, printing presses etc.
Differential compound wound DC motor	Almost constant torque and constant speed Rise in speed with load	For experimental and research work

4.26 Selection of Electric Motor

The proper choice of motor power for a drive is an important aspect for the economical, efficient, and reliable operation of a machine.

a. ***Selection of power rating***: Installing a motor with a higher rating (HP or kW capacity of motor is more than pickup load) than is necessary for a given drive load to extra energy loss during operation of the machine and will incur unnecessary capital expenditure. On the other hand, the installation of a motor of insufficient power (the size of motor is less than the pickup load) reduces the efficiency of the working machine and makes it unreliable; also, the electric motor itself may easily get damaged due to over load.
A motor should be chosen for maximum power utilization. During operation, it should heat up to the maximum permissible temperature, but should not overheat. In addition, the motor should operate normally under possible temporary overloads and develop the starting torque required by the given working machine. Accordingly, the motor is usually chosen based on heating conditions.

b. ***Characteristic of the motor***: For the selection of a reliable and efficient motor, the conditions of service must be well known. It is not sufficient to simply specify the output power in kW and the speed but it is also necessary to know the characteristic of a mechanical load and motor so that work can be quickly, easily, and efficiently carried out without any breakdown. The followings are the particulars of the work
    i. Torque at the shaft during running, starting, and at different loads.
    ii. Accelerating torque and braking torque.
    iii. Duty cycle: continuous, short-term, intermittent, shock loading.
    iv. Switching frequency.
    v. The efficiency of motor at different loads.
    vi. Temperature at the work place
    vii. Environmental condition
    viii. Other working requirements.

In studying the behavior of a motor selected for a particular drive unit, one of the first problems involved is to determine whether the speed-torque characteristic of the motor suits the requirements imposed by the speed-torque characteristic of the driven unit. Drive behavior during the transient period of a start-up, braking, or speed change- over also depends upon how the speed-torque characteristics of the motor and the driven unit vary with speed. It is, therefore, imperative to study these characteristics to be able to select correctly the motor and obtain an economical drive.

4.27 Speed Control of DC Motor

D.C. motors are most suitable for wide-range speed control and are, therefore, indispensable for many adjustable speed drives. The speed of a DC motor is given by the relationship

$$N = \frac{V - I_a R_a}{k\phi} \tag{4.117}$$

Equation (4.117) Shows that speed is dependent upon supply voltage V, the armature circuit resistance R_a and field flux ϕ. In practice, the variation of these factors is used for speed control. Thus, there are three general methods of speed control of DC motor.

a. Variation of resistance in the armature circuit.
b. Variation of the field flux.
c. Variation of the armature terminal voltage.

It is advisable to understand the basic concept of base speed, speed regulation, speed range, constant power drive, and constant torque drive before studying the speed control techniques for DC motors.

I. ***Base Speed*** is defined as the speed at which a motor runs at rated armature voltage and rated field current. Base speed is equal to the rated speed or nameplate speed of the motor.
II. If the speed-change from no load to full load is $\Delta\omega_m$ then ***speed regulation*** is defined as the ratio of $\Delta\omega_m$ to rated speed or base speed ω_m.
III. ***Speed range*** is defined as the ratio of the maximum allowable speed to the minimum allowable speed of the motor. When the speed range of a motor is specified, it must be mentioned whether this speed range is at no-load, full load, or a fraction of full load.
IV. If the motor shaft power remains constant over a given speed range, the system is called a ***constant power drive.*** Note that in constant power drive, higher torques are available at lower speeds and lower torques at higher speeds. The motor size is always decided by the highest torque requirement at the lowest speed.
V. If the motor shaft torque remains constant over a given speed range, the system is called a ***constant torque drive.*** Note that in constant torque drive, shaft power varies as the speed varies. Load exhibits different types of variation of load torque with respect to speed. For example, load torque may be constant or proportional to speed or square of speed (fan type load) or inversely proportional to speed (traction load).

4.27.1 Armature resistance control or rheostatic control

In the armature resistance control method, a variable resistor R_c is connected in series with armature circuit. Fig. 4.41 shows the method of connection in DC shunt and series motor.

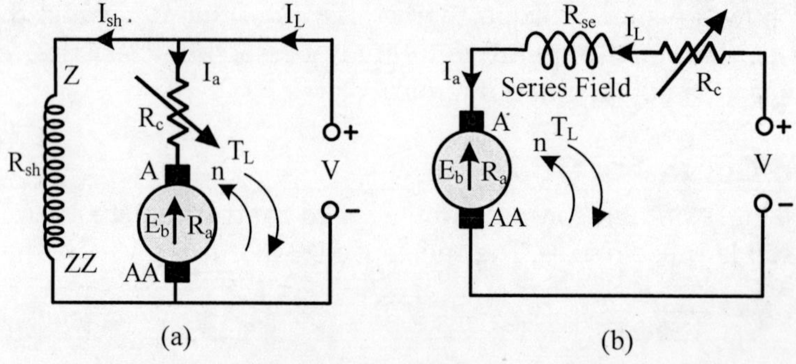

Fig. 4.41 Speed control of (a) DC shunt motor (b) DC Series Motor

In a shunt machine, flux (ϕ) will not be affected. But in DC series machine, flux (ϕ) will be affected when motor is loaded. In both DC motor, according to equation (4.117) flux is constant when applied terminal voltage and shunt field resistance are constant. Therefore, the speed of the motor is directly proportional to induced *emf* $N \propto E_b$ and $E_b = V - I_a R_a$ (equations 4.114 and 4.115). The value of E_b depends upon the drop in the armature circuit. When a variable resistance is connected in series with the armature as shown in fig. 4.41(a) and fig. 4.41(b) the induced emf $[E_b = V - I_a(R_a + R_c)]$ is reduced and hence the speed is reduced.

Thus, the motor runs at a speed lower than the normal speed as shown in fig. 4.42(a) and fig. 4.42(b). In both cases, the motor runs at a lower speed as the value of R_c increased. Since R_c carries full armature current, it must be designed to carry continuously the full armature current.

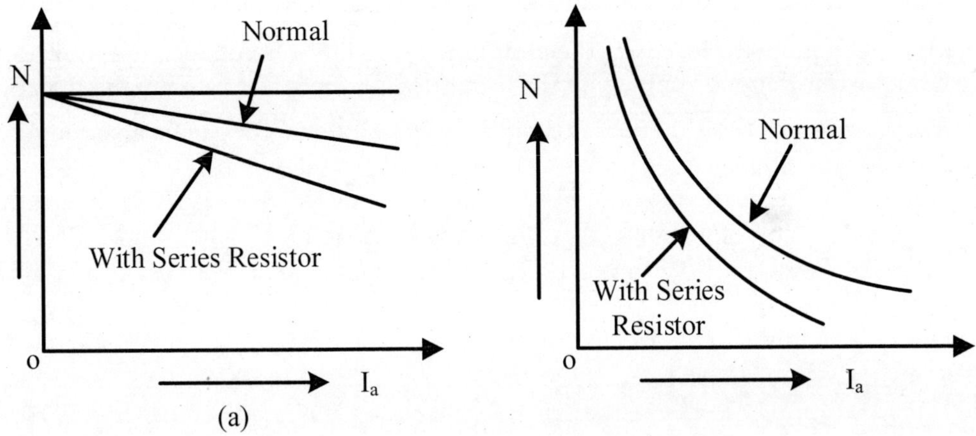

Fig. 4.42 Speed current characteristic of (a) Shunt motor (b) Series motor

Let I_{a0} be the armature current at no load and armature resistance be R_a. Let I_{a1} be the armature current when the controller resistance R_c is added to armature resistance. Let N_0 be the no-load speed and N_1 be the speed of armature current I_{a1}.

$$N_0 \propto (V - I_{a0} R_a); \qquad i.e., N_0 \propto E_b \tag{4.118}$$

$$N_1 \propto (V - I_{a1}(R_a + R_c); \qquad i.e., N_1 \propto E_{b1} \tag{4.119}$$

Since flux is constant for shunt motor, we have

$$\frac{N_0}{N_1} = \frac{V - I_a R_a}{V - I_{a1}(R_a + R_c)} \tag{4.120}$$

$$N_1 = N_0 \left(\frac{V - I_{a1}(R_a + R_c)}{V - I_{a0} R_a} \right) \tag{4.121}$$

Neglecting no-load current I_{a0}

$$N_1 = N_0\left(1 - \frac{I_{a1}(R_a + R_c)}{V}\right) \quad (4.122)$$

Equation (4.122) shows the linear relation between $(R_a + R_c)$ and speed. The load current I_{a1} for which the speed N_1 become zero is obtained by putting $N_1 = 0$ in equation (4.122).

$$0 = N_0\left(1 - \frac{I_{a1}(R_a + R_c)}{V}\right) \quad (4.123)$$

$$I_{a1} = \frac{V}{(R_a + R_c)} \quad (4.124)$$

The equation (4.124) shows the required maximum current for which speed becomes zero. This current is known as stalling current. In this method, a wide range of speeds (below normal) can be obtained. Moreover, the motor develops any desired torque over its operating range since torque depends only upon the armature current (flux remaining unchanged).

The major advantage of this method is that the speed of the motor can be also reduced to any low value and creeping speed (only a few rpm) can also be developed.
The major disadvantage of this method of speed control is

- A large amount of power loss in the control rheostat.
- For a given value of R_c the speed reduction is not constant but varies with the motor load.

This method is suitable for small DC motors.

4.27.2 Field flux control

In this method, current through the field winding is controlled by using a variable resistor in series with the field coil in the DC shunt motor and DC series motor. The flux produced by the shunt field winding depends upon the current flowing through it i.e., $\phi \propto I_{sh}$ and $I_{sh} = \frac{V}{(R_{sh})}$.

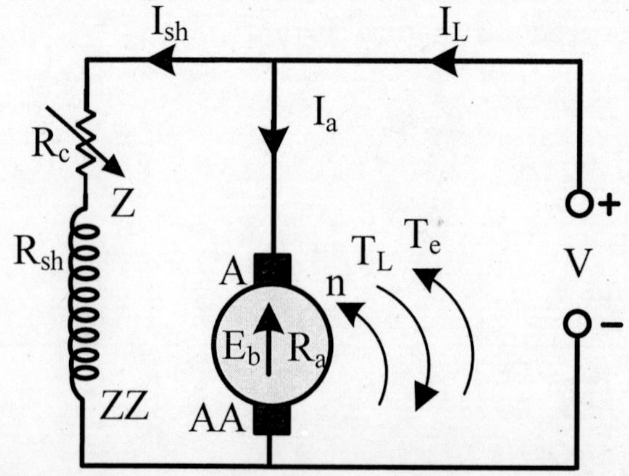

Fig. 4.43 Speed control of DC shunt motor by variation of field flux

In the shunt motor, current through the field winding is controlled by connecting a variable resistor R_c in series with the shunt field winding as shown in fig. 4.43. the resistor R_c is called the shunt field regulator.

The shunt field current is given by $I_{sh} = \frac{V}{R_{sh} + R_c}$. The connection of R_c in the field reduces the field current and hence the flux ϕ is also reduced. The reduction in flux will result in an increase in the speed (according to equation 4.117). The amount of increase in speed depends on the value of resistance R_c. Consequently, the motor runs at a speed higher than normal speed. For this reason, this method of speed control is used to give motor speeds above normal or to correct for a fall in speed due to load. This method is most common since very little power ($I_{sh}^2 R_c$) is wasted in the shunt field variable resistance due to the relatively small value of I_{sh}.

In DC series motor the variation of field current is done by any one of the following methods:

a. A variable resistance R_d is connected in parallel with the series winding as shown in fig. 4.44. The parallel resistor is called the diverter. A portion of the main current is diverted through R_d. Thus, the diverter reduces the current flowing through the field winding. This reduces the flux and increases the speed. The method was in use with DC series motor-operated traction locomotive.
b. The second method uses a tapped field control as shown in fig. 4.45. Here the ampere turns are varied by varying the number of field turns. Tapped field control is used in domestic mixers and food processors.

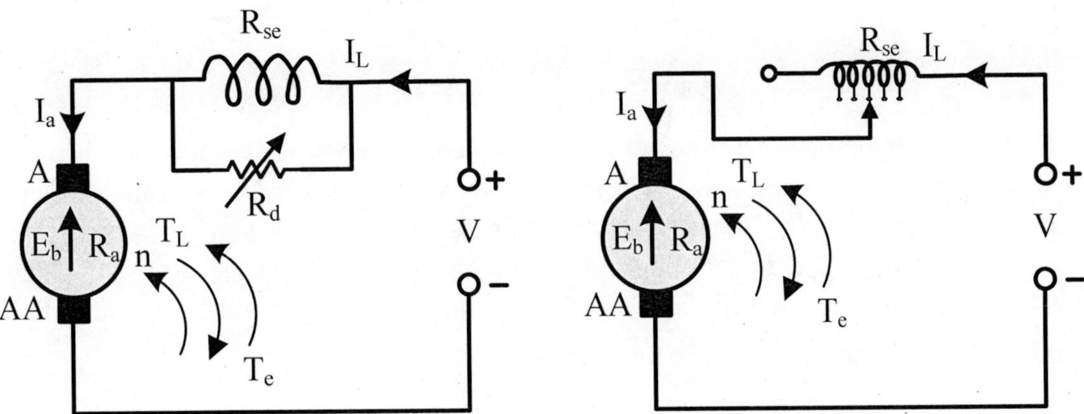

Fig. 4.44 Diverter in parallel with the series of DC Motor

Fig. 4.45 Tapped series field on DC series Motor

Fig. 4.46(a) and fig. 4.46(b) shows the typical speed/ torque curves for shunt and series motors respectively, whose speeds are controlled by the variation of the field flux.

- The flux or field can also be controlled by controlling the reluctance of the magnetic circuit known as the **Reluctance control method**, but this can only be obtained by employing some special mechanical features which increase the cost of machine. Therefore, practically this method is not employed.

- The flux or field can also be controlled by supplying variable voltage to the field winding known as the **Field-Voltage control method**. This can only be achieved by disconnecting the field winding from the armature and supplying variable voltage from some other source. Then the motor is treated as a separately excited DC motor.

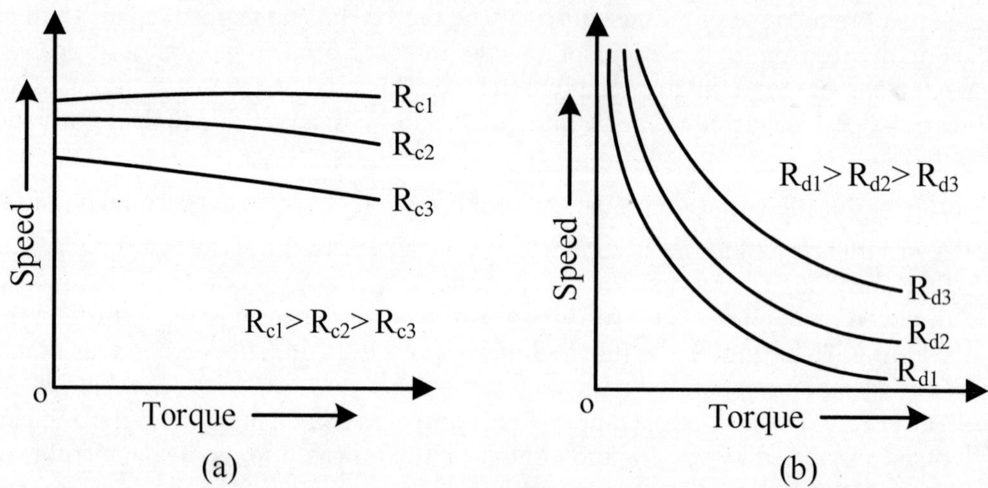

Fig. 4.46 Speed/torque characteristic of (a) DC shunt motor (b) DC Series Motor

The advantages of field control are as follows:

a. This method is easy and convenient.
b. The shunt field current I_{sh} is very small, and the losses in the shunt field are small.
c. The flux cannot usually be increased beyond its normal value due to iron core saturation. The speed control by flux is limited to weakening, which gives an increase in speed. It is applicable over a limited range only, as because if the field is weakened too much, torque would be reduced and there could be loss of stability.

The limitations and drawbacks of this method are:

a. Speeds below normal cannot be obtained.
b. The speeds above normal can be obtained by weakening the field. The advantage of high speed cannot be considered for increase in power output rather to compensate the power, armature has to draw extra current from the mains.
c. To obtain high speed, the field is very weak, to obtain certain load torque, armature draws extra current which may causes overheating of armature winding; poor commutation, sparking and instability.

4.27.3 Armature voltage control

This method of speed control requires an adjustable source of voltage separately from the source supplying the field current. The adjustable voltage for the armature is obtained from an adjustable voltage generator or an adjustable electronics rectifier. Speed below base speed is possible as shown in fig. 4.48. This method gives a wide range of speeds with any desired number of speed points. It is essentially a constant-torque system because the output of the motor decreases with a decrease in applied voltage and a corresponding decrease in speed. This method avoids the disadvantage of poor speed regulation and low efficiency. This method is not applied to any great extend; generally, an account of higher cost of the generating equipment.

Ward-Leonard System

Ward Leonard system of speed control is based on the principle of armature voltage control. This method was introduced in 1891. The drawbacks of the earlier methods can be overcome by this method. When large motors are to be controlled (rotating by starting and speed reversal), a separate motor-driven generator of suitable rating is used. The method is useful in rolling mills, where the slab of molten metal is pressed between two rollers, which are rolled in forward and reverse directions so that sheet metal is finally obtained. In this process, the distance between rollers is also adjusted automatically. The schematic diagram of the Ward Léonard method of speed control of a DC shunt motor is shown in fig. 4.47.

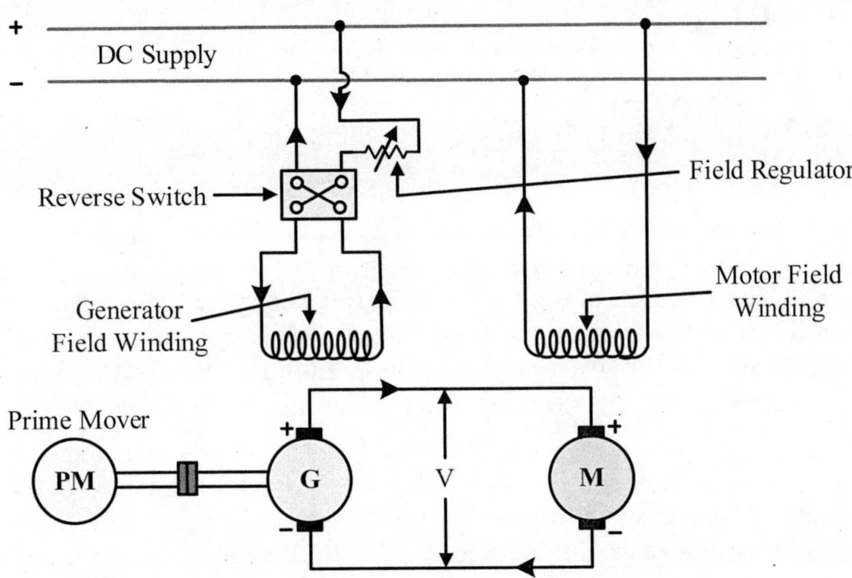

Fig. 4.47 Ward-Leonard method of speed control for DC motor

In this system, M is the main DC motor whose speed is to be controlled, and G is a separately excited DC generator. The DC motor *M* is fed from the generator *G* and its field winding is connected directly to a constant DC supply line. The generator G is driven by a 3-phase driving motor or prime mover (PM) which may be an induction motor or a synchronous motor. The combination of AC driving motor and the DC generator is called the motor-generator (M-G) set. The field winding of the DC generator is connected to a constant voltage DC supply line through a field regulator and reversing switch.

The voltage of the generator fed to the motor, can be varied from zero to its maximum value by means of its field regulator. By reversing the direction of the field current by means of the reversing switch, the polarity of the generated voltage can be reversed at zero speed and field current is increased in reverse direction and hence the direction of rotation of motor *M*. Hence, by this method, the speed and direction of rotation both can be controlled very accurately. When speed control over a wide range is required, combination of armature voltage control and field flux control is used. In armature voltage control method constant torque and variable power drive is obtained from speed below the base speed. This is shown in fig. 4.48.

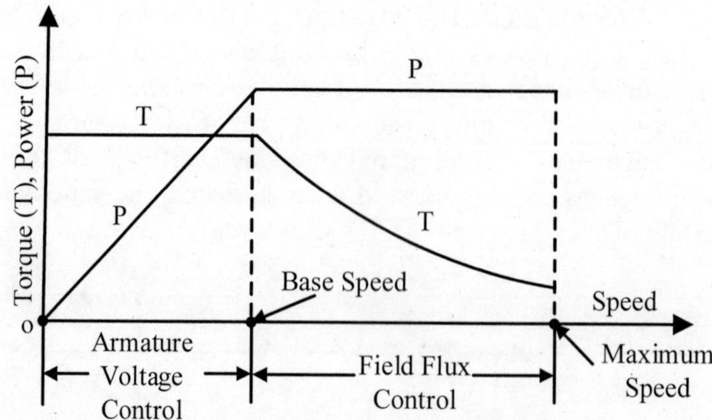

Fig. 4.48 Torque and power characteristic for combined armature voltage and field control

As mentioned, earlier the driving AC motor can be an induction motor or synchronous motor. An induction motor operates at a lagging power factor. The synchronous motor running at fixed speed may be operated at a leading power factor by over -excitation of its field. Leading reactive power taken by over-excited synchronous motor compensates for the lagging reactive power taken by other inductive loads in the plant. Thus, the power factor of the plant is improved.

When the load is heavy and intermittent, a slip-ring induction motor is used as a prime mover. A flywheel is mounted on its shaft. This scheme is known as Ward-Leonard-Ilgener Scheme. It prevents heavy fluctuations in supply current. When the driving AC motor is a synchronous motor, its supply current fluctuations cannot be reduced by mounting a flywheel on its shaft, because a synchronous motor operates only at a constant speed.

In another form of Ward- Leonard drive, non- electrical prime movers can also be used to drive the DC generator either by a diesel engine or a gas turbine. In this system regenerative braking is not possible because energy cannot flow in the reverse direction in the prime mover.

Advantages of Ward- Leonard Drive

1. It is a very smooth speed control of DC motors over a wide range (from zero to normal speed).
2. The speed can be quickly controlled in both directions of rotation of the motor. This is also referred to as the to and fro drive.
3. Motor can run with a uniform acceleration.
4. In this type of armature voltage-controlled starting, starters are not required.
5. It has inherent regenerative braking capacity.
6. The speed regulation of DC motor in this method is high.
7. By using an overexcited synchronous motor as the drive for DC generator, the lagging reactive volt-amperes of the plant are compensated. Therefore, the overall power factor of the plant improves.
8. When the load is intermittent as in rolling mills, the drive motor used is an induction motor with a flywheel mounted on its shaft to smooth out the intermittent electrical loading to a low value.

Disadvantages of the classical Ward- Leonard System

1. Higher initial cost due to the use of two additional machines (M-G set) of the same rating as the main DC motor.
2. The overall efficiency of the system is not sufficient especially if it is not lightly loaded.
3. Larger size and weight.
4. Requires more floor area and costly foundation.
5. Frequent maintenance is needed.
6. The drive produces more noise.

Application Ward Leonard method

This **Ward Leonard method of speed control system** is used where a very wide and very sensitive speed control of a DC motor in both the direction of rotation is required. When the generator field current is zero the motor is rest. When the generator field current is increased in one direction using generator field at rheostat, motor starts in one direction. When generator field current is brought to zero with motor still running in the forward direction, the direction of current reverses. Momentarily generator acts as a motor. The induction motor (prime mover) - generator speed rises beyond the synchronous speed of the induction motor. A certain amount of power is fed back to AC supply - regenerative action of the induction motor in super synchronous speed. The generator field is increased in reverse direction- initially motor is at rest. The generator develops reverse polarity increasing voltage and motor starts in the opposite direction. This cycle continues- making the motor run in forward and reverse direction continuously or to and fro motion.

This speed control system is mainly used in colliery winders, cranes, electric excavators, mine hoists, elevators, steel rolling mills, paper machines, diesel locomotives, etc.

4.27.4 Solid State Control

The DC Motor speed can be controlled through power semiconductor switches. The power semiconductor switches are SCR, MOSFET, IGBT etc., this type of speed control is called **static Ward Leonard Drive**. The static ward-Leonard drive is being used these days because of the drawback of the classical method. Rotating motor-generator sets have been replaced by a solid-state converter to control the speed of DC motor. The converter used is a controlled rectifier or chopper.

In case of ac supply, controlled-rectifier are used to convert fixed AC supply voltage into a variable ac voltage as shown in fig. 4.49. When the supply is DC, a chopper is used to obtain variable DC voltage from the fixed-voltage DC supply as shown in fig. 4.50.

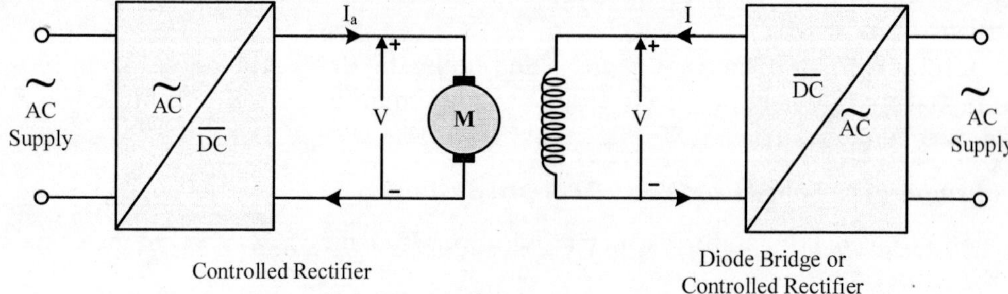

Fig.4.49. Controlled rectifier fed DC drive

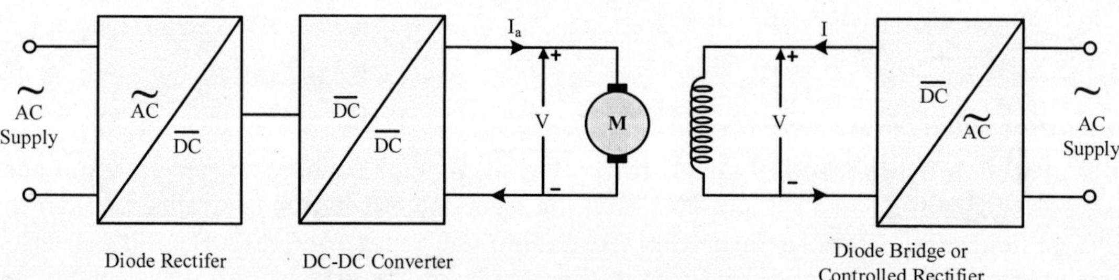

Fig.4.50. Chopper fed DC drive

Drawback of static Ward-Leonard drive

The main drawback of static ward-Leonard drive is as follows:

i. They are not suitable for intermittent loads because fluctuations produced large fluctuations of supply voltage and current. There is no provision for load equalisation in static Ward-Leonard system.

ii. Harmonics are generated in the system which affects the quality of supply.

iii. Such a system operates at low power factor, particularly at low speeds.

In general, static Ward-Leonard drive are used in most applications. However, conventional Ward-Leonard drive are used in large size intermittent loads. In the case of non-electrical prime mover conventional Ward-Leonard system can only be useful.

4.28 Speed Regulation

The speed regulation of DC motor is defined as the change in speed from full load to no load and is expressed as a percentage of the full load speed.

$$\%\text{Speed regulation} = \frac{\text{No load speed} - \text{Full load speed}}{\text{Full load speed}} \times 100$$

$$\%\text{Speed regulation} = \frac{N_0 - N}{N} \times 100 \tag{4.125}$$

4.29 Losses in DC Motor

The losses in DC motor are same as DC generator. Various losses are as

i. ***No-Load Reactional Loss:*** This loss is made up of two losses
   - The iron loss at working flux and speed.
   - The mechanical loss i.e., friction and windage loss at operating speed.

ii. ***I^2R_a Loss:*** It consists of armature circuit loss $I_a^2R_a$, where R_a include the resistance of brush contact, armature winding, interpole winding and compensating winding. It is also accounts for series field winding losses.

iii. ***Stray load Losses:*** These are produced by distortion of the air gap flux due to the armature reaction and currents in the commutated coil. They are assumed to be 1% of the rated output.

4.30 Efficiency of a DC Motor

The electrical motor efficiency is given by

$$\eta_m = \frac{Power\ output}{Power\ input} \tag{4.126}$$

When machine working as a motor

Power input

$$= VI_L$$

Power Output=power input-variable losses-constant losses

$$= VI_L - I_L^2R_a - P_c$$

$$\eta_m = \frac{VI_L - I_L^2R_a - P_c}{VI_L} \tag{4.127}$$

The efficiency of DC motor can also be summed as

Electrical efficiency (η_{elect})

$$\eta_{elect} = \frac{mechanical\ Power\ developed}{Total\ electrical\ power\ input} = \frac{E_b I_b}{VI_L} \tag{4.128}$$

Mechanical efficiency (η_{mech})

$$\eta_{mech} = \frac{Useful\ mechanical\ Power\ output}{Mechanical\ power\ developed} \tag{4.129}$$

$$\eta_{mech} = \frac{BHP\ of\ motor\ \times 745.7}{E_b I_b} \tag{4.130}$$

Overall or commercial efficiency ($\eta_{ov} = \eta_{mech} \times \eta_{elect}$)

$$\eta_{ov} = \frac{Useful\ mechanical\ Power\ output}{Total\ electrical\ power\ input} \tag{4.131}$$

$$\eta_{mech} = \frac{BHP\ of\ motor\ \times 745.7}{VI_L} \tag{4.132}$$

Overall or commercial efficiency (η_{ov}) can also be expressed as follows:

$$\eta_{ov} = \frac{Useful\ Power\ output}{Total\ power\ input} \tag{4.133}$$

$$\eta_{ov} = \frac{Total\ Power\ input - total\ losses}{Total\ power\ input} \tag{4.134}$$

$$\eta_{ov} = \frac{VI_L - total\ losses}{VI_L} \tag{4.135}$$

Condition for Maximum efficiency

Condition of maximum efficiency for DC motor and generator are same. According to Section 4.15

$$I_L^2 R_a = P_c \tag{4.136}$$

i.e., Variable losses = constant losses

Hence, the efficiency of a DC machine will be maximum when the line current is such that the constant loss is equal to the variable loss.

The load current at which the efficiency of DC machine is maximum.

$$I_L = \sqrt{\frac{P_c}{R_a}} \tag{4.137}$$

Example.4.22 A 250 V, 10 kW with armature resistance of 0.7-ohm DC shunt motor rum at 1250 rpm. Its field is adjusted at no load so that its armature carries a current of 1.6 A. When load is applied, the current drawn by it increases to 40 A and speed falls to 1150 rpm. How much flux per pole reduces due to armature reaction?

Solution: -

$$At\ no\text{–}load; \qquad E_{bo} = V - I_{a0}R_a = 250 - 1.6 \times 0.7 = 248.88\ V$$

$$At\ no\text{–}load; \qquad E_b = V - I_aR_a = 250 - 40 \times 0.7 = 222\ V$$

$$\text{Now, } N_0 \propto \frac{E_{bo}}{\phi_0}\ and\ N \propto \frac{E_b}{\phi}$$

Therefore,

$$\frac{N}{N_0} = \frac{E_b}{\phi} \times \frac{\phi_o}{E_{bo}}$$

$$\phi = \frac{E_b}{E_{bo}} \times \frac{N_o}{N} \times \phi_0$$

$$\phi = \frac{222}{248.88} \times \frac{1250}{1150} \times 1 = 0.96956$$

Reduction in flux=1-0.96956=3.044% (Ans.)

Example.4.23 A 250 V shunt motor runs at 1500 rpm at full load with an armature current of 15 A. the total resistance of the armature and brushes is 0.6 ohm. If the speed is to be reduced to 1200 rpm with the same armature current, calculate the amount of resistance to be connected in series with the armature and power lost in this resister.

Solution: -

During normal operating conditions;

$$Armature\ current, I_{a1} = 15\ A$$

$$Back\ emf,\ E_{b1} = V - I_{a1}R_a = 250 - 15 \times 0.6 = 241\ V$$

$$Speed\ , N_1 = 1500\ rpm$$

Let a resistance R ohm be connected in series with armature, then

$$Speed\ , N_2 = 1200\ rpm$$

$$Back\ emf,\ E_{b2} = V - I_{a2}(R + R_a) = 250 - 15 \times (0.6 + R)$$

$$E_{b2} = 241 - 15R$$

$$Now, \quad \frac{E_{b2}}{E_{b1}} = \frac{N_2}{N_1} \quad since\ flux\ is\ constant$$

$$\frac{241 - 15R}{241} = \frac{1200}{1500}$$

$$241 - 15R = 192.8$$

$$R = 3.213\ ohm$$

$$\text{power lost} = I_a^2 R = (15)^2 \times 3.213 = 722\ Watt$$

Example.4.24 A 240 V DC shunt motor has a field resistance of 400 ohms and an armature resistance of 0.1 ohms. The armature current is 50 A and the speed is 1000 rpm. Calculate the additional resistance in the field to increase the speed to 1200 rpm. Assume that armature current remains the same and the magnetization curve is a straight line.

Solution: -

When no resistance is added in the shunt field winding

$$Shunt\ field\ current, I_{sh} = \frac{V}{R_{sh}} = \frac{200}{400} = 0.6\ Amp$$

$$E_{b1} = E_{b2} = V - I_a R_a = 240 - 50 \times 0.1 = 235\ V$$

Now we know that, $\quad N \propto \frac{E_b}{\phi}$

Now, $\quad N_1 \propto \frac{E_{b1}}{\phi_1}$ and $N_2 \propto \frac{E_{b2}}{\phi_2}$

$$or \quad \frac{N_2}{N_1} = \frac{E_{b2}}{\phi_2} \times \frac{\phi_1}{E_{b1}} = \frac{\phi_1}{\phi_2}$$

As the magnetization curve is a straight line, $\phi_1 \propto I_{sh1}\ and\ \phi_2 \propto I_{sh2}$

$$\frac{N_2}{N_1} = \frac{I_{sh1}}{I_{sh2}}$$

$$I_{sh2} = \frac{N_1}{N_2} \times I_{sh1} = \frac{1000}{1200} \times 0.6 = 0.5\ A$$

$$\frac{240}{400 + R} = 0.5$$

$$R = 80\ Ohm\ (Ans.)$$

Example.4.25 A 230 V shunt motor is taking a current of 50 A. Resistance of shunt field is 46 ohm and the resistance of the armature of 0.02 Ohm. There is a resistance of 0.6 ohm in series with the armature and the speed is 900 rpm. What alternation must be made in the armature circuit to raise the speed to 1000 rpm for the same torque.

Solution: -

$$Load\ current, I_{L1} = 50\ Amp\ and\ speed\ \ N_1 = 900\ rpm$$

$$Shunt\ field\ current, I_{sh} = \frac{V}{R_{sh}} = \frac{230}{46} = 5\ Amp$$

$$\text{Armature current, } I_{a1} = I_L - I_{sh}$$

$$= 50 + 5 = 45\ Amp.$$

$$Back\ emf,\ E_{b1} = V - I_a(R + R_a)$$

$$= 230 - 45 \times (0.6 + 0.02) = 202.1\ V$$

Let the series resistance be reduced from 0.6 ohms to R_1 ohm to rise speed to 1000 rpm

Since load torque is constant

$$T_2 = T_1\ or\ I_{a2}\phi_2 = I_{a1}\phi_1$$

$$I_{a2} = I_{a1} \times \frac{\phi_1}{\phi_2} = I_{a1} = 45\ A; Assuming\ flux\ unchanged$$

$$speed\ \ N_2 = 900\ rpm$$

$$Back\ emf,\ E_{b2} = V - I_{a2}(R_a + R_1)$$

$$= 230 - 45 \times (R_1 + 0.02)$$

$$E_{b2} = 229.1 - 45R_1$$

$$Now,\quad \frac{E_{b2}}{E_{b1}} = \frac{N_2}{N_1}\ since\ flux\ is\ constant$$

$$\frac{229.1 - 45R_1}{202.1} = \frac{900}{1000}$$

$$R_1 = 1.0491\ ohm\ (Ans.)$$

Hence, Additional resistance of 0.6 ohm will have to be 1.0491 ohm in order to raise the motor speed from 900 rpm to 1000 rpm. (*Ans.*)

Example.4.26 A 200 V Shunt motor has armature resistance of 0.1-ohm, shunt field resistance 240 ohm, and rotational loss is 236 W. On full load, the line current is 9.8 A with the motor running at 1450 rpm. Determine the (i) mechanical power developed (ii) power output (iii) load torque (iv) full load efficiency.

Solution: -

$$Load\ current, I_L = 9.8\ Amp;\ speed, N = 1450\ rpm$$

$$Shunt\ field\ current, I_{sh} = \frac{V}{R_{sh}}$$

$$= \frac{200}{240} = 0.833\ Amp$$

Armature current, $I_a = I_L - I_{sh}$

$$= 9.8 - 0.833 = 8.967\ Amp$$

$$Back\ emf,\ E_b = V - I_a R_a$$

$$= 200 - 8.967 \times 0.1 = 199.10\ V$$

(i) *Mechanical Power developed,*

$$E_b I_a = 199.10 \times 8.967 = 1785.33\ Watt$$

(ii) Power output=Mech. Power developed− rotational loss

=1785.33−236=1549.33 Watt

(iii) Power output $= \frac{2\pi NT}{60}$

$$\text{Load torque, T} = \frac{\text{Power output } \times 60}{2\pi N}$$

$$\text{Load torque, T} = \frac{1549.33 \times 60}{2\pi \times 1450}$$

$$= 10.2\ Nm\ (Ans.)$$

$$Input\ to\ motor, V\ I_L$$

$$= 200 \times 9.8 = 1960\ W$$

(iv) $full\ load\ efficiency, \eta = \frac{Power\ output}{Input} \times 100$

$$\eta = \frac{1549.30}{1960} \times 100$$

$\eta = 79.04\ \%$ (Ans)

Example.4.27 ***A 220 V DC series motor takes 50 A. Armature resistance 0.1-ohm, series field resistance 0·08 ohm. If the iron and friction losses are equal to copper losses at this load, find the B.H.P. and efficiency.***

Solution: -

Armature current,

$$I_a = I_L = 50\ Amp$$

$$copper\ loss\ = I_a^2(R_a + R_{se})$$

$$= (50)^2 \times (0.1 + 0.08)$$

$$= 450\ Watt$$

Iron and friction losses=copper losses=450 Watt

Total losses= Iron and fraction losses+ copper losses

$$= 450 + 450 = 900\ Watt$$

$$Input\ Power = VI_L$$

$$= 220 \times 50 = 11000\ Watt$$

$$Output\ Power = Input\ Power - Total\ Losses$$

$$= 10000 - 900 = 9100\ Watt$$

$$BHP = \frac{Output\ Power}{745.7}$$

$$= \frac{9100}{745.7}$$

$$= 12.20\ H.P\ (Ans)$$

$$Efficiency\ (\eta)$$

$$\eta\ = \frac{Power\ output}{Input} \times 100$$

$$\eta = \frac{9100}{10000} \times 100$$

$$\eta = 91\ \%\ \text{(Ans)}$$

Example.4.28 ***A 400 V shunt generator has full-load current of 200 A. Its armature resistance is 0·06-ohm, field resistance is 100 ohm and the stray losses are 2000 watt. Find the h.p. of prime-mover when it is delivering full load.***

Solution: -

Load current,

$$I_L = 200\ Amp$$

Shunt field current,

$$I_{sh} = \frac{V}{R_{sh}} = \frac{400}{100} = 4\ Amp$$

Armature current

$$I_a = I_L + I_{sh} = 200 + 4 = 204\ Amp$$

Armature copper loss

$$= I_a^2 R_a = (204)^2 \times 0.06 = 2497\ Watt$$

Shunt field copper loss

$$= I_{sh}^2 R_{sh} = (4)^2 \times 100 = 1600\ Watt$$

Total losses

$$= Armature\ copper\ loss + Shunt\ field\ copper\ loss + stray\ loss$$

$$= 2497 + 1600 + 2000 = 6097\ Watt$$

Output Power= VI_L

$$= 400 \times 200 = 80000\ Watt$$

Input Power

$$= Output\ Power + Total\ Losses$$

$$= 80000 + 6097 = 86097\ Watt\ Watt$$

$$BHP = \frac{Output\ Power}{745.7}$$

$$= \frac{86097}{745.7} = 115.46\ H.P\ (Ans)$$

4.31 Testing of DC Machine

The tests that are performed on DC machines are as follows:

- Open-circuit characteristic,
- Load characteristics,
- Determination of efficiency curve and
- Temperature rise test

The open-circuit characteristics and load characteristics have already been discussed. There are three methods to determine the efficiency of DC motors:

- ***Direct method:*** In this method, the generator or motor is put at full load and the total power developed by it is wasted. Direct test is generally used for small machines because it is inconvenient to apply full load to large-size machines. Therefore, this method is restricted only to determine the efficiency of small-sized machines.
  Brake test comes under this category. The belt must be tight before starting a series motor if brake is applied in the series motor. If this is not done, the armature may get damaged and flywheel broken to pieces.

- ***Indirect method:*** in this method measurement of losses and calculation of efficiency are carried out without actually loading the machine. There is no difficulty in applying this method to large machines. Accordingly, this method is usually employed to determine the efficiency of large DC shunt and compound wound machines. The disadvantage of this method is that the machine is run light during the test and so it does not give any indication of temperature rise at load or of commutating qualities of the machine. Swinburne test is the simplest example of the indirect method. Stray load loss is not accounted for in this method.

- ***Regenerative method:*** In this method, two identical, mechanically coupled machines are required. One of them will act as a motor to drive the other as a generator. The machine working as a generator feeds back power to the supply. The total power drawn from the supply is only used for supplying the internal losses of the two machines. It is possible to test very large machines as the power required is very small. Hopkinson test is a regenerative test for determining efficiency of a DC machine.

4.31.1. Brake test

This test is a kind of direct method and is used to determine the efficiency of comparatively small motors. In direct method, the DC machine is subjected to rated load, and entire output power is wasted. The motor is loaded directly using a mechanical brake or by using an eddy current brake or a calibrated air fan. Fig. 4.51 shows a common type of mechanical brake used in testing – the rope or belt brake. The brake is applied to a pulley mounted on the motor shaft. The load on the motor is increased by tightening the belt mounted on the pulley. The electrical connections are shown in the circuit diagram.

Let, spring balanced reading on the tight side $= W_a$ Kg

Spring balanced reading on the loose side $= W_b$ Kg

$$\text{Motor Speed} = N \text{ rpm}$$

$$\text{Radius of pully} = r \text{ metre}$$

$$\text{Thickness of belt} = t \text{ metre}$$

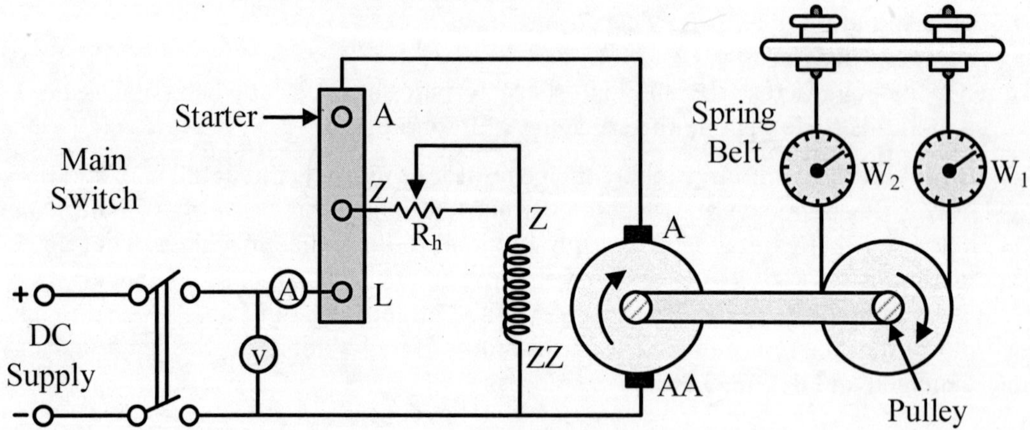

Fig.4.51. Circuit arrangement for brake test

$$\text{Motor output} = T \times \frac{2\pi N}{60} \tag{4.138}$$

$$= (W_a - W_b)\left(r + \frac{t}{2}\right)\frac{2\pi N}{60} \tag{4.139}$$

Since T=effective pull ×effective radius

$$= (W_a - W_b)r \times \frac{2\pi N}{60} \text{ Kg m/s} \tag{4.140}$$

$$= (W_a - W_b)r \times \frac{2\pi N}{60} \times 9.81 \text{ Nm/s or watt} \tag{4.141}$$

If voltmeter reading is V volt and ammeter reading is I ampere

Motor input=VI Watt

Then, the efficiency of motor,

$$\eta = \frac{output}{Input}$$

$$\eta = \frac{(W_a - W_b)r \times \frac{2\pi N}{60} \times 9.81}{VI} \tag{4.142}$$

This method of measuring efficiency has the following disadvantages:

- It is not possible to measure the output power directly.
- It is not possible to use this method for determining internal losses and efficiency of large motors because such facilities of loading are not available.

- The output measured by this method is not accurate because belt is not offering a constant load, because usually belt slips over the pulley.

Belt encircles half the rotating hollow drum. Output power is wasted in rubbing of belt with the rotating drum causing the rise in temperature. Water is normally poured inside rotating hollow drum for cooling purpose. At high temperatures and continuous rubbing action, the belt becomes smooth resulting in slippage of belt over drum. Sprinkling of sand is required for creating friction between belt and drum.

A special precaution is to be observed while performing this test on a series motor. If the brakes applied fails, the motor may obtain a dangerously high speed; therefore, this test is usually applied only on shunt and compound machines.

Example.4.29 In brake test, on DC Shunt motor the tension on the two side of the brake were 2.9 kg and 0.17 kg. radius of the pully was 7 cm. input current was 1 amp at 230 volts. The motor speed was 1500 rpm. Find the torque, power-output and efficiency.

Solution:

Net force on pully,

$$290 - 0.17 = 2.73\ kg$$

$$2.73\ \times 9.81 = 26.78\ kg$$

Net torque,

$$= force\ \times Radious\ = 26.78\ \times \frac{7}{100} = 1.8746\ Nm$$

Power output,

$$= Torque\ \times Radians/sec\ = 1.8746\ \times \frac{2\pi \times 1500}{60} = 294\ Watt$$

Efficiency,

$$= \frac{294}{(230 \times 2)} \times 100 = 63.9\%$$

4.31.2. Swinburne's test

This method is an indirect method of testing a DC machine. It is named after Sir James Swinburne. **Swinburne's test** is the most commonly used and simplest method of testing shunt and compound wound DC machines which have constant flux. This test is performed to determine the constant losses and efficiency at any desired load. In this test, the machine is operated as a motor on no-load or as a generator. The iron and friction losses are determined by measuring the input. A voltmeter and two ammeters A_1 and A_2 are connected in the circuit as shown in fig. 4.52. The normal rated voltage V is applied to the motor terminals. The ammeter A_1 and A_2 measure the no-load line current I_{L0} and shunt field current I_{sh} respectively. The voltmeter measures the applied voltage. As there is no output at no-load, all the power supplied to the motor, given by the product of current I_{L0} and voltage V is being utilized to meet losses only. The speed of the machine is adjusted to the rated speed with the help of the shunt regulator R_d as shown in the figure.

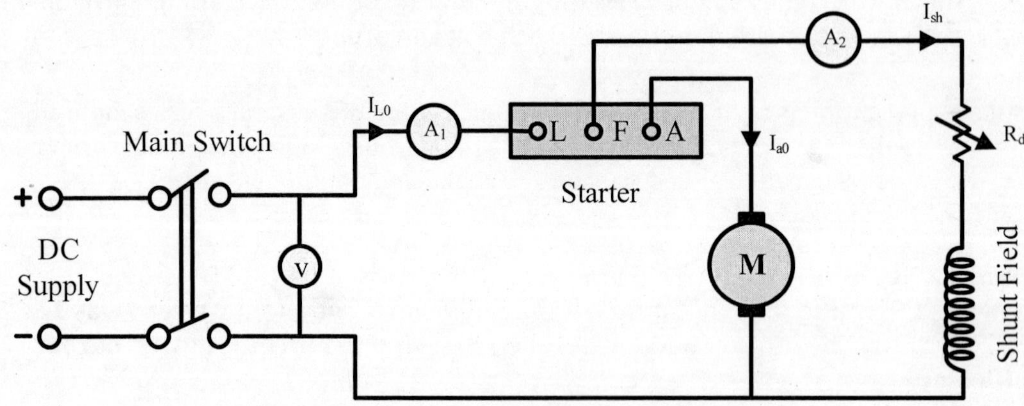

Fig.4.52. Circuit arrangement for Swinburne's test

Let,

V= Supply Voltage

I_{L0}= No load line current

I_{sh}= Shunt field current

The following are the losses at no-load:

a) Shunt field copper losses.
b) Armature copper losses at no-load (very small)
c) Iron losses in the core
d) Windage and friction losses at bearing and commutator.

$$\text{Armature current, } I_{a0} = I_{L0} - I_{sh} \tag{4.143}$$

$$\text{No load input power } = VI_{L0} \quad \text{Watt} \tag{4.144}$$

The resistance of armature circuit including the inter pole winding, etc., is measured by disconnecting one end of the shunt field circuit. Let its value be R_a.

$$\text{Then, variable losses} = I_{a0}^2 R_a \tag{4.145a}$$

$$\text{Shunt field copper} = I_{sh}^2 R_{sh} \tag{4.145b}$$

$$\text{Constant losses, } P_c = \left(VI_{L0} - I_{a0}^2 R_a - I_{sh}^2 R_{sh}\right) \tag{4.146}$$

After determining the constant losses, the efficiency of the machine, when it is working as a motor or generator can be calculated at any load, as discussed below:

Let I_L be the line current at which efficiency is to be calculated.

(i) When the machine is working as a motor:

Armature current, $I_a = I_L - I_{sh}$ (4.147)

Variable or armature copper loss at load $= I_a^2 R_a$ (4.148)

Total losses $= P_c + I_a^2 R_a + I_{sh}^2 R_{sh}$ (4.149)

Input value $= VI_L$ (4.150)

Output power = Input power – total losses

$$= VI_L - \left(P_c + I_a^2 R_a + I_{sh}^2 R_{sh}\right) \quad (4.151)$$

$$\text{Efficiency}, \eta = \frac{output}{Input} = \frac{VI_L - \left(P_c + I_a^2 R_a + I_{sh}^2 R_{sh}\right)}{VI_L} \quad (4.152)$$

(ii) When the machine is working as a generator

Armature current, $I_a = I_L + I_{sh}$ (4.153)

Variable or armature copper loss at load $= I_a^2 R_a$ (4.154a)

Shunt field copper = $I_{sh}^2 R_{sh}$ (4.154*b*)

Total losses $= P_c + I_a^2 R_a + I_{sh}^2 R_{sh}$ (4.155)

Output power= VI_L (4.156)

Input power = output power + total losses

$$= VI_L + \left(P_c + I_a^2 R_a + I_{sh}^2 R_{sh}\right) \quad (4.157)$$

$$\text{Efficiency}, \eta = \frac{VI_L}{VI_L + \left(P_c + I_a^2 R_a + I_{sh}^2 R_{sh}\right)} \quad (4.158)$$

Advantages

Following are the advantages of Swinburne's test

- The power required for the testing of large machines is very small, therefore it is an economical and convenient method of testing DC machines.
- As the constant losses are known, thus the efficiency can be pre-determined at any load.

Disadvantages

The main disadvantages of Swinburne's test are

- Since Swinburne's test is performed on no-load, thus it does not indicate whether the commutation on full load is satisfactory and whether the temperature rise would be within specified limits.
- This test cannot be performed with DC series motors because at no-load series motors obtain dangerously high speeds.

- The change in iron losses is not considered from no-load to full load. At full load, due to the armature reaction, the flux is distorted which increases the iron losses.
- Stray load loss is not considered.

Limitations of the Swinburne's Test

The Swinburne's test has the following limitation –

- The Swinburne's test is only applicable to those DC machines in which the flux is practically constant, which are shunt machines and level compound generators.
- The series DC machines cannot be tested by Swinburne's test since they cannot be run on no-load and their flux and speed vary greatly.

Example.4.30 When running at no load, a 400-V shunt motor takes 5A. Armature resistance is 0.5 ohm and field resistance 200 ohm. Find the output of the motor and efficiency when running on full load and taking a current of 50 A. also, find the percentage change in speed from no load to full load.

Solution:

No-load input,

$$400 \times 5 = 2000\ Watt$$

This input goes to meet all kind on no load losses that is armature cu loss and constant losses

$$I_{sh} = \frac{400}{200} = 2\ Amp.$$

No load current, $I_a = 5 - 2 = 3\ Amp.$

No load armature copper loss, $3^2 \times 0.5 = 4.5\ Watt$

No load shunt field copper loss, $2^2 \times 200 = 800\ Watt$

Constant loss, 2000-4.5$-800 = 1195.5\ Watt$

When armature current is 50 Amp.

$I_a = 50 - 2 = 48\ Amp.$

Armature copper loss, $48^2 \times 0.5 = 1152\ Watt$

Total loss on full load $= 1152 + 1195.5 + 800 = 3147.5\ Watt$

Input, $50 \times 400 = 20000\ Watt$

Output$= 20000 - 3147.5 = 16852.5\ Watt = 16.8\ kW$

Full load efficiency

$$= \frac{16852.5}{20000} = 0.8426\ or\ 84.26\%$$

Now, $E_{b1} = 400 - (3 \times 0.5) = 398.5\ V$

$E_{b1} = 400 - (48 \times 0.5) = 376\ V$

$$\frac{N_1}{N_2} = \frac{E_{b1}}{E_{b2}} = \frac{398.5}{376}$$

$$\frac{N_1 - N_2}{N_2} = \frac{22.5}{376}$$

Percentage change in speed=5.98%

4.31.3. Hopkinson's test (Back-to-Back test)

Hopkinson's test is a method of testing the efficiency of DC machines. Hopkinson's test is known as the *regenerative test* or *back-to-back test* or *heat-run test.* This test requires two identical shunt machines which are mechanically coupled and also connected electrically in parallel. One machine act as a motor and the other as a generator. The motor takes its input from the supply and the mechanical output of the motor drives the generator. The electrical output of the generator is used in supplying the input to the motor. Therefore, the output of each machine is fed as input to the other.

In the process, the two machines draw electrical power or energy to meet the losses of the two machines. Since the mechanics are identical, the losses in each machine are determined by dividing the input into two equal parts. Usually, this test is performed on large-size machines at full-load for longer duration.

Circuit diagram and procedure for performing the test

The connection diagram of Hopkinson's test is shown in fig. 4.53. In the connection diagram, the machine M acts as a motor and is started from the supply with the help of starter. The switch S is kept open. The field current of the machine M is adjusted with the help of field rheostat R_m to make the motor to run at its rated speed. Machine G acts as a generator.

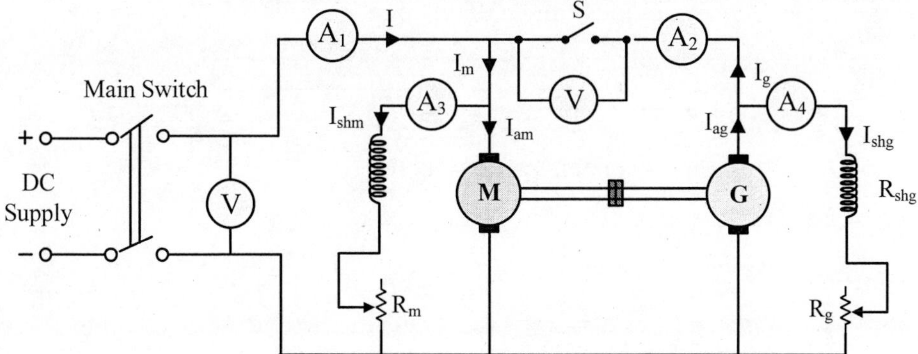

Fig. 4.53 Circuit arrangement for Hopkinson's test

As the G is driven by the machine M, hence it runs at rated speed of M. The field current of machine G is so adjusted with the help of its field rheostat R_g that the armature voltage of the generator G is somewhat higher than the supply voltage. When the voltage of the generator is equal to and of the same polarity of the busbar voltage (voltmeter across Switch S reads zero), the switch S is closed and the generator is connected to the busbar.

Now, both the machines are connected in parallel across the supply voltage. Under this condition, the generator neither taking any current nor giving any current to the supply, thus it is said to float. Now, by adjusting the excitation of the machines with the help of the field rheostats, any load can be thrown on the machines.

Here, Input power from the supply $= VI$ (4.159)

This input power from the supply is equal to the total losses of both machines.

$$\text{Armature Cu loss of motor} = I_{am}^2 R_a \tag{4.160}$$

$$\text{Field Cu loss of the motor} = I_{shm}^2 R_{shm} \tag{4.161}$$

$$\text{Armature Cu loss of generator} = I_{ag}^2 R_a \tag{4.162}$$

$$\text{Field Cu loss of the generator} = I_{shg}^2 R_{shg} \tag{4.163}$$

As the two machines are identical, the constant losses of both the machines P_C are assumed to be equal and given by,

$$P_C = (Power\ input\ from\ supply) - (Armature\ and\ shunt\ cu\ losses\ of\ both\ machines)$$

$$P_C = VI - (I_{am}^2 R_a + I_{shm}^2 R_{shm} + I_{ag}^2 R_a + I_{shg}^2 R_{shg}) \tag{4.164}$$

It was assumed that the constant losses are being equally divided between the two machines.

$$\therefore Constant\ loss\ per\ machine = \frac{P_C}{2} \tag{4.165}$$

Now, the efficiency of two machines can be determined as follows-

$$\text{Generator output} = VI_{ag} \tag{4.166}$$

$$\text{Constant losses for generator} = \frac{P_C}{2} \tag{4.167}$$

$$\text{Armature Cu loss of generator} = I_{ag}^2 R_a \tag{4.168}$$

$$\text{Field Cu loss of generator} = I_{shg}^2 R_{shg} \tag{4.169}$$

Therefore, the efficiency of the generator is given by,

$$\eta_g = \frac{Output}{Output + Losses} = \frac{VI_{ag}}{VI_{ag} + I_{ag}^2 R_a + I_{shg}^2 R_{shg} + \frac{P_C}{2}} \tag{4.170}$$

Efficiency of motor

$$\text{Motor input} = VI_m = V\,(I_{am} + I_{shm}) \tag{4.171}$$

$$\text{Constant losses for motor} = \frac{P_C}{2} \tag{4.172}$$

$$\text{Armature Cu loss of motor} = I_{am}^2 R_a \quad (4.173)$$

$$\text{Field Cu loss of motor} = I_{shm}^2 R_{shm} \quad (4.174)$$

Therefore, the efficiency of the motor given by,

$$\eta_m = \frac{Input - Losses}{Input}$$

$$= \frac{[V\,(I_{am} + I_{shm})] - \left[I_{am}^2 R_a + I_{shm}^2 R_{shm} + \frac{P_C}{2}\right]}{V\,(I_{am} + I_{shm})} \quad (4.175)$$

Advantages of Hopkinson's test

The advantages of Hopkinson's test for determination of efficiency of DC machines are –

- This method is very economical as the power required from the supply is very low.
- The commutation conditions and the temperature rise can be studied under rated load conditions properly.
- Since the test is conducted at full load, any change in iron losses due to flux distortion at full load is taken into account.
- Efficiency at different loads can be determined.
- Large DC machines can be tested without wasting much amount of power.
- The stray losses are taken into account since both the machines are operated under rated load conditions.

Disadvantages of Hopkinson's test

- The main disadvantage of this test is the necessity of two practically identical DC machines.

Example.4.31 In a Hopkinson's test on two 200V, 100 Kw, the circulating current is equal to the full load current and in addition 90 A are taken from the supply. Obtain the efficiency of each machine.

Solution:

Output current of generator,

$$I_g = \frac{100{,}000}{220} = 454.4\ Amp., I = 90\ Amp.$$

Assuming equal efficiencies, we have calculated according to fig. 4.53

$$\eta = \sqrt{\frac{I_g}{I + I_g}} = \sqrt{\frac{454.5}{454.5 + 90}} = 0.914\ or\ 91.4\%\ (\text{Ans.})$$

Example.4.32 The Hopkinson's test on two shunt machines gave the following result for full load: Line voltage=250 V; current takes from supply system excluding field current=50 A; motor armature current=380 V; field current 5A and 4.2 A. calculate the efficiency of the machine working as a generator. Armature resistance of each machine is 0.2 ohm.

Solution:

Motor armature copper loss, $= 380^2 \times 0.02 = 2888\ Watt$

Armature current=380-50=330

Generator armature copper loss, $= 330^2 \times 0.02 = 2178\ Watt$

Power drawn from supply, $250 \times 50 = 12500\ Watt$

Stray losses for two machines, $=12500 - (2888 + 2178) = 7434\ Watt$

Stray losses per machine, $= 7434/2 = 3717\ Watt$

Motor efficiency,

Armature copper loss, $= 2888\ Watt$

No-load input, $= 250 \times 4.2 = 1050\ Watt$

Stray losses, $= 3717\ Watt$

Total Losses $= 2888 + 1050 + 3717 = 7655\ Watt$

Motor input, $= 250 \times 380 + 250 \times 4.2 = 96050\ Watt$

Motor output, $= 96050 - 7655 = 88395\ Watt$

$$\eta = \frac{88395}{96050} = 0.9203\ or\ 92.03\%$$

Generator efficiency

Armature copper loss, $= 2178\ Watt$

Field copper loss, $= 250 \times 5 = 1250\ Watt$

Stray losses, $= 3717\ Watt$

Total Losses $= 2178 + 1250 + 3717 = 7145\ Watt$

Generator Output, $= 250 \times 330 = 82500\ Watt$

Generator input,$= 82500 + 7145 = 89645\ Watt$

$$\eta = \frac{82500}{89645} = 0.9202\ or\ 92.02\%$$

4.32 Field Test for Series Motor

Small DC series machines can be tested by brake test but large DC series machines cannot be tested by brake test because neither it is convenient nor possible to develop a mechanism to apply load on such large machines directly. Moreover, DC series machines cannot be tested by Swinburne's test, because at no-load these machines obtain dangerously high speeds. In view of this *Field Test* is considered to be most suitable for determining efficiency of these machines.

This test is applicable to two similar series motor. Series motor which are mainly used for traction work are easily available in pairs. The two machines are coupled mechanically. One machine runs normally as a motor and drive generator whose output is wasted in a variable load R as shown in fig. 4.54. iron and friction losses of two machines are made equal (i) by joining the series field winding of the generator in the motor armature circuit so that both machines are equally excited and (ii) by running them at equal speed. Load resistance R_L is varied till the motor current reaches its full load value indicated by ammeter A_1. After this adjustment for full load current, different ammeter and voltmeter reading are noted.

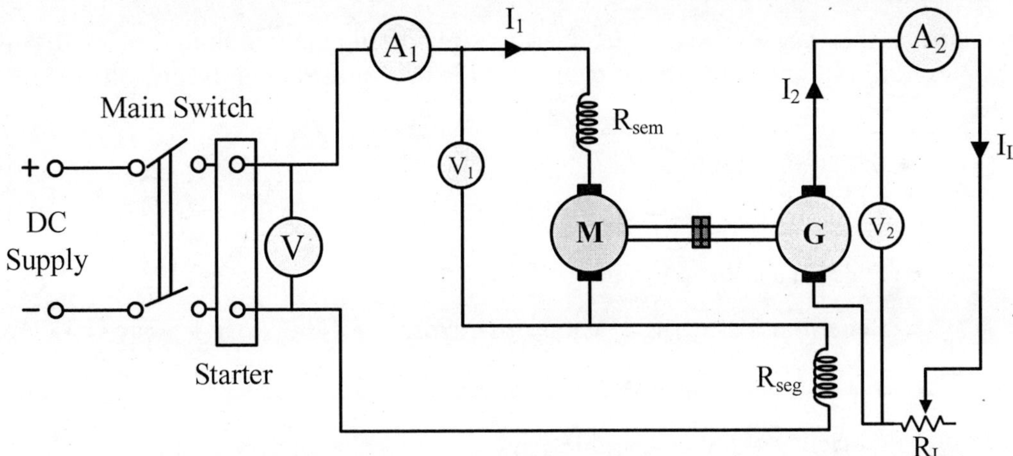

Fig. 4.54 Circuit arrangement for field test of DC series motor

Let V= Supply Voltage; I_1= Motor Current;

V_2= Terminal Voltage of generator; I_2= Load Current

R_a= Armature resistance of each machine;

R_{se}= series field resistance of each machine;

Now

Input to whole set= VI_1 (4.176)

output = V_2I_2 (4.177)

Total losses in the set, $W_t = VI_1 - V_2I_2$ (4.178)

Armature and field copper losses, $W_{cu} = (R_a + 2R_{se})I_1^2 + I_2^2R_a$ (4.179)

Stray losses for the test $W_s = W_t - W_{cu}$

$$\text{Stray losses per Machine} = \eta = \frac{W_t - W_{cu}}{2} \tag{4.180}$$

Stray losses are equally divided between the machine because of their equal excitation and speed.

Efficiency of motor

$$\text{Motor input} = V_1 I_1 \tag{4.181}$$

$$\text{Motor losses; } W_m = \text{Armature Cu loss} + \text{Field Cu loss} + \text{Stray Loss}$$

$$= (R_a + R_{se}) I_1^2 + W_S \tag{4.182}$$

Therefore, the efficiency of the motor given by,

$$\eta_m = \frac{Input - Losses}{Input}$$

$$= \frac{V_1 I_1 - W_m}{V_1 I_1} \tag{4.183}$$

Efficiency of Generator

The generator efficiency will be little use because it is running under abnormal conditions of separate excitation. However, the efficiency under this unusual condition can be found if desired.

$$\text{Generator output} = V_2 I_2 \tag{4.184}$$

$$\text{Armature Cu loss of generator} = I_2^2 R_a \tag{4.185}$$

$$\text{Field Cu loss of generator} = I_1^2 R_{se} \tag{4.186}$$

$$\text{Generator losses; } W_g = \text{Armature Cu loss} + \text{Field Cu loss} + \text{Stray Loss}$$

$$= I_2^2 R_a + I_1^2 R_{se} + W_s \tag{4.187}$$

Therefore, the efficiency of the motor given by,

$$\eta_m = \frac{Output}{Output - Losses}$$

$$= \frac{V_2 I_2}{V_2 I_2 - W_g} \tag{4.188}$$

It should be noted that although the two machines are mechanically coupled yet it not a regenerative method, because the generator output is wasted instead of being fed back into the motor as in Hopkinson's (Back-to-Back test).

Disadvantages of Field Test Method

- Even for a small error in the measurement of the input to motor or output of generator may cause a relatively large error in computed efficiency.
- Whole of the power supplied to the set is wasted.

Example.4.33 A test on two coupled similar motor, with their field connected in series, gave the following result when one machine acted as a motor and other as a generator.

Motor: Armature current=56 A; Armature voltage=590 V; voltage drop across field winding=40 V.

Generator: Armature current=44 A; Armature voltage=400 V; Voltage drops across field winding=40 V; Resistance of each armature=0.3 ohm.

Calculate the efficiency of the motor and gearing at this load.

Solution:

$$\text{Total input,} = 590 \times 56 = 33040\ Watt$$

$$\text{Output,} = 400 \times 44 = 17600\ Watt$$

$$\text{Total Losses in two machines are} = 33040 - 17600 = 15440\ Watt$$

$$\text{Series field resistance, } R_{se} = \frac{40}{56} = 0.714\ ohm$$

$$\text{Total copper loss,} = (0.3 + 2 \times 0.714) \times 56^2 + 44^2 \times 0.3 = 6006\ Wat$$

$$\text{Stray losses for two machines,} =15440 - 6006 = 9434\ Watt$$

$$\text{Stray losses per machine,} = 9434/2 = 4717\ Watt$$

Motor efficiency,

$$\text{Armature copper loss,} = (0.3 + 0.714) \times 56^2 = 3180\ Watt$$

$$\text{Motor input,} = 590 \times 56 = 33040\ Watt$$

$$\text{Stray losses,} = 5837 - find\ above$$

$$\text{Total Losses} = 3180 + 5837 = 9017\ Watt$$

$$\text{Motor output,} = 33040 - 9017 = 24023\ Watt$$

$$\eta = \frac{24023}{33040} = 0.727\ or\ 72.70\%$$

Generator efficiency

$$\text{Armature copper loss,} = 44^2 \times 0.3 = 581\ Watt$$

$$\text{Series field copper loss,} = 40 \times 56 = 2240\ Watt$$

$$\text{Stray losses,} = 5837\ Watt$$

$$\text{Total Losses} = 581 + 2240 + 5837 = 8658\ Watt$$

$$\text{Generator Output,} = 400 \times 44 = 17600\ Watt$$

$$\eta = \frac{17600}{17600 + 8658} = 0.67\ or\ 67\%$$

4.33 Necessity of a Starter

The current drawn by a motor armature is given by the relation

$$I_a = \frac{(V - E_b)}{R_a} \tag{4.189}$$

Where V is supply voltage, E_b the back emf and R_a the armature resistance.

When the motor is at rest, there is, as yet, obviously no back emf developed in the armature ($E_b \propto N$). If, now, full supply voltage is applied across the stationary armature, it will draw a very large current ($I_a = V/R_a$) because armature resistance is relatively small. *Consider the case of a 230-V, 3 H.P. (2.2371 kW) motor having an armature resistance of 0.15 Ω and a full-load current of 30 A. If this motor is started from the line directly, it will draw a starting current of 230/0.15 = 1533.33 A which is 1533.33/30 = 51.11 times its full-load current.* This heavy starting current has the following effects:

- It will blow out the fuses and prior to that it may damage the insulation of armature winding due to excessive heating effect if starting period is more.
- Excessive voltage drop will occur in the lines to which the motor is connected. Thus, the operation of the appliances connected to the same line may be impaired and, in some cases, they may refuse to work.

To avoid this, a resistance is introduced in series with the armature (for the duration of starting period only, say 5 to 10 seconds) as shown in fig. 4.55 which limits the starting current to safe value $I_a = \frac{(V - E_b)}{R_a + R}$. The starting resistance is gradually cut out as the motor gains speed and develops the back emf which then regulates its speed.

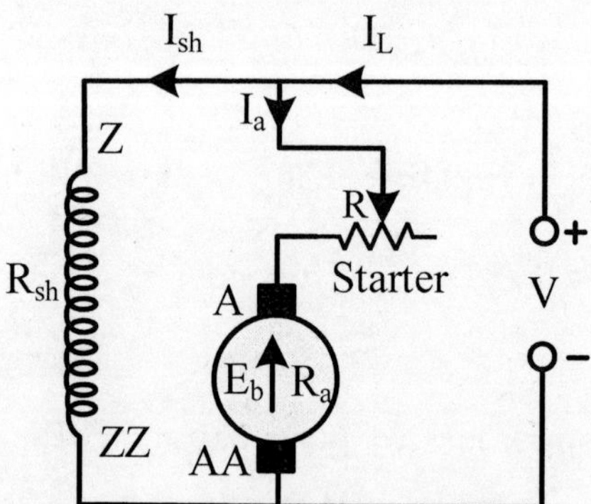

Fig. 4.55 DC Shunt motor starter

Very small motors may, however, be started from rest by connecting them directly to the supply lines. It does not result in any harm to the motor for the following reasons:

- Such motors have a relatively high armature resistance than large motors, hence their starting current is not so high.
- Being small, they have low moment of inertia, hence they speed up quickly.

- The momentary large starting current taken by them is sufficient to produce a large disturbance in the voltage regulation of the supply lines.

Another important feature of a starter is that it contains protective devices such as an overload protection coil (or relay) which provides necessary protection to the motor against overloading and no-volt release coil.

4.34 Starters for DC Shunt and Compound Wound Motors

Starter is limiting the current in the armature circuit during starting or accelerating period. Starters are always rated based on the output power and voltage of the motor with which they are to be employed. The simplest type of starter is just a variable resistance (a rheostat) connected in series with the armature alone as shown in fig. 4.55. It may be noted that shunt field is kept independent of starting resistance. It is because when supply is connected, it receives normal rated voltage and sets-up maximum flux. A higher value of flux results in a low operating speed and a higher motor torque for a particular value of starting current since speed is inversely proportional to flux per pole ($N \propto 1/\phi$) whereas motor torque is proportional to the product of flux per pole and armature current ($T \propto \phi I_a$) Thus, the heating effect on armature winding is reduced.
For all practical applications, this starter is further modified which includes protective devices such as Over Load Release (OLR) and No-Volt Release Coil (NVRC). The conventional starters used for starting shunt and compound motors of ordinary industrial capacity are of two kinds known as three-point and four-point starters respectively.

4.35 Three Point Starter

The schematic connection diagram of a shunt motor starter is shown in fig. 4.56. It consists of starting resistance R divided into several sections. The tapping points of starting resistance are connected to number of studs. The last stud of the starting resistance is connected to terminal A to which one terminal of the armature is connected. The +ve supply line is connected to the line terminal L through main switch. From the line terminal, the supply is connected to the starting lever SL through Over Load Release (OLR) coil $OLRC$. A spring S is placed over the lever to bring it to the OFF position when supply goes OFF. A soft iron piece SI is attached with the starting lever which is pulled by the no-volt release coil under normal running conditions. The far end of the brass strip BS is connected to terminal Z through a No Volt Release Coil (NVRC). One end of the shunt field winding is connected to Z terminal of the starter. An iron piece is lifted by $OLRC$ under abnormal condition to short circuit the NVRC. The negative supply line is connected directly to the other ends of shunt field winding and armature of the DC shunt motor.

Operation

First of all, the main switch is closed with starting lever resting in OFF position. The handle is then turned clockwise to the first stud and brass strip. As soon as it comes in contact with first stud, whole of the starting resistance R is inserted in series with the armature and the field winding is directly connected across the supply through a brass strip. As the handle is turned further the starting resistance is cut OFF from the armature circuit in steps and finally entire starting resistance is removed from armature circuit.

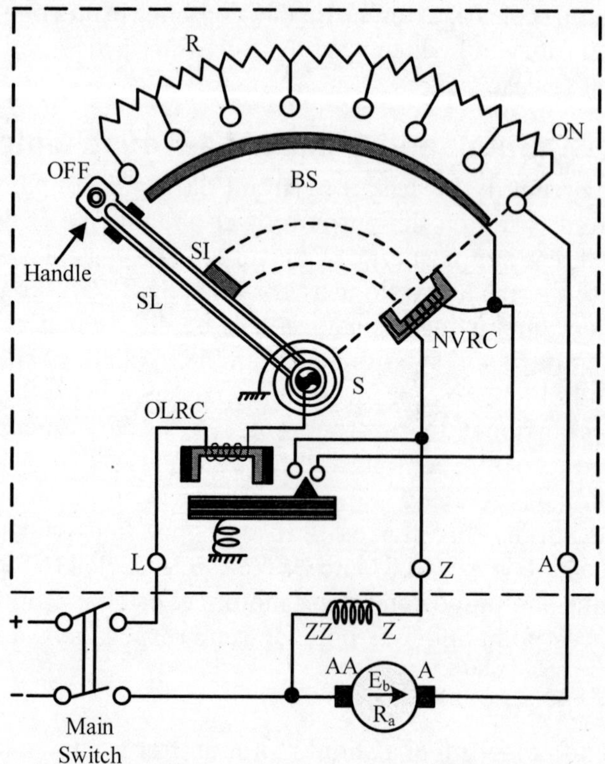

Fig. 4.56. Three Point Starter

No-volt Release Coil (NVRC) function

A No Volt Release Coil (NVRC) is a small electromagnet having many turns of fine wire. It is connected in series with shunt field winding and therefore, carries a small field current. When the handle is turned to ON position, the NVRC is magnetized by the field current and holds the starting lever at ON position. In case of field-failure or disconnection of the supply, this coil is demagnetized and the lever comes to the OFF position due to spring tension. Consequently, the motor is disconnected from the supply. If the spring with the NVRC is not used the lever would remain in *ON*-position in case of supply failure. Again, when the supply comes, the motor would be connected directly to the lines without starter. The other important advantage of connecting the NVRC in series with the shunt field winding is that due to an accident if the circuit of field winding becomes open, the NVRC will be demagnetized and the starting lever is immediately pulled back to OFF position by the spring. Otherwise, the motor would have attained dangerously high speed. In three-point starter, NVRC protects the No-Volt condition as well as Field failure protection. However, in case of Four Point starter, it provides only protection against No-Volt conditions.

Over-load Release Coil (OLRC)function
An Over-Load Release Coil (OLRC) is a feeble electromagnet having small number of turns of thick wire. It is connected in series with the motor and carries the line current. When the motor is overloaded (or short-circuited), a heavy current of more than predetermined value will flow through it. Then, the iron piece (armature or plunger) is lifted and short circuits the NVRC. Hence the starting lever is released and pulled back to the OFF position due to spring tension. Thus, the motor is disconnected from the supply and is protected against overloading.

Drawback of Three Point Starter
The 3-point starter suffers from a serious drawback for motors with a large variation of speed by adjustment of the field rheostat. For increasing the speed of shunt motor - field circuit resistance is increased, which results in decrease in shunt field current. In three-point starter, the holding electromagnet coil of NVRC is connected in series with field circuit.

When field current becomes very low which results in holding electromagnet too weak to overcome the force exerted by the spring. The holding magnet may release the arm of the starter during the normal operation of the motor and thus disconnect the motor from the line. This is not desirable. A 4-point starter is thus used instead, which does not have this drawback.

For speed control of DC shunt or compound motors, a rheostat (variable resistor R_b) is connected in series with the field winding, as shown in fig. 4.57. In this case, if a three-point starter is used and the value of R_b is so adjusted that the current flowing through the shunt field winding is very small. It may be seen that the same current flows through the No-Volt Release Coil, then the magnetic strength of the coil may be insufficient to hold the plunger at its *ON* position. This is an undesirable feature of a three-point starter. This feature makes a three-point starter unsuitable for such applications.

4.36 4-Point Starter

A three-point starter provides inherent Field- Failure protection, but it is not suitable for applications where speed of shunt motor is to be increased by field weakening. Accordingly, a four-point starter is designed, as shown in fig. 4.57 for shunt motor, in which the current flowing through the No-Volt Release Coil is made independent of the shunt field circuit. A circuit used for 4-point starter with a compound wound machine has been shown in fig. 4.58.

Operation
The working of a four-point starter is similar to a three-point starter with slight changes. In this case, when the plunger touches the first stud, the line current is divided into the following three parts:

- First part passes through starting resistance and armature (as well as in series field for compound motors).
- Second part passes through the field winding (and speed control resistance if applied) and
- The third part passes through NVRC and protective resistance is connected in series with the coil.

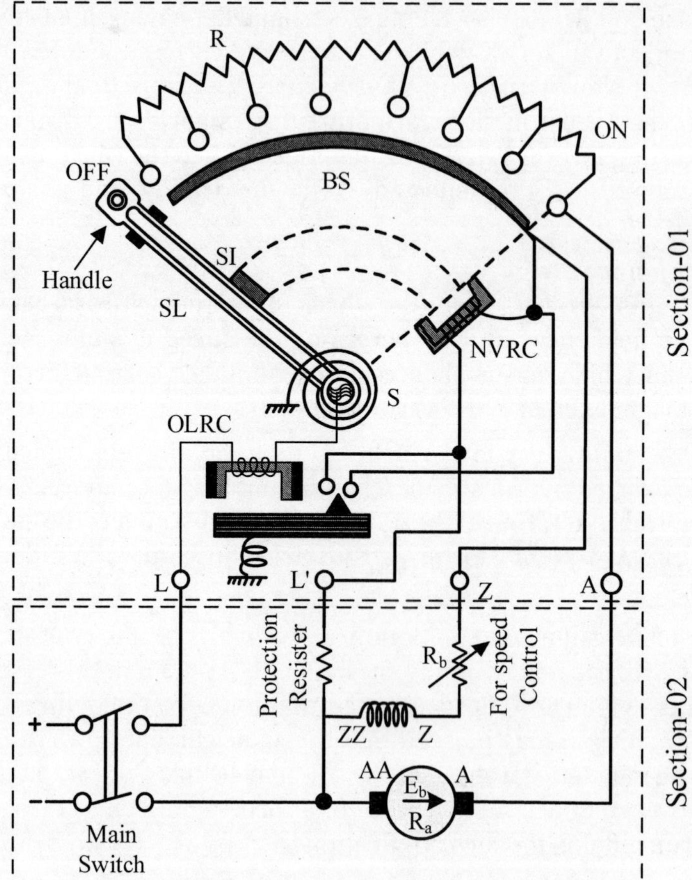

Fig. 4.57 Four-point starter for DC shunt motor

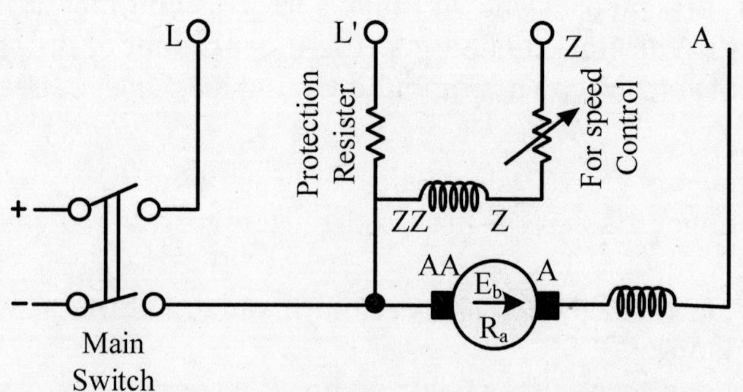

Fig. 4.58 Four-point starter circuit for DC compound motor

It is evident that as a four-point starter, the current through NVRC circuit is independent of the field circuit current. Therefore, the change of current in the field circuit does not affect the pull exerted by the holding coil which remains always sufficient to prevent the spiral spring from restoring the plunger to its *OFF* position. While starting a motor with a four-point starter it is necessary to ensure that the field circuit is closed and the rheostat connected in series with the shunt field winding must be at zero resistance position. Moreover, the whole of the starting resistance must come in series with the armature. Whenever a shunt motor is required to be stopped, it must be stopped by opening the line switch. In fact, this switch can be opened without any appreciable arc since the motor develops back emf nearly equal to the applied voltage and the net voltage across the switch contacts is very small. The electromagnetic energy stored in the field does not appear at the switch but it gradually discharges through the armature. The motor should never be stopped by bringing the plunger (starting arm) back to the *OFF* position, because in such cases when the field circuit breaks at the last stud placed near the *OFF* position, a heavy spark occurs owing to the inductive nature of the field. Usually, this sparking burns the contact. Moreover, while stopping the motor, the value of resistance connected in the field circuit should always be reduced to zero so that speed of motor falls to its normal value. This ensures that, when the motor is started next time, it must start with a strong field and higher starting torque.

4.37 Grading of Starting Resistance for Shunt Motors

Starting torque would be small in designing shunt motor starters, it is usual to allow an overload of 50% for starting and to advance the starter a step when armature current has fallen to definite lower value. For starting a motor from standstill to its rated speed, it is normally desirable to increase the speed gradually to maintain the angular acceleration constant during the starting period. The angular acceleration is proportional to the net torque, which is in turn nearly proportional to the product of flux (ϕ) and armature current I_a i.e., $T \propto \phi I_a$. The flux ϕ will remain constant provided line voltage remains constant. Hence it follows that substantially constant angular acceleration calls for constant armature current during starting period.

We know that, $$N \propto \frac{E_b}{\phi} \propto \frac{V - I_a R}{\phi} \tag{4.190}$$

where R = resistance of armature plus the resistance of starter.

Since I_a and ϕ are to be kept constant, therefore, for increasing the speed (N) gradually, R should be varied (reduced) in such a way that the above relation must be satisfied. For different values of armature current, the value of R is given by the expression,

from Equation (4.190)

$$N\phi \propto V - I_a R \tag{4.191}$$

$$KN \propto V - I_a R \tag{4.192}$$

$$R = \frac{V - KN}{I_a}, \text{Where K is constant} \tag{4.193}$$

In starters, usually, the value of R is changed in steps and, therefore, armature current will change in two extreme values. Accordingly, the steps of the starter are designed in such a way that armature current varies in between these limits so that torque may not change to a greater degree. The fuse or *MCB* placed in the motor circuit is usually not larger than 150% of the motor full load current. Let I_1 and I_2 be the maximum and minimum value of the current drawn by the motor during starting. Let the starter have n-sections each having a resistance as r_1, r_2, r_3...r_n as shown in fig.4.59. Let the total resistance of armature circuit when the starting arm is kept at stud No. 1, 2, 3, ... n and n + 1 be R_1, R_2, R_3, ...R_n and R_{n+1} (where R_{n+1} = R_a), respectively.

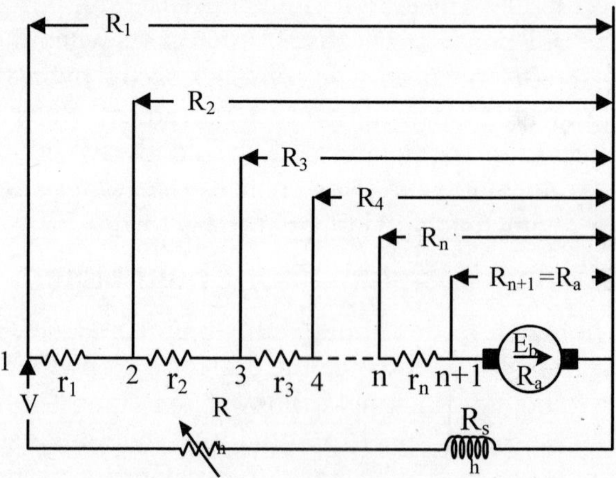

Fig.4.59 Design of shunt motor starter

I_1 the maximum permissible armature current at the start (I_{max}), when the starting arm is brought in contact with stud-1, the motor is stationary and there is no emf. Usually, current is limited to 1.5 times the full-load current of the motor. Hence, the motor develops 1.5 times its full-load torque and accelerates very rapidly. The maximum current is given by

$$I_1 = \frac{V}{R_1} \tag{4.194}$$

when the armature current has fallen to some predetermined value I_2 (also called I_{min}) arm is moved to stud No. 2. Let the value of back e.m.f. be E_{b1} at the time of leaving stud No. 1. Then

$$I_2 = \frac{V - E_{b1}}{R_1} \tag{4.195}$$

It should be carefully noted that I_1 *and* I_2 [(I_{max}) *and* (I_{min})] are respectively the maximum and minimum currents of the motor. When arm touches stud No. 2, then due to diminution of circuit resistance, the current again jumps up to its previous value I_1. Since speed had no time to change, the back e.m.f. remains the same as initially.

$$I_1 = \frac{V - E_{b1}}{R_2} \tag{4.196}$$

From (4.195) and (4.196), we get

$$\frac{I_1}{I_2} = \frac{R_1}{R_2} \tag{4.197}$$

When the arm is held on stud No. 2 for some time, then speed and hence the back emf increases to a value E_{b2}, thereby decreasing the current to the previous value I_2, so that

$$I_2 = \frac{V - E_{b2}}{R_2} \tag{4.198}$$

Similarly, on first making contact with stud No. 3, the current is

$$I_1 = \frac{V - E_{b2}}{R_3} \tag{4.199}$$

From (4.198) and (4.199), we again get

$$\frac{I_1}{I_2} = \frac{R_2}{R_3} \tag{4.200}$$

When the arm is held on stud No. 3 for some time, the speed and hence back emf increases to a new value E_{b3}, thereby decreasing the armature current to value I_2, such that

$$I_2 = \frac{V - E_{b3}}{R_3} \tag{4.201}$$

On making contact with stud No. 4, the current jumps to I_1 given by

$$I_1 = \frac{V - E_{b3}}{R_a} \tag{4.202}$$

From (4.201) and (4.202), we get

$$\frac{I_1}{I_2} = \frac{R_3}{R_a} \tag{4.203}$$

From (4.197), (4.200), and (4.203), it is seen that

$$\frac{I_1}{I_2} = \frac{R_1}{R_2} = \frac{R_2}{R_3} = \frac{R_3}{R_a} = K \tag{4.204}$$

Obviously,

$$R_3 = KR_a;\ R_2 = KR_3 = K^2R_a \tag{4.205}$$

$$R_1 = KR_2 = K.K^2R_a = K^3R_a \tag{4.206}$$

In general, if n is the number of live studs and therefore (n-1) the number of sections in the starter resistance, then from equations (4.205) and (4.206)

$$R_1 = K^{n-1}.R_a \tag{4.207}$$

$$\frac{R_1}{R_a} = K^{n-1} \tag{4.208}$$

$$\left(\frac{I_1}{I_2}\right)^{n-1} = \frac{R_1}{R_a} \tag{4.209}$$

Since I_1 would be given, R_1 can be found from $R_1 = V/I_1$.

Since n is known, K can be found from $R_1/R_a = K^{n-1}$ and the lower current limit I_2 from $I_1/I_2 = K$.

4.38 Electric Braking of DC Motor

The DC motors can be stopped using one of the following methods –

- Mechanical (Friction) Braking
- Electric Braking

In mechanical braking, the motor is stopped due to friction between the moving parts of the motor and the brake shoe. Mechanical braking has several disadvantages as non-smooth stop, wear and tear of moving parts, braking power wasted as heat and greater stopping time etc. In electric braking, the kinetic energy of moving parts of the motor is converted into electrical energy which is either dissipated in a resistance or returned to the supply source.

Types of Electric Braking

There are three types of electric braking methods of a DC motor

- Rheostat Braking or Dynamic Braking
- Regenerative Braking
- Plugging or Counter (reverse) Current Braking

4.38.1 Rheostat Braking or Dynamic Braking

In dynamic braking, the armature of running DC motor is disconnected from the supply and is connected across a braking resistance R_b as shown in fig. 4.60. However, the field winding is left connected to the supply. Hence, the motor now works as a generator and produces a braking torque. This method is also known as rheostat braking since an external resistance R_b is connected across the armature for the electric braking. During dynamic electric braking when the motor works as a generator, the kinetic energy of moving parts of the motor is converted into electrical energy and is dissipated in the form of heat in the braking resistance R_b and the armature circuit resistance R_a. As a result, the motor is brought to standstill quickly. In case of Dynamic Braking, the current changes its direction, the Braking torque is opposite to direction of rotational. Lower is the Dynamic Braking resistance, larger is Braking torque - motor quickly comes to rest. Dynamic braking or rheostat braking is an inefficient method of braking since all the generated energy is dissipated in the form of heat in the resistance.

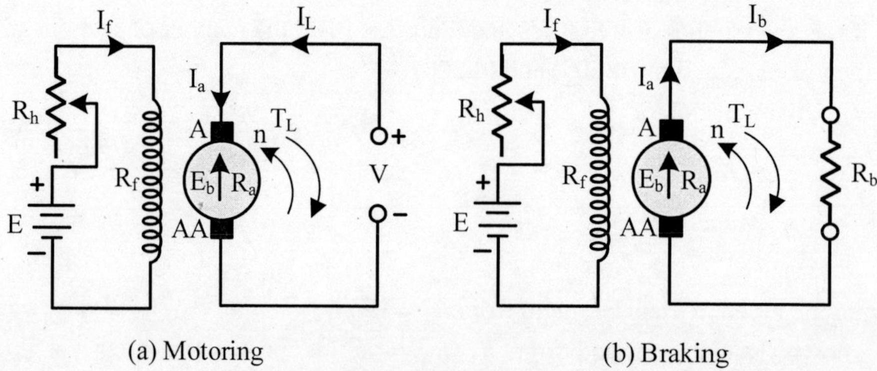

(a) Motoring (b) Braking

Fig.4.60 Rheostat Braking or Dynamic Braking

4.38.2 Regenerative Braking

In regenerative braking, the motor is operated as a generator so the kinetic energy of the moving parts of the motor is converted into electrical energy so that the load drives the motor above its nominal speed. This electrical Energy is then fed back to supply source. The negative regenerative torque produced restricts the overspeeding of motor beyond nominal speed.

Regenerative braking is only possible when the driven load forces the motor to run at a speed greater than the no-load speed with a constant field excitation. Under this condition, the back EMF (E_b) of the motor is more than the supply voltage, which reverses the armature current of the motor. Therefore, the motor now begins to operate as a generator and the generated electrical energy is transferred to the supply source.

Regenerative braking cannot be used for stopping the motor. It is just used for controlling the speed above the no-load speed of the DC motors. The necessary condition for regenerative braking is that the back emf of the motor must be greater than the supply voltage so that the armature current is reversed and the motoring operation is changed to the generating operation.

Regenerative braking is mainly used to control the speed of DC motors driving the loads such as electric locomotives, elevators, cranes and hoists etc.

4.38.3 Plugging or Reverse Current Braking

In the plugging or reverse current braking, the connections of the armature are reversed so that the motor tends to rotate in the opposite direction and provides the necessary braking effect. When the motor comes to rest, the supply source must be disconnected otherwise the motor will start rotating in the opposite direction.

When the armature connections are reversed, the supply voltage and the back EMF will act in the same direction. Hence, during the braking, the resultant voltage across the armature will be equal to ($V + E_b$), which is approximately double the supply voltage. This reverses the armature current and hence a high braking torque is produced. To limit the armature current to a safe value, a current-limiting resistor is connected in series with the armature.

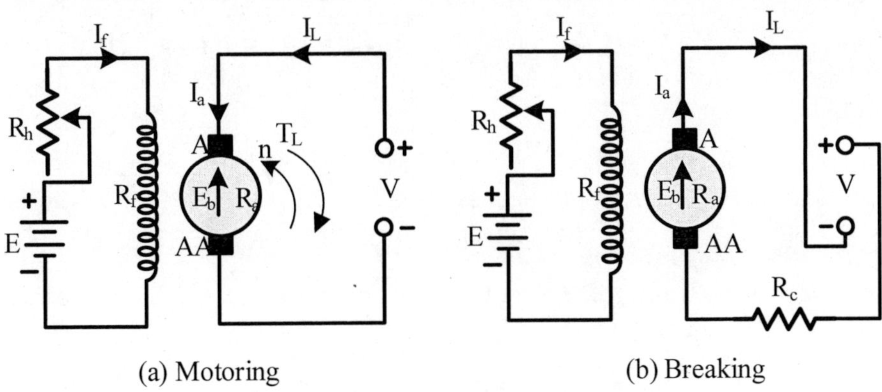

Fig.4.60 Plugging Reverse Braking

In plugging the power supplied by the moving parts as well as power supplied by the source is wasted in resistances. Plugging is mainly used in controlling rolling mills, elevators, machine tools, and printing presses etc.

4.39 Inspection/maintenance of DC Machines

The basic purpose of inspection/maintenance is to ensure uninterrupted and efficient service from the machine. The frequency of routine inspection/maintenance depends upon the conditions under which a machine is operating. The normal routine inspection/maintenance tasks are discussed below:

a. Mechanical maintenance

- ***Bearings:*** Check sound and excessive play if any. Check for proper lubrication-avoid excessive lubrication.
- ***General***: Check all the fixing fixtures – tight them if required.
- Check that all the cover plates and enclosures are clean and correctly in place – if not fix them.

b. Electrical maintenance

Always keep the insulation dry and in good condition for better reliability.

- ***Commutator:*** Trouble may be caused due to dirt deposits-, carbon dust in the grooves between segments, rough surface or eccentricity. Check for these and clean the commutator surface with a dry soft fluff free cloth if required. In large machinery with a commutator and brush arrangement, commutator's carbon dust should be removed by blowing of compressed air. It should be frequently clean using CTC (Carbon Tetra Chloride) liquid.
- ***Brushes:*** Trouble may be caused due to less spring tension, wrong brush positioning, insufficient contact surface or too short a brush. Check for there and do the adjustments or changes as per the need.
- ***Winding***: Trouble may be caused due to deterioration of insulation. Check the insulation resistance of armature winding and field winding periodically – if faulty, it should be repaired.
- ***Insulation resistance***: The insulation resistance of the whole machine be checked periodically which should not be less than 1 MΩ. If its value is found to be less than this then the machine should be cleaned and dried.
- ***Earthing***: Outer frame of the motor must be properly earth, the earth resistance should be checked, if found more pour water in earthing pit.
- ***Starter***: Check the functioning of the starter, its relay must be sensitive enough to protect the machine.

4.40 Fault in DC Machine

In DC machines faults may occur in field windings or in armature winding.

a. ***Faults in field winding***: The faults in field winding may

- ***Open circuit fault***: Check the winding for open circuit, if found faulty, check the opening at the terminals if so tight-up the terminals otherwise replace the field winding.
- ***Short circuit fault:*** Check for the voltage drop in each turn and locate the fault and replace the turn or repair the field winding.
- ***Earth fault***: Check for the earth fault, if found remove it or replace the coil.

b. ***Faults in armature winding:*** In armature winding, following faults may develop:

- ***Short-circuit in coils***: The turns placed in the same slot of the armature may get short circuited due to insulation failure between the turns. Heavy local current flows and heat is produced. This may lead to open circuit. The heavy current is purely local and it does not over-load the brushes.
- ***Open-circuit in coils:*** Due to open circuiting of armature coil, half of the armature which carries sound coils carries double the normal current and the other portion carries no current. This produces heavy sparking at the brushes which may damage it. Therefore, the machine has to be stopped and repaired immediately.
- ***Earthed coil:*** When a coil in the armature comes in direct contact with armature stampings due to insulation rupturing, the fault is called earth fault. If this fault occurs at one place, the effect is not that serious but if this fault occurs at two places, the two coils are being short-circuited causing a serious short circuit fault that produces heavy sparking due to unbalancing of armature currents.

Solved Numerical Examples

Q1. A Long-Shunt compound generator delivers a load current of 50 A at 500 V and has armature, series field and shunt field resistance of 0.05 Ω, 0.03 Ω and 250 Ω respectively. Calculate the generated voltage and the armature current. Allow 1 V per brush for contact drop.

Solution:

$$I_{sh} = 500/250 = 2\ A$$

Current through armature and series winding is

$$= 50 + 2 = 52\ A$$

Voltage drops on series field winding

$$= 52\ \times 0.03 = 1.56\ V$$

Armature voltage drop

$$I_a R_a = 52\ \times 0.05 = 2.6\ V$$

Drop at brushes

$$= 2\ \times 1 = 2\ V$$

$$E_g = V + \ I_a R_a + Series\ drop + Brush\ drop$$

$$= 500 + 2.6 + 1.56 + 2 = 506.16\ V$$

Q2. A Short-Shunt compound generator delivers a load current of 30 A at 220 V and has armature, series field and shunt field resistance of 0.05 Ω, 0.30 Ω and 200 Ω respectively. Calculate the induced emf and the armature current. Allow 1.0 V per brush for contact drop.

Solution:

Voltage drops in series winding $= 30\ \times 0.3 = 9\ V$

Voltage across shunt winding $= 220 + 9 = 229\ V$

$$I_{sh} = \ 229/200 = 1.145\ A$$

$$I_a = \ 30\ + 1.145 = 31.145\ A$$

$$I_a R_a = \ 31.145\ \times 0.05 = 1.56\ V$$

Brush Drop $= 2\ \times 1 = 2\ V$

$$E_g = V + \ I_a R_a + Series\ drop + Brush\ drop$$

$$= 220 + 1.56 + 9 + 2 = 232.56\ V$$

Q3. An 8-pole DC shunt generator with 778 wave-connected armature conductors and running at 500 rpm. supplies a load of 12.5 Ω resistance at terminal voltage of 50 V. The armature resistance is 0.24 Ω and the field resistance is 250 Ω. Find the armature current, the induced emf and the flux per pole.

Solution:

Load current

$$= V/R = 250/12.5 = 20\ A$$

Shunt current

$$= 250/250 = 1\ A$$

$$\text{Armature current} = 20 + 1 = 21\ A$$

$$\text{Induced emf} = 250 + (21\ \times 0.24) = 255.04\ V$$

Now,

$$E_g = \frac{ØZN}{60} \times \left(\frac{P}{A}\right) = 255.04 = \frac{Ø \times 778\ \times 500}{60}\left(\frac{8}{2}\right)$$

$$Ø = 9.83\ mWb$$

Q4. A 4-pole, DC shunt generator with a shunt field resistance of 100 Ω and an armature resistance of 1 Ω has 378 wave-connected conductors in its armature. The flux per pole is 0.02 Wb. If a load resistance of 10 Ω is connected across the armature terminals and the generator is driven at 1000 r.p.m., calculate the power absorbed by the load.

Solution:

Induced e.m.f in the generator is

$$E_g = \frac{ØZN}{60} \times \left(\frac{P}{A}\right)\ volt = \frac{0.02\ \times 378\ \times 1000}{60}\left(\frac{4}{2}\right) = 252\ volt$$

Now, let V be the terminal voltage i.e., the voltage available across the load as well as the shunt resistance.

Load current $= V/10\ A$ and Shunt current $= V/100\ A$

Armature current

$$= \frac{V}{10} + \frac{V}{100} = \frac{11\ V}{100}$$

$$V = E_g - Armature\ drop$$

$$V = 252 - 1 \times \frac{11\ V}{100} = 227\ volt$$

Q5. A shunt generator delivers 195A at terminal p.d. of 250 V. The armature resistance and shunt field resistance are 0.02 Ω and 50 Ω respectively. The iron and friction losses equal 950 W. Find (i) E.M.F. generated (ii) Cu losses (iii) Output of the prime motor (iv) Commercial, Mechanical and electrical efficiencies.

Solution:

$$I_{sh} = 250/50 = 5\ A;$$

$$I_a = 195 + 5 = 200\ A$$

Armature voltage drop

$$= I_a R_a = 200 \times 0.02 = 4\ V$$

Generated e.m.f.

$$= 250 + 4 = 254\ V$$

Armature Cu loss

$$= I_a^2 R_a = 200^2 \times 0.02 = 800\ W$$

Shunt Cu loss $= V\ I_{sh} = 250 \times 5 = 1250\ W$

Total Cu loss $= 1250 + 800 = 2050\ W$

Stray losses $= 950\ W; Total\ losses = 2050 + 950 = 3000\ W$

Output $= 250 \times 195 = 48750\ W$

$Input = 48750 + 3000 = 51750\ W$

Output of prime mover $= 51750\ W$

Generator input $= 51750\ W$

Stray Loss$= 950\ W$

Electrical power produced in armature $= 51750 - 950 = 50800$

$$\eta_m = (50800/51750) \times 100 = 98.2\ \%$$

Electrical or Cu losses $= 2050\ W$

$$\eta_e = \frac{48750}{48750 + 2050} \times 100 = 95.9\%$$

$$\eta_c = (48750/51750) \times 100 = 94.2\%$$

Q6. A D.C. series motor takes 40 A at 220 V and runs at 800 r.p.m. If the armature and field resistance are 0.2 Ω and 0.1 Ω respectively and the iron and friction losses are 0.5 kW, find the torque developed in the armature. What will be the output of the motor?

Solution:

Armature torque is given by

$$T_a = 9.55 \frac{E_b\, I_a}{N} \; N - m$$

$$E_b = V - I_a\,(R_a + R_{se})$$

$$= 220 - 40\,(0.2 + 0.1) = 208\,V$$

$$T_a = 9.55 \times 208 \times 40/800$$

$$= 99.3\,N - m$$

Cu loss in armature and series-field resistance $= 40^2 \times 0.3 = 480\,W$

Iron and friction losses = 500 W

Total losses = 480 + 500 = 980 W

Motor power input = 220 X 40 = 8,800 W

Motor output = 8800 -980 = 7,820 W = 7.82 kW

Q7. A 220-V D.C. shunt motor runs at 500 r.p.m. when the armature current is 50 A. Calculate the speed if the torque is doubled. Given that $R_a = 0.2\,\Omega$.

Solution: -

$$T_a \propto \emptyset I_a\,,since\ \emptyset\ is\ constant, T_a \propto I_a$$

$$T_{a1} \propto I_{a1}\ and\ T_{a2} \propto I_{a2}$$

$$T_{a2}/T_{a1} = I_{a2}/I_{a1}$$

$$2 = I_{a2}/50\ or\ I_{a2} = 100\,A$$

Now, $N_2/N_1 = E_{b2}/E_{b1}$ Since ∅ remains constant.

$$E_{b1} = 220 - (50 \times 0.2) = 210\,V$$

$$E_{b2} = 220 - (100 \times 0.2) = 200\,V$$

$$N_2/500 = 200/210$$

$$N_2 = 476\,r.p.m.$$

Q8. A 4-pole, 220-V shunt motor has 540 lap-wound conductors. It takes 32 A from the supply mains and develops output power of 5.595 Kw. The field winding takes 1 A. The armature resistance is 0.09 Ω and the flux per pole is 30 mWb. Calculate: -

(i) The speed and

(ii) The torque developed in newton-meter

Solution: -

$$I_a = 32 - 1 = 31\ A;$$

$$E_b = V - I_a R_a$$

$$= 200 - (0.09 \times 31) = 217.2\ V$$

$$E_b = \frac{\emptyset ZN}{60}\left(\frac{P}{A}\right)$$

$$217.2 = \frac{30 \times 10^{-3} \times 540 \times N}{60}\left(\frac{4}{4}\right)$$

$$N = 804.4\ r.p.m.$$

$$T_{sh} = 9.55\ \frac{Output}{N}$$

$$= 9.55 \times \frac{5{,}595}{804.4} = 66.5\ N - m$$

Multiple Choice Questions

Q1. The brushes for commutator are made of

a. Copper
b. Aluminum
c. Cast Iron
d. Carbon

Q2. In DC generator, dummy coil is provided

a. To reduced eddy current losses
b. To enhance flux density
c. To amplify voltage
d. To mechanically balanced the rotor

Q3. Hysteresis loss in a DC machine depends upon

a. Volume and grade of iron
b. Maximum value of flux density
c. Frequency of magnetic reversals
d. All of the above

Q4. The main requirement of a DC armature winding is that, it must be

a. A wave winding
b. A lap winding
c. A closed one
d. Drum winding

Q5. The OCC of a self-excited DC generator starts with minimum voltage instead of zero due to

a. Residual pole flux
b. High armature speed
c. Magnetic inertia
d. High field circuit resistance

Q6. In DC generator, compensating winding is provided to

a. Compensate for decrease in main flux
b. Neutralize armature mmf
c. Neutralized cross magnetizing flux
d. Maintain uniform flux distribution

Q7. In a DC generator, the effect of armature reaction on the main field is to

a. Reverse it
b. Distort it
c. Reduced it
d. Both (b) and (C)

Q8. To reduce the sparking in a loaded DC generator, brushes have to be shifted.

a. Clockwise
b. Counter clockwise
c. Either (a) or (b)
d. In the direction of rotation

Q9. In DC generator, armature reaction is produced by

a. Its field current
b. Armature conductors
c. Field pole winding
d. Load current in armature

Q10. Shunt generators are most suitable for parallel operation because of their voltage characteristic

a. Identical
b. Dropping
c. linear
d. Rising

Q11. Which of the following types of generator does not need equalizers for satisfactory parallel operation?

a. Series
b. Over-compound
c. Flat compound
d. Under compound

Q12. The main function of an equalizer bar is to make the parallel operation of two over compound DC generator

a. Stable
b. Possible
c. Regular
d. Smooth

Q13. The external characteristic of a shunt generator can be obtained directly from its

a. Internal characteristic
b. Open characteristic
c. Load saturation characteristic
d. Performance

Q14. The voltage build-up process of a DC generator is

a. Difficult
b. Delayed
c. Cumulative
d. Infinite

Q15. Which of the following DC generator cannot build-up on open circuit?

a. Shunt
b. Series
c. Short shunt
d. Long shunt

Q16. In a DC motor, unidirectional torque is produced with the help of

a. Brushes
b. Commutator
c. End-plate
d. Both (a) and (b)

Q17. The normal value of armature resistance of a DC motor is

a. 0.005
b. Depending on the capacity of motor
c. 10
d. 100

Q18. The mechanical power developed by the armature of a DC motor is equal to

a. Armature current multiplied by back emf
b. Power input minus losses
c. Power output multiplied by efficiency
d. Power output plus iron losses

Q19. The induced emf in the armature conductor of a DC motor is

a. Sinusoidal
b. Trapezoidal
c. Rectangular
d. Alternating

Q20. A DC motor can be locked upon as DC generator with the power flow

a. Reduced
b. Reversed
c. Increased
d. Modified

Q21. In a DC motor, the mechanical output power actually comes from

a. Field system
b. Air gap flux
c. Back emf
d. Electrical input power

Q22. The maximum torque of DC motors is limited by

a. Commutation
b. Heating
c. Speed
d. Armature current

Q23. Under constant load condition, the speed of a DC motor is affected by

a. Field flux
b. Armature current
c. Back emf
d. Both (b) and (c)

Q24. A DC shunt motor is found suitable to drive fans because they require

a. Small torque at startup
b. Large torque at high speeds
c. Practically constant voltage
d. Both (a) and (b)

Q25. As the load is increased, the speed of a DC shunt motor

a. Increase proportionately
b. Remains constant
c. Increased slightly
d. Reduced slightly

Q26. A series motor is best suited for driving

a. Lathes
b. Cranes and hoist
c. Shears and punches
d. Machine tools

Q27. The DC series motor should never be switched ON at no load because

a. The field current is zero
b. The machine does not pickup
c. The speed becomes dangerously high
d. It will take too long to accelerate

Q28. A shunt DC motor works on AC mains

a. Unsatisfactory
b. Satisfactory
c. Not at all
d. None of the above

Q29. The speed of a DC motor can be controlled by

a. Its flux per pole
b. Resistance of armature circuit
c. Applied voltage
d. All of the above

Q30. In the rheostatic method of speed control for a DC shunt motor, use if armature divertor makes the method

a. Less wasteful
b. Less expensive
c. Unsuitable for changing loads
d. Suitable for rapidly changing loads

Q31. Ward-Leonard system of speed control is not recommended for

a. Wide speed range
b. Constant speed operation
c. Frequent motor reversals
d. Very low speeds

Q32. Thyristor chopper circuit are employed for

a. Lowering the level of DC voltage
b. Rectifying the AC voltage
c. Frequency conversing
d. Providing commutation circuitry

Q33. The inverter circuit is employed to convert

a. AC to DC voltage
b. DC to AC voltage
c. DC to DC voltage
d. Frequency conversion

Q34. A solid-state chopper converts a fixed voltage DC supply into a

a. Variable voltage AC supply
b. Variable voltage DC supply
c. Higher voltage DC supply
d. Lower voltage AC supply

Q35. The most economical method of finding no load losses of a large DC shunt motor is

a. Hopkinson's test
b. Swinburne's
c. Retardation test
d. Field's test

Keys to multiple choice questions

1.	d	2.	b	3.	d	4.	c	5.	a	6.	c
7.	d	8.	d	9.	d	10.	b	11.	d	12.	a
13.	b	14.	c	15.	b	16.	d	17.	b	18.	a
19.	a	20.	b	21.	d	22.	a	23.	a	24.	d
25.	d	26.	b	27.	c	28.	a	29.	d	30.	d
31.	b	32.	a	33.	b	34.	b	35.	b	36	a

Short type questions

Q1. What are the conditions to be fulfilled for a DC shunt generator to build-up emf?

Q2. Why a DC series generator is not considered suitable for general electric supply?

Q3. What type of DC generator can be used for battery charging?

Q4. What causes sparking at brushes?

Q5. What kind of generator can be expected to have no-load voltage and full load voltage to be the same?

Q6. What is critical speed in a DC shunt generator?

Q7. What is meant by build-up of a generator?

Q8. What is the critical resistance of field circuit?

Q9. What are the different types of excitations employed in DC machines?

Q10. What is meant by OCC of a DC generator?

Q11. What are the types of commutation possible in a DC machine?

Q12. What type of current flows in field winding and armature winding of a DC shunt generator?

Q13. What cause heating of armature?

Q14. How can you differentiate between the excitation system of a DC series generator and that of a DC shunt generator?

Q15. Describe the function of interpoles in a DC machine.

Q16. If the terminal voltage of a DC generator at full load is more than its open-circuit voltage. What is the mode of excitation of the generator?

Q17. Why transformer rating in kVA?

Q18. In what respect the separately excited DC generator is superior to self-excited DC generator?

Q19. How may the direction of rotation of a DC motor be reversed?

Q20. What will happened if field of the DC motor is opened while running?

Q21. How does a DC motor differ from a DC generator in construction?

Q22. What do you understand by self-excitation mode of DC machine? Name two DC machines working in this mode.

Q23. What do you mean by speed regulation of a DC motor?

Q24. In a DC motor, the brushes are given a backward shift why?

Q25. Why large variable speed DC motors are fitted with compensating windings?

Q26. What is the purpose and location of a series field winding?

Q27. What is the difference between cumulative compound and differential compound wound motors?

Q28. What is the field of application of DC shunt motor and DC series motor?

Q29. What would happen if a DC motor is directly switched ON to the supply, without any starter?

Q30. Why the emf generated in the armature of a DC motor is called the back emf?

EXERCISES

Q1. A 500 V shunt generator has a full load current of 100 A and stray losses being 1.5 kW. Armature and field resistances are 0.3 and 250 ohms respectively. Calculate the input power and efficiency.

(**Ans.** 55.621 kW; 89.89 %)

Q2. A four-pole, lap-wound DC shunt generator has field and armature resistances of 100 ohms and 0.1 ohms respectively. It supplies power to 50 lamps rated for 100 volts, 60 watts each. Calculate the total armature current and the generated emf by allowing a contact drop of 1 volt per brush.

(**Ans.** 31 A; 105.1 V)

Q3. The field and armature resistance of a four-pole shunt generator with lap-connected armature is 50 ohm and 0.1 ohm respectively. It is supplying a 2400 W load at a terminal voltage of 100 V. Calculate the total armature current, the current per armature path and the generated emf.

(**Ans.** 26A; 6.5 A; 102.6 V)

Q4. A DC series motor on full load takes 50 A from 230 V DC mains the total resistance of the motor is 0.22 ohm. If the iron and fraction losses together amount to 5% of the input. Calculate the power delivered by the motor shaft. The total voltage drop due to the brush contact is 2V.

(**Ans.** 10.275 kW)

Q5. A shunt motor running at no load takes 5 A at 200 V. The resistance of the field circuit is 150 ohm and of the armature is 0.1 ohm. Determine the efficiency of the motor when the input current is 120 A.

(**Ans.** 89.8%)

Q6. A 20 kW, 440 V short shunt compound DC generator has a full load efficiency of 80%. If the resistance of the armature and interpole is 0.4 ohm and that of the series and shunt field 0.25 and 240 ohms. Calculate the combined bearing friction, windage, and core loss of the machine.

(**Ans.** 2739.28 Watt)

Q7. The field resistance and armature resistance of a 400 V DC shunt motor is 200 ohm and 0.5 ohm respectively. It draws 5 A at no-load. What will be its output and efficiency when it draws 50 A from the mains? Also, find the percentage change in speed from no-load to full-load.

(**Ans.** 16.852 kW; 84.26%; 5.65%)

Q8. The field and armature resistance of a 220 V DC series motor is 0.3 and 0.5 ohm respectively. When connected to rated voltage, at a certain load it draws 25 A and runs at 1000 rpm. If the current remains constant, calculate the resistance necessary to reduce the speed to 500 rpm.

(**Ans.** *4* ohms)

Q9. A 110 V shunt generator has a full load current of 100 A. Shunt field resistance of 55 ohm and constant losses of 5002 watts. If full load efficiency is 88%, find armature resistance. Assume voltage to be constant at 100 V. Calculate the efficiency at half load, and 50% overload. Find the load current.

(**Ans.** 0.078 ohm; 85.8%; 96.2 A)

Q10. The field and armature resistance of a 250 V DC shunt generator is 50 ohm and 0.02 ohm respectively. It delivers 195 A at rated voltage. The iron and friction losses equal 1050 W. Find (*a*) emf generated (*b*) copper losses (*c*) output of the prime-mover (*d*) commercial, mechanical, and electrical efficiencies.

(**Ans.** 254 V; 2050 W; 51.85 kW; 94.11%, 97.975%, 95.96%)

Q11. A 4-pole machine running at 1500 rpm has an armature conductor with 90 slots and 6 conductors per slot. The flux per pole is 10mWb. Determine the terminal voltage as DC generator if the coil is lap connected. If the current per conductor is 100 A. determine the electrical power.

(**Ans.** 135 V; 54 kW)

Q12. Calculate the flux in 4-pole dynamo with 722 armature conductor generating500 V when running at 1000 rpm when the armature is (a) lap connected (b) wave connected.

(**Ans.** 41.55 mWb; 20.78mWb)

Q13. A DC series motor draws 80 A at 230 V and rotates at 1000 rpm. The armature and series field resistance are 0.11 ohm and 0.14 ohm respectively. Calculate the speed of the motor when it draws 20 A from the mains assuming that the field is reduced to 0.4 times the previous one.

(**Ans.** *2678 rpm*)

Q14. A 12 hp, 230 V shunt motor has an armature circuit resistance of 0.5 ohms and a field resistance of 115 ohms. At no-load and rated voltage, the speed is 1200 rpm and the armature current is 2A. If the load is applied, the speed drops to 1100 rpm. Determine the armature current, the line current and the torque.

(**Ans.** *40.2 A; 42.2 A; 73.2 Nm*)

Q15. A 120 V DC shunt motor has an armature and shunt field winding resistance of 0.2 ohms and 120 ohms respectively. The brush voltage drop is 2 V. The rated full-load armature current is 75 A. Calculate the current at the instant of starting and its value in terms of percentage of full-load armature current.

(**Ans.** *591 A; 78.8%*)

Q16. The field resistance and armature resistance of a 120 V DC shunt motor is 60 ohm and 0.2 ohm respectively. The motor draws 60 A current at full-load, 1800 rpm. If the brush contact drop is 3V, find the speed of the motor at half load.

(**Ans.** *1902 rpm*)

Q.17. A DC series motor draws 80 A at 230 V and rotates at 1000 rpm. The armature and series field resistance are 0.11 ohm and 0.14 ohm respectively. Calculate the speed of the motor when it draws 20 A from the mains assuming that the field is reduced to 0.4 times the previous one.

(**Ans.** *2678 rpm*)

Books for further reading

1 Electric Machinery, Fitzgerald, Kingslay, Umans, Tata McGraw-Hill.

2. Electric Machinery Fundamentals, Chapman, McGraw-Hill Higher Education.

3. Electric Machines, Nagrath and Kothari, Tata McGraw-Hill.

4. Electric Machinery, P.S. Bimbhra, Khanna Publishers.

5. Electrical Machines, R. K. Srivastava, 2/e, Cengage Learning Pvt. Ltd.-2011

6. Electrical Machines, Smarajit Ghosh, 2/e, Pearson edu, 2012

7. Electrical Machines-I, D. K. Palwalia, N K Garg, P Kumar, G Jain. Ashirwad Publishers-2020

REFERENCE BOOKS:

1. Electric Machinery and Transformer, Guru, Hiziroglu, Oxford University press.

3. Basic Electric Machines, Vincent Deltoro, Prentice Hall.

3. Performance and Design of A.C. machines, M. G. Say

For further reading scan to: -

Lecture Notes on DC Machines

Video animation on DC Machines

Video lecture on DC Machines

Video lecture on Principal of operation of DC Machines

5 Transformers

UNIT SPECIFICS

This unit discusses the following aspects of the transformers:

- *Principle, construction, and operation of the transformer,*
- *Equivalent circuit and phasor diagram of the transformer,*
- *Voltage regulation, losses, and efficiency of the transformer,*
- *Testing of transformers,*
- *Autotransformers.*

The unit on transformers contains the fundamental aspects of its operation, control, and testing. The unit covers the working principles, construction, types, and applications of transformers. It covers the topic of transformer losses and efficiency. It illustrates transformer-related problems, such as determining voltage regulation, calculating efficiency, and designing transformer systems for specific power requirements. A number of multiple-choice questions as well as questions of short- and long-answer types are given. A list of references and suggested readings are given in the unit so that one can go through them for practice. Some QR codes have been provided in different sections which can be scanned for relevant supportive knowledge.

RATIONALE

This unit is designed to provide a comprehensive understanding of the principles, construction, and operation of transformers, both in theory and practice. Transformers play a vital role in electrical power systems, enabling efficient transmission and distribution of electrical energy by stepping up or stepping down voltage levels.

Throughout this unit, the fundamental concepts and advanced topics related to transformers is explored. It starts by examining the principle of electromagnetic induction and its application in transformers. It explores the construction of single-phase transformers, discussing their essential components. Next, it will explore the equivalent circuit and phasor diagram representation of transformers, which will aid in understanding their electrical behavior and voltage regulation characteristics. We will investigate the various losses in transformers, such as copper and iron losses, and their impact on transformer efficiency.

Testing is an integral part of transformer analysis, and it will cover open circuit tests, short circuit tests, polarity tests, and back-to-back tests to determine transformer parameters and performance. It will also delve into the separation of hysteresis and eddy current losses, which provides insights into the magnetic core's behavior and efficiency improvement strategies. Further, it explores three-phase transformers, their construction, different types of connections, and their comparative features. It also discusses the parallel operation of transformers. Autotransformers will be examined, including their construction, principle of operation, applications, and a comparison with traditional two-winding transformers.

Throughout this unit, you will have the opportunity to apply theoretical concepts to real-world scenarios through practical exercises and laboratory sessions. By the end of this unit, you will have gained the necessary knowledge and skills to analyze, design, and operate transformers effectively. You will be equipped with the tools to evaluate transformer performance, optimize efficiency, and ensure the reliable and safe operation of transformers in electrical power systems.

PRE-REQUISITES

Mathematics: Vectors, Integrals, Differentiation, Algebra (Class XII)
Physics: Electricity and Magnetism (Class XII)

UNIT OUTCOMES

The list of outcomes of this unit is as follows:

U5-O1: Knowledge of transformer construction, types, and application.

U5-O2: Understanding of transformer operation

U5-O3: Development of Equivalent circuit and phasor diagram of the transformer,

U5-O4: Analyse voltage regulation, losses, and efficiency of the transformer,

Unit-5 Outcomes	EXPECTED MAPPING WITH COURSE OUTCOMES (1-Weak Correlation; 2-Medium correlation; 3-Strong Correlation)			
	CO-1	CO-2	CO-3	CO-4
U5-O1	1	-	-	3
U5-O2	1	-	-	3
U5-O3	-	-	-	3
U5-O4	-	-	-	3

5.1 Introduction

Nowadays, for the generation, transmission, and distribution of electric power, the AC system is widely used over the DC system. Electrical power is generated in different power plants in the three-phase form at the frequency of 50 Hz (in India). The generated voltage at the generating station is 6.6 kV, 11 k V or higher. For economic reasons, high voltages are required for transmission whereas, for safety reasons, low voltages are required for utilization. For transmission purposes, it is required to step up to a voltage of 132 kV or higher. Again, in urban and rural areas it is required to step down to 3.3 kV and 6.6 kV, respectively, and 11 kV at the substation. For domestic purposes, it is required to step it down to 433 V or 230 V. As alternating voltages can be raised or lowered as per requirements, this is possible with a **transformer.** Three-phase transformers are used to step up the generated voltage before transmission of electrical power and also to step down the high voltage before distribution, that is, at the substation. Transformer is an essential part of the power system. Hence, it is rightly said that a **transformer is a backbone of a power system**.

The transformer is a device that transfers electrical energy from one electric circuit to another electrical circuit through the medium of a magnetic field without a change in the frequency. The electric circuit which receives energy from the supply mains is called primary winding and the other circuit which delivers electric energy to the load is called secondary winding. The transformer is an electromagnetic energy conversion device. The primary and secondary circuits are not connected electrically but are coupled magnetically. This coupling magnetic field allows the transfer of energy in either direction of the circuit (high voltage to low voltage or low voltage to high voltage circuit). If the transfer of energy occurs at the same voltage, the purpose of the transformer is merely to isolate the two electric circuits. A transformer can be termed a step-up or step-down transformer only after it has been put into service. Therefore, when referring to the winding of a particular transformer, the terms high voltage winding and low voltage winding should be used instead of the primary and secondary winding.

A transformer is the most widely used device in both low and high-current circuits. As such, transformers are built in an amazing range of sizes.

In general, important tasks performed by transformers are:

a. for decreasing or increasing voltage and current levels from one circuit to another circuit.
b. for matching the impedance of a source and its load for maximum power transfer in electronic and control circuits.
c. for isolating DC while permitting the flow of AC between two circuits or for isolating one circuit from another.
d. As a coupling device in electronic circuits
e. To measure voltage and currents; these are known as instrument transformers.

Transformer is, therefore, an essential apparatus both for high and low current circuits.

5.2 Definition

A **transformer** is an electromagnetic energy conversion device that transfers electrical energy from one electrical circuit to another electrical circuit with a desired change in voltage and current at the same frequency. The winding having less number of turns is called low-voltage winding and the winding having more number of turns is called high voltage winding.

Also, the winding to which the AC supply is connected is called a primary winding and the other one is called a secondary winding to which load is connected.

5.3 Working Principle of Transformer

A transformer works on the principle of *electromagnetic induction* between two (or more) coupled circuits or coils. According to this principle, an emf is induced in a coil if it links a changing flux. Fig. 5.1 shows a basic single-phase transformer having two windings wound on a common magnetic core which are electrically separated and magnetically linked through a common magnetic path. Once the AC supply of voltage V_1 is given to primary winding, an alternating flux is set-up in the magnetic core which links with the primary and secondary winding. Consequently, self-induced emf E_1 and mutually-induced emf E_2 are induced in primary and secondary, respectively. These induced emf's are developed in phase opposition to V_1 as per Lenz's law. The self-induced emf in the primary is also called back emf since it acts in the opposite direction to the applied voltage. If a closed path is provided at the secondary winding, this induced emf at the secondary drives a current.

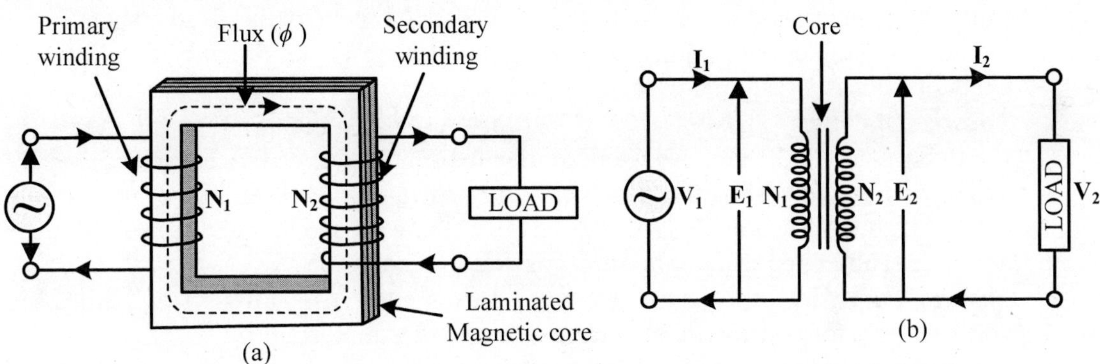

Fig.5.1 (a) Single phase transformer (core and winding) (b) Circuit representation

The magnitude of induced emf in a coil depends upon the rate of change of flux linkages i.e., $e \propto \frac{d\phi}{dt}$. since, the rate of change of flux for both the winding is the same, the magnitude of induced emf $E_1 \propto N_1$ in primary and secondary will depend upon their number of turns, i.e., primary induced emf and secondary induced emf $E_2 \propto N_2$. When $N_2 > N_1$, the transformer is called a step-up transformer, on the other hand, when $N_2 < N_1$ the transformer is called a step-down transformer.

Turn ratio: The ratio of primary to secondary turns is called turn ratio, i.e., turn ratioN_1/N_2.

Transformation ratio: The ratio of secondary voltage to primary voltage is called the voltage transformation ratio of the transformer. It is represented by K.

$$K = \frac{E_2}{E_1} = \frac{N_2}{N_1} \tag{5.1}$$

In brief, we can say the following:

- The transformer is a static device.
- It transfers electrical power from one circuit to another.
- During the transfer of power, there is no change in frequency.
- It uses electromagnetic induction to transfer electrical power.
- The two electrical circuits are in the mutual inductive influence of each other.

Scan to know more on transformers

5.4 Types of Transformer

The classification is carried out based on the ratio of input and output voltages. From the application point of view, the following transformers are most important:

- ***Power and distribution transformer:*** These transformers are used for transmission and distribution of power. Power transformers usually operate at a high average load, which would cause continuous capacity copper loss, thus affecting their efficiency. To have minimum losses during 24 hours, such transformers are designed with low copper losses. Distribution transformers are continuously energized causing iron losses for all 24 hours, Generally, the load on these transformers fluctuates from no-load to full load during this period. To obtain high efficiency, such transformers are designed with low iron losses.
- ***Autotransformer:*** These transformers are used to change the voltage within relatively small limits and are used for starting AC motors.
- ***Transformers for feed installations with static converters:*** These are used for converting AC to DC and also DC to AC. The first one is used for rectification purposes and the second one for inversion purposes.
- ***Testing transformers:*** These are used to conduct tests at high and ultra-high voltages. e.g., for testing the dielectric strength of transformer oil.
- ***Power transformers for special applications:*** These transformers are used in furnaces, welding, tractions etc.
- ***Radio transformers:*** These are used in radio engineering and similar purposes. From a frequency range point of view, transformers can be divided as (50–400 Hz) audio transformers, wide band and narrow band transformers, and pulse transformers. Transformers can also be divided depending on the number of windings such as two-winding (conventional) and single winding known as an autotransformer.
- ***Instrument Transformers:*** To measure high voltages and currents in the power system potential transformer (*P.T.*) and a current transformer (*C.T.*) are used, respectively. The potential transformers are used to decrease the voltage and current

transformers are used to decrease the current up to a measurable value. These are also used with protective devices.

- ***Isolation transformer:*** These transformers are used only to isolate (electrically) the electronic circuits from the main electrical lines, therefore, their transformation ratio is usually one.
- ***Impedance matching transformer:*** These transformers are used at the output stage of the amplifier for impedance matching to obtain maximum output from the amplifiers.

5.5 Construction of Transformer

A single-phase transformer consists of primary and secondary windings put on a magnetic core. The magnetic core is used to confine the flux to a definite path. Transformer cores are made from thin sheets (called laminations) of high-grade silicon steel 0.35 mm thick for 50 Hz transformers. The laminations reduce eddy-current loss and the silicon steel reduces hysteresis loss. The laminations are insulated from one another by heat-resistant enamel insulation coating. For reducing the core losses, nearly all transformers have their magnetic core made from cold-rolled grain-oriented sheet-steel (C.R.G.O.). This material when magnetized in the rolling direction, has low core loss and high permeability and allows flux density as high as 2.8 Wb/sq.m. I-type and E-type laminations are used as shown in fig. 5.2. The laminations are built up into a stack and the joints in the laminations are staggered to minimize airgaps (a continuous air gap in an un-staggered core results in a large magnetizing current). The laminations are tightly clamped. There are two basic types of transformer constructions, the core type, and the shell type.

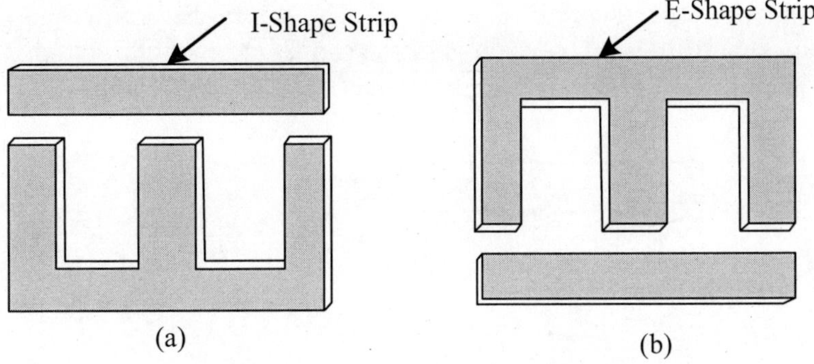

Fig.5.2 Laminations of E and I-shape

5.5.1 Core Type Construction

In core-type transformers, the winding surrounded a considerable part of the steel core which consists of two vertical legs or limbs with two horizontal sections, called yokes. To keep the leakage flux to a minimum, half of each winding is placed on each leg of the core as shown in fig. 5.3. The low-voltage winding is placed next to the core, and the high-voltage winding is placed around the low-voltage winding to reduce the insulating material required. Thus, the two winding are arranged as concentric coils. Such winding is called concentric winding or cylindrical winding. For a given output and voltage rating, core type transformer requires less iron but more conductor material as compared to shell type transformer.

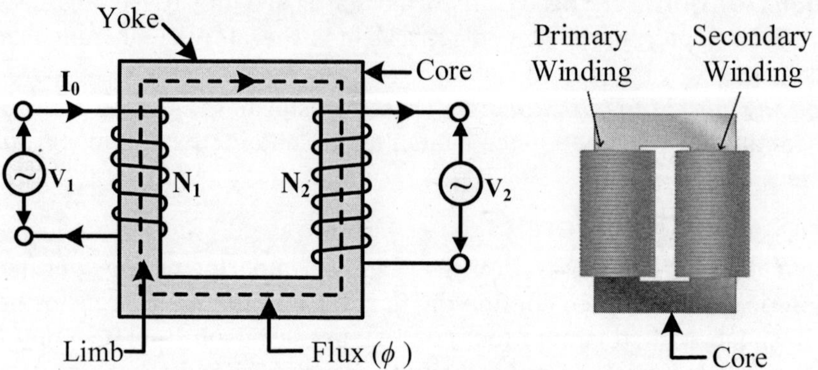

Fig. 5.3 Core-type transformer

5.5.2 Shell Type Construction

The shell-type transformer has a double magnetic circuit and three limbs. Both windings are placed on the central limb. The coils occupy the entire space of windows. The coils are usually multi-layer disc type or sandwich coils. The low-voltage coils are placed nearest to the iron core to reduce the amount of high-voltage insulation, also referred to as insulation coordination. The core is laminated. Special care is taken to arrange the laminations of the core. All the points at alternate layers are staggered properly to avoid a narrow air gap at the joint, right through the cross-section of the core. The joints are known as *overlapped or mitered* joints. The shell-type construction is preferred for a few high-voltage transformers. Since the windings are surrounded by a core, natural cooling does not exist.

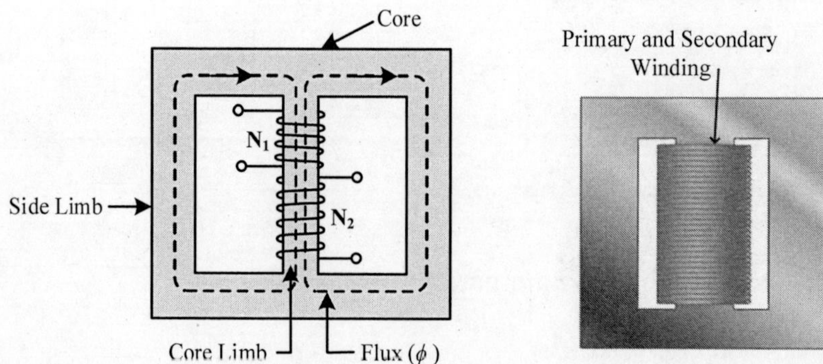

Fig. 5.4 Shell-type transformer

To remove any winding during maintenance, the removal of a large number of laminations is required. Fig.5.4 shows a shell-type transformer. Due to better provision for mechanical support and bracing of coils in the shell-type transformer, better resistance to combat high mechanical force is obtained. High mechanical forces are developed for a high current during a short circuit.

Comparison between Core-type and Shell-type Transformers:

Sr. No.	*Core-type transformer*	*Shell-type transformer*
1.	The windings surround a considerable portion of the core.	The *core* surrounds a considerable portion of the windings.
2.	Core has *two limbs* to carry the windings.	The core has *three or more limbs* but the central limb carries the windings.
3.	The mean length of the coil turn is *shorter*.	The mean length of the coil turn is *longer*.
4.	Windings are of *form*-wound and are of cylindrical-type.	Windings are *sandwich-type*. The coils are first wound in the form of pancakes, and the complete winding consists of stacked discs.
5.	More suitable for *high-voltage* transformers.	More suitable and economical for *low voltage* transformers.

5.5.3 Berry-type Construction

A berry-type transformer is a specially designed shell-type transformer and is named after its designer. The transformer core consists of laminations arranged in groups that radiate from the center as shown (top view) in fig. 5.5.

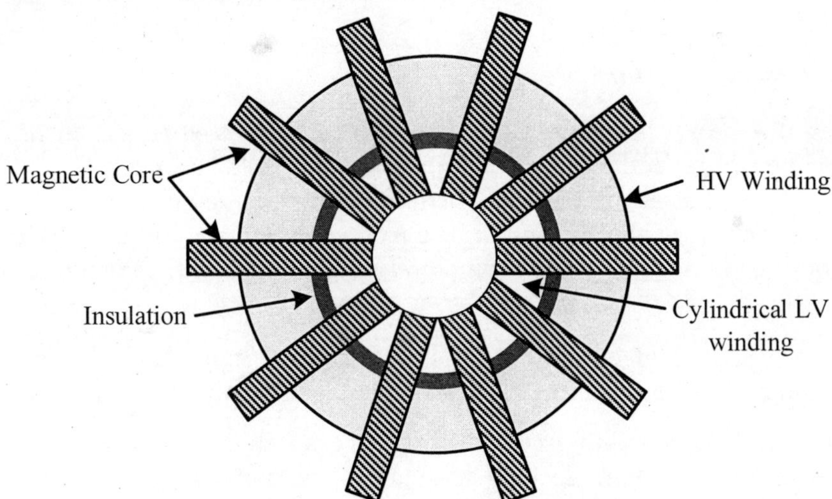

Fig. 5.5 Berry-type transformer

The core of a berry-type transformer looks like the spokes of wheels. The core of this transformer is constructed in such a way that the magnetic circuit finds several paths to pass flux. Berry transformer is also called a toroidal transformer, which is a large type of electronic transformer and has been widely used in home appliances and other electronic equipment with high technical requirements. Its main use is as a power transformer and an isolation transformer. To avoid possible insulation damage due to the movement of strips and winding, the core and coil of the transformer are provided with rigid mechanical bracing. Good bracing reduces vibration and objectionable noise - a humming sound - during operation.

Features of Berry Transformers

1. The core of the Berry transformer is made of continuous-oriented silicon steel strips. After annealing, it will form a highly consistent magnetic flux guidance. At the same time, the winding of the Berry transformer is tightly covered with the Berry core for winding. , the magnetic flux density is higher because the Berry transformer can achieve a high electrical efficiency of 95%.
2. Less vibration noise: The iron core has no air gap, which can reduce the noise of the induced vibration of the iron core, and the windings evenly and tightly wrap the annular iron core, effectively reducing the "hum" sound caused by magnetostriction.
3. Low operating temperature: Because the iron loss can reach 1.1W/kg, the iron loss is very small, the temperature rise of the iron core is low, and the winding has a good heat dissipation on the iron core with a lower temperature, so the temperature rise of the transformer is low.
4. Easy installation: The Berry transformer has only one installation screw in the center core, which is especially easy to quickly install and disassemble in electronic equipment.

Advantages of Berry Transformers

The iron core of the Berry transformer is made of high-quality cold-rolled silicon steel sheets (the sheet thickness is generally below 0.35mm), which is rolled seamlessly, which makes its core performance better than the traditional laminated core. The coil of the Berry transformer is evenly wound on the iron core, and the direction of the magnetic field lines generated by the coil almost completely coincides with the magnetic circuit of the iron core. Describe a series of advantages.

1. The iron core with high electrical efficiency has no air gap, and the stacking coefficient can be as high as 95% or more. The no-load current is only 10% of the laminated type.
2. Small size and light weight Berry transformers can reduce the weight by half compared to laminated transformers. As long as the cross-sectional area of the iron core is kept equal, the Berry transformer can easily change the ratio of the length, width, and height of the iron core, and can be designed to meet the requirements.
3. The magnetic interference is small. The iron core of the Berry transformer has no air gap, and the windings are evenly wound on the Berry iron core. This structure leads to small magnetic leakage and small electromagnetic radiation. High-sensitivity transformers can be used without additional shielding. Electronic equipment, such as low-level amplifiers and medical equipment.

4. Less vibration and noise.
5. The iron core has no air gap to reduce the iron core.

Applications of Berry transformers

1. ***Audio equipment:*** Among the Berry transformer products of this factory, there are first and foremost products that are well-known in audio equipment. Such as high-performance products that can be used as power transformers in Class 3 high-power hi-fi amplifiers and power transformers for medium and small power audio equipment. High-quality Berry transformers can be used for power ranges from 6VA to 1000VA. Since the range is 1000VA consider the dropping of power.
2. ***Electrical control:*** New high-performance magnetic materials such as permalloy soft magnetic materials, amorphous materials, or nanocrystalline alloys can be used as magnetic cores, with precise intelligent instruments and special testing methods. Testing method for high precision, high stability Berry transformer. It can ensure accurate transformer parameters and completely consistent performance.
3. ***Medical equipment:*** Berry transformers are specially designed and manufactured for medical equipment. In addition to high efficiency, high reliability, and high safety requirements. It also especially strengthens the electrical strength and improves the anti-humid heat performance. A thermal fuse is also required to be added inside the transformer to ensure reliability.
4. ***Application in other fields:*** Berry transformers can also be used in power inverters of various sizes, such as solar and wind power generation systems. This high-efficiency Berry transformer with a low-loss iron core can greatly improve the overall efficiency of the inverter power supply.

5.6 Transformer Winding

Barry-type transformers are usually made of high-grade copper conductors. Standard conductors are used for carrying higher currents. To avoid each turn coming in contact with the other, the windings are provided with insulation. In addition to inter-turn insulators, bare copper wires are provided with enamel coating. Usually, single or double-layer cotton is used. Sometimes press board or cotton insulation is also used to support the windings. Usually, additional insulation is provided for line end turns for their protection from lightning and switching overvoltages. During transient disturbances, the distribution of voltage is not uniform along the windings and 80% of voltage at that time appears across the first 10% of turns from the line end. Heat generation occurs due to energy loss, which is proportional to the volume of the material in which the losses occur. The heat dissipation is proportional to the surface area of the same material and the tank. The ratio of heat generated to heat dissipated is approximately proportional to the ratio of the volume of the material for conductors and the core to the surface area of the material for conductors, the core, and the tank, which must approach unity to limit the temperature rise. This can be achieved by corrugating the surface area of the tank. To get effective cooling, radiators are used.

The following are the most important requirements of transformer windings:

- The windings must be economical.
- The heating conditions of the windings should satisfy standard requirements.
- The windings must have good mechanical strength to combat the force that originates due to a short circuit.
- The windings must have the necessary electrical strength during over-voltage.

The following are the two different types of windings: (i) Concentric windings and (ii) sandwich windings. Concentric windings are used in core-type transformers as shown in fig. 5.6(a). Sandwiched windings are almost exclusively used in shell-type transformers as shown in fig. 5.6(b).

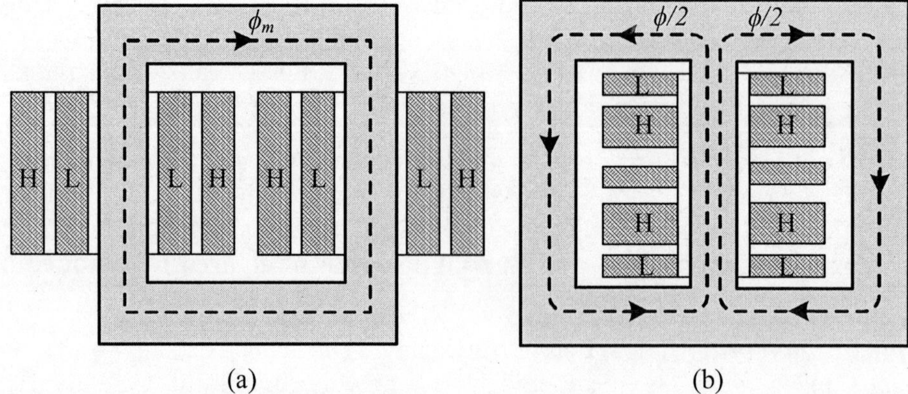

Fig. 5.6 Sectional view of single phase transformer (a) Core type (b) Shell type

The positioning of the *HV* and *LV* windings with respect to the core is also very important from the point of view of insulation requirements. The low-voltage winding is placed nearer to the core in the case of concentric windings and on the outside positions in the case of sandwiched windings as shown in fig. 5.6(a) and fig. (b) on account of less and easier insulation facilities.

5.7 Ideal Transformer

To understand the theory, operation, and applications of a transformer, a transformer is assumed to be an ideal one, the various assumptions are as follows :

a. Its coefficient of *coupling (k)* is *unity.*
b. Its primary and secondary windings are pure inductive (winding resistances are negligible)having an infinitely large value
c. All the flux set up by the primary links the secondary windings, *i.e.* all the flux is confined to the magnetic core.
d. The core losses (hysteresis and eddy current losses) are negligible.

e. Its leakage flux and *leakage inductances* are *zero.*
f. Its self and mutual inductances are zero having *no reactance.*
g. Its efficiency is 100 percent having *no loss* due to resistance, hysteresis, or eddy current.
h. Its transformation ratio (or turn ratio) is equal to the ratio of its secondary to primary terminal voltage and also the ratio of its primary to secondary current.
i. Its core has permeability (μ) of *infinite* value. Hence, less magnetizing current requires for magnetizing their core

It is to be noted that a practical or commercial transformer has none of these properties even though its operation is close to ideal.

An ideal transformer with a secondary side open circuited shown in fig. 5.7. It has no ohmic resistance and leakage reactance. There is no loss in an ideal transformer. In fig. 5.7, alternating voltage (V_1) is applied at the primary and hence alternating current flows in the primary. The primary of transformer draws the magnetizing current (I_μ) only because it is purely inductive. I_μ is small in magnitude and lags behind V_1 by an angle of 90°.

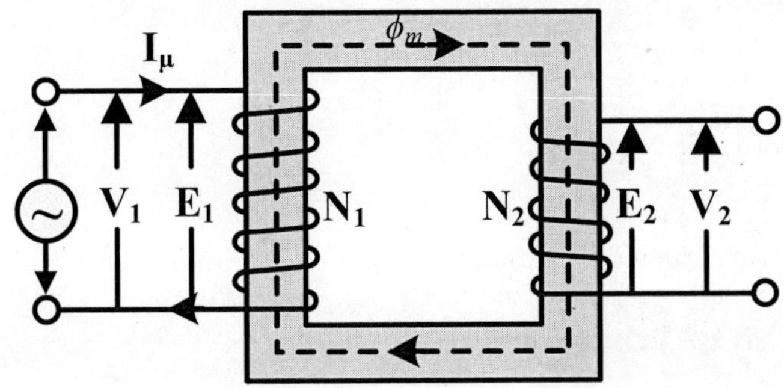

Fig. 5.7 Elementary dtiagram of an ideal transformer

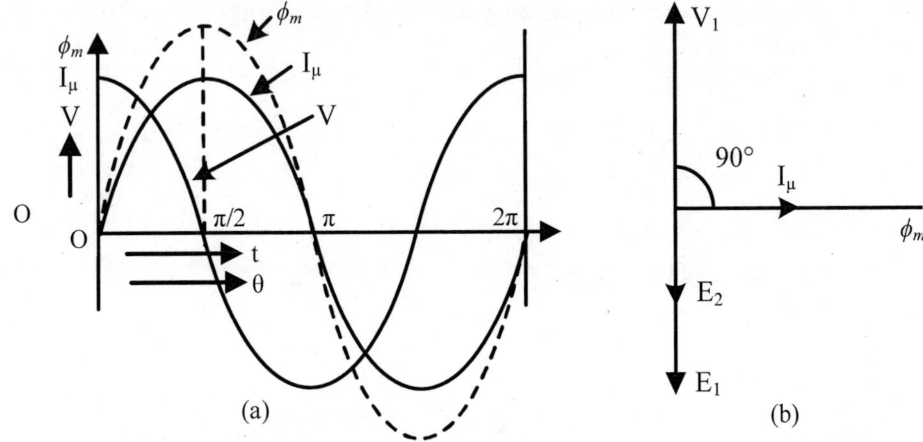

Fig. 5.8 Ideal transformer (a) Wave diagram (b) Phasor Diagram

The function of I_μ is to magnetize the core, and it produces an alternating flux (ϕ), which is proportional to I_μ. The alternating flux (ϕ) is linked with both primary and secondary windings and causes self-induced emf (E_1) in the primary. This self-induced emf (E_1) is equal and opposite to V_1 at any instant of time. This induced emf is known as *back emf* or *counter emf.* Due to mutual induction, an emf E_2 is produced in the secondary. This emf is known as *mutually induced emf.* It is anti-phase with V_1 and its magnitude is proportional to the rate of flux as well as the number of turns of the secondary windings.

Fig. 5.8(a) shows the instantaneous values of applied voltage, induced emf, flux, and magnetizing current by sinusoidal waves, while fig. 5.8(b) shows the vectorial representation of the effective values of the above quantities.

For an ideal transformer, since it has no losses, the output must be equal to the input, therefore.

$$E_2 I_2 cos\theta = E_1 I_1 cos\theta \tag{5.2}$$

$$E_2 I_2 = E_1 I_1 \tag{5.3}$$

$$K = \frac{E_2}{E_1} = \frac{I_1}{I_2} \tag{5.4}$$

Since $E_2 \propto N_2$; $E_1 \propto N_1$ and $E_1 \cong V_1$; $E_1 \cong V_1$

$$K = \frac{V_2}{V_1} = \frac{E_2}{E_1} = \frac{N_2}{N_1} = \frac{I_1}{I_2} \tag{5.5}$$

Hence, primary and secondary currents are inversely proportional to their respective turns.

5.8 EMF Equation of Transformer

When a sinusoidal voltage is applied to the primary winding of a transformer, a sinusoidal flux is set up in the iron core which links with primary and secondary winding as shown in fig. 5.9. Fig. 5.9 shows the representation of alternating flux, varying sinusoidally, which increases from its zero value to maximum value (ϕ_m) in one-quarter of the cycle, that is in the second to one-fourth of a cycle.

Let, ϕ_m =maximum value of flux in Wb.

f = Supply frequency in Hz or c/s

N_1 = Number of turns in the primary winding

N_2 = Number of turns in the secondary winding

The average rate of change of flux is given by

$$\frac{\phi_m}{{}^1/_{4f}} = 4f\phi_m \ Wb/s \ or \ V \tag{5.6}$$

This rate of change of flux per turn is the induced emf of volts.

Therefore, average emf/turn= $4f\phi_m \ Volt$ (5.7)

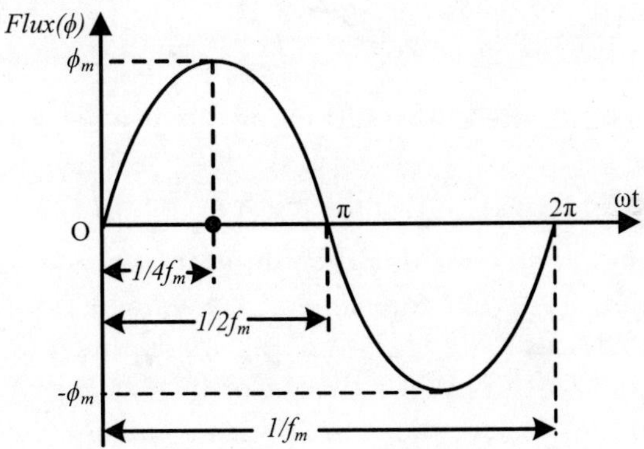

Fig. 5.9 Representation of alternating flux

$$\text{Form factor } = \frac{RMS\ Value}{Average\ Value} = 1.11 \tag{5.8}$$

$$\text{RMS value per turn} = 1.11 \times 4f\phi_m = 4.44\ f\phi_m\ Volt \tag{5.9}$$

The RMS value of induced emf in the primary winding is given by

$$E_1 = (4.44\ f\phi_m) \times N_1 = 4.44\ fB_mA_rN_1 \tag{5.10}$$

Where, $B_m = {}^{\phi_m}/_{A_r}$ is the maximum value of flux density having unit Tesla (T) and A_r is the area of cross-section.

Similarly, the RMS value of induced emf in the secondary winding is given by

$$E_2 = (4.44\ f\phi_m) \times N_2 = 4.44\ fB_mA_rN_2 \tag{5.11}$$

From equations (5.10) and (5.11), we have

$$\frac{E_1}{N_1} = \frac{E_2}{N_2} = 4.44\ f\phi_m = \text{ Constant} \tag{5.12}$$

$$\frac{E_2}{E_1} = \frac{N_2}{N_1} = K \tag{5.13}$$

Where K is the transformation ratio.

Equation (5.12) shows that emf induced per turn in primary and secondary windings is equal.

5.9 Rating of Transformer

Generally, the rating of the machine should indicate the power supplied by it. But in the case of a transformer, the output power is not constant. It keep changing with the load. The output power factor is also a function of load. Hence, the rating of the transformer

$$\text{Rating of transformer} = \text{ Primary voltage } \times \text{ Primary Current}$$

$$= \text{ secondary voltage } \times \text{ Secondary Current}$$

Rating in VA or KVA or MVA $= V_1 \times I_1 = V_2 \times I_2$

$$I_1 = \frac{kVA \times 10^3}{V_1}\ ; \qquad I_2 = \frac{kVA \times 10^3}{V_2}$$

If the rating is specified in MVA, then the multiplying factor will be 10^6.

5.10 Why Rating of Transformer in kVA?

A transformer is an energy-transferring device at different voltage levels. As seen copper loss of the transformer depends on current and iron loss on voltage. Hence, total transformer loss depends on volt-amp (VA). i.e. it depends on the load power factor which is why the transformer rating is in kVA not in kW. Or the VA or kVA rating of the transformer simply indicates how much maximum apparent or total power a transformer can supply. It indicates the active power and the reactive power. The output of total power supplied by the transformer (VA or KVA), a part or full will be useful, which depends on the load. At the time of manufacturing of a transformer, the nature of the load is not known. Hence, the maximum total power supply capacity is given as the rating. Hence the transformer is rated in VA or kVA and not in W or kW.

5.11 Transformer on DC

The basic working principle of a transformer is electromagnetic induction, when flux linking with a coil changes an emf is induced in it. If DC is applied to one of the windings of a transformer, it will set a constant magnetic field in the magnetic core. Hence, no emf will be induced either in primary or secondary. Then electric power cannot be transformed from one circuit to the other. Moreover, if rated DC voltage is applied to its primary, a high current will be drawn by it since there is any counter (self) induced emf that limits the current. Consequently, heaving heat will be produced and winding insulation will burn. *Hence, a transformer cannot work on DC* and it is never put-on rated DC supply.

Example 5.1: A single-phase 2310/220 V transformer has an emf of 13 V per turn approximately. What will be the number of primary and secondary turns?

Solution:

$$\frac{E_1}{N_1} = \frac{E_2}{N_2} = 13\ V\ \text{(Given)};$$

$$E_1 = 2310\ V;\ E_2 = 220\ V$$

Primary turns,

$$N_1 = \frac{E_1}{13} = \frac{2310}{13} = 177.69 \cong 178\ \ (Ans)$$

Secondary turns,

$$N_2 = \frac{E_2}{13} = \frac{220}{13} = 16.92 \cong 17\ (Ans)$$

Example 5.2: A power transformer has 1000 primary turns and 100 secondary turns. The cross-sectional area of the core is 5 sq. cm and the maximum flux density while in operation is 10000 Gauss. Calculate turns per volt for the primary and secondary windings.

Solution:

Here, N_1= 10000; $N_2 = 100$; $A_i = 5\,cm^2 = 5 \times 10^{-4} m^2$

$$B_m = 10000 \text{ gauss} = 10000 \text{ X } 10^{-8} \text{ X } 10^4 = 1 \text{ tesla}$$

We know,

$$E_1 = 4.44\, N_1 f B_m A_i = 4.44 \times 1000 \times 50 \times 1 \times 5 \times 10^{-4} = 111\, V$$

$$E_2 = 4.44\, N_2 f B_m A_i = 4.44 \times 100 \times 50 \times 1 \times 5 \times 10^{-4} = 11.1\, V$$

On the primary side, the number of turns/volt = $\frac{N_1}{E_1} = \frac{1000}{111} = 9\ (Ans)$

On the secondary side, the number of turns/volt = $\frac{N_2}{E_2} = \frac{100}{11.1} = 9\ (Ans)$

The number of turns per volt or voltage per turn on primary and secondary remains the same.

Example 5.3: A 30 kVA transformer has 500 and 40 turns at primary and secondary respectively. If the primary is connected to 3000 V, 50 Hz mains, calculate (i) primary and secondary currents at full load; (ii) The secondary emf, and (iii) The maximum flux in the core. Neglect magnetic leakage, the resistance of the winding, and the primary no-load current to the full load current.

Solution:

(i) At full load, $I_1 = \frac{30 \times 10^3}{3000} = 10\ (Ans)$

Now, $\frac{I_1}{I_2} = \frac{E_2}{E_1} = \frac{N_2}{N_1}$

Secondary current,

$$I_2 = \frac{N_1}{N_2} \times I_1 = \frac{500}{40} \times 10 = 125\, A\ (Ans)$$

(ii) Secondary emf,

$$E_2 = \frac{N_2}{N_1} \times E_1 = \frac{40}{500} \times 3000 = 240\, V\ (Ans)$$

(iii) Using relation, $E_1 = 4.44 N_1 f \phi_m$

$3000 = 4.44 \times 500 \times 50 \times \phi_m$

$$\text{Or } \phi_m = \frac{3000}{4.44 \times 500 \times 50} = 27\, mWb\ (Ans)$$

Example 5.4: A single-phase 11000 /415 V, 50 Hz, core type transformer has emf per turn of 15 V. The maximum flux density in the core is 2.5 T. Find the number of primary and secondary turns and net cross-sectional area of the core.

Solution:

Here, $E_1 = 11000$ V; $E_2 = 415$; f = 50 Hz; $B_m = 2.5$ T

EMF/turns, $= \frac{E_1}{N_2} = \frac{E_2}{N_2} = 15$

Number of primary turns, $N_1 = \dfrac{E_1}{15} = \dfrac{11000}{15} = 733.33\ (Ans)$

Number of secondary turns, $N_2 = \dfrac{E_2}{15} = \dfrac{415}{15} = 27.67\ (Ans)$

Now, $E_1 = 4.44 N_1 f A_i B_m$

$$\frac{E_1}{N_1} = 4.44 f A_i B_m \ or\ 15 = 4.44 \times 50 \times A_i \times 2.5$$

Net area, $A_i = \dfrac{15}{4.44 \times 50 \times 1.5} = 0.045045\ m^2 = 450.45\ cm^2\ (Ans)$

Example 5.5: A single phase 50 Hz core type transformer has rectangular cores 30. 20 cm. and the maximum allowable density is 1.02 tesla. Find the number of turns per limb on the high and low voltage sides for a voltage ratio of 3300/200 volt. Take the iron factor as 0.93.

Solution:

Gross-cross-sectional Area = $30 \times 20 = 600\ cm^2$; $A_{gc} = 600 \times 10^{-4} m^2$

The iron factor is to be taken into consideration as the laminations are insulated from each other and

$$K_i = \frac{Net\ Area\ of\ Cross - Section}{Gross\ Area\ of\ Cross - Section} = \frac{A_i}{A_{gc}}$$

$$A_i = K_i \times A_{gc}$$

$$= 0.93 \times 600 \times 10^{-4} = 558 \times 10^{-4}\ m^2$$

Emf induced per turn, $= 4.44 f B_{max} A_i$

$$= 4.44 \times 50 \times 1.02 \times 558 \times 10^{-4} = 13\ volt\ (Ans)$$

Primary turns $= \dfrac{3300}{13} = 254\ turns\ (Ans)$

Secondary turns $= \dfrac{200}{13} = 16\ turns\ (Ans)$

Example 5.6: A 500 kVA, 4400/400 V, 50 Hz, single-phase transformer has 500 turns in the secondary winding. Determine (i) emf per turn, (ii) primary turns, (iii) secondary full load current (iv) maximum flux (v) gross cross-sectional area of the core for flux density of 1.2 T and iron factor is 0.92 (vi) if the core is of square cross-section find the width of the limb.

Solution:

Here, Rating $= 500$ kVA; $E_1 = V_1 = 4400\ V; E_2 = V_2 = 400\ V; f = 50\ Hz;\ N_2 = 500; B_m = 1.2\ T; k_i = 0.92$

(i) Emf per turn

$$= \frac{E_2}{N_2} = \frac{400}{500} = 0.8\ V/turn\ (Ans)$$

(ii) Emf per turn

$$= \frac{E_2}{N_2} = \frac{E_1}{N_1} = \frac{4400}{N_1} = 0.8$$

Primary turns N_1

$$= \frac{4400}{0.8} = 5500\ (Ans)$$

(iii) Secondary full load current,

$$I_2 = \frac{kVA\ \times 1000}{V_2} = \frac{500\ \times 1000}{400} = 1250\ A\ (Ans)$$

(iv) Maximum flux, ϕ_m

$$= \frac{E_2}{4.44\ \times\ N_2\ \times f} = \frac{400}{4.44\ \times\ 500\ \times 50} = 3.6\ mWb\ (Ans)$$

(v) Iron area of the core,

$$A_i = \frac{\phi_m}{B_m} = \frac{3.6\ \times\ 10^{-3}}{1.2} = 3\ \times\ 10^{-3}\ m^2 = 30\ cm^2$$

Gross area of the core, A_g

$$= \frac{A_i}{k_i} = \frac{30}{0.92} = 32.60\ cm^2\ (Ans)$$

(vi) Width of squared limb

$$= \sqrt{A_g} = \sqrt{32.60} = 5.71\ cm\ (Ans)$$

Example 5.7: A 100 kVA, 4000/200 volt, 50 Hz single phase transformer has 40 turns on the secondary, calculate: (i) the values of primary and secondary currents (ii) the number of primary turns (iii) the maximum value of the flux. If the transformer is to be used on a 25 Hz system, calculate (iv) the primary voltage, assuming that the flux is increased by 10% (v) the kVA rating of the transformer assuming the current density in the windings to be unaltered.

Solution:

(i) Full load primary current,

$$I_1 = \frac{100 \times 1000}{4000} = 25\ A\ (Ans)$$

Full load secondary current,

$$I_2 = \frac{100 \times 1000}{200} = 500\ A\ (Ans)$$

(ii) Number of primary turns,

$$N_1 = N_2 \times \frac{E_1}{E_2} = 40 \times \frac{4000}{200} = 800\ (Ans)$$

(iii) We know, $E_2 = 4.44 f \phi_{max} N_2\ volt$

$200 = 4.44 \times 50 \times \phi_{max} \times 40$

$$\phi_{max} = \frac{200}{4.44 \times 50 \times 40} = 0.0225\ Wb\ (Ans)$$

(iv) As the flux is increased by 10% at 25 Hz

Flux at 25 Hz,

$\phi_m = 0.0225 \times 1.1$

$= 0.02475\ Wb$

Primary Voltage

$= 4.44 \times N_1 \times f \times \phi_m\ volt$

$= 4.44 \times 800 \times 25 \times 0.02475 = 2197.8\ volt\ (Ans)$

(v) For the same current density, the full load primary and secondary current remains unaltered.

kVA rating of the transformer

$$= \frac{25 \times 2197.8}{1000} = 54.95\ kVA\ (Ans)$$

5.12 Transformer on No Load

A transformer is said to be on no-load when the secondary winding is open-circuited as shown in fig. 5.10. The secondary current I_2 is thus zero. Hence, the secondary winding is not causing any effect on the magnetic flux set-up in the core or on the current drawn by the primary. But the losses cannot be ignored. When an alternating voltage is applied to the primary, a small current I_0 flows in the primary. The current I_0, is called the no-load current of the transformer. The current I_0 lags behind the voltage vector V_1 by an angle ϕ_0 (called hysteresis angle of advance) which is less than 90° (around 80° to 85°)., as shown in fig. 5.11(a). The angle of lag depends upon the losses in the transformer. The no-load current I_0 made up of two components I_μ and I_w.

- The component I_μ, is called the magnetizing component. It magnetizes the core. In this current, it sets up flux in the core and therefore I_μ, is in phase with ϕ_m. The other, I_μ is in quadrature with the applied voltage V_1. The current I_μ is also called a reactive, or wattless component of no-load current. It produces flux in the core and does not consume any power.
- The component I_w supplies the hysteresis and eddy-current losses in the core and the negligible I^2R loss in the primary winding. The current I_w is called the active component or wattful component of no-load current. It is in phase with the applied voltage V_1. The no-load current I_0, is small in the order of 3 to 5 percent of the rated current of the primary. Moreover, the secondary copper losses are zero as I_2 is zero.

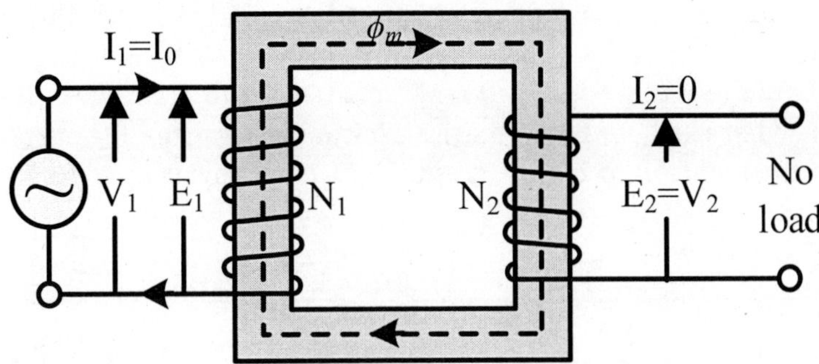

Fig. 5.10 Circuit diagram of the transformer at no-load

$$\text{Active or working or iron loss component, } I_w = I_0 cos\theta_0 \tag{5.14}$$

$$\text{Magnetizing Component, } I_\mu = I_0 sin\,\theta_0 \tag{5.15}$$

$$\text{No}-\text{load current, } I_0 = \sqrt{I_w^2 + I_\mu^2} \tag{5.16}$$

$$\text{Primary Power factor at no}-\text{load, } \quad cos\theta_0 = \frac{I_w}{I_0} \tag{5.17}$$

$$\text{No}-\text{load power input, } P_0 = V_1 I_0 cos\theta_0 \tag{5.18}$$

$$\text{Exiciting resistance,} \qquad R_0 = \frac{V_1}{I_w} \qquad (5.19)$$

$$\text{Exiciting reactance,} \qquad X_0 = \frac{V_1}{I_\mu} \qquad (5.20)$$

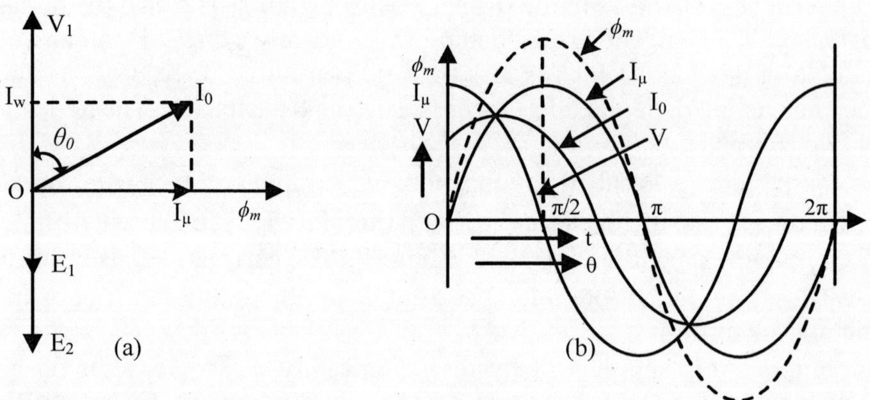

Fig. 5.11 Transformer at no load (a) Phasor diagram (b) Wave diagram

5.13 No-load Equivalent Circuit

The equivalent circuit of a transformer at no-load is shown in fig. 5.12. The actual transformer is replaced by an ideal transformer with resistance R_0 and reactance X_0 in parallel. Here, R_0 represents the exciting resistance of the transformer which carries power loss or core loss component of no-load current, i.e., I_w used to meet with the no-load losses in the transformer, whereas X_0 represents the exciting reactance of the transformer which carries wattless component of no-load current, i.e., $I_\mu$used to set up a magnetic field in the core.

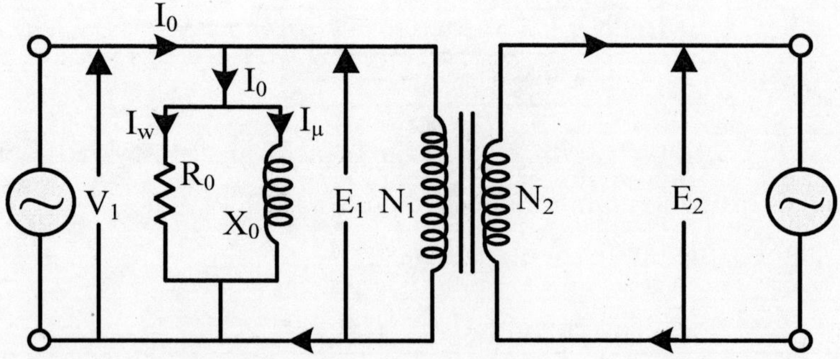

Fig. 5.12 Equivalent circuit of the transformer at no-load

From equivalent circuit

$$\text{The core loss of actual transformer} = I_w^2 R_0 = \frac{V_1^2}{R_o} \qquad (5.21)$$

5.14 Transformer on Load (Neglecting winding resistance and leakage flux)

When a certain load is connected across the secondary, a current I_2 flows through it as shown in fig. 5.13. The magnitude of current I_2 depends upon the terminal voltage V_2 and impedance of the load. The phase angle of the secondary current I_2 with respect to V_2 depends upon the nature of the load i.e., whether the load is resistive, inductive, or capacitive.

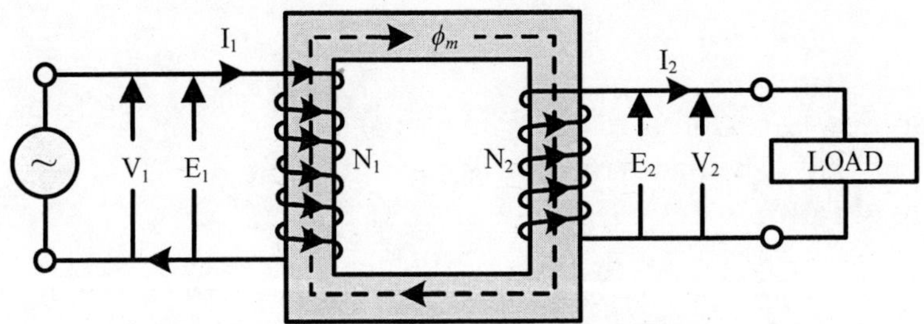

Fig. 5.13 Circuit diagram for loaded transformer

The operation of the transformer on load is explained below with the help of a number of diagrams;

i. When the transformer is on no-load as shown in fig. 5.14(a) it draws no-load current I_0 from the supply mains. The no-load current I_0 produces an mmf $N_1 I_0$ which sets up flux in the core.

ii. When the transformer is loaded, current I_2 flows in the secondary winding. This secondary current I_2 produces an mmf $N_2 I_2$ which sets up flux ϕ_2 in the core. As per Lenz's law, this flux opposes the main flux ϕ, shown in fig. 5.14(b).

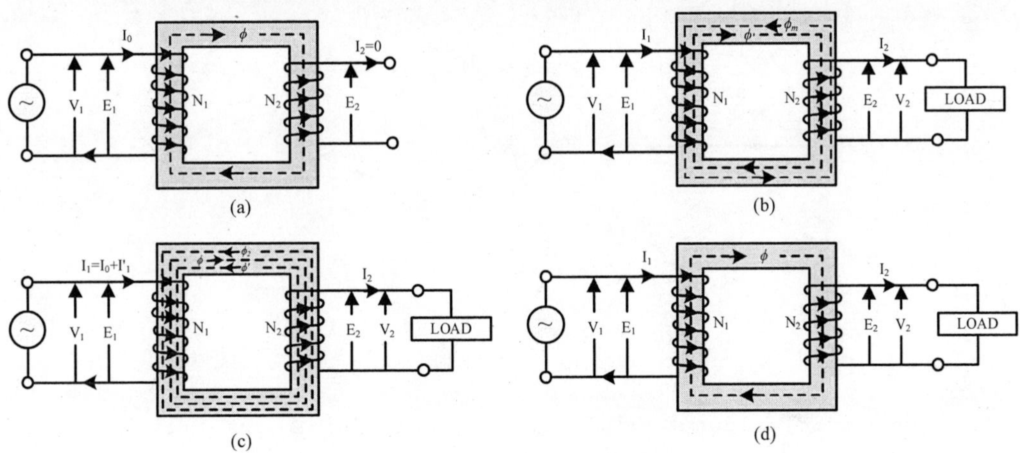

Fig. 5.14 Transformer under loaded condition.

iii. As ϕ_2 is set up in the opposite direction to the main flux, the resultant flux tends to decrease and causes the reduction of self-induced emf E_1 momentarily. Thus V_1 predominates over E_1 causing additional primary current I_1' drawn from the supply mains. The amount of this additional current I_1 is such that the original conditions i.e., flux in the core must be restored. So that $V_1 = E_1$. The current I_1' is in phase opposition with I_2 and is called primary counterbalancing current. This additional current I_1' produces an mmf $N_1\, I_1'$ which sets up flux ϕ, in the same direction as that of ϕ as shown in fig. 5.15(c), and cancels the flux ϕ_2 set up by mmf $N_2\, I_2$.

Now $N_1\, I_1' = N_2\, I_2$ (ampere-turns balance)

$$I_1' = \frac{N_2}{N_1}\, I_2 = K\, I_2 \tag{5.22}$$

iv. Thus, the flux is restored to its original value as shown in fig. 5.14(d). The total primary current I_1 is the vector sum of current I_0 and I_1'.

$$I_1 = I_0 + I_1' \tag{5.23}$$

This shows that flux in the core of a transformer remains the same from no-load to full load; this is the reason why iron losses in a transformer remain the same from no-load to full load.

Phasor Diagram of a Loaded Transformer (Neglecting voltage drops in the windings; ampere-turns balance)

Since the voltage drops in both the windings of the transformer are neglected, therefore,

$$V_1 = E_1 \text{ and } E_2 = V_2 \tag{5.24}$$

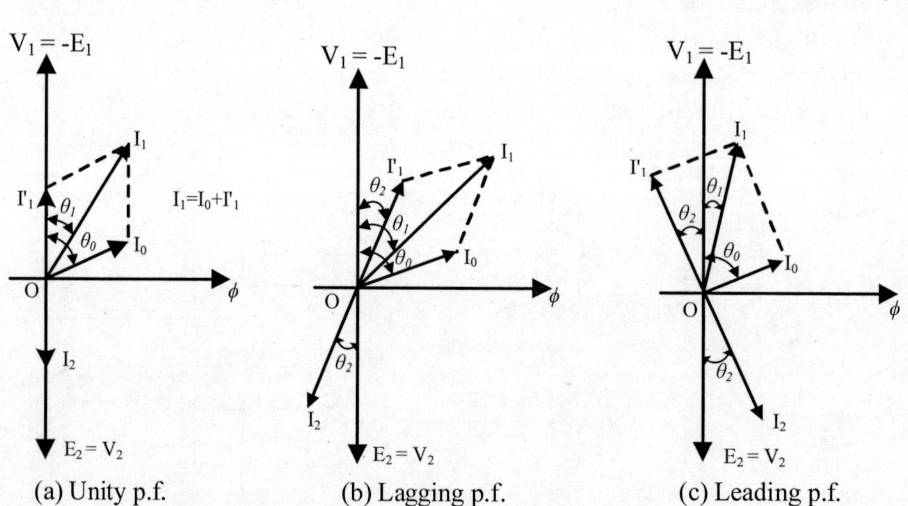

Fig. 5.15 Phasor diagram on-load (neglecting winding resistance and leakage reactance)

While drawing the phasor diagram the following important points are to be considered.

i. For simplicity, let the transformation ratio $K = 1$ be considered, therefore, $E_2 = V_2$.

ii. The secondary current I_2 is in phase, lags behind, and leads the secondary terminal voltage V_2 by an angle θ_2 for resistive, inductive, and capacitive loads, respectively.

iii. The counter balancing current

$$I_1' = \frac{N_2}{N_1} I_2 \tag{5.25}$$

$I_1' = K\, I_2\ here\ K = 1\ \therefore\ I_1' =\ I_2$ and is 180° out of phase with I_2.

iv. The total primary current I_1 is the vector sum of the no-load primary current I_0 and counterbalancing current I_1'.

$$\overline{I_1} = \overline{I_0} + \overline{I_1}'\ or\ \overline{I_1} = \sqrt{(I_0)^2 + (I_1')^2 + 2I_0\ I_1' \cos\theta} \tag{5.26}$$

Where θ is the phase angle between $I_0\ and\ I_1'$.

v. The p.f. on the primary side is $\cos\theta_1$ which is less than the load p.f. $\cos\theta_2$ on the secondary side. Its value is determined by the relation;

$$\cos\theta_1 = \frac{I_0\ \cos\theta_0 + I_1'\ \cos\theta_2}{I_1} \tag{5.27}$$

The phasor diagrams of the transformer for resistance, inductive and capacitive loads are shown in fig. 5.15 (a), (b), and (c), respectively.

Example 5.8: A single-phase 4,400/400 V, 50 Hz transformer has a no-load current of 0.04 A. It consumes power 80 W at no load while the secondary is kept open. Calculate (i) the Power factor of the no-load current (ii) the Iron loss component of the current (iii) Magnetizing component of the current.

Solution:

$W_0 = 80\ W;$

$I_0 = 0.04\ A;$

$V_1 = 4400\ V$

(i) Since $W_0 = V_1 I_0 \cos\theta_0$

$$\cos\theta_0 = \frac{W_0}{V_1 I_0} = \frac{80}{4400\ \times 0.04} = 0.454\ (Ans)$$

The no-load power factor is 0.454 (lagging).

(ii) $I_W = I_0 \cos\theta_0$

$= 0.04\ \times 0.454 = 0.0187\ A\ (Ans)$

(iii) $\sin\theta_0 = \sqrt{1 - cos^2\,\theta_0}$

$= 0.891$

$I_\mu = I_0 \sin\theta_0$

$= 0.04\ \times 0.891 = 0.0356\ A\ \ (Ans)$

Example 5.9: A 2,200/220 V, 10 kVA, single-phase transformer takes a no-load current of 1.3 A when high-voltage winding is kept open. The iron loss component of no-load current is equal to 0.5A. Calculate (i) No-load input power (ii) Magnetizing component and power factor of no-load current.

Solution:

Given $I_0 = 1.3\ A;\ I_W = I_0 \cos\theta_0 = 0.5\ A$

$$V_1 = E_1 = 2200\ V\ and\ V_2 = E_2 = 220\ V$$

(i) No-load input power,

$$(W_0) = V_2\, I_0 \cos\theta_0 = V_2\, I_W = 220\ \times 0.5 = 110\ W\ (Ans)$$

(ii) Now, $\cos\theta_0 = \frac{I_W}{I_0}$

$$= \frac{0.5}{1.3} = 0.385\ (lagging)$$

$$\sin\theta_0 = \sqrt{1 - cos^2\,\theta_0} = 0.923$$

$$I_\mu = I_0\ \sin\theta_0 = 1.3\ \times 0.923 = 1.199\ A$$

Example 5.10: A single phase 6000/400 V transformer has no-load current of 0·7 A at 0·24 power factor lagging. A current of 120 A at a power factor of 0·8 lagging is supplied by its secondary then determine the current drawn by the primary winding.

Solution:

Here, $I_0 = 0.7\ A;\ \cos\theta_0 = 0.24\ lag;\ I_2 = 120\ A;\ \cos\theta_2 = 0.8\ lag$

Transformer ratio, $K = \frac{V_2}{V_1}$

$$= \frac{400}{6600} = \frac{2}{30}$$

Let the primary counterbalance current be I_2.

$$N_1 I_1' = N_2 I_2$$

$$\text{Or}\ \ I_1' = \frac{N_2}{N_1} \times I_2 = K\, I_2$$

$$= \frac{2}{30} \times 120 = 8\ A$$

Now, $\cos\theta_0 = 0.24;\ \theta_0 = cos^{-1}0.24 = 76.11°$

$$\cos\theta_2 = 0.8;\ \theta_2 = cos^{-1}0.8 = 36.87°$$

$$\theta = 76.11° - 36.87° = 39.24°$$

Current drawn by the primary,

$$I_1 = \sqrt{(I_0)^2 + (I_1')^2 + 2\, I_0 I_1' \cos\theta}$$

$$= \sqrt{(0.7)^2 + (8)^2 + 2 \times 0.7 \times 8 \times cos39.24''} = 8.55\ A\ (Ans)$$

Example 5.11: A 230V, 50 Hz transformer has 200 primary turns. It draws 5 A at 0.25 p.f lagging at no-load. Determine (i) The maximum value of flux in the core; (ii) Core loss; (iii) Magnetising current (iv) Exciting resistance and reactance of the transformer.

Solution:

(i) Using the relation, $E_1 = 4.44 N_1 f \phi_m$

$230 = 4.44 \times 220 \times 50 \times \phi_m$

The maximum value of flux $\phi_m = 5.18\ m\,Wb\ (Ans)$

(ii) Core loss, $P_0 = V_1 I_0 \cos\theta_0$

$= 230 \times 5 \times 0.25 = 287.5\ W\ (Ans)$

(iii) No-load p.f., $\cos\theta_0 = 0.25$;

$\sin\phi_0 = \sin cos^{-1} 0.25 = 0.9682$

Magnetizing current component,

$I_m = I_0 \sin\theta_0 = 5 \times 0.9682 = 4.84\ A\ (Ans)$

(iv) Exciting resistance,

$$R_0 = \frac{V_1}{I_W} = \frac{230}{I\cos\theta_0} = \frac{230}{5 \times 0.25} = 184\ \Omega\ (Ans)$$

Exciting reactance,

$$X_0 = \frac{V_1}{I_\mu} = \frac{230}{4.84} = 47.52\ \Omega\ (Ans)$$

5.15 Equivalent Resistance

In an actual transformer, the primary and secondary windings have some resistance represented by R_1 and R_2, respectively as shown in fig. 5.16. To make the calculations easy the resistance of the two windings can be transferred to either side. The resistance is transferred from one side to the other in such a manner that the percentage voltage drop remains the same when represented on either side. It is assumed that there is no leakage of flux in the surrounding medium.

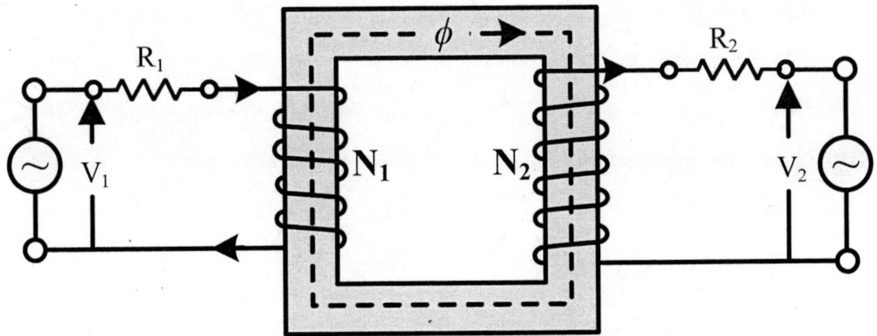

Fig. 5.16 Transformer winding with resistance

Let N_1 and N_2 be the number of turns of primary and secondary windings respectively. Let the turn ratio be 'K' and I_1 and I_2 be the current in primary and secondary windings respectively.

Neglecting $I_0, \frac{I_1}{I_2} = K$. Let the referred value of R_2 be R'_2 when it is transferred to primary.

Now, Copper loss across $R_2 = I_2^2 R_2$ (when R_2 is in secondary) (5.28)

Copper loss across $R'_2 = I_1^2 R'_2$ (when R_2 transferred to primary) (5.29)

These two losses must me equal.

$$I_2^2 R_2 = I_1^2 R'_2 \tag{5.30}$$

$$R'_2 = \left(\frac{I_2^2}{I_1^2}\right) R_2 = \frac{1}{K^2} R_2 \tag{5.31}$$

Total resistance referred to primary

$$R_{01} = R_1 + R'_2 \tag{5.32}$$

$$R_{01} = R_1 + \frac{1}{K^2} R_2 \tag{5.33}$$

R_{01} is known as effective or equivalent resistance referred to as primary. fig. 5.17 show the equivalent circuit of fig. 5.16 when secondary resistance is transferred to primary.

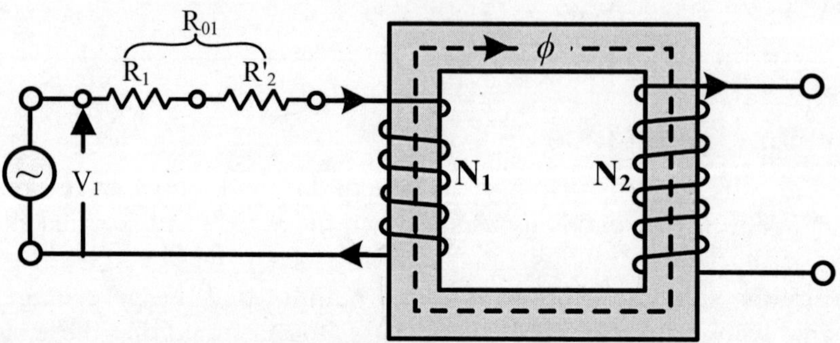

Fig. 5.17 Resistance referred to primary

Let the referred value of R_1 be R'_1 when it is transferred to secondary.

Now, Copper loss across $R_1 = I_1^2 R_1$ (when R_1 is in primary) (5.34)

Copper loss across $R'_1 = I_2^2 R'_1$ (R_1 transferred to secondary) (5.35)

These two losses must me equal.

$$I_1^2 R_1 = I_2^2 R'_1 \tag{5.36}$$

$$R'_1 = \left(\frac{I_1^2}{I_2^2}\right) R_1 = K^2 R_1 \tag{5.37}$$

Total resistance referred to secondary

$$R_{02} = R_2 + R'_1 \tag{5.38}$$

$$R_{02} = R_2 + K^2 R_1 \tag{5.39}$$

R_{02} is known as effective or equivalent resistance referred to as secondary. Fig. 5.18 shows the equivalent circuit of 5.16 when secondary resistance is transferred to secondary.

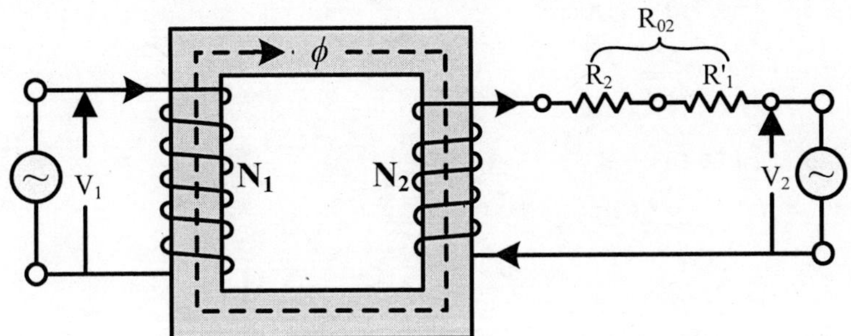

Fig. 5.18 Resistance referred to secondary

5.16 Mutual and Leakage Flux

It is assumed that when the AC supply is connected to the primary winding of a transformer, an alternating flux is set up in the core, and the whole of this flux links with both the primary and secondary windings. However, in an actual transformer, both windings produce some flux. Till now we have assumed that all the flux linked with the primary also links with the secondary. But in practice, the permeability of the core of the transformer is finite. All the fl ux linked with the primary do not link with the secondary. *The flux that links with both the windings of the transformer is called* **mutual flux** *and the flux which links only with one winding of the transformer and not to the other is called* **leakage flux.**

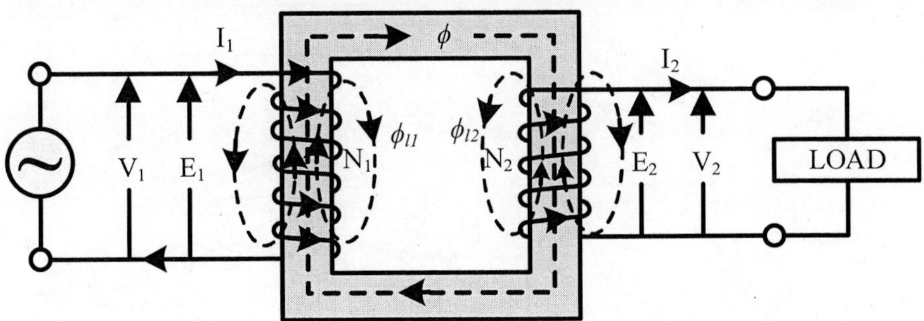

Fig. 5.19 Primary and secondary winding reactance

The primary ampere turns to produce some flux ϕ_{l1} which is set up in the air and links only with the primary winding, as shown in fig. 5.19, is called *primary leakage flux.* Similarly, secondary ampere-turns produce some flux ϕ_{l2} which is set up in the air and links only with secondary winding called *secondary leakage flux.* The primary leakage flux ϕ_{l1} is proportional to the primary current I_1 and secondary leakage flux ϕ_{l2} is proportional to the secondary current I_2. The primary leakage flux ϕ_{l1} produces self-leakage inductance L_{l1} $(= N_1\phi_{l1}/I_1)$

which in turn produces leakage reactance X_1 $(= 2\pi f L_{l1})$. Similarly, secondary leakage flux ϕ_{l2} produces leakage reactance X_2 $(= 2\pi f L_{l2})$.

5.17 Equivalent Reactance

Let the referred value of X_2 be X'_2 when it is transferred to primary. We have

$$I_2^2 X_2 = I_1^2 X'_2 \tag{5.40}$$

$$X'_2 = \left(\frac{I_2^2}{I_1^2}\right) X_2 = \frac{1}{K^2} X_2 \tag{5.41}$$

Total resistance referred to primary

$$X_{01} = X_1 + X'_2 \tag{5.42}$$

$$X_{01} = X_1 + \frac{1}{K^2} X_2 \tag{5.43}$$

X_{01} is known as effective or equivalent reactance referred to as primary. Fig. 5.20 shows the equivalent circuit secondary resistance is transferred to primary.

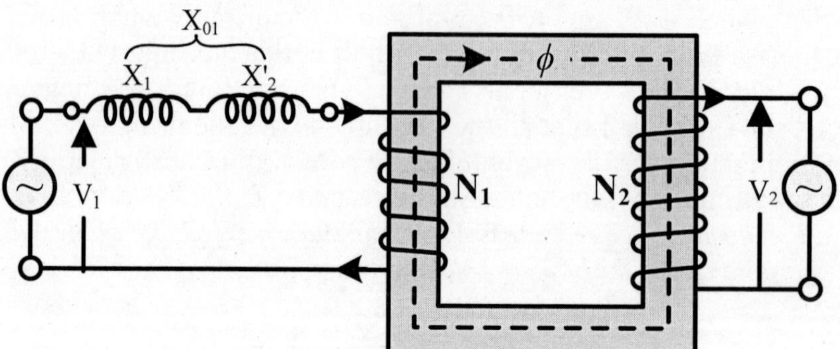

Fig. 5.20 Reactance referred to primary

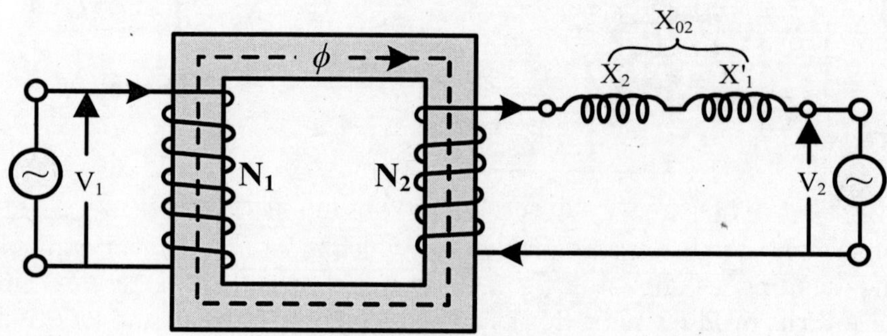

Fig. 5.21 Reactance referred to secondary

Let the referred value of X_1 be X'_1 when it is transferred to secondary. We have

$$I_1^2 X_1 = I_2^2 X'_1 \tag{5.44}$$

$$X'_1 = \left(\frac{I_1^2}{I_2^2}\right) X_1 = K^2 X_1 \tag{5.45}$$

Total reactance referred to secondary

$$X_{02} = X_2 + X'_1 \tag{5.46}$$

$$X_{02} = X_2 + K^2 X_1 \tag{5.47}$$

X_{02} is known as effective or equivalent reactance referred to as secondary. Fig. 5.21 shows the equivalent circuit when primary reactance is transferred to secondary.

Now, total impedance refer to the primary side

$$Z_{01} = R_{01} + jX_{01} \tag{5.48}$$

$$Z_{01} = \sqrt{R_{01}^2 + X_{01}^2} \tag{5.49}$$

Similarly, total impedance refer to the secondary side

$$Z_{02} = R_{02} + jX_{02} \tag{5.50}$$

$$Z_{02} = \sqrt{R_{02}^2 + X_{02}^2} \tag{5.51}$$

The total equivalent circuit refer to either side shown in fig. 5.22

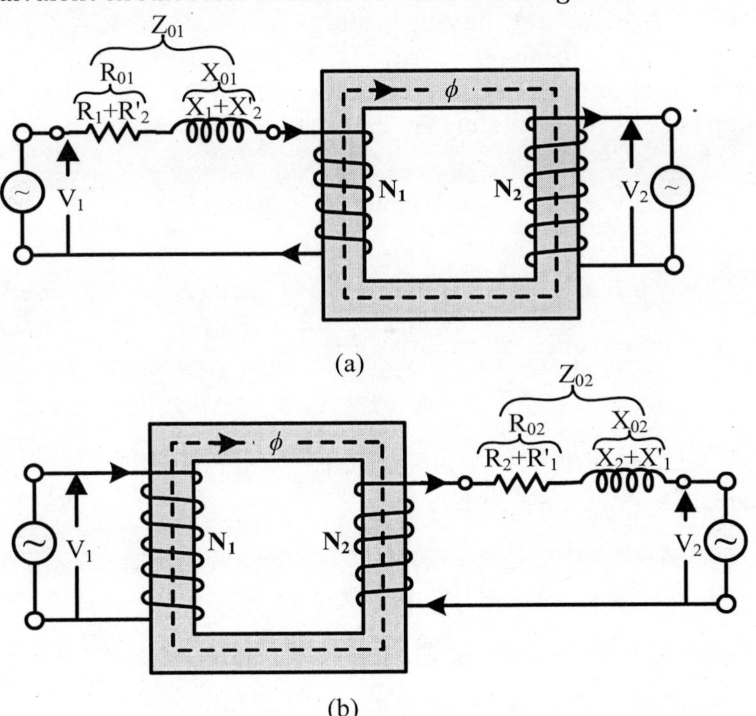

Fig. 5.22 Equivalent circuit referred to (a) primary side (b) secondary side

Example 5.12: Find equivalent resistance and reactance as referred to secondary winding when a 63 kVA, 1100/440 V single-phase transformer has R_1 = 0.18 ohm, X_1 = 0.55 ohm, R_2 = 0.0068 ohm and X_2 = 0.03 ohm.

Solution:

Here, transforming rating $= 63\ kVA;\ V_1 = 1100\ V;\ V_2 = 440\ V;$

$$R_1 = 18\ \Omega;\ X_1 = 0.55\ \Omega;\ R_2 = 0.0068\ \Omega;\ X_2 = 0.03\ \Omega$$

Transformation ratio, $K = \frac{V_2}{V_1} = \frac{440}{1100} = 0.4$

Equivalent resistance referred to the secondary side,

$$R_{02} = R_2 + R_1' = R_2 + R_1 \times K^2$$

$$= 0.0068 + 0.18 \times (0.4)^2 = 0.0356\ \Omega\ (Ans)$$

Equivalent reactance referred to the secondary side,

$$X_{02} = X_2 + X_1' = X_2 + X_1 \times K^2$$

$$= 0.02 + 0.5 \times (0.2)^2 = 0.118\ \Omega\ (Ans)$$

Example 5.13: Determine the total resistance and reactance referred to secondary side. A single-phase transformer having voltage ratio 2000/200V (primary to secondary) has a primary resistance and reactance 1·8 ohm and 4·2 ohm, respectively. The corresponding secondary values are 0·02 and 0·045 ohm. Also, calculate the impedance of the transformer referred to the secondary side.

Solution:

Here, $R_1 = 1.8.\Omega;\ X_1 = 4.2\ \Omega;\ R_2 = 0.02\ \Omega;\ X_2 = 0.045\ \Omega$

Transformation ratio, $K = \frac{V_2}{V_1} = \frac{200}{2000} = 0.1$

Total resistance referred to the secondary side,

$$R_{02} = R_2 + R_1' = R_2 + R_1 \times K^2$$

$$= 0.02 + 0.18 \times (0.1)^2 = 0.038\ \Omega\ (Ans)$$

Total reactance referred to the secondary side,

$$X_{02} = X_2 + X_1' = X_2 + X_1 \times K^2$$

$$= 0.045 + 4.2 \times (0.1)^2 = 0.087\ \Omega\ (Ans)$$

The impedance of the transformer referred to the secondary side,

$$Z_{02} = \sqrt{(R_{02})^2 + (X_{02})^2} = \sqrt{(0.038)^2 + (0.087)^2} = 0.095\ \Omega\ (Ans)$$

5.18 Transformer with Resistance and Leakage Reactance (Actual Transformer)

A transformer having resistances R_1 and R_2 of primary and secondary windings respectively, and leakage reactances X_1 and X_2 of primary and secondary windings respectively. It has iron and copper losses. The equivalent circuit of an actual transformer is shown in fig. 5.23.

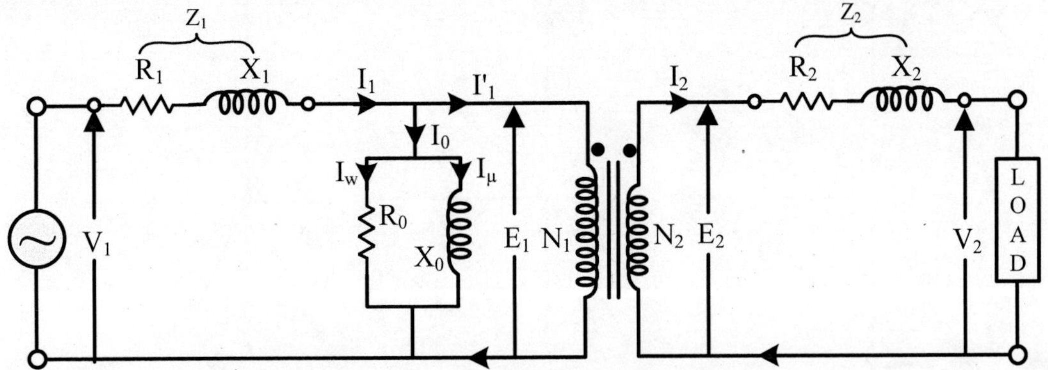

Fig. 5.23 Equivalent circuit for loaded transformer

The primary impedance is given by

$$\overline{Z_1} = R_1 + jX_1 \tag{5.52}$$

$$Z_1 = \sqrt{R_1^2 + X_1^2} \tag{5.53}$$

The secondary impedance is given by

$$\overline{Z_2} = R_2 + jX_2 \tag{5.54}$$

$$Z_2 = \sqrt{R_2^2 + X_2^2} \tag{5.55}$$

The applied voltage V_1 on the primary side is given by

$$\overline{V_1} = \overline{E_1} + \overline{I_1}(R_1 + jX_1) \tag{5.56}$$

$$\overline{V_1} = \overline{E_1} + \overline{I_1}\overline{Z_1} \tag{5.57}$$

According to fig. 5.23

$$\overline{I_1} = \overline{I'_1} + \overline{I_0} \tag{5.58}$$

Similarly, the secondary voltage is given by

$$\overline{V_2} = \overline{E_2} - \overline{I_2}(R_2 + jX_2) \tag{5.59}$$

$$\overline{V_2} = \overline{E_2} - \overline{I_2}\overline{Z_2} \tag{5.60}$$

Fig. 5.24 shows the phasor diagrams of an actual transformer for resistive (unity power factor), inductive (lagging power factor), and capacitive(leading power factor) loads. The drops in resistances are drawn in phase with current vectors and drops in reactances are drawn perpendicular to the current vectors.

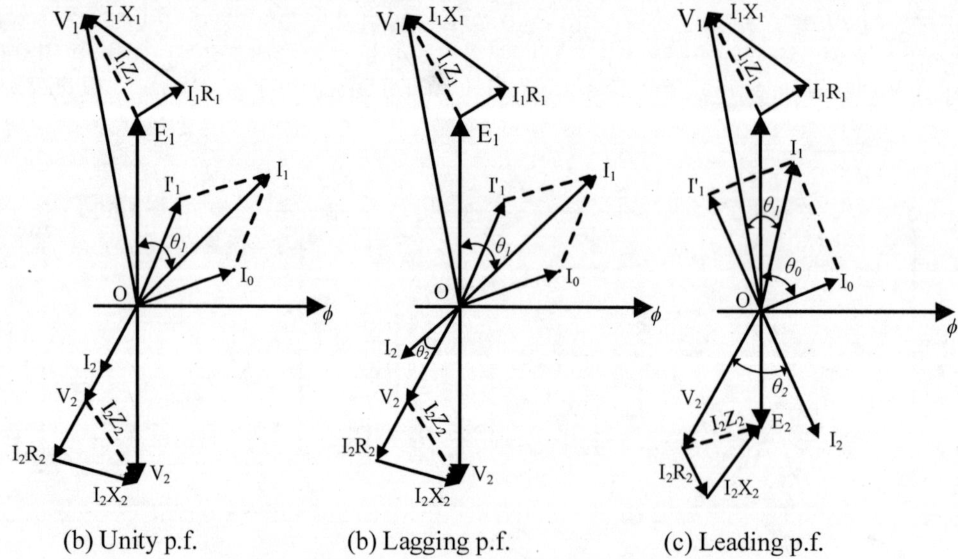

Fig. 5.24 Phasor diagram for loaded transformer

5.19 Equivalent Circuit

The equivalent circuit of a single-phase transformer having load impedance Z_L shown in fig. 5.25. An equivalent circuit in which the resistance and leakage reactance of the transformer are imagined to be external to the winding whose only function then is to transform the voltage. The no-load current I_0 is simulated by pure inductance X_0 taking the magnetizing component I_μ and a non-inductive resistance R_0 taking the working component I_w connected in parallel across the primary circuit.

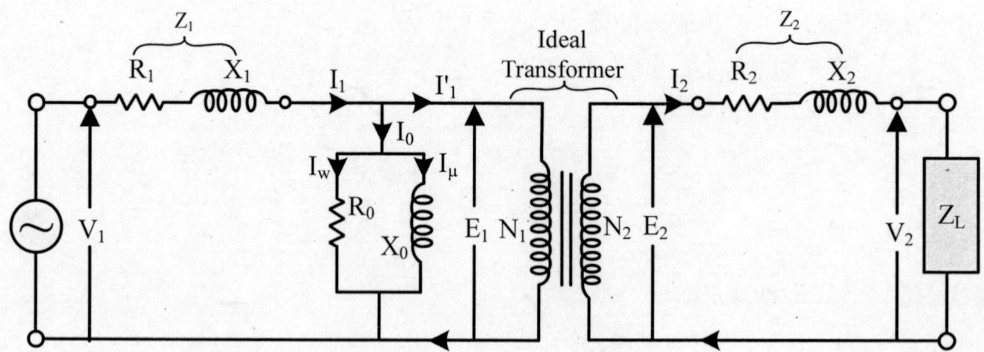

Fig. 5.25 Equivalent Circuit

The value of E_1 is obtained by subtracting vectorially I_1Z_1 from V_1. The value of $X_0 = E_1/I_\mu$ and $R_0 = E_1/I_w$. It is clear that E_1 and E_2 are related to each other by expression

$$\frac{E_2}{E_1} = \frac{N_2}{N_1} = K \tag{5.61}$$

To make transformer calculation simpler. It is preferable to transfer voltage, current, and impedance either to primary or to secondary. In that case, we would have to work in one winding only which is more convenient.

The primary equivalent of the secondary induced voltage is

$$E'_2 = \frac{E_2}{K} = E_1 \tag{5.62}$$

Similarly, the primary equivalent of secondary terminal or output voltage is

$$V'_2 = \frac{V_2}{K} = V_1 \tag{5.63}$$

The primary equivalent of secondary current

$$I'_2 = KI_2 \tag{5.64}$$

For transferring secondary impedance to primary K^2 is used (refer to section 5.14 and 5.17 and fig. 5.17 to fig. 5.22)

$$R'_2 = \left(\frac{I_2^2}{I_1^2}\right) R_2 = \frac{1}{K^2} R_2 \tag{5.65}$$

$$X'_2 = \left(\frac{I_2^2}{I_1^2}\right) X_2 = \frac{1}{K^2} X_2 \tag{5.66}$$

$$Z'_2 = \left(\frac{I_2^2}{I_1^2}\right) Z_2 = \frac{1}{K^2} Z_2 \tag{5.67}$$

Similarly, the relationship is used for shifting an external load impedance to primary. The secondary circuit is shown in fig. 5.26(a) and its equivalent primary values are shown in fig. 5.26(b).

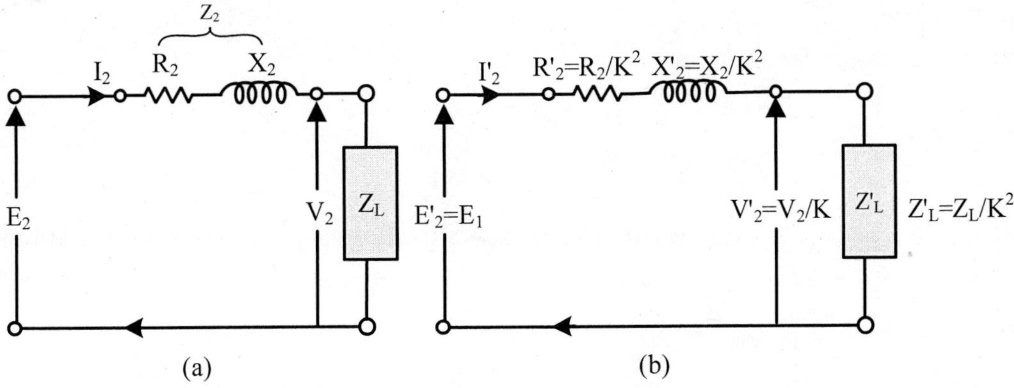

Fig. 5.26 Equivalent Circuit (a) secondary circuit (b) equivalent primary value

The total equivalent circuit of the transformer is obtained by adding in the primary impedance as shown in fig. 5.27. This is known as the exact equivalent circuit but it presents a somewhat harder circuit problem to solve. A simplification can be made by transferring the exciting circuit across the terminals as in fig. 5.28. it should be noted that in this case $X_0 = V_1/I_\mu$.

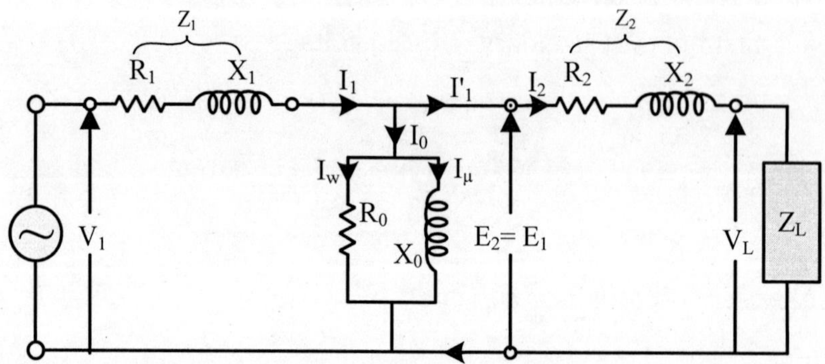

Fig. 5.27 Exact equivalent Circuit

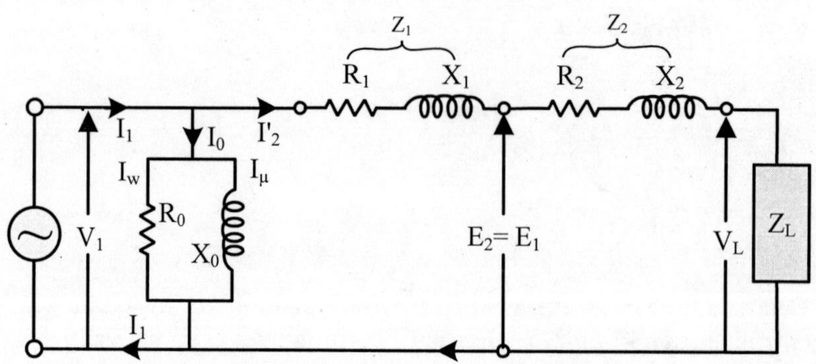

Fig. 5.28 Approximate equivalent Circuit

Further simplification may be achieved by omitting I_0. From fig. 5.26 it is found that the total impedance between the input terminal is

$$Z = Z_1 + Z_m \| (Z'_2 + Z'_L) \tag{5.68}$$

$$Z = Z_1 + \frac{Z_m(Z'_2 + Z'_L)}{Z_m + Z'_2 + Z'_L} \tag{5.69}$$

Where $Z'_L = R'_L + X'_L$ is load impedance and Z_m is the impedance of the exciting circuit.

$$V_1 = I_1 Z \tag{5.70}$$

$$V_1 = I_1 \left[Z_1 + \frac{Z_m(Z'_2 + Z'_L)}{Z_m + Z'_2 + Z'_L} \right] \tag{5.71}$$

R_o, X_o, R_{01}, and X_{01} are the four important parameters of the transformer. From the open circuit and short circuit tests of a transformer, these parameters can be determined.

5.20 Losses in Transformer

Two types of losses occur in a transformer:

- Core loss or iron loss occurs in a transformer because it is subjected to an alternating flux.
- The windings carry current due to loading and hence copper losses occur.

5.31.1 Core or iron loss

The separation of core losses has already been introduced. The alternating flux is set up in the core and it undergoes a cycle of magnetization and demagnetization. Therefore, loss of energy occurs in this process due to hysteresis. This loss is called *hysteresis loss* (P_h) which is expressed by

$$P_h = K_h B_m^{1.6} f\, V \;\; Watt \tag{5.72}$$

where K_h is hysteresis constant depending on the material, B_m is the maximum flux density, f is the frequency and V is the volume of the core.

The induced emf in the core sets up eddy current in the core, and hence eddy current loss (P_e) occur, which is given by

$$P_e = K_e B_m^2 f^2 t^2 V \; Watt \tag{5.73}$$

where K_e is the eddy current constant and t is the thickness of the core.

Since the supply voltage V_1 at rated frequency f is always constant, the flux in the core is almost constant. Therefore, flux density in the core remains constant. Hence, hysteresis and eddy current losses are constant at all loads. Thus, the core loss or iron loss is also known as constant loss. The iron loss is denoted by P_i.

Iron loss is reduced using high-grade core material such as Cold Rolled Grain Oriented silicon steel having a very low hysteresis loop for reducing hysteresis loss and laminated core for reducing the eddy current loss.

5.31.2 Copper loss

The loss of power in the form I^2R due to the resistances of the primary and secondary windings is known as *copper losses.* The copper loss also depends on the magnitude of currents flowing through the windings. The total Cu loss is given by

$$P_{Cu} = I_1^2 R_1 + I_2^2 R_2 = I_1^2 R_{01} = I_2^2 R_{02} \tag{5.74}$$

Copper losses are determined based on R_{01} or R_{02} which is determined from short circuit test. Since the standard operating temperature of an electrical machine is taken as 75°C, it is then corrected to 75°C.

The copper loss due to full-load current is known as full-load Copper loss(Cu loss). The Cu loss is also known as *variable loss.*

There are two other losses known as stray loss and dielectric loss. Since the leakage field is present in a transformer, eddy currents are induced in the conductors, tank walls and bolts etc. Stray losses occur due to these eddy currents. Dielectric loss occurs in the insulation of coil and solid insulation. These two losses are small and hence neglected.

Therefore, the total loss of the transformer= Iron loss+Cu loss= $P_i + P_{Cu}$

5.31.3 Stray losses (leakage flux)

Leakage inductance is by itself largely lossless since energy supplied to its magnetic fields is returned to the supply with the next half-cycle. However, any leakage flux that intercepts nearby conductive materials such as the transformer's support structure will give rise to eddy currents and be converted to heat. There are also radiative losses due to the oscillating magnetic field, but these are usually small and negligible.

5.31.4 Dielectric losses

In the solid insulation or transformer oil i.e., insulation material of the transformer, dielectric loss occurs when the solid insulation gets damaged or the oil gets deteriorated or its quality decreases over time. Hence, the overall efficiency of the transformer may be affected and get damaged due to either excessive voltage or excessive dielectric loss.

5.31.5 Magnetostriction losses

Magnetic flux in a ferromagnetic material, such as the core, causes it to physically expand and contract slightly with each cycle of the magnetic field, an effect known as magnetostriction. This produces the buzzing sound commonly associated with transformers and can cause losses due to frictional heating.

5.31.6 Mechanical losses

In addition to magnetostriction, the alternating magnetic field causes fluctuating forces between the primary and secondary windings. These incite vibrations within nearby metalwork, adding to the buzzing noise, and consuming a small amount of power.

5.21 Separation of Hysteresis and Eddy Current Losses

Core losses of a transformer are constituted by (*i*) hysteresis loss and (*ii*) eddy current loss. According to Steinmetz's empirical relations;

$$P_h = K_h B_m^{1.6} f\, V \;\; Watt \tag{5.75}$$

$$P_e = K_e B_m^2 f^2 t^2 V \; Watt \tag{5.76}$$

If the thickness of laminations and volume of the core is kept constant, these losses will depend upon supply frequency and maximum flux density and hence.

$$P_h = P\, f\, B_m^{1.6} \; Watt \tag{5.77}$$

$$P_e = Q\, f^2 B_m^2 \; Watt \tag{5.78}$$

Here P and Q are the new constants.

For the transformer emf equation is given by the relation;

$$E = 4.44\, N\, f\, B_m\, A_i\, Watt \tag{5.79}$$

Or $B_m \propto \frac{E}{f} \propto \frac{V}{f}$ as other values are constant.

For a particular value of B_m the core losses per cycle may be represented as;

$$\frac{P_i}{f} = A + Bf \tag{5.80}$$

Where A and B are other constants ($A = P\, B_m^{1.6}$ and $B = Q\, B_m^2$)

The value of constants A and B can be determined by performing an open circuit test on the transformer at different frequencies but keeping $\frac{V}{f}$ constant at every instant. While performing this test, the applied voltage V and frequency f are varied together.

At every step, take the reading of frequency meter 'f' and wattmeter 'P_i'. Plot a curve between f and $\frac{P_i}{f}$, it will give a straight line curve as shown in fig. 5.29.

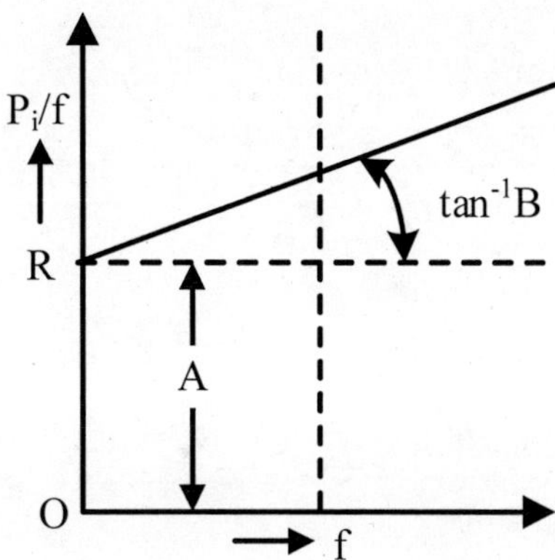

Fig. 5.29 Curve between the ratio of core loss and frequency versus frequency (Keeping V/Hz constant)

Where this line intercepts the vertical axis (R) gives the value of constant A (i.e., A = OR), whereas the slope of the line gives the value of constant B. Knowing the value of A and B we can separate the hysteresis and eddy current losses.

5.22 Efficiency of a Transformer

The efficiency of a transformer is defined as the ratio of output to input power, the two being measured in the same units (either in watts or in kW). Due to the losses in a transformer, its output power is less than the input power.

∴ Power output = Power input – Total losses

∴ Power input = Power output + Total losses = Power output + $P_i + P_{Cu}$

$$\text{Transformer efficiency}, \eta = \frac{Power\ Output}{Power\ Input}$$

$$= \frac{Power\ Output}{Power\ Output\ +\ P_i +\ P_{Cu}} \tag{5.81}$$

The output power of a transformer at full load = $V_2 I_{2fl} \cos\theta_2$,

Where, $\cos\theta_2$ is the power factor of the load,

I_{2fl} is the secondary current at full load,

V_2 is the rated secondary voltage of the transformer.

$I_{2fl}^2 R_{02}$ =full-load copper loss of the transformer

The efficiency of the transformer at full load is given by

$$\eta_{fl} = \frac{V_2 I_{2fl} \cos\theta_2}{V_2 I_{2fl} \cos\theta_2 +\ P_i + I_{2fl}^2 R_{02}} \tag{5.82}$$

$$\%\,\eta_{fl} = \frac{V_2 I_{2fl} \cos\theta_2}{V_2 I_{2fl} \cos\theta_2 +\ P_i + I_{2fl}^2 R_{02}} \times 100 \tag{5.83}$$

The efficiency of the transformer at any load m is given by

$$\eta = \frac{m\ \times Output\ at\ full\ load}{m\ \times output\ at\ full\ load\ \ P_i + m^2\ I_{2fl}^2 R_{02}}\ in\ p.u. \tag{5.84}$$

$$\text{Where } m = \frac{Actual\ Load}{Full\ Load}\ or\ m = \sqrt{\frac{P_i}{P_{Cu\ fl}}}$$

$P_{Cu\ fl}$ is the Cu loss of the transformer at full load

- ***Conditions for maximum efficiency***

For maximum efficiency

$$\frac{d\eta}{dI_2} = 0$$

$$\text{Now,} \quad \eta = \frac{V_2 I_2 \cos\theta_2}{V_2 I_2 \cos\theta_2 +\ P_i + I_2^2 R_2}$$

$$\frac{d\eta}{dI_2} = \frac{d}{dI_2}\left[\frac{V_2 I_2 \cos\theta_2}{V_2 I_2 \cos\theta_2 +\ P_i + I_2^2 R_2}\right] = 0$$

$$\frac{V_2 \cos\theta_2\ (V_2 I_2 \cos\theta_2 +\ P_i + I_2^2 R_2) - V_2 I_2 \cos\phi_2\ (V_2 \cos\theta_2 + 2I_2 R_2)}{(V_2 I_2 \cos\theta_2 + P_i + I_2^2 R_2)^2}$$

$$V_2 I_2 \cos\theta_2 + P_i + I_2^2 R_2 - V_2 I_2 \cos\theta_2 - \ 2I_2 R_2 = 0$$

$$P_i =\ I_2^2 R_2 \tag{5.85}$$

Iron loss = Cu loss or Constant loss = Variable loss

Thus, the efficiency of a transformer will be maximum when copper (or variable) losses are equal to iron (or constant) losses.

Now, $$\eta_{max} = \frac{V_2 I_2 \cos\theta_2}{V_2 I_2 \cos\theta_2 + 2P_i} \times 100 \qquad (5.86)$$

During the working of a transformer at constant voltage and frequency, its efficiency varies with the load. Its efficiency increases as the load increases. At a certain load, its efficiency becomes maximum. If the transformer is further loaded, its efficiency starts decreasing. Fig. 5.30 shows the plot of efficiency versus load current.

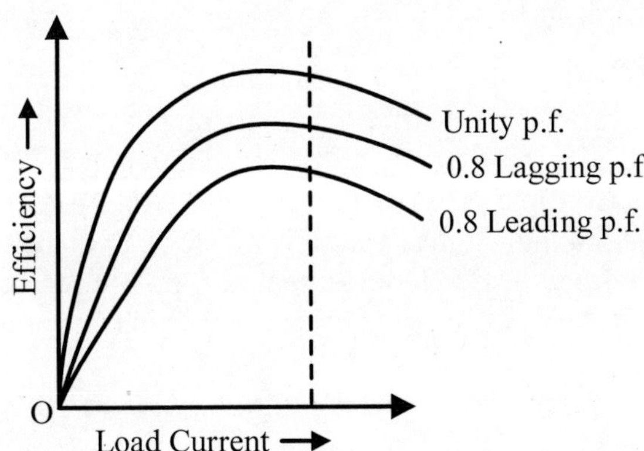

Fig. 5.30 Variation of efficiency versus current for different power factors

To determine the condition of maximum efficiency, let us assume that the power factor of the load remains constant and the secondary terminal voltage (V_2) is constant. Therefore, efficiency becomes only a function of the load current (I_2).

- ***Load Current at Maximum Efficiency***

Let $I_{2m,}$ be the load current at maximum efficiency.

$$I_{2m}^2 R_2 = P_i$$

$$I_{2m}^2 = \frac{P_i}{R_2}$$

$$I_{2m} = \sqrt{\frac{P_i}{R_2}} \qquad (5.87)$$

Or let I_{2fl} be the full load current.

$$I_{2m}^2 = \frac{P_i I_{2fl}^2}{I_{2fl}^2 R_{02}} = I_{2fl}^2 \times \frac{P_i}{P_{Cu\,fl}}$$

$$i.e., I_{2m} = I_{2fl}\sqrt{\frac{P_i}{P_{Cu\,fl}}} \qquad (5.88)$$

Equation (5.88) shows the load current in terms of full-load current at maximum efficiency.

- ***kVA Supplied at Maximum Efficiency***

For constant V_2 , the kVA supplied is the function of load current only.

$$kVA \ at \ \eta_{max} = I_{2m} V_2 \times 10^{-3} = V_2 I_{2fl} \sqrt{\frac{P_i}{P_{Cu\,fl}}} \times 10^{-3}$$

$$kVA \ at \ \eta_{max} = (full \ load \ kVA \ rating) \times \sqrt{\frac{P_i}{P_{Cu\,fl}}} \quad (5.89)$$

5.23 All Day Efficiency

The load on certain transformers fluctuates throughout the day. The distribution transformers are energized for 24 hours, but they deliver very light loads for the major portion of the day. Thus iron losses occur for the whole day but copper losses occur only when the transformer is loaded. Hence, the performance of such transformers cannot be judged by the commercial power efficiency, but it can be judged by all-day efficiency also known as operational efficiency or energy efficiency which is computed based on energy consumed during 24 hours.

*The **all-day efficiency** is defined as the ratio of output in kWh (or Wh) to the input in kWh (or Wh) of a transformer over 24 hours.*

All day efficiency, $\eta_{All-day}$

$$= \frac{Output \ in \ kWh \ over \ 24 \ hours}{Input \ in \ kWh \ over \ 24 \ hours} \quad (5.90)$$

All day efficiency, $\eta_{All-day}$

$$= \frac{Output \ in \ kWh \ over \ 24 \ hours}{Output \ in \ kWh \ over \ 24 \ hours + energy \ loss \ over \ 24 \ hours} \quad (5.91)$$

Example 5.14: A 220/400 V, 10 kVA, 50Hz, single-phase transformer has copper loss of 120 W at full load. If it has an efficiency of 98% at full load, the unity power factor, determine the iron losses. What would be the efficiency of the transformer at half full-load at 0.8 p.f. lagging.

Solution:

$$\eta_x = \frac{x \ kVA \times 1000 \times \cos\theta}{x \ kVA \times 1000 \times \cos\theta + P_i + x^2 P_c} \times 100$$

$$98 = \frac{1 \times 10 \times 1000 \times 1}{1 \times 10 \times 1000 \times 1 + P_i + 1 \times 1 \times 120} \times 100$$

$or \ P_i = 84.08 \ W \ (Ans)$

When x = ½ and $\cos\theta = 0.8$;

$$\eta_x = \frac{0.5 \times 10 \times 1000 \times 0.8}{0.5 \times 10 \times 1000 \times 0.8 + 84.08 + (0.5)^2 \times 120} \times 100 = 97.23\,\% \ (Ans)$$

Example 5.15: A single-phase 440/110 V transformer has primary and secondary winding resistance of 0·3 ohm and 0·02 ohm, respectively. If the iron loss on normal input voltage is 150 W, calculate the secondary current at which maximum efficiency will occur. What is the value of this maximum efficiency for unity power factor load?

Solution:

Primary resistance, $R_1 = 0.3\ \Omega$

Secondary resistance, $R_2 = 0.02\ \Omega$

Iron losses, $P_i = 150\ W$

Load power factor, $\cos\theta = 1$

Primary induced voltage, $E_1 = 440\ V$

Secondary induced voltage, $E_2 = 110\ V$

Transformer ratio,

$$K = \frac{E_2}{E_1} = \frac{110}{440} = \frac{1}{4}$$

Primary resistance referred to secondary,

$$R_1' = K^2 R_1 = \frac{1}{4} \times \frac{1}{4} \times 0.3 = 0.01875\ \Omega$$

Equivalent resistance referred to secondary,

$$R_{02} = R_2 + R_1' = 0.02 + 0.01875 = 0.03875\ \Omega$$

We know the condition for maximum efficiency is,

Copper loss = iron loss i.e., $I_2^2 R_{02} = P_i$

Secondary current at which the efficiency is maximum,

$$I_2 = \sqrt{\frac{P_i}{R_{02}}}$$

$$= \sqrt{\frac{150}{0.03875}} = 62.22\ A\ (Ans)$$

The maximum efficiency,

$$\eta_{max} = \frac{I_2 V_2 \cos\theta}{I_2 V_2 \cos\theta + 2P_i} \times 100$$

$$= \frac{62.22\ \times 110\ \times 1}{62.22\ \times 110\ \times 1 + 2\ \times 150} \times 100 = 95.8\%\ (Ans)$$

Example 5.16: A 50 kVA transformer on full load has a copper loss of 600 watts and iron loss of 500 watts, calculate the maximum efficiency and the load at which it occurs.

Solution:

$$\%\,\eta = \frac{Output}{Output + Iron\ loss + Copper\ loss} \times 100$$

Efficiency will be maximum when;
Copper loss = Iron loss = 500 W
The fraction at which the efficiency is maximum,

$$x = \sqrt{\frac{P_i}{P_c}}$$

$$= \sqrt{\frac{500}{600}} = 0.9128$$

Load at which the efficiency is maximum, i.e.,

$$\text{Output} = \ x\ \times\ kVA = 0.9128\ \times 50 = 45.64\ kVA$$

$$= 45.64\ \times 1 = 45.64\ kW\ (Since\ \cos\phi = 1\ )$$

$$\eta_m = \frac{45.64\ \times 1000}{45.64\ \times 1000 + 500 + 500} \times 100 = 97.85\%\ (Ans)$$

Example 5.17: The iron and full-load copper losses of a 100 kVA single-phase transformer are 1 kW and 1.5 kW, respectively. Calculate the kVA loading at which the efficiency is maximum and its efficiency at this loading: (i) at unit p.f. (ii) at 8 p.f. lagging.

Solution:

Here, Rated capacity = 100 kVA; Iron Loss, $P_i = 1kW$

Full-load copper loss, $P_c = 1.5\ kW$

Output kVA corresponding to maximum efficiency,

$$= x \times rated\ kVA = \sqrt{\frac{P_i}{P_c}} \times rated\ kVA = \sqrt{\frac{1}{1.5}} \times 100 = 81.65\ kVA\ (Ans)$$

(i) At unity p.f.

$$\eta = \frac{81.65 \times 1}{81.65 \times 1 + 1 + 1} \times 100 = \frac{81.65}{83.65} \times 100 = 97.6\%\ (Ans)$$

(ii) At 0.8 p.f. lagging

$$\eta = \frac{81.65 \times 0.8}{81.65 \times 0.8 + 1 + 1} \times 100 = \frac{65.32}{67.32} \times 100 = 97.03\%\ (Ans)$$

Example 5.18: A transformer has a maximum efficiency of 98% at 15 kVA at unity p.f. It is loaded as follows: 12 hrs –2 kW at p.f. 0.5; 6 hrs – 12 kW at p.f. 0.8; 6 hrs – 18 kW at p.f. 0.9, calculate the all-day efficiency of the transformer.

Solution:

We know,

$$\eta_{max} = \frac{kVA\ \cos\phi}{kVA\ \cos\phi + 2\,P_i} \qquad (\because\ P_i = P_c\,)$$

$$\frac{98}{100} = \frac{15\ \times 1}{15\ \times 1 + 2\,P_i}$$

$$15 +\ 2\,P_i = \frac{15\ \times 100}{98} = 15.306$$

Iron losses, $P_i = 0.153\ kW$

Full load copper losses, $P_c = \ P_i = 0.153\ kW$

During 24 hrs. the transformer is loaded as under:

Hrs.	Load in kW	P.f.	Load in kVA $\frac{kW}{p.f.}$	Fraction of load $X = \frac{given\ load\ in\ kVA}{full\ load\ in\ kVA}$
12	2	0.5	2/0.5 =4	4/15 = 0.267
6	12	0.8	12/0.8 = 15	15/15 = 1
6	18	0.9	18/0.9 = 20	20/15 = 1.333

kWh output in 24 hrs

= 2 X 12 + 12 X 6 + 18 X 6 = 204 kWh

Iron losses for 24 hrs

= 0.153 X 24 = 3.672 kWh

Copper losses for 24 hrs

$$= \ (0.267)^2\ \times 0.153\ \times 12 +\ (1)^2\ \times 0.153\ \times 6 +\ (1.333)^2\ \times 0.153\ \times 6$$

$$= 2.68\ kWh$$

Input in 24 hrs

= 204 + 3.672 + 2.68 = 210.352 kWh

All-day efficiency,

$$\eta_{all-day} = \frac{204}{210.352} \times 100 = 96.98\%\ (Ans)$$

Example 5.19: A 30 kVA, 3,000/300 V, 50 Hz, single-phase transformer has the following winding resistances (R's) and leakage reactances (X's):
$R_1 = 2.5\ \Omega;\ R_2 = 0.018\ \Omega; X_1 = 3.8\ \Omega;\ X_2 = 0.052\ \Omega$
Calculate the following:

(i) Equivalent resistance, leakage reactance, and impedance referred to as high voltage side.
(ii) Equivalent resistance, leakage reactance, and impedance referred to as low-voltage side.
(iii) Total Cu loss of the transformer at full load condition.

Solution:

Here, $R_1 = 2.5\ \Omega;\ R_2 = 0.018\ \Omega; X_1 = 3.8\ \Omega;\ X_2 = 0.052\ \Omega$

$$\text{Turn ratio (K)} = \frac{3000}{300} = 10$$

(i) Equivalent resistance referred to as HV side:

$$R_{01} = R_1 + K^2 R_2 = 2.5 + (10)^2 \times 0.018 = 4.3\ \Omega$$

Equivalent leakage reactance referred to as HV side:

$$X_{01} = X_1 + K^2 X_2 = 3.8 + (10)^2 \times 0.052 = 9\ \Omega$$

Equivalent impedance referred to as HV side:

$$Z_{01} = \sqrt{R_{01}^2 + X_{01}^2} = \sqrt{(4.3)^2 + (9)^2} = 9.974\ \Omega$$

(ii) Equivalent resistance referred to as LV side:

$$R_{01} = \frac{R_1}{K^2} + R_2 = \frac{2.5}{10^2} + 0.018 = 0.043\ \Omega$$

Equivalent reactance referred to as LV side:

$$X_{01} = \frac{X_1}{K^2} + X_2 = \frac{3.8}{10^2} + 0.052 = 0.09\ \Omega$$

Equivalent impedance referred to as LV side:

$$Z_{02} = \sqrt{R_{02}^2 + X_{02}^2} = \sqrt{(0.043)^2 + (0.09)^2} = 0.09974\ \Omega$$

(iii) Total Cu loss at full load:

$$\text{Primary current, } I_1 = \frac{30 \times 10^3}{3000} = 10\ A$$

$$\text{Secondary current, } I_2 = \frac{30 \times 10^3}{300} = 100\ A$$

$$\text{Total Cu loss} = I_1^2 R_{01} = (10)^2 \times 4.3 = 430\ W$$

$$\text{Also, total Cu loss} = I_2^2 R_{02} = (100)^2 \times 0.043 = 430\ W$$

$$\text{Also, Total Cu loss} = I_1^2 R_1 + I_2^2 R_2 = 10^2 \times 2.5 + 100^2 \times 0.018 = 430\ W$$

5.24 Voltage Regulation

Voltage regulation is a measure of change in the voltage magnitude between the sending and receiving end of a component. It is commonly used in power engineering to describe the percentage voltage difference between no-load and full-load voltages of distribution lines, transmission lines, and transformers.

When a transformer is loaded, with a constant supply voltage, the terminal voltage changes due to voltage drop in the internal parameters of the transformer i.e., primary and secondary resistances and leakage reactances. The voltage drop at the terminals also depends upon the load and its power factor. The change in terminal voltage from no-load to full-load at constant supply voltage with respect to no-load voltage is known as voltage regulation of the transformer.

An electrical power transformer is open-circuited, meaning that the load is not connected to the secondary terminals. In this situation, the secondary terminal voltage of the transformer will be its secondary induced emf E_2.

Whenever a full load is connected to the secondary terminals of the transformer, the rated current I_2 flows through the secondary circuit and voltage drop comes into the picture. In this situation, the primary winding will also draw an equivalent full load current from the source. The voltage drop in the secondary is $V_2 = I_2Z_2$ where Z_2 is the secondary impedance of the transformer. The voltage V_2 across load terminals which is less than no load secondary voltage E_2 and this is because of I_2Z_2 voltage drop in the transformer.

The equation for the voltage regulation of the transformer, represented in per unit, is

$$\text{Voltage Regulation (\%)} = \frac{E_2 - V_2}{E_2} \text{ (Per Unit)} \tag{5.92}$$

Represented in percentage, is

$$\text{Voltage Regulation (\%)} = \frac{E_2 - V_2}{E_2} \times 100\% \tag{5.93}$$

When all the quantities are referred to the primary side of the transformer;

$$\text{Voltage Regulation (\%)} = \frac{V_1 - E_1}{V_1} \times 100\% \tag{5.94}$$

5.31.1 Approximate expression for voltage regulation

- ***Voltage Regulation of Transformer for Lagging Power Factor (for inductive load)***

The lagging power factor of the load is $\cos\theta_2$, which means the angle between the secondary current and voltage is θ_2. The phasor diagram for the lagging power factor or inductive load is shown in fig. 5.31.

Here, from the above diagram,

$$OC = OA + AB + BC$$

$$\text{Here, } OA = V_2$$

Here, $AB = AE \cos\theta_2 = I_2R_2 \cos\theta_2$

and, $BC = DE \sin\theta_2 = I_2X_2 \sin\theta_2$

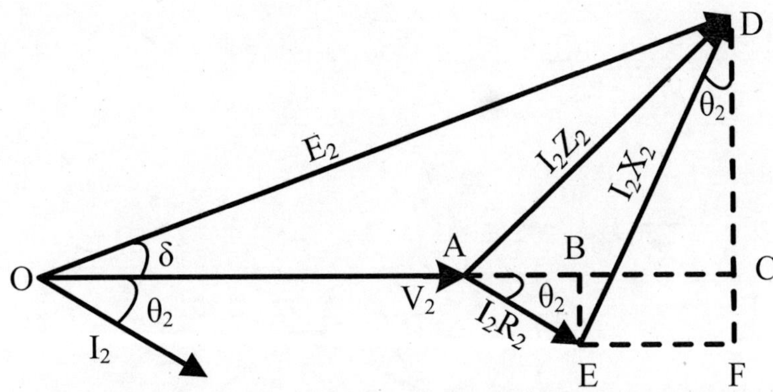

Fig. 5.31 Voltage regulation at the lagging power factor

The angle between OC and OD may be very small, so it can be neglected and OD is considered nearly equal to OC i.e.

$$E_2 = OC = OA + AB + BC$$

$$E_2 = OC = V_2 + I_2R_2 \cos\theta_2 + I_2X_2 \sin\theta_2$$

$$E_2 - V_2 = I_2R_2 \cos\theta_2 + I_2X_2 \sin\theta_2 \quad (5.95)$$

From equation (5.93)

$$\%\text{Reg.} = \frac{E_2 - V_2}{E_2} \times 100 = \frac{I_2R_2 \cos\theta_2 + I_2X_2 \sin\theta_2}{E_2} \times 100 \quad (5.96)$$

$$\%\text{Reg.} = \frac{I_2R_2 \cos\theta_2}{E_2} \times 100 + \frac{I_2X_2 \sin\theta_2}{E_2} \times 100 \quad (5.97)$$

Where,

$\frac{I_2R_2}{E_2} \times 100$, Percentage resistance drop

$\frac{I_2X_2}{E_2} \times 100$, Percentage reactance drop

$$\%\text{Reg.} = \%\ \text{drop} \cos\theta_2 + \%\ \text{drop} \sin\theta_2 \quad (5.98)$$

- ***Voltage Regulation of Transformer for Leading Power Factor (for capacitive load)***

The leading power factor of the load is $\cos\theta_2$, which means the angle between the secondary current and voltage is θ_2. The phasor diagram for the leading power factor or capacitive load is shown in fig. 5.31.

Here, from fig.5.32.

$$OC = OA + AB - BC$$

Here, $OA = V_2$

Here, $AB = AE\ \cos\theta_2 = I_2R_2\ \cos\theta_2$

$and, BC = DE\ \sin\theta_2 = I_2X_2\ \cos\theta_2$

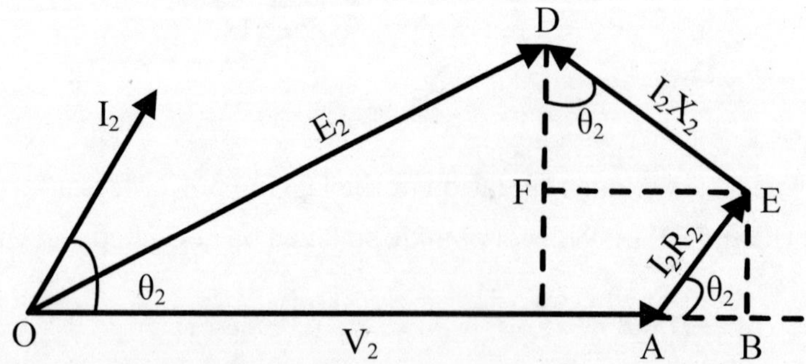

Fig. 5.32 Voltage regulation at leading power factor

Angle between OC and OD may be very small, so it can be neglected and OD is considered nearly equal to OC i.e.

$$E_2 = OC = OA + AB - BC$$

$$E_2 = OC = V_2 + I_2R_2\ \cos\theta_2 - I_2X_2\ \sin\theta_2$$

$$E_2 - V_2 = I_2R_2\ \cos\theta_2 - I_2X_2\ \sin\theta_2 \qquad (5.99)$$

From equation (5.93)

$$\%\text{Reg.} = \frac{E_2 - V_2}{E_2} \times 100 = \frac{I_2R_2\ \cos\theta_2 - I_2X_2\ \sin\theta_2}{E_2} \times 100 \qquad (5.100)$$

$$\%\text{Reg.} = \frac{I_2R_2\cos\theta_2}{E_2} \times 100 - \frac{I_2X_2\ \sin\theta_2}{E_2} \times 100 \qquad (5.101)$$

Similarly from inductive load

$$\%\text{Reg.} = \ \%\ \text{drop}\ \cos\theta_2 - \%\ \text{drop}\ \sin\theta_2 \qquad (5.102)$$

- ***Voltage Regulation of Transformer for Zero Power Factor (for resistive load)***

For resistive load: % Reg = % resistance drop (5.103)

5.31.2 Zero voltage regulation of a transformer

Zero voltage regulation' indicates that there is no difference between its 'no-load voltage' and its 'full-load voltage'. This means that the voltage regulation is equal to zero. This is not practical – and is only theoretically possible in the case of an ideal transformer.

From equation (5.102), we can derive the condition at which the regulation of a transformer becomes zero, it means when the load is no more, the terminal voltage remains the same.

$$0 = \quad \% \text{ resistance drop } \cos\theta_2 - \% \text{ reactance drop } \sin\theta_2$$

$$\% \text{ resistance drop } \cos\phi_2 = \% \text{ reactance drop } \sin\phi_2$$

$$\frac{I_2R_2\cos\theta_2}{E_2}\times 100 = \frac{I_2X_2\ \sin\theta_2}{E_2}\times 100$$

$$R_2\cos\theta_2 = X_2\ \sin\theta_2$$

$$\tan\theta_2 = \frac{R_2}{X_2}, \qquad \theta_2 = \tan^{-1}\frac{R_2}{X_2} \tag{5.104}$$

$$\text{Load power factor, } \cos\theta_2 = cos\tan^{-1}\frac{R_2}{X_2} \tag{5.105}$$

The above expression reveals that the regulation of a transformer will become zero only at the leading *pf* of the load that too when

$$\cos\theta_2 = cos\tan^{-1}\frac{R_2}{X_2} \tag{5.106}$$

5.31.3 Condition for maximum regulation

This condition is never suggested for any transformer, but let us see under what condition it may occur. Regulation will be maximum if

$$\frac{d}{d\theta_2}(\text{regulation}) = 0$$

$$\frac{d}{d\theta_2}\left(\frac{I_2R_2\ \cos\theta_2 - I_2X_2\ \sin\theta_2}{E_2}\right) = 0$$

$$-\frac{I_2R_2}{E_2}\sin\theta_2 + \frac{I_2X_2}{E_2}\cos\theta_2 = 0$$

$$\frac{I_2R_2}{E_2}\sin\theta_2 = \frac{I_2X_2}{E_2}\cos\theta_2$$

$$\tan\theta_2 = \frac{X_2}{R_2}, \qquad \theta_2 = \tan^{-1}\frac{X_2}{R_2} \tag{5.107}$$

This condition will occur at a lagging *Pf* that too when,

$$\cos\theta_2 = cos\tan^{-1}\frac{X_2}{R_2} \tag{5.108}$$

Example 5.20: Determine its regulation when the power factor is (i) 0·8 lagging (ii) 0·8 leading (iii) unity. If the ohmic loss of a transformer is 2% of the output and its reactance drop is 4% of the voltage.

Solution:

Ohmic loss or resistance drop = 1%; Reactance drop = 5%

(i) When p.f., $\cos\theta_2 = 0.8\ lagging$

$$\sin\theta_2 = \sin cos^{-1}0.8 = 0.6$$

$$\%\ Regulation = \%\ Resistance\ drop \times \cos\phi_2 + \%\ Reactance\ drop \times \sin\phi_2$$

$$= 2 \times 0.8 + 4 \times 0.6 = 4\%\ (Ans)$$

(ii) When p.f., $\cos\phi_2 = 0.8\ leading$; $\sin\phi_2 = \sin cos^{-1}0.8 = 0.6$

$$\%\ Regulation = \%\ Resistance\ drop \times \cos\theta_2 - \%\ Reactance\ drop \times \sin\theta_2$$

$$= 2 \times 0.8 - 4 \times 0.6 = -0.8\%\ (Ans)$$

(iii) When p.f. is unity % regulation = % Resistance drop = 2% (Ans)

5.25 Testing of Transformer

All the transformers are tested before placing them in the field. By performing these tests, we can determine the parameters of a transformer to compute its performance characteristics (like voltage regulation and efficiency etc.).

Large transformers cannot be tested by direct loading because of the following reasons:

- It is impossible to arrange a large load required for direct loading.
- While performing a direct load test, there is huge power wastage.
- It is very inconvenient to handle the power equipment.

Therefore, to furnish the required information open circuit and short circuit tests are conducted conveniently without actually loading the transformer. The other important tests which are conducted on a transformer are the polarity test, voltage ratio test, and Back-to-back test.

5.26 Open Circuit and Short Circuit Test

Open-circuit and short-circuit tests are performed to determine the circuit constants, efficiency, and regulation without actually loading the transformer. These tests give more accurate results than those obtained by taking measurements on fully loaded transformers. Also, the power consumption in these tests is very small compared with the full-load output of the transformer.

5.31.1 Open circuit test

An open circuit test is carried out at the rated voltage to determine the no-load loss or core loss or iron loss. It is also used to determine the no-load current I_0 which helps find the no-load parameters i.e., exciting resistance R_0 and exciting reactance X_0 of the transformer. This test is performed on the low-voltage side of the transformer, fig. 5.33(a) shows the connection diagram for the open circuit test. The high voltage (hv) side is left open. A voltmeter V, an armature A, and a low power factor (LPF) wattmeter 'W' are connected in the low-voltage (LV) side (primary) which is supplied at rated voltage and frequency as given on the nameplate of the transformer. Thus, the voltmeter V reads the rated voltage V_1 of the primary.

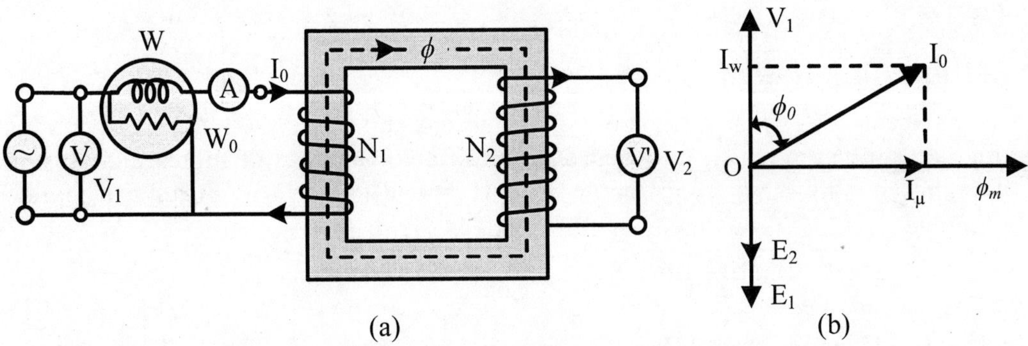

Fig. 5.33 Open circuit test (a) circuit diagram (b) phasor diagram

Since the secondary is open-circuited, a very small current I_0, called the no-load current, flows in the primary. The ammeter A, therefore, reads the no-load current I_0. Since the no-load current I_0 is very small (usually 2 to 5 percent of the full-load primary current). The power loss in the transformer is due to core loss and a very small I^2R (copper loss) loss in the primary. The I^2R loss in the primary winding can be neglected. There is no I^2R (copper loss) loss in the secondary since it is open and $I_2 = 0$. The core loss depends upon the flux. Since the rated voltage V_1 is applied, the flux set up by it will have a normal value so that normal core losses will occur. This core loss is the same at all loads. Therefore, the wattmeter which is connected to measure input power reads the core loss (iron loss) P_i only. The voltmeter V' if connected on the secondary side measures the secondary-induced voltage V_2. The ratio of voltmeter readings gives the transformation ratio of the transformer. The phasor diagram of the transformer at no-load is shown in fig. 5.33(b).

Voltmeter reading $= V_1$ (primary rated voltage)

Ammeter reading $= I_0$ (No load current)

Wattmeter reading $= W_0$

From these measured values the components of the no-load equivalent circuit can be determined.

$$\text{Iron loss of the transformer, } P_i = V_1 I_0 \cos\theta_0 \tag{5.109}$$

$$\text{No} - \text{load power factor}, \cos\theta_0 = \frac{P_i}{V_1 I_0} \tag{5.110}$$

$$\text{Working Component}, I_w = I_0 \cos\theta_0 \tag{5.111}$$

$$\text{Magnetising Component}, I_\mu = I_0 \sin\theta_0 \tag{5.112}$$

$$\text{Equivalent exciting resistance}, R_0 = \frac{V_1}{I_w} \tag{5.113}$$

$$\text{Equivalent exciting reactance}, X_0 = \frac{V_1}{I_\mu} \tag{5.114}$$

The Iron losses measured by this test are used to determine transformer efficiency and parameters of the exciting circuit of a transformer shown in fig. 5.12, section 5.13.

5.31.2 Short circuit test

In the short-circuit test, usually, the low-voltage side is short-circuited by a thick conductor or through an ammeter which may serve an additional purpose of indicating rated load current as shown in fig. 5.34. A wattmeter *W*, voltmeter *V*, and an ammeter *A* are connected in high-voltage winding (now the high voltage side is say primary).

The reasons for short-circuiting the *LV* side and taking measurements on the *HV* side are as follows:

- The rated current on the *HV* side is lower than that on the *LV* side. This current can be safely measured with the available laboratory ammeters.
- Since the applied voltage is less than 5 percent of the rated voltage of the winding, greater accuracy in the reading of the voltmeter is possible when the *HV* side is used as the primary.

This test is carried out to determine the following:

- Copper losses at full load (or at any desired load). These losses are required for the calculations of the efficiency of the transformer.
- Equivalent impedance (Z_{01} or Z_{02}), resistance (R_{01} or R_{02}) and leakage reactance (X_{01} or X_{02}) of the transformer referred to as the winding in which the measuring instruments are connected. Knowing equivalent resistance and reactance, the voltage drop in the transformer can be calculated and hence regulation of the transformer is determined.

The high-voltage winding is supplied at low voltage with the help of an autotransformer. The supply voltage is gradually increased until full-load primary current flows. When the rated full-load current flows in the primary winding rated full-load current will flow in the secondary winding by transformer action.

Readings of the ammeter, voltmeter and wattmeter are noted. The ammeter reading I_{Isc} gives the full-load primary current. The voltmeter reading V_{Isc} gives the value of the primary applied voltage when full-load current is flowing in the primary and secondary. Since the applied voltage is low (usually about 5 to 10 percent of the normal rated supply voltage), the flux ϕ produced is low (1/30 to 1/8 of normal flux). Also, since the core loss is nearly proportional to the square of the flux, the core loss is so small that it can be neglected. However, the windings are carrying normal full-load currents, and therefore the input is supplying the normal full-load copper losses. Thus, the wattmeter gives the full-load copper losses P_{cfl}.

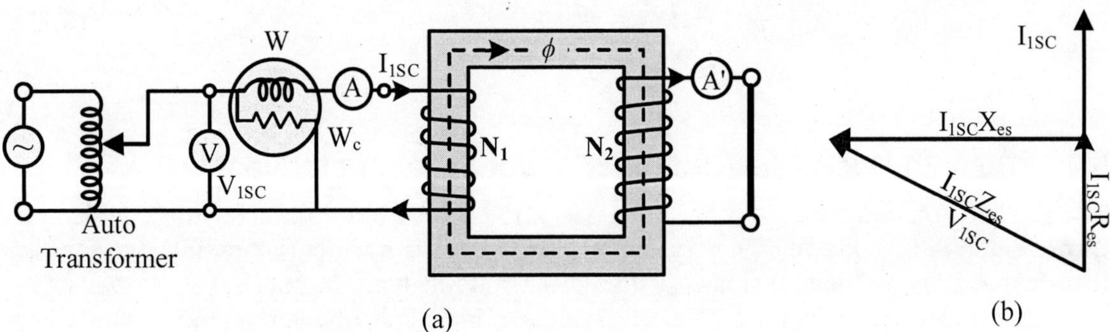

Fig. 5.34 (a)Short circuit test circuit diagram (b)Phasor diagram

The readings of the instruments in a short-circuit test are as follows:

Voltmeter reading = V_{1sc} (short circuit voltage)

Ammeter reading = I_{1sc} (full load primary current)

Wattmeter reading = W_c (full load copper loss of the transformer)

The output voltage V_2 is zero because of the short circuit. Consequently, the whole of the primary voltage is used in supplying the voltage drops in the total impedance Z_{01} referred to the primary.

$$V_{Isc} = I_{Isc}\, Z_{01} \tag{5.115}$$

If $\cos\phi_{sc}$ = power factor at short circuit, then full load copper loss of the transformer is

$$P_{cfl} = V_{ISC}\, I_{ISC}\, \cos\theta_{sc} \tag{5.116}$$

$$\cos\theta_{sc} = \frac{R_{01}}{Z_{01}} \tag{5.117}$$

Or $$P_{cfl} = I_{1sc}^2 R_{01} \tag{5.118}$$

The equivalent resistance of the transformer referred to primary

$$R_{01} = \frac{P_{cfl}}{I_{1sc}^2} \tag{5.119}$$

Equivalent impedance referred to primary

$$Z_{01} = \frac{V_{1sc}}{I_{1sc}} \tag{5.120}$$

Equivalent reactance referred to primary

$$X_{01} = \sqrt{Z_{01}^2 - R_{01}^2} \tag{5.121}$$

With short-circuit test performed only on one side the equivalent circuit constants referred to another side can also be calculated as follows:

$$Z_{02} = Z_{01}\left(\frac{N_2}{N_1}\right)^2 = K^2 Z_{01} \tag{5.122}$$

$$R_{02} = R_{01}\left(\frac{N_2}{N_1}\right)^2 = K^2 R_{01} \tag{5.123}$$

$$X_{02} = X_{01}\left(\frac{N_2}{N_1}\right)^2 = K^2 X_{01} \tag{5.124}$$

5.27 Back to Back Test (Sumpner's test or regenerative test)

To determine the maximum temperature rise, it is necessary to conduct a full-load test on a transformer. For small transformers, the full-load test is conveniently possible, but for large transformers, the full-load test is very difficult. A suitable load to absorb the full-load power of a large transformer may not be easily available. It will also be very expensive as a large amount of energy will be wasted in the load during the test. Large transformers can be tested for determining the maximum temperature rise by the back-to-back test. This test is also called the **Regenerative test or Sumpner's test.**

The back-to-back test on single-phase transformers requires two identical transformers. Fig. 5.35, shows the circuit diagram for the back-to-back test on two identical single-phase transformers T_{r_1} and T_{r_2}. The primary windings of the two transformers are connected in parallel and supplied at rated voltage and rated frequency. A voltmeter, an ammeter, and a wattmeter are connected to the input side as shown in fig. 5.35.

The secondaries are connected in series with their polarities in phase opposition, which can be checked by the voltmeter V_2. The range of this voltmeter should be double the rated voltage of a transformer secondary. To check that secondaries are connected in series opposition, any two terminals (say B and C) are joined together and the voltage is measured between the remaining terminals A and D.

If the voltmeter V_2 reads zero, the two secondaries are in series opposition, and terminals A and D are used for the test. If the voltmeter reads a value approximately equal to twice the rated secondary voltage of either transformer, then the secondaries are acting in the same direction. Then terminals A and C are joined and the terminals B and D are used for the test.

If the primary circuit is now closed, the total voltage across the two secondaries in series will be zero. There will be no current in the secondary windings. The transformers will behave as if their secondary windings are open-circuited. Hence, the reading of the Wattmeter W_1 gives the iron losses of both the transformers when rated low voltage V_1 is applied across parallel connected primaries.

A small voltage is injected into the secondary circuit by a regulating transformer T_r excited by the main supply. The magnitude of the injected voltage is adjusted till the armature A_2 reads full-load secondary current.

The secondary current produces full-load current to flow through the primary windings. This current will follow a circulatory path through the main busbars as shown by the dotted line in fig. 5.35. The reading of the Wattmeter W_2 will not be affected by this current. Thus, wattmeter W_2 gives the sum of full-load copper losses of the two transformers.

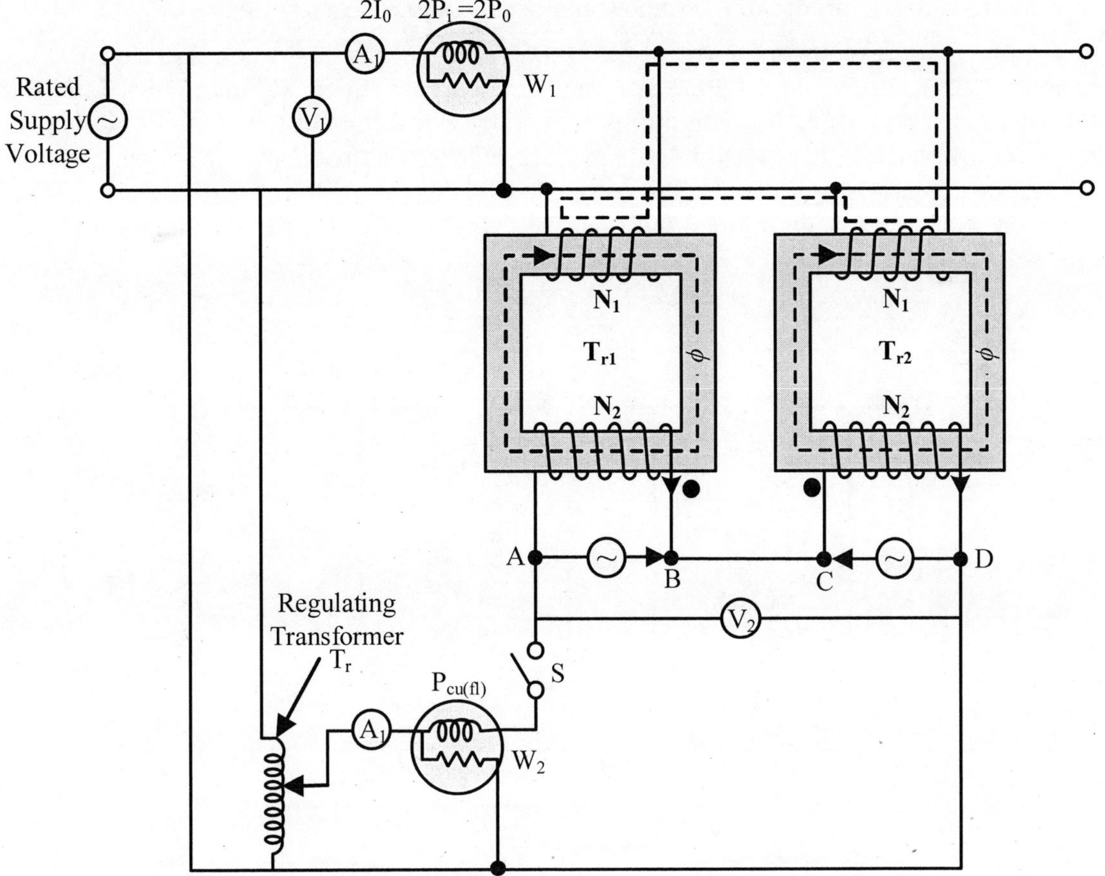

Fig. 5.35 Back-to-back test on two identical single-phase transformer

The armature A_1 gives the total no-load current of the two transformers. Thus, in this method, we have loaded the two transformers to full load but the power taken from the supply is that necessary to supply the losses of both transformers.

The temperature rise of the transformers can be determined by operating these transformers back-to-back for a long time, say 48 hours, and measuring the temperature of the oil at periodic intervals of time, say every hour.

5.28 Polarity Test

A polarity test is performed to determine the terminals with the same instantaneous polarity of the two windings when terminals are not being marked. The relative polarities of the primary and secondary terminals are required to be known for

- Interconnecting two or more transformers in parallel.
- Connecting three single-phase transformers while doing the poly-phase transformation of power.
- Connecting windings of the same transformer in parallel or series.

For determining the relative polarity of the two windings of a transformer, the two windings are connected in series and a voltmeter is connected across them as shown in fig. 5.36. One of the winding (preferably *HV* winding) is excited from a suitable AC voltage (less than the rated value). If the polarities of the windings are as marked on the diagram, then the windings will have a subtractive polarity and the voltmeter will read the difference of E_1 and E_2 $(E_1 - E_2)$. If the voltmeter reads $E_1 + E_2$the polarity marking of one of the windings must be reversed.

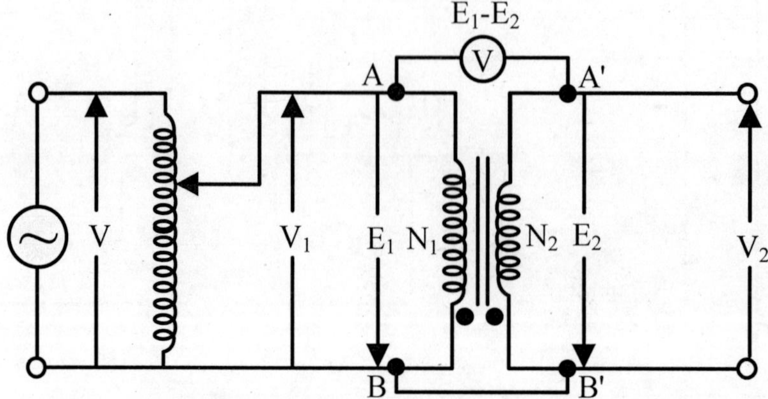

Fig. 5.36 Circuit test for polarity test

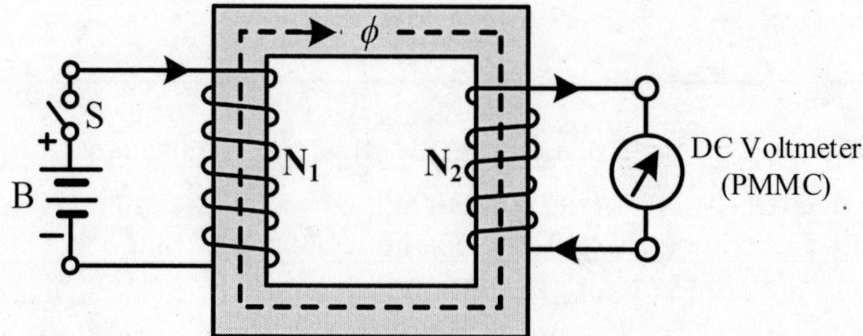

Fig. 5.37 Polarity test using PMMC

While performing the polarity test, the subtractive polarity method is preferred over the additive polarity method, because in this case, the voltage between *A* and *A′* or that between

B and A' is reduced. The leads connected between these terminals and two windings are not subjected to high voltage stresses. Whereas, in the case of additive polarity the two windings and leads connected between AA' and BB' are subjected to high voltage stresses. When the transformer is placed in the field, it may not be convenient to perform the above test to check the polarity. In such cases, polarity may be checked by using a battery, a switch, and a DC voltmeter (*PMMC* type) which are connected in the circuit as shown in fig. 5.37. When the switch (S) is closed, the primary current increases which increases the flux linkages with both the windings inducing emf in them. The positive polarity of this induced emf in the primary is at the end to which the battery is connected. The end of the secondary which simultaneously acquires positive polarity which is indicated by the deflection in the (*PMMC*) voltmeter. If deflection does not occur, then open the switch. At this instant, if deflection occurs the polarity of the secondary is opposite.

5.29 Voltage Ratio Test

The true voltage ratio is based on the turn ratio of the two windings of a transformer. In case the two voltages are measured at no-load, their ratio is almost equal to the true value. Similarly, if the primary and secondary currents are measured on short circuit, their ratio gives true-ratio particularly if the transformer has little leakage flux and low core reluctance.

$$\text{voltage ratio } \frac{V_2}{V_1} = \frac{I_1}{I_2}.$$

5.30 Parallel Operation of Single-Phase Transformer

When the primaries and secondaries of two or more transformers are connected separately to the same incoming and outgoing lines to share the load, the transformers are said to be connected in parallel.

The two single-phase transformers A and B are placed in parallel as shown in fig. 5.38. Here the primary windings of the two transformers are joined to the supply bus-bars and the secondary windings are joined to the load through load bus-bars. Under these conditions;

$$V_1 = \text{Primary applied voltage}$$

$$V_2 = V = \text{Secondary load voltage}$$

5.30.1 Need for parallel operation in single-phase transformers

There are three principal reasons for connecting transformers in parallel.

1. Firstly, if one transformer fails, the continuity of supply can be maintained through other transformers.
2. Secondly, when the load on the substation becomes more than the capacity of the existing transformers, another transformer can be added in parallel.
3. Thirdly, any transformer can be taken out of the circuit for repair/routine maintenance without interrupting the supply to the consumers.

Some more reasons that necessitate parallel operation of transformers are as follows:

- Non-availability of a single large transformer to meet the total load requirement.

- The power demand might have increased over time necessitating augmentation of the capacity.
- To ensure improved reliability. Even if one of the transformers gets into a fault or is taken out for maintenance/repair the load can be continued to be serviced.
- To reduce spare capacity. If many smaller-size transformers are used one machine can be used as a spare. If only one large machine is feeding the load, a spare of a similar rating has to be available. The problem of spares becomes more acute with fewer transformers in service at a location.
- When transportation problems limit the installation of large transformers at a site, it may be easier to transport smaller ones to the site and work them in parallel.

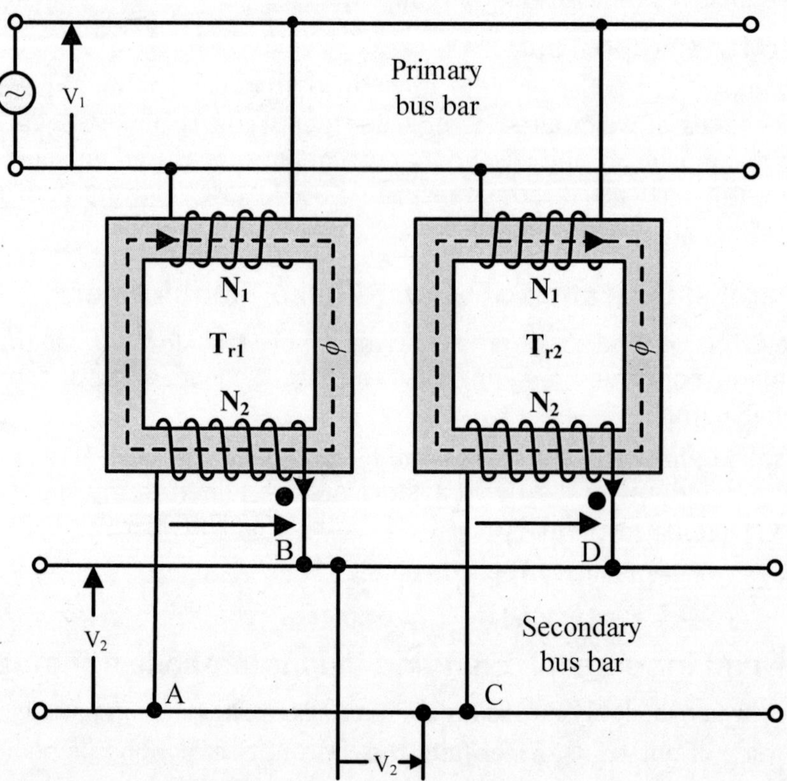

Fig. 5.38 Parallel operation of two single-phase transformer

5.30.2 Conditions for parallel operation of single-phase transformer

In order that the transformers to work satisfactorily in parallel, the following conditions should be satisfied:

1. Transformers should be properly connected with regard to their polarities.
2. The voltage ratings and voltage ratios of the transformers should be the same.

3. The per unit or percentage impedances of the transformers should be equal.
4. The reactance/resistance ratios of the transformers should be the same.

1. Connection with regard to Polarity

The first condition for the successful parallel operation of single-phase transformers is that the transformers should be properly connected with regard to their polarities. This condition is absolutely essential because wrong connections may result in a dead short circuit. Fig.5.39 (a) shows the correct method of connecting two single-phase transformers in parallel. It will be seen that around the loop formed by the secondaries, the two secondaries e.m.f. E_A *and* E_B oppose and there will be no circulating current.

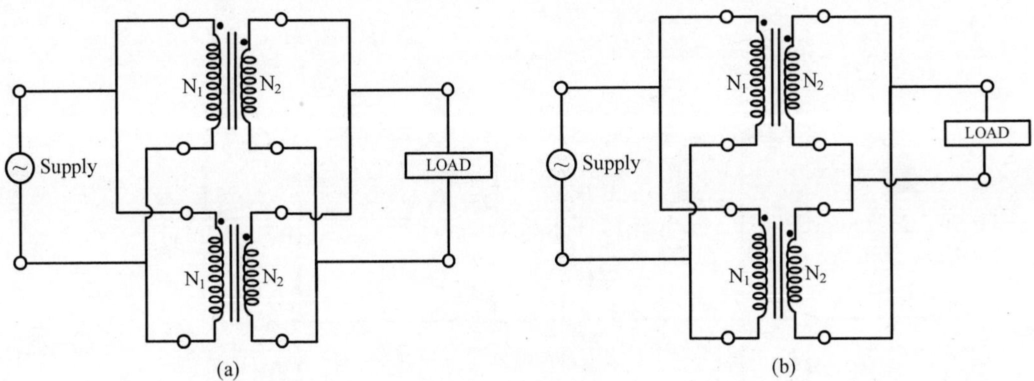

Fig. 5.39 Transformer Connection Based on Polarity (a) Correct Connections (b) Wrong Connections

Fig.5.39(b) shows the wrong method of connecting two single-phase transformers in parallel. Here the two secondaries are so connected that their emf E_A *and* E_B are additive. This may lead to short-circuiting conditions and a very large circulating current will flow in the loop formed by the two secondaries. Such a condition may damage the transformers unless they are protected by fuses and circuit breakers.

2. Same Voltage Rating and Voltage Ratio

The next condition for parallel operation of a single-phase transformer is the same voltage rating and voltage ratio. This condition is desirable for the satisfactory parallel operation of transformers. If this condition is not met, the secondary emf will not be equal and there will be a circulating current in the loop formed by the secondaries even at no load. This will result in the unsatisfactory parallel operation of transformers. Consider two single-phase transformers A and B operating in parallel as shown in fig.5.40.

Let E_A and E_B be their no-load secondary voltages and Z_A and Z_B be their impedances referred to the secondary. Then at no-load, the circulating current in the loop formed by the secondaries is Circulating current,

$$I_C = \frac{E_A - E_B}{Z_A + Z_B} \quad \text{Assuming } E_A > E_B \qquad (5.125)$$

Even a small difference in the induced secondary voltages can cause a large circulating current in the secondary loop because the impedances of the transformers are small. This secondary circulating current will cause current to be drawn from the supply by the primary of each transformer. These currents will cause copper losses in both primary and secondary. This creates heating with no useful output. When the load is connected to the system, this circulating current will tend to produce unequal loading conditions i.e., the transformers will not share the load according to their kVA ratings. It is because the circulating current will tend to make the terminal voltages of the same value for both transformers. Therefore, the transformer with a smaller voltage ratio will tend to carry more than its proper share of the load. Thus, one transformer would tend to become overloaded than the other and the system could not be loaded to the summation of transformer ratings without overloading one transformer.

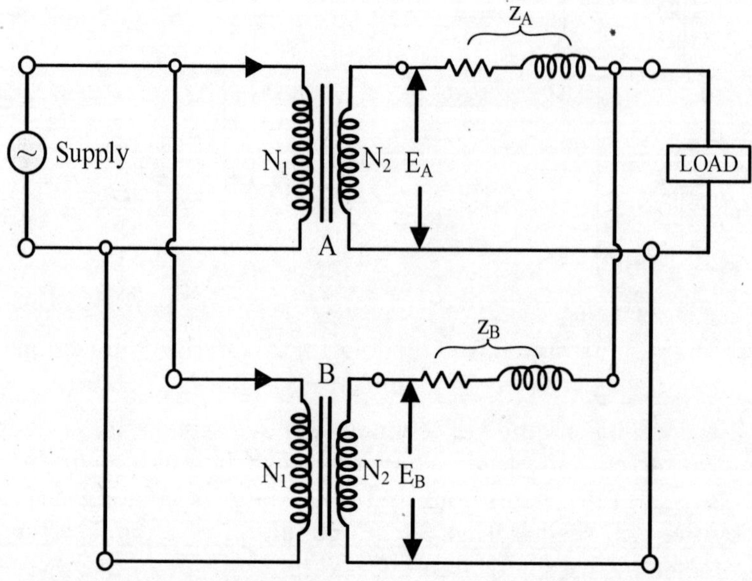

Fig. 5.40 Parallel operation of two single-phase transformer

3. Equal Percentage Impedance

Another condition for parallel operation of a single-phase transformer is all transformers should have equal percentage impedance. This condition is also desirable for the proper parallel operation of single-phase transformers. If this condition is not met, the transformers will not share the load according to their kVA ratings. Sometimes this condition is not fulfilled by the design of the transformers. In that case, it can be corrected by inserting the proper amount of resistance or reactance or both in series with either primary or secondary circuits of the transformers where the impedance is below the value required to fulfill this condition.

4. Same Reactance/Resistance Ratio

If the reactance/resistance ratios of the two transformers are not equal, the power factor of the load supplied by the transformers will not be equal. In other words, one transformer will be operating with a higher and the other with a lower power factor than that of the load.

5.30.3 Load sharing between two transformers connected in parallel

The load sharing between two transformers connected in parallel depends upon the various conditions as discussed below: -

Case-01: When the two transformers have the same voltage ratios and their impedance voltage triangles are identical in size and shape:

The condition for which the transformers have the same voltage ratio and impedance voltage triangles is known ideal condition or ideal case. Let E be the no-load secondary voltage of each transformer and V the terminal voltage. Fig. 5.41 shows the equivalent circuit in which I_A and I_B are the currents supplied by transformers A and B, respectively. I_L be the total load current lagging behind the voltage V by an angle ϕ. The impedance voltage triangles of the individual transformers are identical in shape and size and are, therefore, represented by a single triangle V_{AB} with the resistance drop (side VA) parallel to the load current vector OI as shown in fig. 5.42. The current I_A and I_B in the individual transformers are both in phase with the load current I_L and are inversely proportional to their respective impedance.

$$I_L = I_A + I_B \tag{5.126}$$

$$I_A Z_A = I_B Z_B \tag{5.127}$$

Or

$$\frac{I_A}{I_B} = \frac{Z_B}{Z_A} \tag{5.128}$$

$$I_A = \frac{Z_B}{Z_A} . I_B \tag{5.129}$$

Substituting the value of I_A in eq. (5.126), we get,

$$I_L = I_B \frac{Z_B}{Z_A} + I_B = I_B \left[\frac{Z_B + Z_A}{Z_A}\right]$$

$$I_B = \frac{Z_A}{Z_A + Z_B} \times I_L \tag{5.130}$$

Similarly,

$$I_A = \frac{Z_B}{Z_A + Z_B} \times I_L \tag{5.131}$$

Multiplying both sides by the common terminal voltage V, we get,

$$I_A V = \frac{Z_B}{Z_A + Z_B} V I_L \tag{5.132}$$

Similarly, $$I_B V = \frac{Z_A}{Z_A + Z_B} V I_L \tag{5.133}$$

$$kVA_A = \frac{Z_B}{Z_A + Z_B} \times kVA \tag{5.134}$$

and $$kVA_B = \frac{Z_A}{Z_A + Z_B} \times kVA \tag{5.135}$$

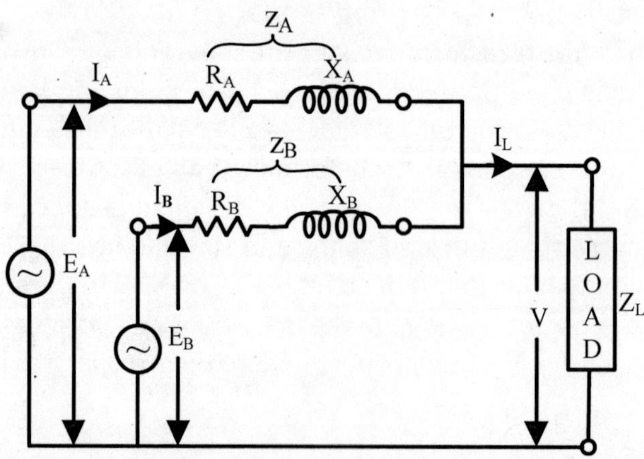

Fig. 5.41 Circuit diagram of two transformer connected in parallel

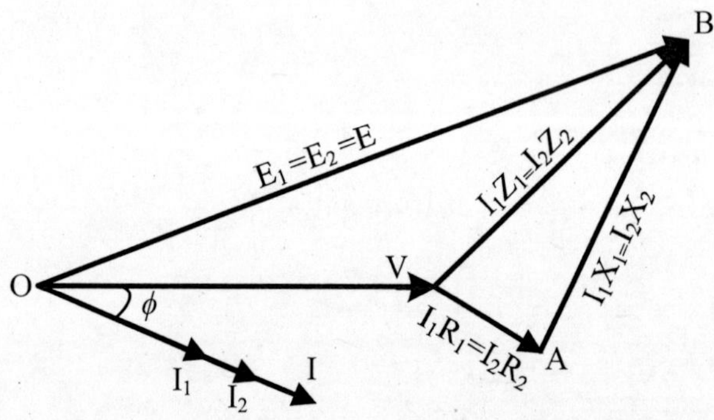

Fig. 5.42 Phasor diagram (impedance voltage triangle are identical)

Case-02: **When the two transformers have the same voltage ratios but different voltage triangles:**

In this case, no-load voltages of both secondary are equal in magnitude as well as in phase i.e., there is no phase difference between E_A and E_B which will only be possible if the magnetizing currents of the two transformers are not very different from each other or nearly the same. Under these conditions, both sides of two transformers can be connected in parallel, and no current will circulate between them on no-load. Fig. 5.41 shows the equivalent circuit diagram when the parallel connected transformers are sharing the load current I_L, and it represents two impedances in parallel. The impedance voltage triangle is now represented by two triangles $VAB, VA'B'$ having common hypotenuse VB as shown in fig. 5.43. The resistance drops sides of the triangles VA, VA' are parallel to the phasors OI_A *and* OI_B of the respective secondary currents. The sum of these vectors OI_A *and* OI_B represents the load current OI.

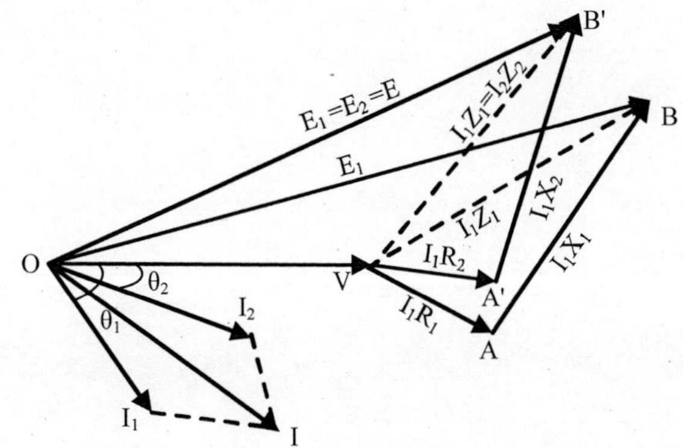

Fig. 5.43 Phasor diagram (voltage triangle are diffrent)

Z_A and Z_B = impedance of the two transformers

I_A and I_B = the currents of the two transformers

I_L = total load current

V = common terminal voltage

The total current I_L is given by

$$I_L = I_A + I_B \tag{5.136}$$

From the equivalent circuit diagram, it is seen that voltage ratio is same

$$E_A = E_B = E \tag{5.137}$$

the common terminal voltage

$$I_A Z_A = I_B Z_B = V \tag{5.138}$$

Since Z_A and Z_B are in parallel, we have the equivalent impedance Z

$$Z = \frac{Z_A Z_B}{Z_A + Z_B} \tag{5.139}$$

The total combined current

$$I_L = \frac{V}{Z} = \frac{V(Z_A + Z_B)}{Z_A Z_B} \tag{5.140}$$

Or, $$V = \frac{Z_A Z_B}{Z_A + Z_B} \times I_L \tag{5.141}$$

$$I_A Z_A = \frac{Z_A Z_B}{Z_A + Z_B} \times I_L \quad (5.142)$$

Or, $$I_A = \frac{Z_A}{Z_A + Z_B} \times I_L \quad (5.143)$$

Similarly, $$I_B = \frac{Z_B}{Z_A + Z_B} \times I_L \quad (5.144)$$

Multiplying both sides by the common terminal voltage V, we get,

$$I_A V = \frac{Z_A}{Z_A + Z_B} \times V I_L \quad (5.145)$$

$$I_B V = \frac{Z_B}{Z_A + Z_B} \times V I_L \quad (5.146)$$

Let $V \times I_L \times 10^{-3} = kVA$, the combined load in kVA

$V \times I_A \times 10^{-3} = kVA_1$, load shared by transformer "A" in kVA

$V \times I_B \times 10^{-3} = kVA_2$, load shared by transformer "B" in kVA

$$kVA_A = \frac{Z_B}{Z_A + Z_B} \times kVA \quad (5.147)$$

$$kVA_B = \frac{Z_A}{Z_A + Z_B} \times kVA \quad (5.148)$$

$$\frac{kVA_A}{kVA_B} = \frac{Z_B}{Z_A} \quad (5.149)$$

The load shared by each transformer is inversely proportional to their impedances.

Again, from equations (5.143) and (5.144), we get,

$$I_A = \frac{1}{\frac{Z_A}{Z_B} + 1} \times I_L \quad (5.150)$$

Or, $$I_B = \frac{1}{\frac{Z_B}{Z_A} + 1} \times I_L \quad (5.151)$$

Equation (5.150) shows that the load shared by each transformer depends upon the ratio of impedances, so the unit in which they are measured does not matter.

***Case-03:* When the two transformers have different voltage ratios and different voltage triangles: -**

In this case, the voltage ratios or transformation ratios of the two transformers are different. It means these no-load secondary voltages are unequal. Let E_A, E_B be the no-load secondary emf of the two transformers and Z be the load impedance across the secondary. The vector diagram is shown in Fig. 5.44. It is seen that even when secondaries are on no-load, there will be some circulating current in the secondaries because of inequality in their induced emf's. This circulating current I_C is given by

$$I_C = (E_A - E_B)/(Z_A + Z_B) \tag{5.152}$$

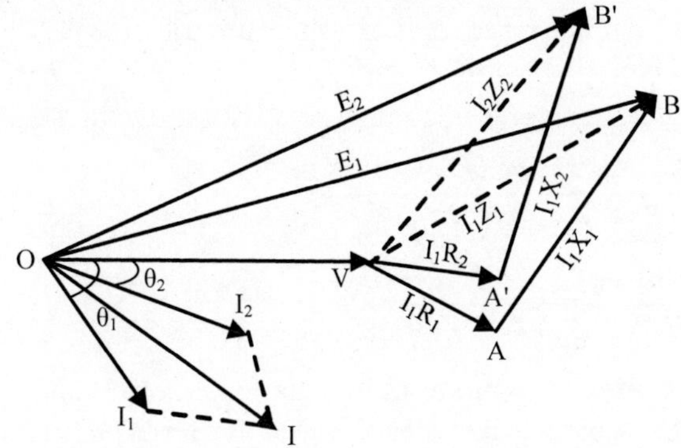

Fig. 5.44 Phasor diagram (having different voltage ratios and different voltage triangles)

As the induced emf's of the two transformers are equal to the total drops in their respective circuits.

$$E_A = I_A Z_A + V_B \tag{5.153}$$

$$E_B = I_B Z_B + V_B \tag{5.154}$$

$$V_B = I.Z_L = (I_A + I_B)Z_L \tag{5.155}$$

Where Z_L = load impedance

$$E_A = I_A Z_A + (I_A + I_B)Z_L \tag{5.156}$$

$$E_B = I_B Z_B + (I_A + I_B)Z_L \tag{5.157}$$

$$(E_A - E_B) = I_A Z_A - I_B Z_B \tag{5.158}$$

$$I_A = \frac{(E_A - E_B) + I_B Z_B}{Z_A} \tag{5.159}$$

Substituting this value of I_A in equation (iii), we get,

$$E_B = I_B Z_B + \left[\frac{(E_A - E_B) + I_B Z_B}{Z_A} + I_B\right] Z_L \tag{5.160}$$

$$I_B = \frac{E_B Z_A - (E_A - E_B)Z_L}{Z_A Z_B + Z_L\,(Z_A + Z_B)} \tag{5.161}$$

From the symmetry of the expression, we get

$$I_A = \frac{E_A Z_B + (E_A - E_B)Z_L}{Z_A Z_B + Z_L\,(Z_A + Z_B)} \tag{5.162}$$

The two equations (5.161) and (5.162) then give the values of secondary currents shared by the two transformers. By the division of transformation ratio i.e., K, and by addition (if not negligible) of the no-load current the primary current may be obtained. Usually, E_A and E_B have the same phase (as assumed above) but there may be some phase difference between the two due to some difference of internal connection viz. for the connections in parallel of a star/star and star/delta 3-phase transformers.

If Z_A and Z_B are small as compared to Z_L i.e., when the transformer is not operated near short-circuit conditions, then equations for I_A *and* I_B can be put in a simpler and more easily understandable form. Neglecting $Z_A Z_B$ in comparison with the expression $Z_L\ (Z_A + Z_B)$ we have

$$I_A = \frac{E_A Z_B}{Z_L\ (Z_A + Z_B)} + \frac{(E_A - E_B)}{Z_A + Z_B} \tag{5.163}$$

$$I_B = \frac{E_B Z_A}{Z_L\ (Z_A + Z_B)} - \frac{(E_A - E_B)}{Z_A + Z_B} \tag{5.164}$$

The physical interpretation of the second term in equations (5.163) and (5.164) is that it represents the cross-current (circulating current) between the secondaries. The first term shows how the actual load current divides between the loads. If $E_A = E_B$, then ratios of the currents are inverse of the impedance.

Example 5.21: A load of 500 A, at 0·8 power (lagging), at a terminal voltage of 400 V is supplied by two transformers that are connected in parallel. The equivalent impedances of the two transformers referred to as the secondary sides are (2 + j3) ohm and (2.5 + j5) ohm, respectively. Calculate the current and kVA supplied by each transformer and the power factor at which they operate.

Solution:

Here, $Z_A = 2 + j3 = \sqrt{(2)^2 + (3)^2}\ \angle tan^{-1}\frac{3}{2} = 3.606\ \angle 56.31°$

$Z_B = 2.5 + j5 = \sqrt{(2.5)^2 + (5)^2}\ \angle tan^{-1}\frac{5}{2.5} = 5.59\ \angle 63.44°$

$$Z_A + Z_B = 2 + j3 + 2.5 + j5 = 4.5\ j8 = \sqrt{(4.5)^2 + (8)^2}\ \angle tan^{-1}\frac{8}{4.5}$$

$$= 9.178\angle 60.64°$$

$I = I\ \angle cos^{-1}\ 0.8 = 500\ \angle - 36.87°$

$$I_1 = \frac{Z_B}{Z_A + Z_B} \times I = \frac{5.59\ \angle 63.44°}{9.178\ \angle 60.64°} \times 500\ \angle - 36.87°$$

$$= 30.45\ \angle - 34.07°$$

$I_1 = 304.5\ A\ (Ans)$

Power factor, $cos\ \theta_1 = \cos(-34.07°) = 0.8184\ lag\ (Ans)$

$$I_2 = \frac{Z_B}{Z_A + Z_B} \times I = \frac{3.606\ \angle 56.31°}{9.178\angle 60.64°} \times 500\ \angle - 36.87°$$

$$= 196.45\ \angle - 41.2°$$

$I_1 = 196.45\ A\ (Ans)$

Power factor, $cos\ \theta_2 = \cos(-41.2°) = 0.75524\ lag\ (Ans)$

Example 5.22: A transformer 'A' having an open circuit emf of 6600 V with impedance (0.3+j3) ohm referred to as secondary is connected in parallel with transformer 'B' having an open circuit emf of 6400 V with impedance (0.2+j1) ohm referred to the secondary side. Calculate the current delivered by each transformer to a load impedance of (8+j6) ohm.

Solution: Here, $E_A = 6600\angle 0° = 6600\ \pm j0$; $E_B = 6400\angle 0° = 6400\ \pm j0$

$$Z_A = 0.3 + j3 = 3.015\angle 84.29°;\ Z_B = 0.2 + j1 = 1.02\angle 78.69°$$

$$Z_A + Z_L = 0.3 + j3 + 0.2 + j1 = 0.5 + j4 = 4.031\angle 82.87°$$

$$Z_L = 8 + j6 = 10\angle 36.87°\ ;$$

$$I_1 = \frac{E_A\, Z_B + (E_A - E_B)Z_L}{Z_A\, Z_B + \overline{Z_L}\,(Z_A + Z_B)}$$

$$= \frac{6600\angle 0° \times 1.02\angle 78.69°\ (6600 - 6400)\angle 0° \times 10\angle 36.87°}{3.015\angle 84.29° \times 1.02\angle 78.69° + 10\angle 36.87° \times 4.031\angle 82.87°}$$

$$= \frac{6732\angle 78.69° + 2000\angle 36.87°}{3.0753\angle 162.98° + 40.31\angle 119.74°}$$

$$= \frac{(1320\ + j6601) + (1600 + j1200)}{(2.94 + j0.9) + (-20 + j35)}$$

$$= \frac{2920 + j7801}{-22.94 + j35.9}$$

$$= \frac{8329.6\angle 69.48°}{42.6\angle -57.42°} = 195.53\angle 126.9°$$

$$I_A = 195.53\ A\ (in\ magnitude)\ (Ans)$$

Similarly,

$$I_B = \frac{E_B Z_A - (E_A - E_B)Z_L}{Z_A Z_B + \overline{Z_L}\,(Z_A + Z_B)}$$

$$= \frac{6400\angle 0° \times 3.015\angle 84.29°\ (6600 - 6400)\angle 0° \times 10\angle 36.87°}{42.6\angle -57.42°}$$

$$= \frac{19296\angle 84.29° + 2000\angle 36.87°}{42.6\angle -57.42°}$$

$$= \frac{1920 + j19200 - 1600 - j1200}{42.6\angle -57.42°}$$

$$= \frac{320 + j18000}{42.6\angle -57.42°}$$

$$= \frac{18003\angle 88.98°}{42.6\angle -57.42°} = 422.6\angle 146.4°$$

$$I_B = 422.6\ A\ (in\ magnitude)\ (Ans)$$

Example 5.23: Two single-phase transformers *A* and *B* of equal voltage ratio are running in parallel and supply a load of 800 A at 0.8 power factor lagging having equivalent impedances (1.5 + *j*3) Ω and (2 + *j* 4.5) Ω respectively. Find the current supplied by each transformer and the ratio of the kW output of the two transformers.

Solution:

$$Z_A = (1.5 + j3)\Omega\,;\ Z_B = (2 + j\,4.5)\,\Omega$$

$$\frac{I_A}{I_B} = \frac{Z_B}{Z_A}$$

$$= \frac{2 + j\,4.5}{1.5 + j3}$$

$$= 1.47 + j0.067$$

$$I_A = (1.47 + j0.067)I_B$$

Let us take the secondary terminal voltage as a reference.

$$I = 800\,(0.8 - j0.6)$$

$$= (640 - j480)A$$

Again, $I = I_A + I_B$

$$= (1.47 + 0.067)I_B + I_B$$

$$= (2.47 + j0.067)I_B$$

$$I_B = \frac{I}{2.47 + j0.067}$$

$$= \frac{640 - j480}{2.47 + j0.067}$$

$$= 253.65 - j201.2 = 323.76\angle - 38.42°$$

$$I_A = (1.47 + j0.067)I_B$$

$$= (1.47 + j0.067) \times (253.65 - j201.2)$$

$$= 386.34 - j287.77$$

$$= 481.74\angle - 36.68°\ A$$

The ratio of kW outputs is equal to the ratio of in-phase components of the two currents.

$$\frac{Output\ of\ A}{Output\ of\ B} = \frac{386.34}{253.65} = 1.523$$

Example 5.24: Two transformers *A* and *B* are connected in parallel and supply a common load. Open circuit emf of *A* and *B* are 5,500 V and 5,400 V, respectively. Equivalent impedance in terms of secondary of *A* and *B* are (0.4 + *j*4) Ω and (0.1 + *j* 1.5) Ω respectively. The load impedance is (10 + *j* 6) Ω. Find the current supplied by each transformer.

Solution:

$I_A = 5500\,V;\ E_B = 5400\,V;\ Z_L = (10 + j6)\Omega;\ Z_A = (0.4 + j4)\Omega;\ Z_B = (0.1 + j1.5)\Omega$

$$I_A = \frac{E_A Z_B + (E_A - E_B)Z_L}{Z_A Z_B + Z_L\,(Z_A + Z_B)}$$

$$= \frac{5500(0.1 + j1.5) + (5500 - 5400) \times (10 + j6)}{(0.4 + j4) \times (0.1 + j1.5) + (10 + j6) \times (0.4 + j4 + 0.1 + j1.5)}$$

$$= \frac{1550 + j8850}{-33.96 + j59} = (101.31 - j84.58)A$$

Similarly,

$$I_B = \frac{E_B Z_A - (E_A - E_B)Z_L}{Z_A Z_B + Z_L\,(Z_A + Z_B)}$$

$$= \frac{5400(0.4 + j4) + (5500 - 5400) \times (10 + j6)}{(0.4 + j4) \times (0.1 + j1.5) + (10 + j6) \times (0.4 + j4 + 0.1 + j1.5)}$$

$$= \frac{1160 + j21000}{-33.96 + j59} = (258.85 - j168.66)A$$

5.31 Auto-Transformer

In two-winding transformers, the two windings are electrically isolated and emf is induced in the secondary winding due to mutual induction. The transformers in which a part of the winding is common to both the primary and secondary circuits are termed as *autotransformers.* In an autotransformer, the two windings are electrically connected and it works on the principle of induction and conduction.

5.31.1 Construction

The core of an auto-transformer is rectangular, a single winding is wound around one or two limbs of the rectangular core as shown in fig. 5.45. Terminal "Y" is taken as a common point from which one terminal for the primary and one terminal for the secondary is taken out. The second terminal of the secondary is connected to point "Z" which may be fixed or movable as shown in fig. 5.45.

The number of turns between XY is taken as N_1 and the number of turns between YZ is taken as N_2 as shown in fig. 5.46(a). Thus, one section of the same windings acts as primary and the other section of the same windings acts as secondary. When the number of secondary turns N_2 is less than the primary turns N_1 (i.e., $N_2 < N_1$) as shown in fig. 5.46(a), the auto-transformer works as a step-down transformer, whereas, it works as a set-up transformer if the number of secondary turns N_2 is more than primary turns N_1 as shown in fig. 5.46(b). Fig. 5.46(c) shows the auto-transformer having combined step-up and step-down operations. Here, tapping 1-4 operates on the step-down operation and 5-6 operates on the step-up operation.

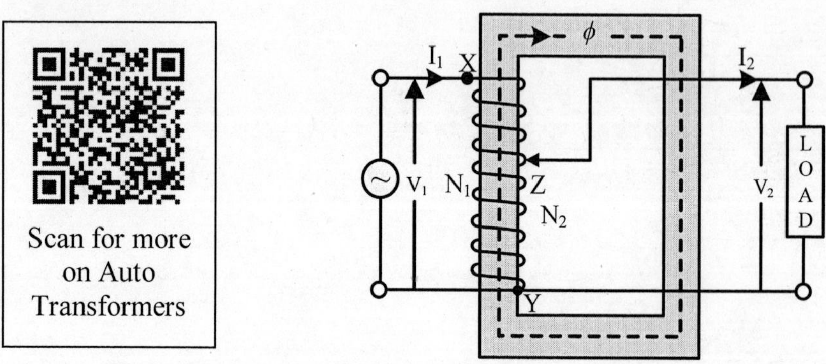

Fig. 5.45 Single winding Rectangular core auto-transformer

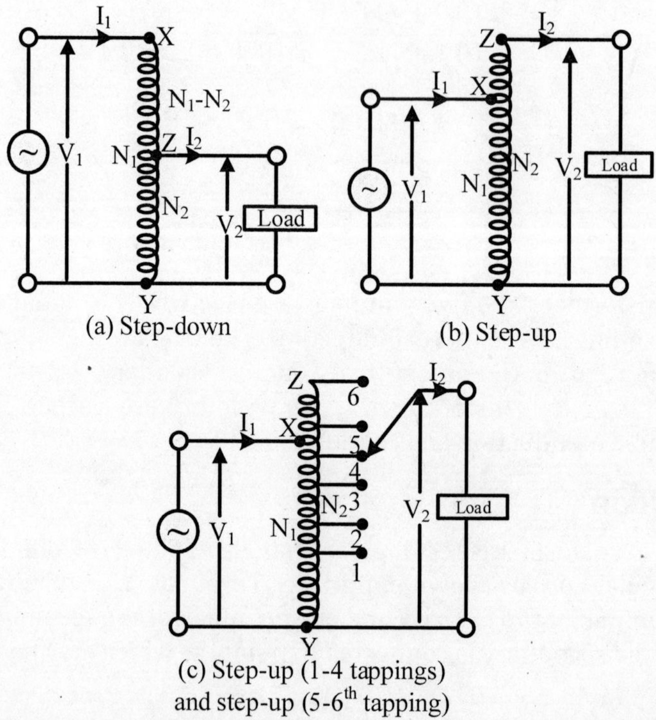

Fig. 5.46 Circuit of single phase rectangular core auto-transformer

The core of an auto-transformer may also be ring-type in shape, a single winding is wound over the rings as shown in fig. 5.47. Fig. 5.48 shows the ring type step-up and step-down autotransformer.

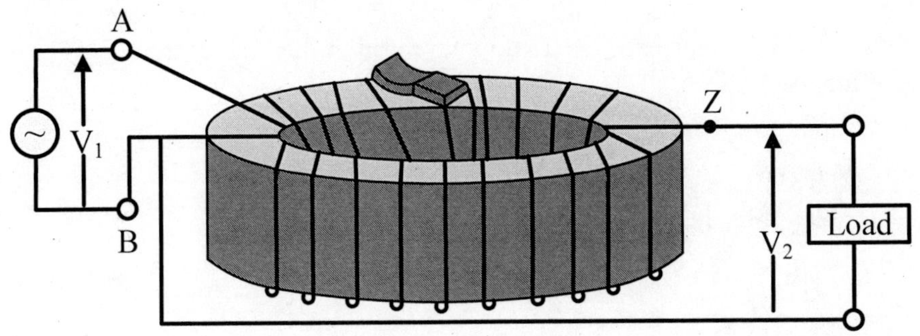

Fig. 5.47 Ring-type autotransformer (core and winding)

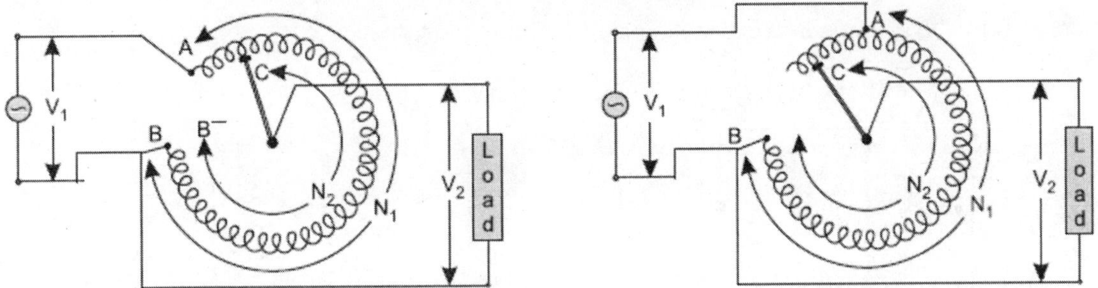

Fig. 5.48 Circuit diagram of ring type autotransformer (a) Step down (b) Step up

The pictorial view of a single-phase auto-transformer used in labs is shown in fig. 5.49. Here, point C is attached to a moveable arm that carries a carbon brush. The brush moves over the number of turns wound over a toroidal laminated core and its position determines the output voltage. Autotransformer and CT cores are made using ribbon or strips rolled to form the toroidal core.

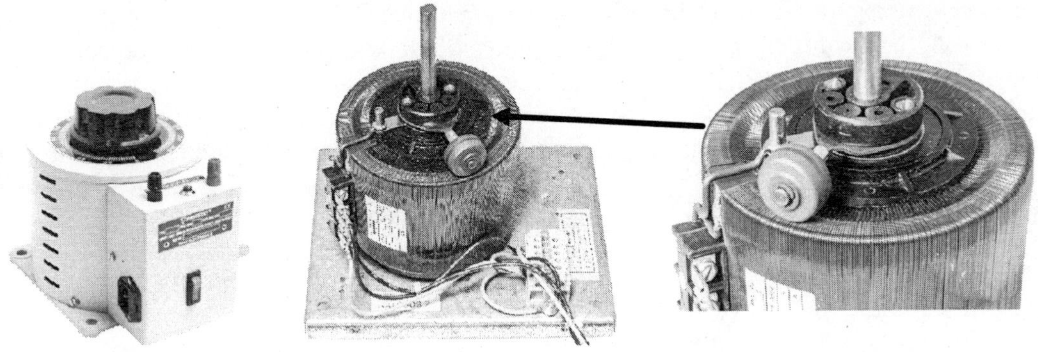

Fig. 5.49 Single-phase circular core autotransformer

5.31.2 Working

As shown in fig.5.50 when AC voltage V_1 is applied to winding XY, an exciting current start flowing through the full winding XY if the internal impedance drops is neglected, then the voltage per turn in windings XY is V_1/N_1 and therefore, the voltage across YZ is $(V_1/N_1)\,N_2$. When switch S is closed, a current I_2 start flowing through the load current I_1 is drawn from the source. Neglecting losses,

Input power = output power

$$V_1 I_1 \cos\theta_1 = V_2 I_2 \cos\theta_2 \tag{5.165}$$

If internal (or leakage) impedance drops and losses are neglected, then

$$\cos\theta_1 = \cos\theta_2 \tag{5.166}$$

$$V_1 I_1 = V_2 I_2 \tag{5.167}$$

$$\frac{V_2}{V_1} = \frac{I_1}{I_2} = \frac{N_2}{N_1} = K \tag{5.168}$$

Here K is less than unity. The expression is identical to a two-winding transformer.

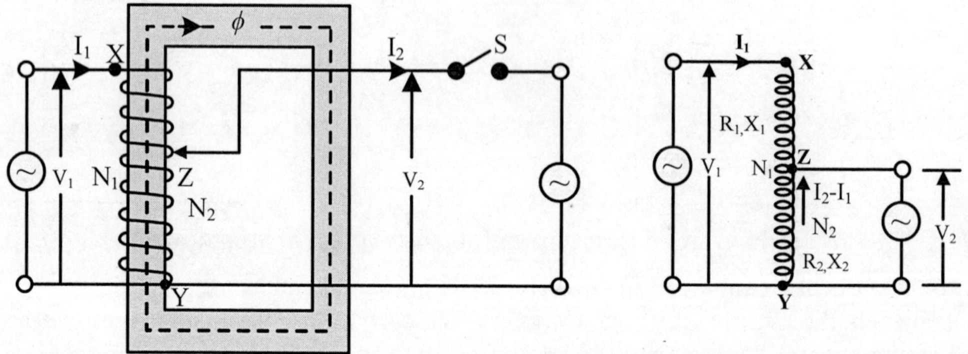

Fig. 5.50 Single phase autotransformer (a) core and winding (b) Equivalent circuit

Fig. 5.50 shows a step-down autotransformer. Let the point X be positive with respect to Z. At no load the exciting current flows from X to Z and it produces a working mmf vectorially downwards, i.e., from X to Z. During the presence of load at the secondary, the current flows from Z to Y and it weakens the produced working magnetomotive force (mmf), and the transformer draws extra current from the primary. Ultimately, it will maintain the same working mmf. In winding XZ, the current is I_1 (from X to Y) whereas in winding YZ the current is I_2 (from Z to Y). $V_2 < V_1$, $I_2 > I_1$ and the net current through YZ is $I_2 - I_1$ (from Z to Y). Therefore,

$$I_{YZ} = I_2 - I_1 \tag{5.169}$$

Mmf of winding XZ

$$= I_1\ (N_1 - N_2) = I_1 N_1 - I_1 N_2$$

$$= I_2 N_2 - I_1 N_2 \ (\text{Since; } I_1 N_1 = I_2 N_2)$$

$$= (I_2 - I_1)N_2 = \text{mmf of winding ZY } (= I_{ZY}N_2) \quad (5.170)$$

It is, therefore, seen that the transformer action takes place between winding, section XZ, and winding section YZ. In other words, the volt-amperes across winding XZ are transferred by transformer action to the load connected across winding YZ.

Power transformer in VA

$$= V_{XY}I_{XY} = (V_1 - V_2)I_1 \quad (5.171)$$

Total power to be transferred or input power in VA

$$= V_1I_1 \quad (5.172)$$

Hence,
$$\frac{\text{Transformer power in } VA}{\text{Input Power } VA} = \frac{(V_1 - V_2)I_1}{V_1I_1}$$

$$= 1 - \frac{V_2}{V_1} = (1 - K) \quad (5.173)$$

$$\boldsymbol{Power\ transformed = (1 - K) \times Power\ input} \quad (5.174)$$

Out of the input volt-amperes V_1I_1, only $V_{XY}I_{XY} = (V_1 - V_2)I_1$ is transformed to the output by transformer action. The remaining power in volt-ampere required for the output, is conducted directly to the secondary from the primary (due to an electrical connection).

Power conducted in VA = Total power input in VA – Transformed power in VA

$$= V_1I_1 - (V_1 - V_2)I_1 = V_2I_1$$

$$\frac{\text{Power conducted in } VA}{\text{Total power input } VA} = \frac{V_2I_1}{V_1I_1} = K \quad (5.175)$$

$$\boldsymbol{Power\ conducted = K \times Power\ input} \quad (5.176)$$

5.31.3 Saving of copper in an auto-transformer

The total length of the winding is proportional to the number of turns (N), while its cross-section is proportional to the current (I) for any winding. Therefore, its copper weight is proportional to the product of N and I.

$$\text{Weight of copper} \propto NI \quad (5.177)$$

where I is current in winding and N is the number of turns of the winding.

Fig. 5.51(a) and fig. 5.51(b) shows a step-down autotransformer and a two-winding transformer respectively. Let W_{tw} and W_{ta} be the total weight of copper in the two-winding transformer and in the autotransformer. let N_1 and N_2 be the number of turns of primary and secondary windings, respectively.

Now, with reference to fig. 5.51(a), the total weight of copper (Cu) required in two winding transformers.

$W_{tw} \propto$ Total weight of copper (Cu) on its primary
+ Total weight of copper (Cu) on its secondary

$$W_{tw} \propto I_1N_1 + I_2N_2 \quad (5.178)$$

Total weight of copper required in the auto-transformer from fig. 5.50(b).

$$W_{ta} \propto \text{Total weight of copper (Cu) in section XZ} + \text{Total weight of copper (Cu) in section ZY}$$

$$W_{ta} \propto I_1(N_1 - N_2) + (I_2 - I_1)\, N_2$$

$$W_{ta} \propto I_1N_1 + I_2N_2 - 2I_1N_2 \tag{5.179}$$

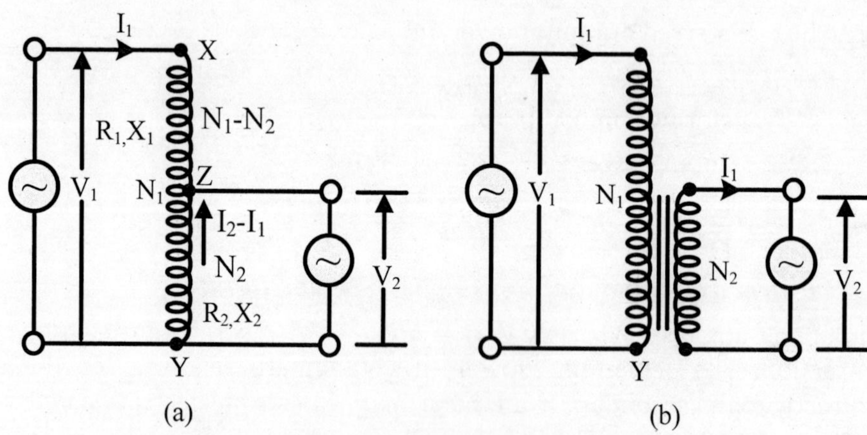

Fig. 5.51 Circuit for two winding and autotransformer

Now, the ratio of the weight of copper in an auto-transformer to the weight of copper in an ordinary transformer.

$$\frac{W_{ta}}{W_{tw}} = \frac{I_1N_1 + I_2N_2 - 2I_1N_2}{I_1N_1 + I_2N_2}$$

$$= \frac{I_1N_1 + I_2N_2}{I_1N_1 + I_2N_2} - \frac{2I_1N_2}{I_1N_1 + I_2N_2}$$

$$= 1 - \frac{2I_1N_2/I_1N_1}{I_1N_1/I_1N_1 + I_2N_2/I_1N_1} = 1 - K$$

$$W_{ta} = (1 - K)W_{tw} \tag{5.180}$$

Now Saving of copper

= weight of copper required in an ordinary transformer – the weight of copper required in an auto-transformer

$$= W_{tw} - W_{ta}$$

$$= W_{tw} - (1 - K)W_{tw}$$

$$\text{Saving of copper} = K \times W_{tw} \tag{5.181}$$

Saving of copper $= K \times$ weight of copper required for two winding transformers

Hence, saving in copper increases as the transformation ratio approaches unity, therefore, autotransformers are used when K is nearly equal to unity.

5.31.4 Conversion of two winding transformer into single phase auto-transformer

A conventional two-winding transformer with its voltage ratings and polarity markings is shown in fig. 5.52. It can be converted to a step-up autotransformer by connecting the two windings electrically in series with additive or subtractive polarities. With additive polarity between the high-voltage and low-voltage sides, a step-up transformer is obtained. Whereas, with subtractive polarity a step-down transformer is obtained. Let us consider a conventional 30 kVA, 2000/200 V transformer to be connected in an autotransformer configuration.

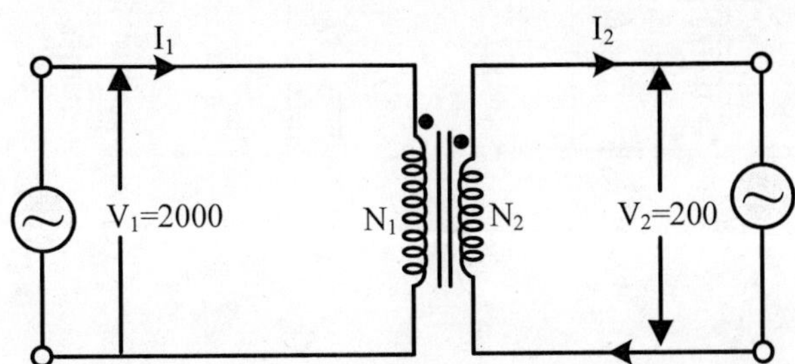

Fig. 5.52 Two-winding transformer

Additive Polarity

Fig. 5.53(*a*) shows the series connections of the windings with additive polarity. The circuit is redrawn in fig. 5.53(*b*) showing the common terminal of the autotransformer at the top. Fig. 5.53(*c*) shows the same circuit with the common terminal at the bottom. Since the polarity is additive, V_{HV} = 2000+ 200 = 2200 V and V_{LV} = 2000 V, the transformer acts as a step-up autotransformer.

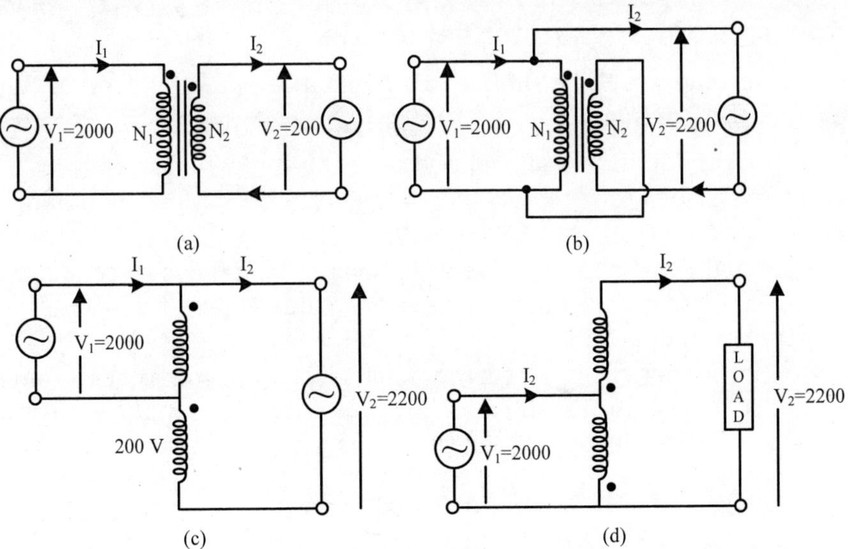

Fig. 5.53 Additive polarity

Subtractive polarity

Fig. 5.54(a) shows the series connections of the windings with subtractive polarity. The circuit is redrawn in fig. 5.54(b) with the common terminal at the top. Fig.5.54(c) shows the same circuit with the common terminal at the bottom. Since the polarity is subtractive V_{HV} = 2000 V and V_{LV} = 2000 – 200 = 1800 V, the transformer acts as a step-down autotransformer.

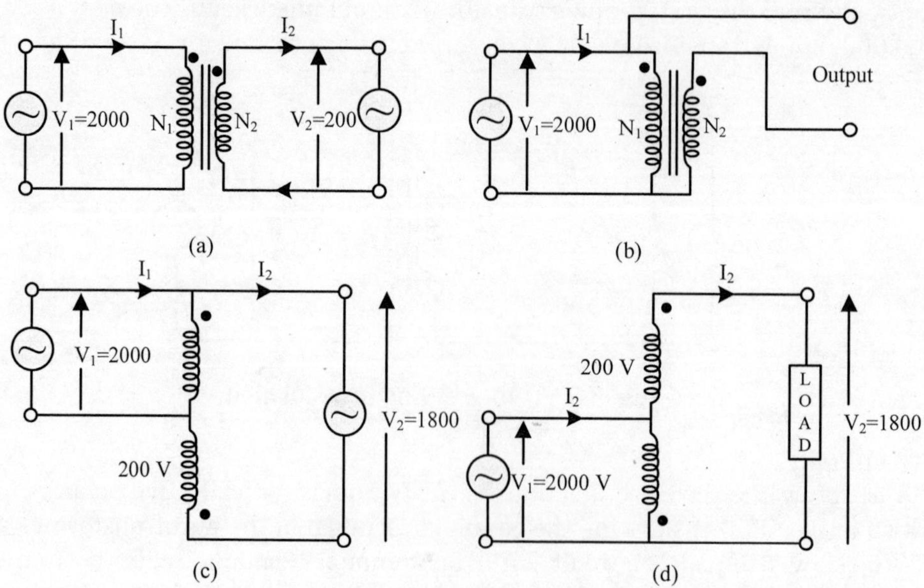

Fig. 5.54 Subtractive polarity

5.31.5 Auto-transformer vs potential divider

An auto-transformer appears to be similar to a resistance potential divider. But this is not because of the following reasons:

- The potential divider is less efficient because it has more losses.
- An autotransformer can step up the voltage, Whereas a resistive potential divider can't.
- In a potential divider, almost the entire power to load flows by conduction, whereas in auto-transformer, a part of the power is conducted and the rest is transferred to load by transformer action.
- In a potential divider, the input current must always be more than the output current, this is not so in an auto-transformer. If the output voltage in the auto-transformer is less than the input voltage, the load current is more than the input current.

5.31.6 Advantages of autotransformers

Autotransformers have the following advantages:

- Less amount of copper is required compared to two winding transformers.
- Cost is less compared to a two-winding transformer due to its compact size.
- It has good voltage regulation because the resistance and reactance are less compared to a two-winding transformer
- Copper (Cu) loss is less.
- The volt-ampere rating is more compared to a two-winding transformer.
- Greater efficiency due to less loss.
- Smooth and continuous variation of voltage.

5.31.7 Disadvantages of autotransformers

Despite having several advantages, autotransformers also possess the following disadvantage.:

- There is the possibility of high short circuit currents for short circuits on the secondary side due to low impedance.
- The full primary current will appear across the secondary causing higher voltage on the secondary resulting in danger of accidents if the common winding is open-circuited.
- Since there is no electrical isolation between primary and secondary, risk factor appears at high voltage levels.
- It is economical only if the voltage ratio is less than 2.

5.31.8 Applications of Autotransformers

Autotransformers have the following applications:

- It is used as a starter for safely starting machines (providing up to 50% to 60% of full voltage) such as induction motors and synchronous motors.
- It is used as a booster to give a small boost to a distribution cable for compensating the voltage drop.
- It can be used as a furnace transformer to supply power to the furnaces at the required supply voltage.
- In a laboratory or when a continuously variable voltage across a wide range is required, an autotransformer is applied as a variac.

Example 5.25: An auto-transformer having 1500 turns is connected across a 500 V AC supply. What secondary voltage will be obtained if a tap is taken at the 900th turn?

Solution:

$$\text{Supply voltage, } V_1 = 500\ V$$

$$\text{Total turns, } N_1 = 1500$$

$$\text{Secondary turns, } N_2 = 900$$

Voltage per turn

$$= \frac{V_1}{N_1} = \frac{500}{1500} = \frac{1}{3}\ V$$

Secondary voltage, $V_2 = Voltage\ per\ turn\ \times\ N_2$

$$= \frac{1}{3} \times 900 = 300\ V\ (Ans)$$

Example 5.26: Determine the core area, the number of turns, and the position of the tapping point for a 500 kVA, 50 Hz single phase, 6600/5000-volt auto-transformer, assuming the following approximate values: emf per turn 8 volt, maximum flux density 1.3 T.

Solution:

We know, $E = 4.44 f B_m A_i N$

$$\frac{E}{N} = 4.44 f B_m A_i$$

$$8 = 4.44 \times 50 \times 1.3 \times A_i$$

$$A_i = \frac{8}{4.44 \times 50 \times 1.3} = 0.02772\ m^2$$

$$= 277.2\ cm^2\ (Ans)$$

Turn on the primary side,

$$N_1 = \frac{6600}{8} = 825$$

Turn on the primary side,

$$N_2 = \frac{5000}{8} = 625$$

Hence tapping should be 200 turns from the high voltage end or 625 turns from the common end.

Example 5.27: Determine the values of the currents flowing in the various branches of a 3-phase, star-connected auto-transformer loaded with 400 kW at 0.8 power factor lagging and having a ratio of 440/550 volt. Neglect voltage drops, magnetizing current, and all losses in the transformer.

Solution:

$$\text{Primary current} = \frac{kW \times 1000}{\sqrt{3} \times E_{L1} \times \cos\phi}$$

$$I_1 = \frac{400 \times 1000}{\sqrt{3} \times 440 \times 0.8}$$

$$= 656\ A$$

$$\text{Secondary line current} = \frac{kW \times 1000}{\sqrt{3} \times E_{L2} \times \cos\phi}$$

$$I_1 = \frac{400 \times 1000}{\sqrt{3} \times 550 \times 0.8}$$

$$= 525\ A$$

5.32 Three-Phase Transformer

A three-phase system is adopted for the generation, transmission, and distribution of electrical power due to economical reasons. The generation of electric power is three-phase in nature and the generated voltage is 13.2 kV, 22 kV or higher. Transmission of power is carried out at high voltages like 132 kV, 433 kV, and up to 1200 kV. Similarly, at the distribution substation, the voltage must be stepped down and it is necessary to reduce the voltage level up to 433 V or 230 V in domestic applications. Thus, three-phase transformers are required to step-up or step-down voltages in various stages of a power system network. Therefore, it is economical to use three-phase transformers for transmission and utilization purposes.

Transformers for 3-phase circuits can be constructed in one of the following ways:

1. Three separate single-phase transformers are suitably connected for 3-phase operation. Such an arrangement is called a 3-phase bank of transformers.
2. A single three-phase transformer in which the cores and windings for all three phases are combined in a single structure.

5.33 Advantage of a Bank of Three Single-Phase Transformer

The voltage level in a three-phase system at the generating stations and at the receiving stations can be changed either by employing a bank of three single-phase transformers (interconnecting them in star or delta) or by employing one three-phase transformer. A transformer bank of three single-phase transformers has the following advantages over a unit three-phase transformer of the same kVA rating:

1. One single-phase transformer of a bank may be provided with a higher kVA rating than the others to supply an unbalanced load.

2. When one single-phase transformer of a bank is damaged and removed from the service, the remaining two units may be used in open-delta or V-V at reduced capacity.
3. Where single units are concerned only one spare single-phase transformer is needed as a standby instead of a complete spare 3-phase transformer. The provision of a spare standby 3-phase transformer is more costly than the provision of a single-phase spare transformer for a three-phase transformer bank. Thus standby requirement is lesser.
4. The transportation of sinlge-phase transformer is more convenient.
5. It is common practice to use single-unit 3-phase transformers. However, 3-phase banks are also used depending on the requirements.

5.34 Advantages of Three-Phase Transformer

Generally, one three-phase transformer is preferred over a bank of three single-phase transformers because of the following advantages over three single-phase transformer bank of the same kVA rating:

1. It is lighter, smaller, and cheaper.
2. The core is of smaller size and hence less material is required.
3. It occupies less space for the same rating, compared to a bank of three single-phase transformers.
4. It requires a smaller quantity of iron and copper. Hence, its cost is nearly 15% lesser than a bank of three single-phase transformers of equal rating.
5. It operates at slightly better efficiency and regulation.
6. The costly high voltage terminals (Bushings) to be brought out of the transformer housing are reduced to three rather than the six necessary for three separate single-phase transformers.
7. Since only one unit is required to be handled, it is easy for the operator.
8. It has smaller size and can be accommodated in smaller tank and hence needs smaller quantity of oil for cooling.
9. The busbar structure, switchgear, and wiring for a single 3-phase transformer installation are simpler than those for a transformer bank of three 1-phase transformers.

These transformers suffer from the following disadvantage.

1. In a single 3-phase transformer, however, if one of the phase windings breaks down, the whole transformer has to be removed from service for repairs, thereby disturbing the continuity of the power supply.
2. It is more difficult and costly to repair three-phase transformers.
3. It is more difficult to transport a single large unit of a three-phase transformer than to transport three single-phase transformers individually.

The advantages of the three-phase transformer (such as lower cost, lower weight, lower space requirement etc.) overweighs its disadvantages and hence are invariably employed in the power system to step-up or step-down the voltage level.

5.35 Construction of Three-Phase Transformer

Three-phase transformers, from the construction point of view are also classified as Core type transformers and Shell type transformers.

Core Type Transformers

Construction of the magnetic core of a 3-phase core-type transformer may be understood by considering three single-phase core-type transformers positioned at 120° to each other as shown in fig. 5.55(a). For simplicity, only the primary windings are shown. If balanced 3-phase sinusoidal voltage are applied to the windings, the fluxes ϕ_a, ϕ_b and ϕ_c will also be sinusoidal and balanced. If the three legs carrying these fluxes are merged, the total flux in the merged leg is zero. This leg can therefore be removed because it carries no flux, as shown in fig. 5.55(b).

This structure is not convenient to build. Usually the structure of fig. 5.56 with the three limbs of an equal area of the cross-section in the same plane is used. Three limbs are joined by two horizontal (top and bottom) members called yokes. The area of the cross-section of all the limbs and yokes is the same since at every instant magnitude of flux set-up in each part is the same. Each leg carries both low-voltage and high-voltage windings. Since it is easier to insulate the LV windings from the core than the HV windings, the LV windings are placed next to the core with suitable insulation between the core and the LV windings. The HV windings are placed over the LV windings with suitable insulation between them. The core consists of laminations of silicon steel material having oxide film coating on both sides for insulation. The laminations are usually of E and I shape and are staggered alternately to decrease the reluctance of the magnetic path and increase the mechanical strength. This type of transformer is usually wound with circular cylindrical coils.

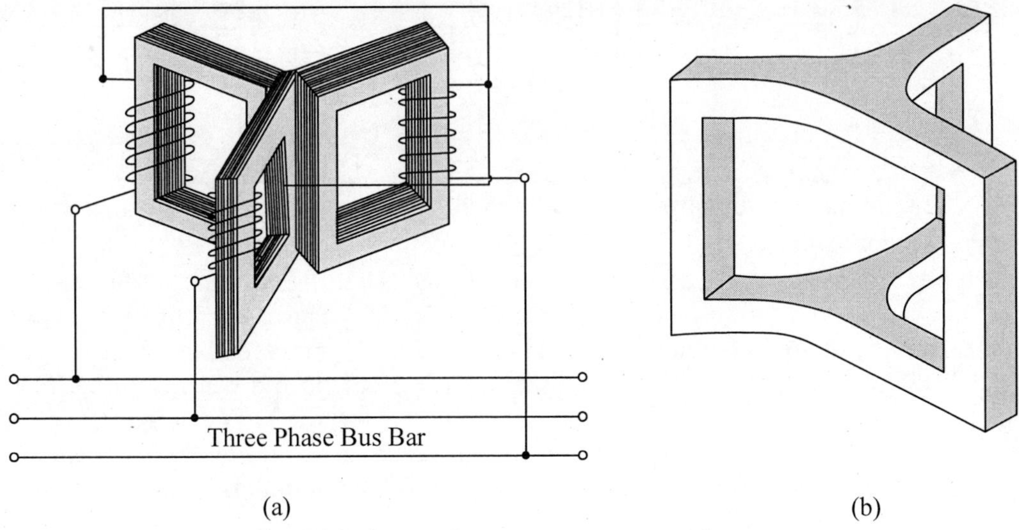

Fig. 5.55 Three-phase core-type transformer

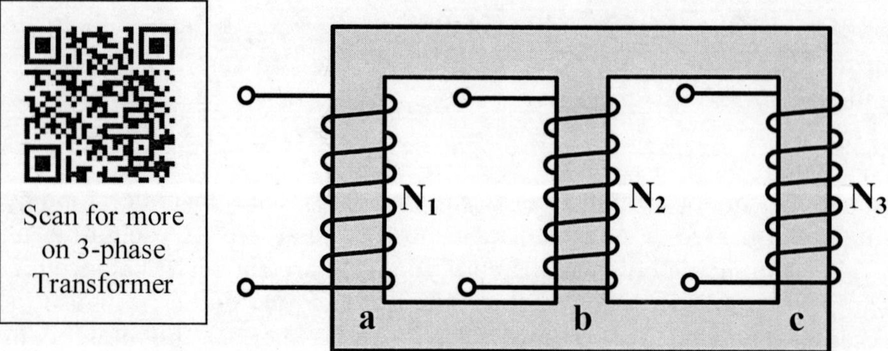

Fig. 5.56 Core and winding of the core-type three-phase transformer

The core construction for very large capacity three-phase transformers is slightly changed. In this case, the core consists of three main limbs on which windings are arranged and two additional limbs at the sides without winding are formed as shown in fig. 5.57. This arrangement allows for decreasing the height of the yoke and consequently decreases the overall height of the core. However, the length increases. This facilitates the transportation of transformers by rail. In this arrangement, the magnetic circuits for each phase are virtually independent. In either case, the magnetic circuits of the three phases are somewhat unbalanced, the middle phase having less reluctance than the outer two. This causes the magnetizing current of the middle phase slightly less than the outers. But during operation, the magnetizing current is so small that it does not produce any noticeable effect.

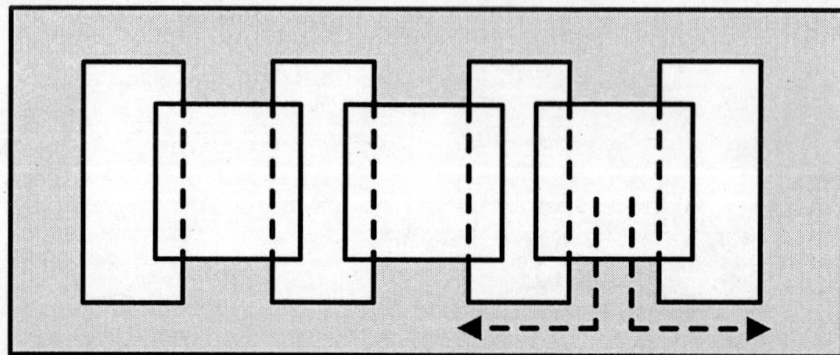

Fig. 5.57 Core and winding of very large capacity core-type three-phase transformer

Shell Type Transformers

In shell-type transformers, the core construction is such that the windings are embedded in the core instead of surrounding the iron as shown in fig. 5.58. The area of the cross-section of the central limbs is double to that of the side limbs and horizontal members. The low voltage and high voltage windings of the three phases are wound on the central limbs. The low voltage (*LV*) winding is always placed nearer to the core and the high voltage (*HV*) winding is placed over the low voltage winding for economic reasons. To obtain uniform distribution of flux in the core, usually second winding placed on the central limb is wound in the reverse direction as shown in fig. 5.58.

The winding direction of the central unit b is made opposite to that of units a and c. If the system is balanced with phase sequence a-b-c, the fluxes will also be balanced.

That is,

$$\phi_a = \propto \phi_b = \propto^2 \phi_c \quad (5.182a)$$

Where $\propto = 1\angle 120°$,

$\propto^2 = 1\angle 240°$

The adjacent yoke sections of units a and b carry a combined flux of

$$\frac{1}{2}\phi_a + \frac{1}{2}\phi_b = \frac{1}{2}\phi_a\,(1\angle 0° + 1\angle 240°)$$

$$= \frac{1}{2}\phi_a \angle 60° \quad (5.182b)$$

Thus, the magnitude of this combined flux is equal to the magnitude of each of its components. In this way, the cross-sectional area of the combined yoke sections may be reduced to the same value as that used in the outer legs and in the top and bottom yokes. The slight unbalance in the magnetic paths among the three phases has very little effect on the performance of the 3-phase shell-type transformer. Its behavior is essentially the same as that of a bank of three single-phase transformer.

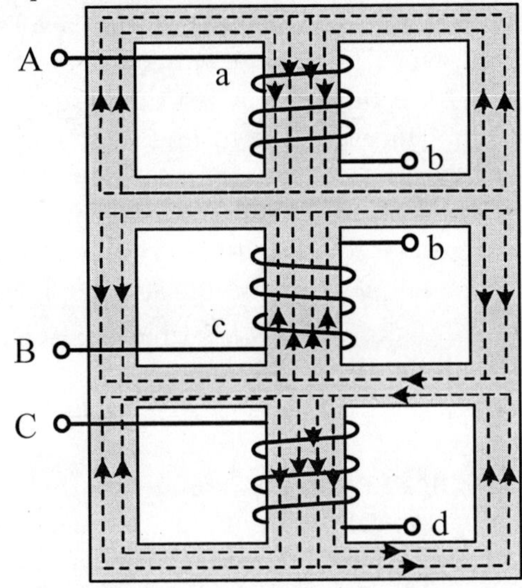

Fig. 5.58 Construction of three-phase shell-type transformer

5.36 Phasor Representation of Alternating Quantities in Three-Phase Transformer Connection

It is necessary to learn the characteristics of balanced three-phase systems as well as the conventions followed to designate currents and voltages of a three-phase system before studying the connections of three-phase transformers. If the supply voltage is balanced, the voltages can be represented by a voltage triangle shown in fig. 5.59(a), where A, B, and C are the nomenclatures of the three lines of the system and N stands for the neutral or star point of the system. A, B, and C are the three vertices of the equilateral triangle ABC and N is the circumcentre of the triangle.

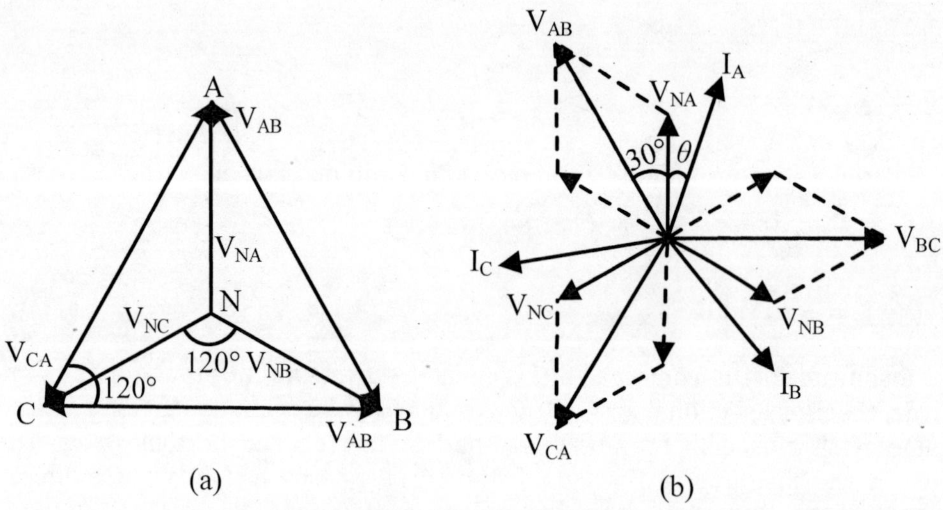

Fig. 5.59 Balanced three-phase system phasor

The voltages and currents with double subscripts notation are represented by phasors. For example, V_{AB}, I_{AB} represent the voltage of point A with respect to point B and the current flowing from point A to point B, respectively. With the arrow pointing towards A the line represents voltage phasor V_{AN}. Thus the line-to-line voltage (also called as line voltage) phasor and line-to-neutral voltage (also known as phase voltage) phasors can be drawn from the voltage triangle shown in fig. 5.59(a). The phasor diagram is shown in fig. 5.59(b) where phase voltage phasors that are equal in magnitude are displaced from each other by 120°. The line voltage phasors are also equal in magnitude and displaced from each other by 120°, but the magnitude of a line voltage phasor is $\sqrt{3}$ times the magnitude of a phase voltage phasor. Further, it may be noted that the set of line voltage phasors is displaced from the set of phase voltage phasors by 30°.

As per Indian Standards Specifications, the terminals of a three-phase transformer are marked as U, V, W with numerics 1, 2, 3, etc.

If the lines carry a balanced load then the magnitudes of the line currents I_A, I_B and I_C are equal and displaced with respect to each other by 120°. They are equally displaced from the corresponding phase voltages by an angle ϕ as shown in fig. 5.59(b), the power factor being $cos\phi$. In this case, the power factor is lagging as currents lag behind the phase voltages.

Relation in three-phase connected system:

a. ***Star Connection:***

$$V_L = \sqrt{3}V_{ph} \tag{5.183}$$

$$I_L = I_{ph} \tag{5.184}$$

$$\text{Power, } P = \sqrt{3}V_L I_L cos\theta = 3V_{ph}I_{ph}cos\theta \tag{5.185}$$

b. *Delta Connection:*

$$V_L = V_{ph} \tag{5.186}$$

$$I_L = \sqrt{3} I_{ph} \tag{5.187}$$

$$\text{Power, } P = \sqrt{3} V_L I_L cos\theta = 3 V_{ph} I_{ph} cos\theta \tag{5.188}$$

c. *In Three Phase Transformer*

Primary Side:

$$E_{ph1} = 4.44\, f\, \phi_m N_1 \tag{5.189}$$

$$E_{ph1} = 4.44\, f\, B_m A_i N_1 \tag{5.190}$$

Secondary Side:

$$E_{ph2} = 4.44\, f\, \phi_m N_2 \tag{5.191}$$

$$E_{ph2} = 4.44\, f\, B_m A_i N_2 \tag{5.192}$$

Transformarion Ratio:

$$\text{Trandformation ratio, K} = \frac{\mathrm{E_{ph2}}}{\mathrm{E_{ph1}}} = \frac{\mathrm{N_2}}{\mathrm{N_1}} \tag{5.193}$$

Example 5.28: Find the turn-ratio (primary to secondary) of 11000/400 volt, delta/star connected, three-phase transformer.

Solution:

Primary phase voltage,

$E_{ph1} = E_{L1} = 11000\, V$ (Delta-connected primary)

Secondary phase voltage,

$$E_{ph2} = \frac{E_{L2}}{\sqrt{3}} = \frac{400}{\sqrt{3}} = 231\, V \quad \text{(Star − connected secondary)}$$

Turn-ratio (Primary to secondary),

$$\frac{N_1}{N_2} = \frac{E_{ph1}}{E_{ph2}} = \frac{11000}{231} = 47.62\ (Ans)$$

Example 5.29: A 50 Hz, three-phase core type transformer is to be built for an 11 kV /440 V ratio, connected in delta star. The cores are to have a square section and the coils are of circular. Taking an induced emf of 15 V per turn and maximum core flux density of about 1.1 T. Find the primary and secondary number of turns and cores cross-sectional area neglecting insulation thickness.

Solution:

Here, Connections- Delta/Star

$$E_{L1} = 11\ kV = 11000\ V;\ E_{L2} = 440\ V;$$

$$B_m = 1.1\ T$$

$$EMF/turn = 15\ V;$$

Primary phase voltage,

$$E_{ph1} = E_{L1} = 11000\ V$$

Primary turns/ Phase,

$$N_1 = \frac{11000}{15} = 733.3\ (Ans)$$

Secondary phase voltage,

$$E_{ph2} = \frac{E_{L2}}{\sqrt{3}} = \frac{440}{\sqrt{3}} = 254\ V$$

Secondary turns/ Phase,

$$N_2 = \frac{254}{15} = 17\ (Ans)$$

Now, $E_{ph1} = 4.44 f B_m A N_1$

$$11000 = 4.44 \times 50 \times 1.1 \times A \times 733.3$$

Cross-sectional area,

$$\mathrm{A} = 0.06143\ m^2 = 614.3\ cm^2\ (Ans)$$

Example 5.30: A three-phase, 50 Hz transformer of shell type has cross-sectional area of the core as 400 cm2. If the flux density is limited to 1.2 Tesla, find the number of turns per phase on high voltage and low voltage side. The voltage ratio is 11000/400 V, the higher voltage side being connected in star and low voltage side in delta. Also determine the transformation ratio.

Solution:

Here, Connections – Star/Delta

$f = 50\ Hz;\ A = 400\ cm^2 = 400\ \times\ 10^{-4}\ m^2;\ B_m = 1.2\ T$

$E_{L1} = 11000\ V;$

$E_{L2} = 400\ V;$

Primary phase voltage,

$$E_{ph1} = \frac{E_{L1}}{\sqrt{3}} = \frac{11000}{\sqrt{3}} = 6351\ V$$

Now, $E_{ph1} = 4.44 f B_m A N_1$

Primary turns/ Phase,

$$N_1 = \frac{E_{ph1}}{4.44 f B_m A}$$

$$= \frac{6351}{4.44\ \times 50\ \times 1.2\ \times 400\ \times\ 10^{-4}} = 596\ (Ans)$$

Transformation, ratio,

$$K = \frac{E_{ph2}}{E_{ph1}} = \frac{400}{11000/\sqrt{3}} = \frac{\sqrt{3}\ \times 400}{11000} = 0.06298\ (Ans)$$

Secondary turns/phase,

$$N_2 = KN_1 = \ 0.06298\ \times 596 = 37.54\ (Ans)$$

5.37 Three-Phase Transformer Connection

A three-phase transformer consists of three transformers, either separate or combined on one core. When a transformer is to be placed in a power system to step-up or step-down the voltage, it is selected as per its connections The primaries and secondaries of any 3-phase transformer can be independently connected in either a star (Y) or delta (Δ).

Thus, there are four possible connections for a 3-phase transformer bank:

1. Δ-Δ (Delta primary – Delta secondary)
2. Y-Y (Star primary – Star secondary)
3. Δ-Y (Delta primary – Star secondary)
4. Y-Δ (Star primary – Delta secondary)

Here it is assumed that all transformers in the bank have the same kVA rating.

Factors affecting the choice of connections

Some of the factors governing the choice of connections are as follows:

1. Availability of a neutral connection for grounding, protection, or load connections.
2. Insulation to the ground and voltage stress.
3. Availability of a path for the flow of third harmonic (exciting) currents and zero-sequence (fault) currents.
4. Need for partial capacity with one circuit out of service.
5. Parallel operation with other transformers.
6. Operation under fault conditions.
7. Economic considerations.

5.37.1 Delta-Delta (Δ-Δ) connection

Fig. 5.60 shows the Δ-Δ connection of three identical single-phase transformers or three identical windings on each of the primary and secondary sides of the three-phase transformer. The secondary winding $a_1\, a_2$ corresponds to the primary winding $A_1\, A_2$, the terminal A_1 and $a_1$have the same polarity. The polarity of the terminal a connecting a_1 and c_2 is the same as that of A connecting A_1 and C_2.

Fig. 5.61 shows the phasor diagrams for the lagging power factor $\cos\theta$. Magnetizing current and voltage drops in impedance have been neglected. Under balanced conditions, in the Δ-Δ configuration the corresponding line and phase voltages are identical in magnitude on both primary and secondary sides (Equation. 5.186). The line currents are $\sqrt{3}$ times the phase (winding) currents and displaced behind the phase currents (Equation. 5.187).

The secondary line-to-line voltage V_{ab}, V_{bc} and V_{ca} are in phase with primary line-to-line voltages V_{AB}, V_{BC} and V_{CA} with voltage ratios equal to the turns ratio:

$$\frac{V_{AB}}{V_{ab}} = \frac{V_{BC}}{V_{bc}} = \frac{V_{CA}}{V_{ca}} = a \tag{5.194}$$

The current ratios when the magnetizing current is neglected are:

$$\frac{I_{AB}}{I_{ab}} = \frac{I_{BC}}{I_{bc}} = \frac{I_{CA}}{I_{ca}} = \frac{I_A}{I_a} = \frac{I_B}{I_b} = \frac{I_C}{I_c} = \frac{1}{a} \tag{5.195}$$

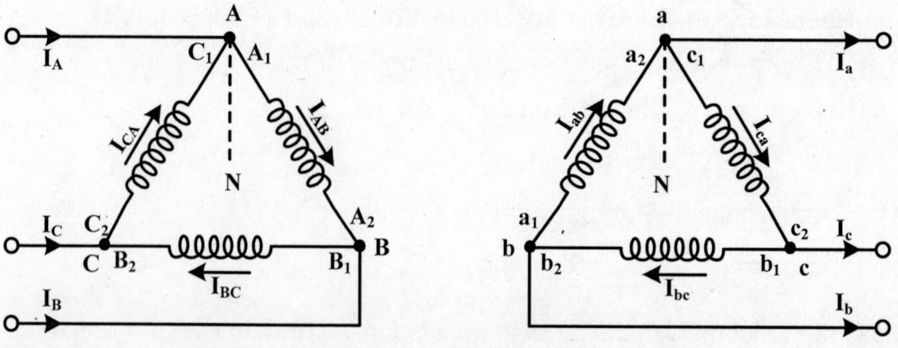

Fig. 5.60 Delta-delta connection of transformer

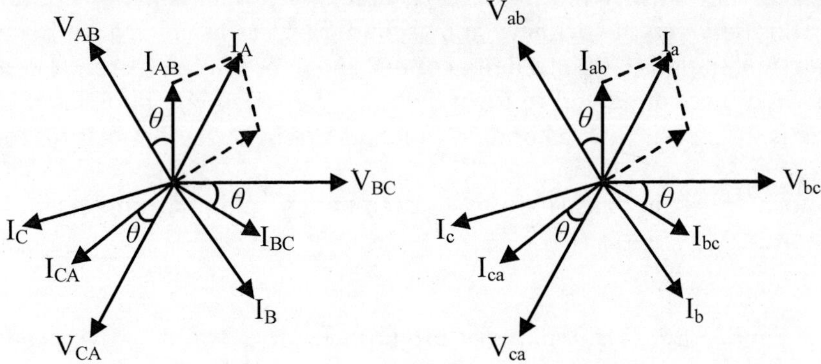

Fig. 5.61 Phasor diagram (0° phase shift)

It is to be noted that in fig. 5.60 each winding is drawn along the line of the phasor of its induced voltage. The voltage and current phasors are determined very easily from the windings drawn in this manner. It is seen from the phasor diagram as shown in fig. 5.61, the primary and secondary line voltages are in phase. This connection is called 0°-connection.

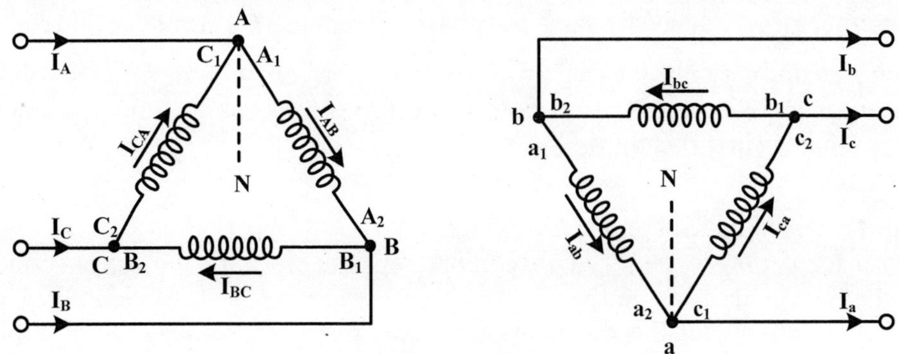

Fig. 5.62 Delta-delta connection of transformer (180° phase shift)

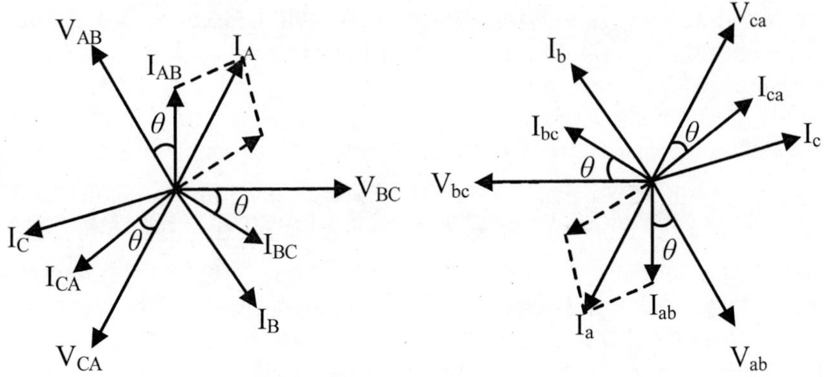

Fig. 5.63 Phasor diagram (180° phase shift)

If connections of the phase windings are reversed on either side, we obtain the phase difference of 180° between the primary and secondary systems. Such a connection is known as 180°-connection. In fig. 5.62 delta-delta connection with 180° phase shift is shown. Hence b_1c_2, c_1a_2 and a_1b_2 are connected to form delta on the secondary side. Fig. 5.63 shows the phasor diagrams. It is seen that secondary voltages are in phase opposition to the primary voltages.

The Δ-Δ transformer has no phase shift associated with it, and no problems with unbalanced loads or harmonics.

Advantages

i. The Δ-Δ connection is satisfactory for both balanced and unbalanced loading. No difficulty is experienced even though the load is unbalanced on the secondary side.
ii. There is no phase displacement between the primary and secondary voltages.
iii. There is no distortion of flux, since the third harmonic component of magnetizing current flows in the delta-connected primary winding without flowing in the line wires.
iv. If a third harmonic is present, it circulates in the closed path and therefore does not appear in the output voltage wave.
v. For winding, conductors with smaller diameters are required as the cross-section of the conductor is reduced because the phase current is $1/\sqrt{3}$ times of line current.
vi. If one transformer fails, the remaining two transformers will continue to supply three-phase power. This is called an open-delta (or V - V) connection. The operation is discussed later in this chapter.

Disadvantage

i. No star point (neutral point) is available.
ii. More insulation is required in comparison to star-star connections since phase voltage is equal to line voltage.
iii. It is not suitable for the three-phase four-wire system because the neutral point is absent.
iv. This connection is generally used for low-voltage transformers.
v. In these connections, the star point is absent, if one line gets earthed due to fault, the maximum voltage between windings and core will become full line voltage
vi. Δ-Δ connection is useful when neither primary nor secondary requires a netural and the voltages are low and moderate.

5.37.2 Star- Star (*Y-Y*) connection

The star-star connections of a three-phase transformer are shown in fig. 5.64. The phasor diagrams are drawn in a similar manner as done in Δ-Δ connection. The phase current is equal to the line current and they are in phase (Equation 5.184). The line voltage is $\sqrt{3}$ times the phase voltage (Equation 5.183). There is a phase separation of 30° between line and phase voltage. Fig. 5.64 shows the *Y-Y* connection for 0° phase shift and in fig. 5.66 there is a phase shift of 180° between primary and secondary systems. Their phasor diagrams are also shown in fig. 5.65 and fig. 5.67.

For an ideal transformer, the voltage ratios are

$$\frac{V_{AN}}{V_{an}} = \frac{V_{BN}}{V_{bn}} = \frac{V_{CN}}{V_{cn}} = a \tag{5.196}$$

And current ratios are

$$\frac{I_A}{I_a} = \frac{I_B}{I_b} = \frac{I_C}{I_c} = \frac{1}{a} \tag{5.197}$$

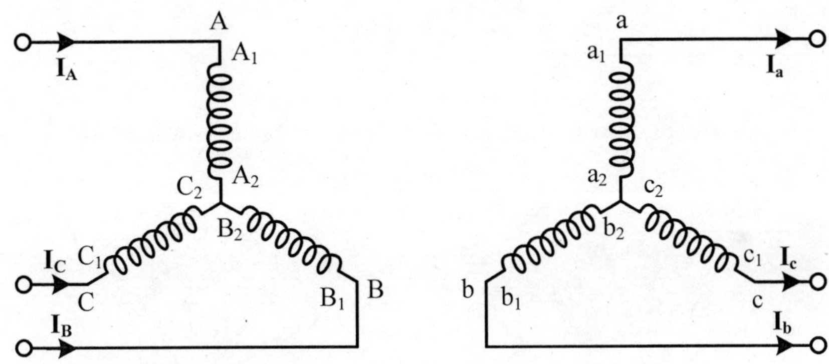

Fig. 5.64 Star-Star connection of transformer

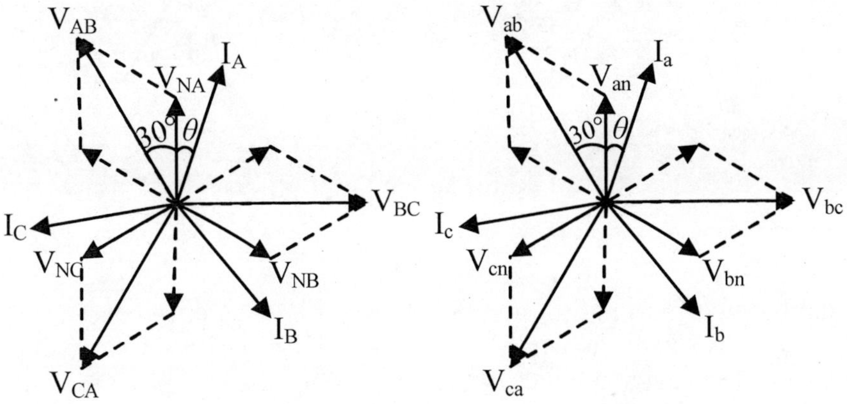

Fig. 5.65 Phasor diagram (0° phase shift)

Star-star connected three-phase transformers are operated with grounded neutrals, i.e., the neutral of the primary is connected to the neutral of the power source. If the neutral is kept in isolation, the unbalanced load on the secondary will shift the position of neutral which changes the magnitude of phase voltages. A grounded neutral in the primary prevents this unsatisfactory operation. With an isolated neutral the third-harmonic components in the magnetizing currents of the three primary windings are in phase and as such they have no path. As the path for the third harmonic current is absent, the phase voltages become non-sinusoidal though the line voltages are sinusoidal.

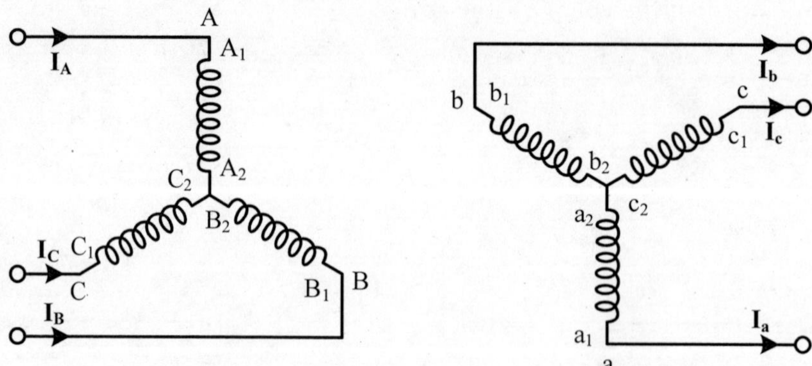

Fig. 5.66 Star-Star connection of transformer (180° phase shift)

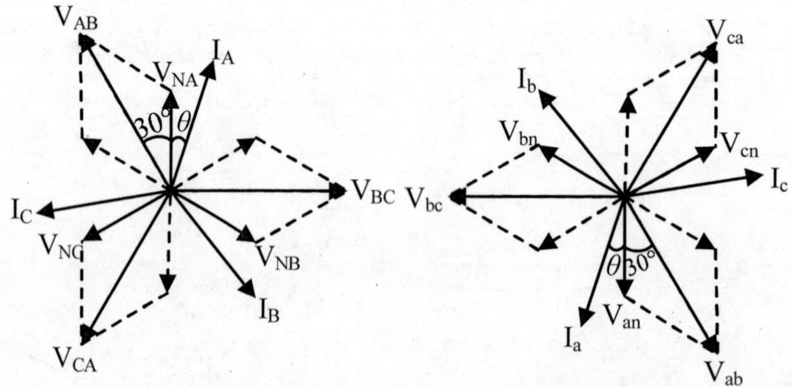

Fig. 5.67 Phasor diagram (180° phase shift)

Advantages

i. Less number of turns and less quantity of insulation is required because $V_{Ph} = \sqrt{3}\ V_L$.

ii. There is no phase displacement between the primary and secondary voltages.

iii. It is possible to provide a neutral connection since the star point is available on both sides.

Disadvantage

The *Y-Y* connection has two very serious problems:

1. If the neutral is not provided, the phase voltages tend to become severely unbalanced when the load is unbalanced. Therefore, the *Y-Y* connection is not satisfactory for unbalanced loading in the absence of a neutral connection.
2. The magnetizing current of any transformer is very nonsinusoidal and contains a very large third harmonic, which is necessary to overcome saturation in order to produce a sinusoidal flux. In a balanced three-phase system, the third harmonic components in the magnetizing currents of three primary windings are equal in magnitude and in

phase with each other. Therefore, they will be directly additive. Their sum at the neutral of a star connection is not zero. Since there is no path for these current components in an ungrounded star connection these components will distort the flux wave which will produce a voltage having a third harmonic in each of the transformers, both on the primary and secondary sides. The third harmonic component of induced voltage may be nearly as large as the fundamental voltage. When this voltage is added to the fundamental, the peak voltage is nearly two times the normal value.

Both the unbalance and third harmonic problems of the *Y-Y* connection can be solved using one of the following methods:

1. **Solid grounding of neutrals:** By providing a solid (low impedance) connection between the star point of the primary transformer and the neutral point of the alternator allows third harmonic currents to flow in the neutral instead of building up large voltages. The triple-frequency currents in the neutral wire may interfere with nearby telephone and other communication circuits. The neutral also provides a return path for unbalanced currents due to unbalanced loads.
2. **Providing tertiary windings:** when it is necessary to have a *Y-Y* connection without neutral, each transformer is provided with a third winding in addition to primary and secondary. The third winding is called tertiary. It is connected to delta. This connection is called *Y-Δ-Y* connection. It is discussed later in this chapter.

Star-star connections are rarely used because of the difficulties associated with the exciting current although these are more economical.

5.37.3 Delta- Star (Δ-Y) connection

A three-phase transformer having delta and star connections on the primary and secondary windings, respectively Shown in fig. 5.68. It may be noted that the secondary system phase voltages i.e., V_{an} etc., lag the primary system phase voltages V_{AN} etc., by 30°. This connection is called -30° connection. In Δ-Y connection, the primary line voltage is equal to the primary phase voltage $V_{LP} = V_{phP}$. The relationship between secondary voltage is $V_{LS} = \sqrt{3}\ V_{phS}$. No difficulty arises due to third harmonic currents as a delta connection allows a path for these currents. Therefore, the line-to-line voltage ratio of this connection is

$$\frac{V_{LP}}{V_{LS}} = \frac{V_{phP}}{\sqrt{3}\ V_{phS}} \tag{5.198}$$

$$But \quad \frac{V_{phP}}{V_{phS}} = a \tag{5.199}$$

Therefore

$$\frac{V_{LP}}{V_{LS}} = \frac{a}{\sqrt{3}} \tag{5.200}$$

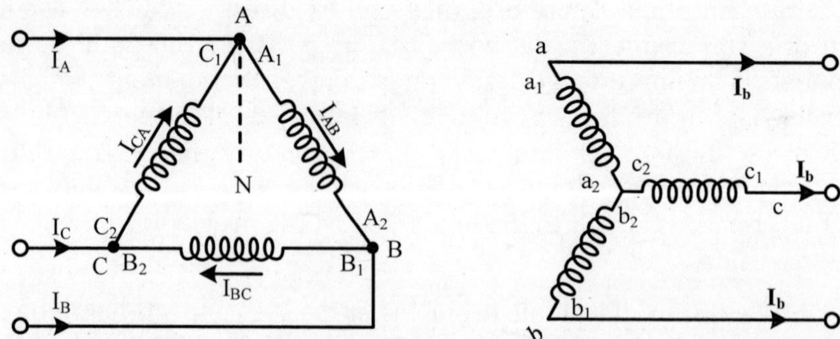

Fig. 5.68 Delta-star connection of transformer (30° phase shift lag)

The phasor diagrams for the Δ-Y connection supplying a balanced load at a power factor cos θ leading is shown in fig. 5.69. It is seen from the phasor diagram that the secondary phase voltage V_{an} lags the primary phase voltage V_{AN} by 30°. Similarly, V_{bn} lags V_{BN} by 30° and V_{cn} lags V_{CN} by 30°. This is also the phase relationship between the respective line-to-line voltages. This connection is called -30° connection.

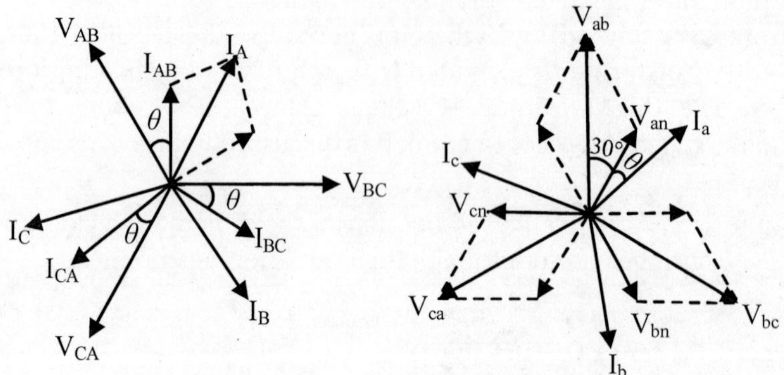

Fig. 5.69 Phasor diagram (30° phase shift lag)

The use of such connections permits a grounded neutral on the secondary side to provide a three-phase four-wire supply system. By reversing the connections on either side, the secondary system voltage can be made to lead the primary system by 30° as shown in fig. 5.70.

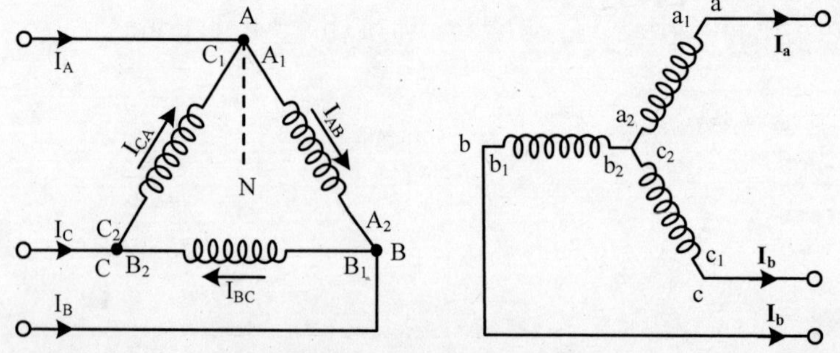

Fig. 5.70 Delta-star connection of transformer (30° phase shift lead)

The phasor diagrams for the Δ-Y connection supplying a balanced load at a power factor cos θ lagging is shown in fig. 5.71. It is seen from the phasor diagram that the secondary phase voltage V_{an} leads the primary phase voltage V_{AN} by 30°. Similarly, V_{bn} leads V_{BN} by 30° and V_{cn} leads V_{CN} by 30°. This is also the phase relationship between the respective line-to-line voltages. This connection is called +30° connection.

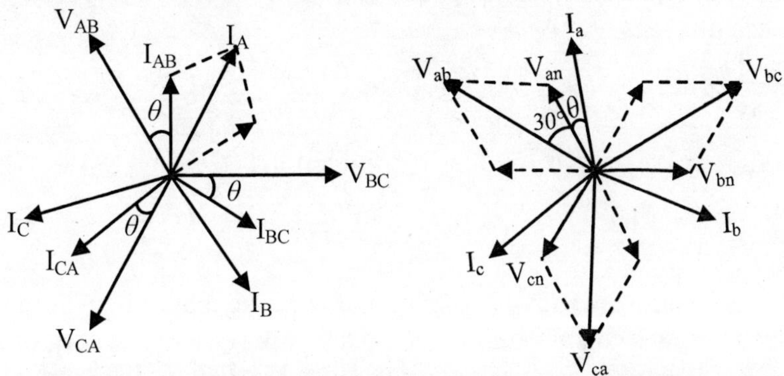

Fig. 5.71 Phasor diagram (30° phase shift lead)

Advantages:

i. The neutral is available on the secondary side, three-phase and four-wire supply can be carried out.
ii. In this case, the neutral point is stable and will not "float" when the load is unbalanced.
iii. On the high voltage side of the transformer, insulation is stressed only to the extent of 57.7% of line voltage.
iv. Saving in the cost of insulation is possible due to the availability of star connection on one side.
v. Since the primary is delta connected, the winding cross-section is small.
vi. There is no distortion of flux because the primary is delta connected which allows a path for the 3rd harmonic current.
vii. No difficulty occurs due to a large unbalanced load

Disadvantages:

i. In this scheme of connection the line voltage ratio is $\sqrt{3}$ times of transformer turn-ratio.
ii. The secondary line voltages have a phase shift of ± 30° with respect to primary line voltages.

The neutral of the secondary is grounded to provide a three-phase, four-wire system, and this scheme of connections is widely used in distribution systems because it can be used to serve both the three-phase power equipment and single-phase lighting circuits. 30° lagging voltage required for the generation of three phase thyristor controlled rectifier uses such transformer connection at low voltage.

Star–delta or delta–star connected transformers cannot be operated is parallel with star-star or delta-delta connected transformers even though the voltage ratios are correctly adjusted as there will be a 30° phase difference between corresponding voltages on the secondary side.

5.37.4 Star-Delta (Y-Δ) connection

The Y-Δ connection of three-phase transformers is shown in fig. 5.72. In this connection, the primary line voltage is equal to $\sqrt{3}$ times primary phase voltage ($V_{LP} = \sqrt{3}\ V_{phP}$). The secondary line voltage is equal to the secondary phase voltage ($V_{LS} = V_{phS}$). When operated in Y-Δ, the primary neutral is sometimes grounded to connect it to a four-wire system. The voltage ratio of each phase is

$$\frac{V_{phP}}{V_{phS}} = a \tag{5.201}$$

Therefore line-to-line voltage ratio of a $Y - \Delta$ connection is

$$\frac{V_{LP}}{V_{LS}} = \frac{\sqrt{3}\ V_{phP}}{V_{phS}} = \sqrt{3}\ a \tag{5.202}$$

The phasor diagrams can be drawn with the help of winding diagrams. There is a phase shift of 30° lead between respective line-to-line voltages. Similarly, a phase shift of 30° lead exists between respective phase voltages. This connection is called + 30° connection. Fig. 5.73 shows the star-delta connection of the transformer for a phase shift of 30° lag. This connection is known as - 30° connection.

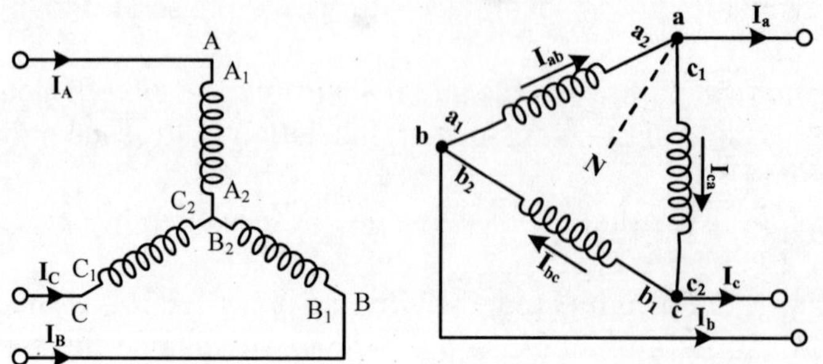

Fig. 5.72 Star-delta connection of transformer (30° phase shift lead)

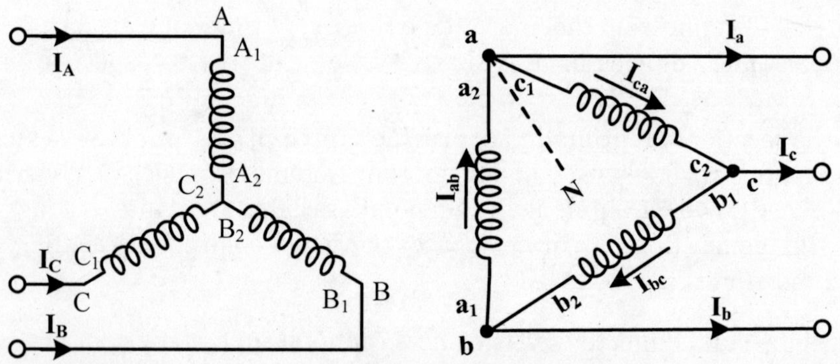

Fig. 5.73 Star-delta connection of transformer (30° phase shift lag)

It is to be noted that a Y-Δ connection is simply obtained by interchanging the primary and secondary roles in Δ-Y connection.

The Δ-Y connection or Y-Δ connection has no problem with unbalanced load and third harmonics. The delta connection assures balanced phase voltages on Y side and provides a path for the circulation of the third harmonics and their multiples without the use of a neutral wire.

Usually, the transformers with these connections are used where the voltage is to be stepped down. For example, at the receiving end of a transmission line. In this type of transformer connection, the neutral of the primary winding is earthed. In this system line voltage ratio is $^1/_{\sqrt{3}}$ times of transformer turn-ratio and secondary line voltages have a phase shift of ± 30° with respect to primary line voltages. On the *HV* side of the transformer, insulation is stressed only to the extent of 57.7% of the line voltage and, therefore, there is some saving in the cost of insulation.

Advantages:

i. The primary is star-connected, fewer numbers of turns are required in the primary, which makes it economical for high-voltage, step-down power transformers.
ii. The available neutral point on the primary side can be earthed to avoid distortion.
iii. It is possible to handle large, unbalanced load.

Disadvantages:

i. Since the secondary voltage is not in phase with the primary, it is not possible to make it parallel with star–star and delta–delta transformers.

Table 5.1 Comparision of voltage and current relationship for different types of connection

Types of Connection	Primary Side				Secondary Side			
	Line voltage	Phase voltage	Line current	Phase current	Line voltage	Phase voltage	Line current	Phase current
Delta-Delta	V_L	V_L	I_L	$\frac{I_L}{\sqrt{3}}$	$\frac{V_L}{a}$	$\frac{V_L}{a}$	aI_L	$a\left(\frac{I_L}{\sqrt{3}}\right)$
Star-Star	V_L	$\frac{V_L}{\sqrt{3}}$	I_L	I_L	$\frac{V_L}{a}$	$\frac{V_L}{a\sqrt{3}}$	aI_L	aI_L
Delta-Star	V_L	V_L	I_L	$\frac{I_L}{\sqrt{3}}$	$\frac{\sqrt{3}V_L}{a}$	$\frac{V_L}{a}$	$a\left(\frac{I_L}{\sqrt{3}}\right)$	$a\left(\frac{I_L}{\sqrt{3}}\right)$
Star-Delta	V_L	$\frac{V_L}{\sqrt{3}}$	I_L	I_L	$\frac{V_L}{a\sqrt{3}}$	$\frac{V_L}{a\sqrt{3}}$	$\sqrt{3}aI_L$	aI_L

5.38 Parallel Connection of Three-phase Transformers

When the primaries and secondaries of two or more transformers are connected separately to the same incoming and outgoing lines to share the load, the transformers are said to be connected in parallel.

The two three-phase transformers *A* and *B* are placed in parallel as shown in fig. 5.74. Here the primary windings of the two transformers are connected to the supply bus-bars and the secondary windings are connected to the load through load bus-bars. Under these conditions

V_{1L} = Primary applied voltage

$V_{2L} = V_L$ = Secondary load voltage

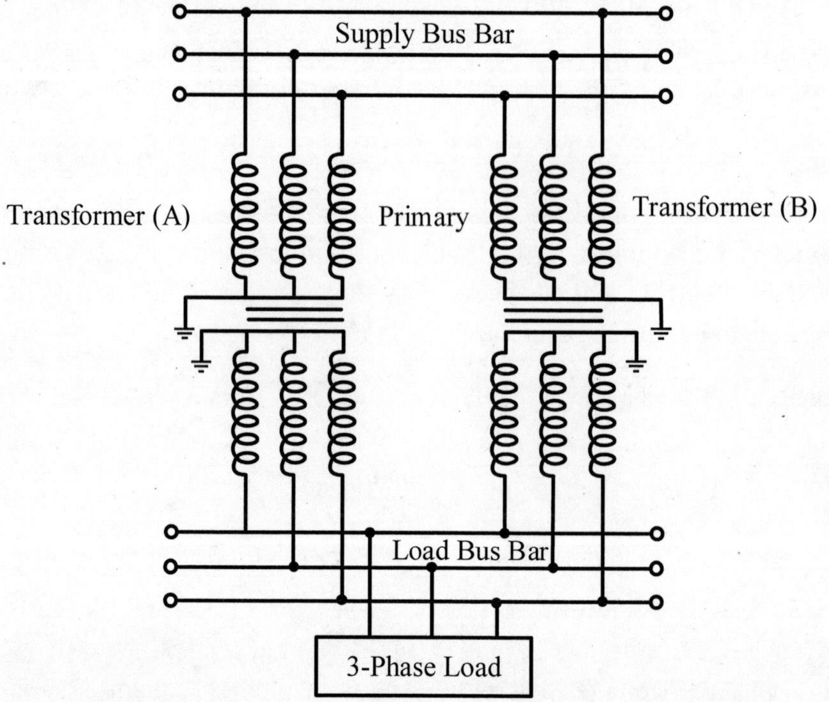

Supply Bus Bar
Transformer (A)
Primary
Transformer (B)
Load Bus Bar
3-Phase Load

Fig. 5.74 Parallel connection of two three-phase transformers

5.38.1 Necessity of parallel connection of three-phase transformers

As explained earlier, the following are the reasons why transformers are put in parallel.

1. It is desirable to place another transformer in parallel when the electrical load on the existing transformer increases beyond its rated capacity.
2. Parallel operation of transformers is necessary when the amount of power to be transformed is much more than that which can be handled by a single unit (transformer).
3. It is desirable to do parallel operation of transformers if we want to keep the spare transformer of smaller size. At the grid substations, spare transformers are always necessary to ensure the continuity of supply in case of breakdown.

5.38.2 Conditions for parallel operation of three-phase transformers

The conditions for two transformers connected in parallel depend upon the various conditions as discussed in Art No. 5.30.2. When two or more Three-phase transformers are to be operated successfully in parallel to deliver a common load the following conditions are to be fulfilled.

1. **Both the transformers should have the same transformation ratio i.e., the voltage ratings of both primaries and secondaries must be identical.**

When the transformation ratios are unequal the induced emfs in the secondary windings will not be equal. Due to inequality of induced emfs in the secondary windings, there will be some circulating current flowing from one secondary winding (having higher induced emf) to the other secondary windings (having lower induced emf), even at no-load. In other words, there will be circulating currents between the secondary windings. For satisfactory parallel operation, the circulating current should not exceed 10% of the normal load current.

2. **Both transformers should have the same percentage (or per unit) impedance.**
3. **Both three-phase transformers must have the same phase-sequence i.e., the transformers must be properly connected with regard to their phase-sequence.**
4. **In the case of three-phase transformers, the two transformers must have such connections that there should not be any phase displacement between the secondary line voltages.**

The primaries and secondaries of three-phase transformers may be connected in different forms of connections. These connections produce various magnitudes and phase displacements. The magnitudes can be adjusted by changing the tappings but phase displacement cannot be compensated. Therefore, the following types of connections are permissible for connecting three-phase transformers in parallel.

Transformer-I:	Yy	Dd	Yy	Yd	Yd
Transformer-II:	Yy	Dd	Dd	Dy	Yz

However, transformers with (α + 30°) and (–30°) angle can also be connected in parallel but only after reversing the connections of either primary or secondary.

5.38.3 Load sharing between three-phase transformers connected in parallel

The load sharing between two transformers connected in parallel depends upon the various conditions as discussed in Art No. 5-30.2 The only difference is that, in case of three-phase transformers per phase impedance is to be considered while determining the load sharing. If percentage impedances of the two transformers having different ratings are given, their values have to be converted as per base kVA for calculating load sharing, i.e.,

$$\% \text{ Resistance or } \% \text{ Reactance at base kVA}$$

$$= \frac{Base \text{ kVA}}{Rated \text{ kVA}} \times \%\text{R or } \%\text{ X at rated kVA.}$$

Example 5.31: Two 1000 kVA and 500 kVA, three-phase transformers are operating in parallel. The transformation ratio is same for both i.e., 6600/400, delta-star. The equivalent secondary impedances of the transformers are (0.001 + j0.003) ohm and (0.0028 + j0.005) ohm per phase respectively. Determine the load shared and pf of each transformer if the total load supplied by them is 1200 kVA at 0.866 pf lagging.

Solution:

The impedance of 1000 kVA transformer,

$$Z_A = 0.001 + j0.003$$

$$= 0.003162\angle 71.57° \; ohm$$

Impedance of 500 kVA transformer,

$$Z_B = 0.0028 + j0.005$$

$$= 0.00573\angle 60.75° \; ohm$$

$$Z_A + Z_B = (0.001 + j0.003) + (0.0028 + j0.005)$$

$$= (0.0038 + j0.008)$$

$$= 0.00886\angle 64.59° \; ohm$$

Load supplied, $S = 1200 \angle - cos^{-} 0.866$

$$= 1200 \angle - 30° \; kVA$$

Load shared by 1000 kVA transformer,

$$S_A = S \times \frac{Z_B}{Z_A + Z_B}$$

$$= 1200\angle - 30° \times \frac{0.00573\angle 60.75°}{0.00886\angle 64.59°}$$

$$= 776\angle - 33.84° \; kVA$$

$$= 645 \; kW \; at \; pf \; 0.83 \; lagging \; (Ans)$$

Load shared by 500 kVA transformer,

$$S_B = S \times \frac{Z_A}{Z_A + Z_B}$$

$$= 1200\angle - 30° \times \frac{0.003162\angle 71.57°}{0.00886\angle 64.59°}$$

$$= 428\angle - 23° \; kVA$$

$$= 394 \; kW \; at \; pf \; 0.92 \; lagging \; (Ans)$$

5.39 Three-Phase Power with Two Single-Phase Transformers

Three-phase power can be transformed by means of only two single-phase transformers or by means of only two windings placed on the primary and secondary of a three-phase transformer. This can be done by two methods, namely

- Open-delta or V-V connection
- Scott connection or T-T connection

Both of these methods result in slightly unbalanced output voltage under load because of the unsymmetrical relations. However, this problem is not considered to be a serious problem in commercial transformers.

5.39.1 Open-delta or V-V connection

The delta-delta connections of three transformers for a three-phase system are shown in fig. 5.60. Let us suppose that in one of the transformers, a fault develops, damage or accidentally open the system will continue to supply three-phase power. If this defective transformer is removed as shown in fig. 5.75. the remaining two transformers continue to function as a three-phase bank. If the primaries are connected to a three-phase supply as shown in fig. 5.75 then three equal phase voltages will be available at the secondary terminals at no-load. This method of transforming three-phase powers by means of only two one-phase transformers is called **open delta** or **V-V** connections.

The basis of operation of open-delta connections is because of the fact that the vector sum of any two of the line voltages in a balanced three-phase system is equal to the third line voltage. Thus, even though one of the transformers has been removed, the voltage between the terminals of the secondary to which the load has been connected remains unchanged.

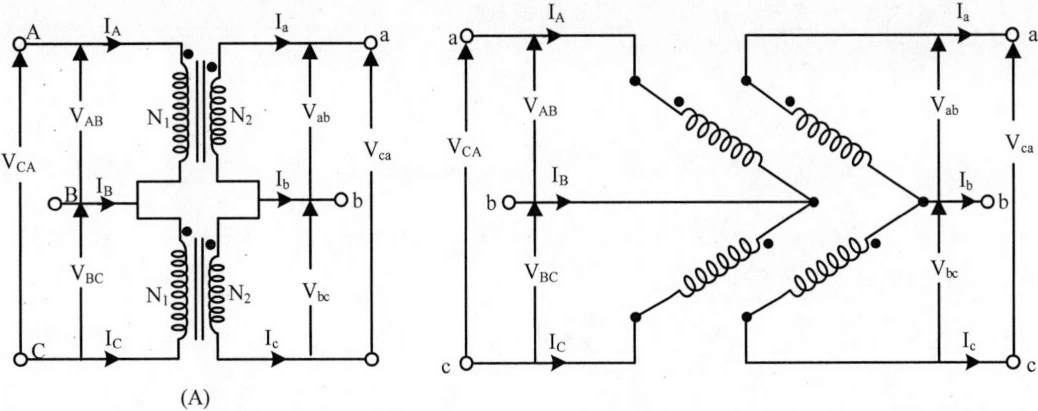

Fig. 5.75 Open delta or V-V connection (a) Common physical arrangement (b) Schematic diagram

Let V_{AB}, V_{BC} and V_{CA} be the applied voltage of the primary. The voltage induced in the transformer secondary or LV winding-I is V_{ab}. The voltage induced in the LV winding-II is V_{bc}. There is no winding between a and c, but there is a potential difference a and c. this voltage may be found by applying KVL around a closed path made up of points a,b, and c. Thus,

$$V_{ab} + V_{bc} + V_{ca} = 0 \tag{5.203}$$

$$V_{ca} = -V_{ab} - V_{bc} \tag{5.204}$$

Let $V_{AB} = V_{LP}\angle 0°$

$$V_{BC} = V_{LP}\angle -120°$$

$$V_{CA} = V_{LP}\angle +120°$$

Where, V_{LP} is the magnitude of line voltage on the primary side.

If the leakage impedance of the transformer is negligible, then

$$V_{ab} = V_{LS}\angle 0°$$

$$V_{bc} = V_{LS}\angle -120°$$

Where, V_{LS} is the magnitude of line voltage on the secondary side.

Substituting the value of V_{ab} and V_{bc} in Equation. (5.204)

$$V_{ca} = -V_{LS}\angle 0° - V_{LS}\angle -120°$$

$$= -V_{LS} - (0.5V_{LS} - j0.866V_{LS})$$

$$= -0.5V_{LS} + j0.866V_{LS}$$

$$V_{ca} = V_{LS}\angle -120° \tag{5.205}$$

It is seen that V_{ca} is equal in magnitude to the secondary transformer voltage and 120° apart in time from both of them. Thus balanced three-phase line voltage applied to the V-V primaries produces a balanced three-phase voltage on the secondary side if leakage impedance is negligible.

It is important to note that the total load that can be handled by *V-V* connections is not two-third of the capacity of a delta-delta bank but is only 57.7% of it. Mathematically, it is proved as below:

Power in the delta arrangement

$$= \sqrt{3}V_L I_L \cos\theta \tag{5.206}$$

$$= \sqrt{3}V_L \sqrt{3}I \cos\theta \quad \text{Where, } I_L = \sqrt{3}I \tag{5.207}$$

$$= 3V_L I \cos\theta$$

Where V_L is the line or phase voltage and $\cos\phi$ is the p.f.

Power in V-V connection $= \sqrt{3}V_L\, I \cos\theta$

$$\frac{\text{Power in V} - \text{V connection or Open delta}}{\textit{Power in closed delta}} = \frac{\sqrt{3}V_L\, I \cos\theta}{3V_L I \cos\theta} = \frac{1}{\sqrt{3}} = 0.577$$

$$= 57.7\%$$

Therefore, the output power of the open delta is $\frac{1}{\sqrt{3}}$ or 57.7% of the output of the closed delta.

The following are the points that favor the use of open-delta or vee-vee connections.

1. When the three-phase load is comparatively small the installation does not warrant a three-phase transformer bank.

2. When one of the transformers in a Δ-Δ bank fails, the service may be continued until the faulty transformer is repaired or a good one is substituted.
3. In a new installation, advantage is taken of the open-delta or V-V connections by installing initially only two transformers of the capacity to meet the present maximum demand. When the load on the system increases to the expected full load, a third transformer is added to close the delta.

Hence open-delta or V-V connections are used when it is anticipated that in future load will increase necessitating the closing of the open-delta at some later stage. The phasor diagram of an open delta or V-V connection is shown in fig. 5.76.

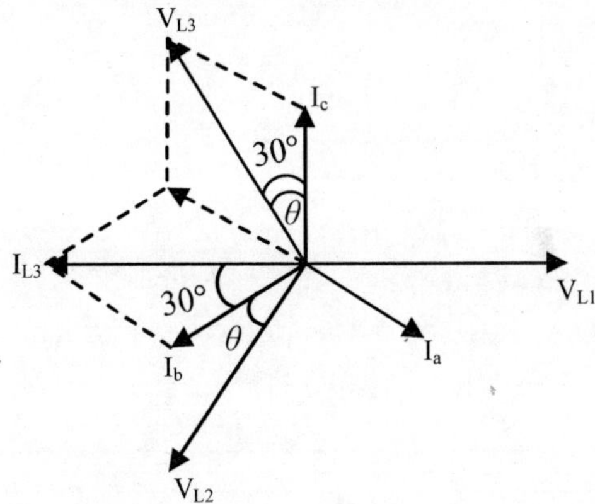

Fig. 5.76 Phasor diagram for open delta connection

Comparison of Delta and Open Delta Connections

1. The ratio of the power in V-V to Power in delta connection is 0.577.
2. When similar transformers are used, the voltages given by the delta and V-V connections are the same and their outputs are proportional to their line currents.
3. In balanced delta connection the line current is $\sqrt{3}$ of the phase current whereas in V-V-connection the line current is the same as the phase current.
4. With non-inductive balanced load, each transformer of the delta connection carries one-third of the total load at unity power factor. Under the same conditions each transformer of V-V-connection carries one half of the load at a p.f. 0.866.

Applications of Open-Delta System

The open-delta system is used in one of the following circumstances:

- As a temporary measure when one transformer of a Δ-Δ system is damaged and removed for repair and maintenance.
- To provide service in a new development area where the full growth of load may require several years. In such cases a V-V system is installed in the initial stage. This reduces the initial cost. Whenever the need arises at a future date to accommodate the growth in the power demand, a third transformer is added for Δ-Δ operation. The addition of one transformer increases the capacity of the total bank by 73.2%.
- To supply a combination of large single-phase and smaller three-phase loads.

Example 5.32: A balanced three-phase load of 1000 kW and 0.8 p.f. lagging is supplied by Vee-connected transformers. Calculate the line and phase currents and the power factor at which each transformer is working. The working voltage is 3.3 kV.

Solution:

Line current, $I_L = \dfrac{kW \times 1000}{\sqrt{3} \times 3.3 \times 1000 \times 0.8}$ 218.8 A

Phase current of the transformer is equal to the line current for Vee connections

$$I_1 = I_2 = I_{ph} = 218.8\ A$$

Load pf, $\cos\theta = 0.8$;

$$\phi = cos^{-1}0.8 = 36.87°$$

The current I_1 will make an angle of $\theta_1 = 30 + \theta = 30 + 36.87° = 66.87°$ with phase voltage of the respective transformer.

$cos\ \theta_1 = cos\ 66.87° = 0.393\ lagging\ (Ans)$

Current I_2 will make an angle of $\theta_2 = 30° - \theta = 30° - 36.87°$
$= -6.87°$ with phase voltage of the respective transformer

$cos\ \theta_2 = cos(-6.87°) = 0.993\ (Ans)$

Example 5.33: The primary and secondary windings of two transformers each rated 220 kVA, 11/2 kV and 50 Hz are connected in open delta. Find (i) the kVA load that can be supplied from this connection; (ii) currents on HV side if a delta connected three phase load of 250 kVA, 0.8 pf (lag) 2 kV is connected to the LV side of the connections.

Solution: -

(i) The kVA load that can be supplied by two transformers, each having rating of 220 kVA

$= 2 \times kVA\ rating\ of\ each\ transformer \times 0.866$

$= 2 \times 220 \times 0.866 = 381\ kVA\ (Ans)$

Secondary line current, $I_{L2} = \dfrac{kVA \times 1000}{\sqrt{3} \times V_{L2}}$

$$= \frac{220 \times 1000}{\sqrt{3} \times 2000} = 63.5\ A$$

Secondary phase current, I_{ph2} = Secondary line current, $I_{L2} = 63.5\ A$

(ii) Primary phase current (Current on HV side)

= Secondary phase current X Transformation ratio

$$= 63.5 \times \frac{2}{11} = 11.54\ A\ (Ans)$$

5.39.2 Scott connection or T-T connection

A Scott connection or *T* –*T* connection, which was originally proposed by Charles F. Scott shown in fig.5.77. This connection needs two transformers on each side instead of three transformers and accomplishes three-phase to three-phase transformations. The transformer, which is a horizontal member of the connection having center taps both on the primary and the secondary, is known as the main transformer. The other transformer of the primary and secondary whose one end is connected to the main transformer has a 0.866 tap and it is called the teaser transformer. The current ratings of the two transformers should be the same. The connections are shown in fig. 5.77.

It may be noted that one end of both the primary and secondary of the teaser transformer are connected to the center taps on both the primary and secondary of the main transformers respectively whereas, the two ends of the main transformer (*A* and *B*) and 86.6% tapping point (*C*) on teaser transformer are connected to the three-phase supply on the primary side. The two ends (*a* and *b*) of the secondary of the main transformer and 86.6% tapping point (*c*) of the secondary of the teaser transformer are taken to connect the load, as shown in fig. 5.77.

Since the primary and secondary of the teaser transformer are connected to the center tap of the primary and secondary of the main transformer respectively giving a shape of English letter '*T*', as shown in fig. 5.77 the connections are known as *T*-*T* connections.

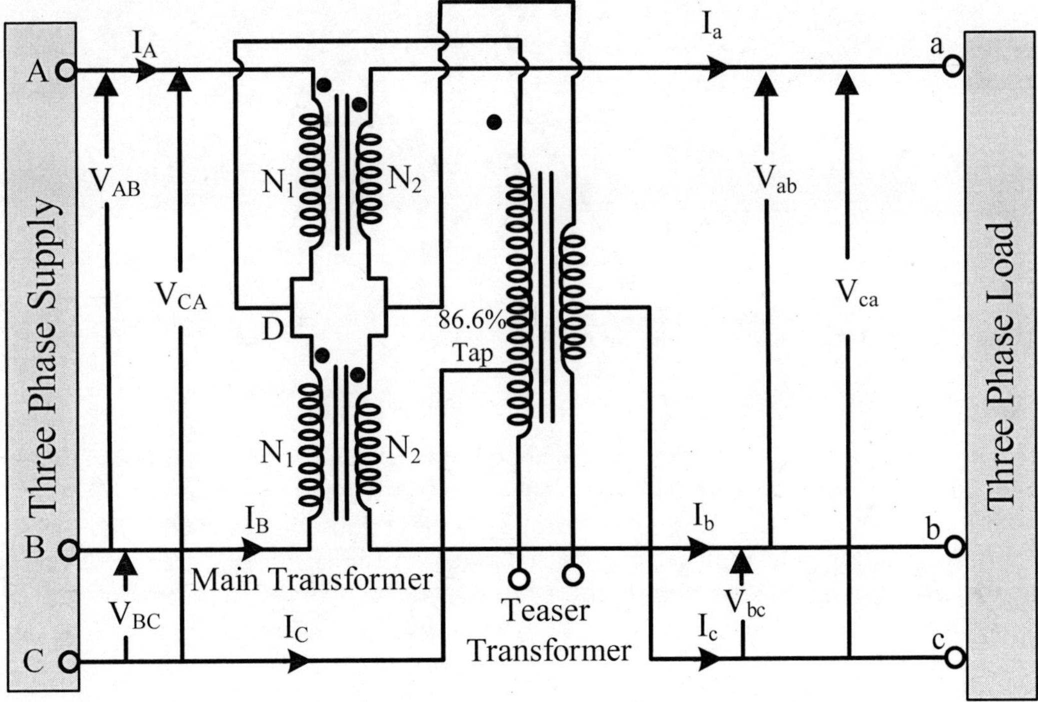

Fig.5.77 Scott connection or T-T connection

Let V_{AB}, V_{BC} and V_{CA} be the applied voltage of the primary. Then

$$V_{CD} = V_{CA} + V_{AD} \tag{5.208}$$

$$V_{CD} = V\angle 120° + V\angle 0°$$

$$= -(0.5V_{LS} - j0.866V_{LS}) + 0.5V_{LS}$$

$$= j0.866V_{LS}$$

$$V_{CD} = V\angle 90° \tag{5.209}$$

The above expression shows that the voltage applied across the teaser is 0.866 times that applied across the main and has a phase difference of 90°. If both the primary and secondary of the teaser transformer have 0.866 times the respective turns on the main transformer, then the induced voltages in the secondary circuit will have the same phase and magnitude relationship as that of applied voltages on the primary. Thus, the voltage induced in the teaser will be 0.866 times that of the induced voltage in the main transformer and has a phase difference of 90°. Consequently, a balanced three-phase system of voltages across points *a*, *b*, and *c* will be available.

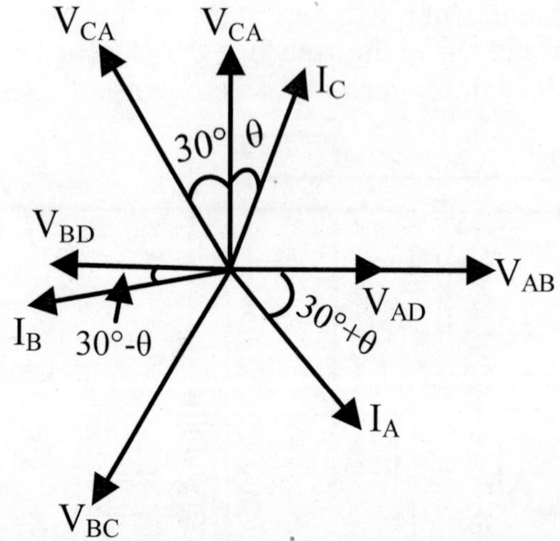

Fig. 5.78 Phasor diagram for T-T connection

The phasor diagram for T-T connection is shown in fig. 5.78. For any balanced load of power factor cos θ (lag), one of the two halves of the main transformer operates at a pf of cos (30° + θ) and the other at cos (30° – θ). This is similar to the condition as in open Δ. Mainly because of this (i.e., different loading effects) the voltages on the secondary side tend to become unbalanced to a greater extent with the increase in load. It may be noted that this arrangement provides the three-phase, four-wire system. Three-phase power loads may be connected between lines *a*, *b*, and *c* while lighting load may be connected between *ad* and *bd*. There is a further advantage of the availability of neutral on the teaser transformer. This permits a true three-phase four-wire system with the use of two transformers, which we could not get in open Δ system.

Example 5.34: Two single-phase Scott-connected transformers supply a three-phase four-wire 50 Hz distribution system with 400 V between lines. The hv windings are connected to a two-phase 6600 V (per phase) system. The core area is 200 cm2, while the maximum allowable flux density is 1.2 T. Determine the number of turns on each winding and the point to be tapped for the neutral wire on the three-phase side.

Solution: -

Maximum flux density,

$$B_m = 1.2\,T$$

Maximum allowable flux,

$$\phi_m = B_m \times core\ area$$

$$= 1.2 \text{ X } 0.02 = 0.024 \text{ Wb}$$

Voltage on the primary (HV) side,

$$V_1 = 6600\,V$$

Number of turns on the HV side of both the transformers,

$$N_1 = \frac{V_1}{4.44 f \phi_m}$$

$$= \frac{6600}{4.44 \times 50 \times 0.024}$$

$$= 1239\ (Ans)$$

Number of turns on the LV side of main transformers,

$$N_2 = \frac{V_2}{4.44 f \phi_m}$$

$$= \frac{400}{4.44 \times 50 \times 0.024}$$

$$= 75\ (Ans)$$

Number of turns on LV side of teaser transformer $0.866\ N_2$

$$= 0.866 \times 75 = 65\ (Ans)$$

Number of turns between CN

$$= \frac{2}{3} \times 65 = 43\ (Ans)$$

5.40 Three-Phase to Two-Phase Conversion

In some cases such as for electric furnaces, it is desirable to work with two-phase currents. From the power supply system, a three-phase AC supply is available, therefore, it is necessary to convert three-phase supply to two-phase supply. To convert three phases to two phases or vice versa, the connection is shown in fig. 5.79 is used, which is nothing but a Scott connection having two transformers of different ratings. If both transformers are identical, they should have suitable tappings.

The points s and r of the secondary are connected as shown in fig. 5.78(b), which gives the two-phase three-wire system. Here, $V_{sp} = 86.6\,V$ whereas $V_{sq} = 100\,V$, producing two unequal voltages. To get the same volt/turn both in the primary and the secondary, one line of three-phase supply is connected to point P_1, where SP_1 is 86.6 percent of the teaser primary turns as shown in fig. 5.79(a). The secondary voltages will be equal in magnitude if and only if the secondary transformers have the same number of turns, which results in a symmetrical two-phase three-wire system instead of an unsymmetrical two-phase three-wire system.

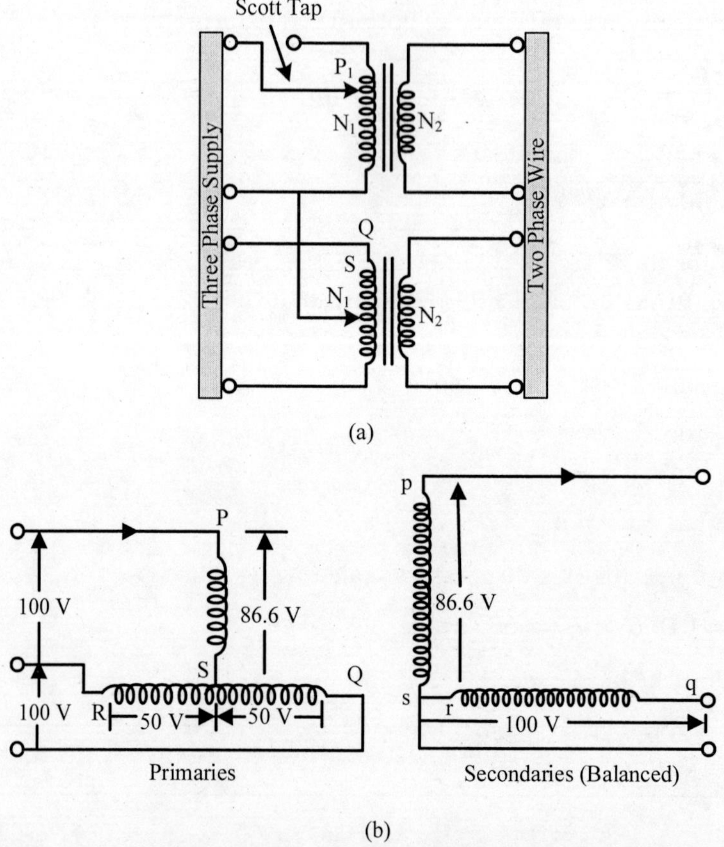

Fig. 5.79 Three-phase to two-phase conversion

Fig. 5.80 shows a different connection where the primary has N_1 number of turns and is connected between Q and R of the main transformer. Now, $V_{PQ} = V_{QR} = V_{RS} = V_1$ if the supply line voltage is V_1 and $V_{PS} = \frac{\sqrt{3}}{2} V_1$.

The number of turns between P and S is $\frac{\sqrt{3}}{2} N_1$ to give the same volt/turn in both primaries. The secondary terminal voltages will be equal in magnitude if the secondaries have an equal number of turns. But the secondary terminal voltages will be quadrature to each other. Since $V_{PS} = \frac{\sqrt{3}}{2} V_1$ and it is not equal to $\frac{V_1}{\sqrt{3}}$, S is not the neutral point. To get the position of the neutral point, let us take N as the neutral point. N will be the neutral point only if $V_{PN} = \frac{V}{\sqrt{3}}$. Now $V_{NS} = \frac{\sqrt{3}}{2} V_1 - \frac{V_1}{\sqrt{3}} = 0.288\ V_1$. This means that N is above S by a number of turns equal to 28.8 percent of N_1. N divides the teaser winding in the ratio of 2:1 because $\frac{0.866}{3} = 0.288$.

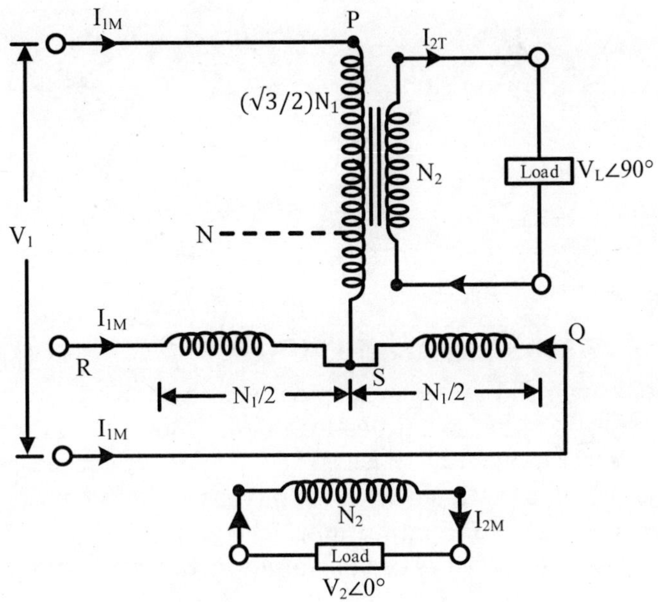

Fig. 5.80 A different connection

Let the current supplied by the teaser secondary be I_2T at unity power factor. The teaser primary current becomes

$$I_{IT} = I_{2T} \times \frac{N_2}{\frac{\sqrt{3}N_1}{2}} = 1.15 \left(\frac{N_2}{N_1}\right) I_{2T} = 1.15\, K\, I_{IT} \qquad (5.210)$$

where $K = \frac{N_2}{N_1}$ is the turns ratio. The current I_{M1} of each half of the main transformer of the primary has the following two components:

- One part is required to balance the secondary current (I_{2M}), which is $I_{2M} \times \frac{N_2}{N_1} = KI_{2M}$.
- The other part is equal to one half of the teaser primary current or (1/2) I_1T because the main transformer primary serves as the return path for teaser primary current and it divides itself into two halves at the mid-point *S*.

The condition due to a balanced two-phase load at a lagging power factor of 0.866 has been shown in fig. 5.81(a). The three-phase side is balanced here. The main transformer rating is 15 percent greater than the rating of the teaser transformer. This is because its voltage is 15 percent greater. But its current remains the same. Fig.5.81(b) shows the condition due to an unbalanced two-phase load having different currents and power factors.

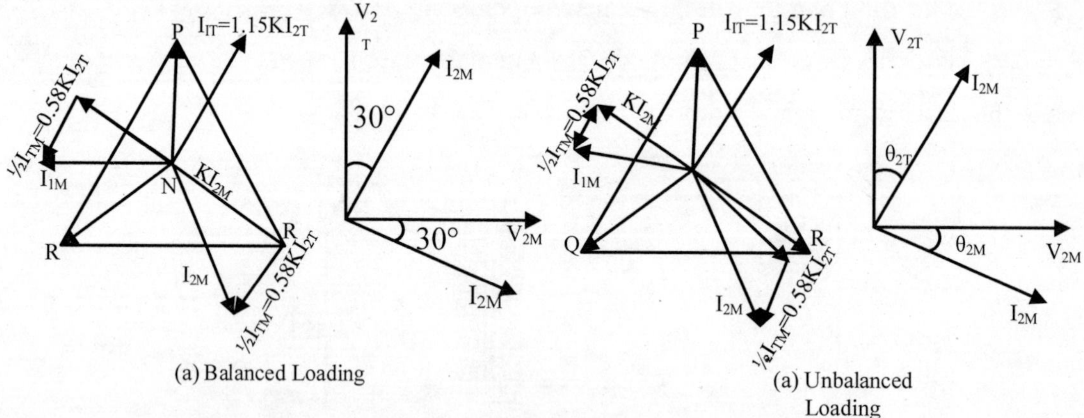

Fig. 5.81 Vector diagram for Balanced and Unbalanced loading

5.41 Three-Phase to Six-Phase Conversion

Rectifiers convert AC power to DC power. A smoother waveform is obtained on the DC side as the number of phases is increased. Objectionable harmonics in alternating currents are also reduced with a greater number of phases. The efficiency of conversion from AC to DC by rectifier and thyristor circuits increases with the increase in number of phases. Six phase is, therefore, preferable to 3-phase for rectification. Since 12-phase units are more complex and costlier than 6-phase units, 12-phase is used in larger installation units.

The following are commonly used schemes for three-phase to six-phase conversion:

1) Double-star connection
2) Double-delta connection
3) Six-phase star connection
4) Diametrical connection

5.41.1 Double-Star Connection

A double-star connection of transformers for three-phase to six-phase conversion, where three identical single-phase transformers are used shown in fig. 5.82. The three primaries are connected in delta, whereas each transformer in the secondary unit splits into two equal connections. Each set of secondaries is connected in star.

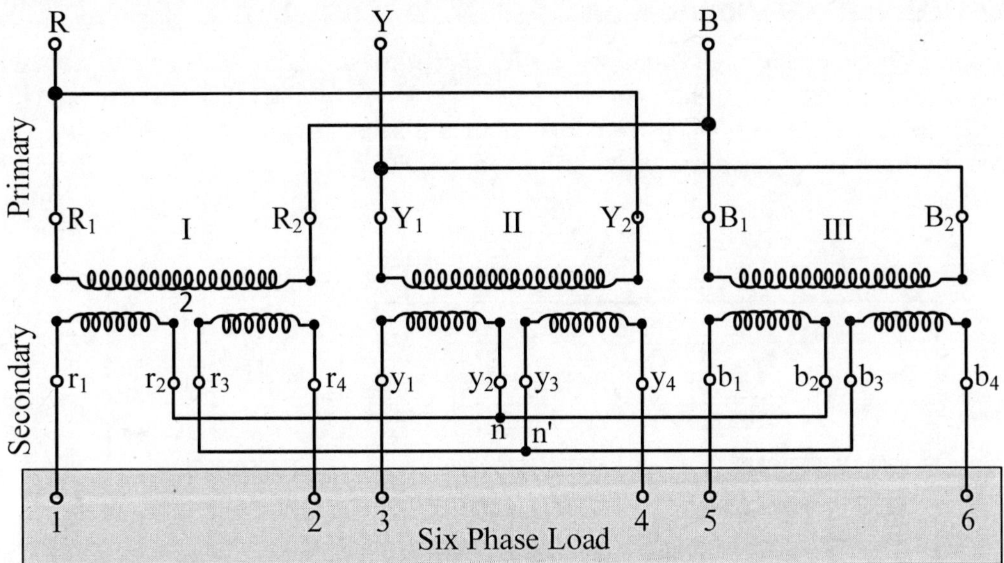

Fig. 5.82 Double-Star Connection

The output terminals for the first star connection are $r_1y_1b_1$ with $r_2y_2b_2$ connected together to form the neutral terminal n. The output terminals for the second star connection (or reversed star connection) are $r_4y_4b_4$ with $r_3y_3b_3$ connected together to form neutral n_1. The connection diagram of the double star connection for 3-to-6 phase transformation is shown in fig. 5.81. Here, the terminals $r_1y_1b_1$ and $r_4y_4b_4$ are connected to the 6-phase load terminals 1, 3, 5, 4, 6, 2 respectively. The two neutral terminals n and n_1 may be connected together. This neutral point serves as the neutral point of the DC supply from the rectifier. In such a way, a true 6-phase star-connected system with a neutral terminal is obtained.

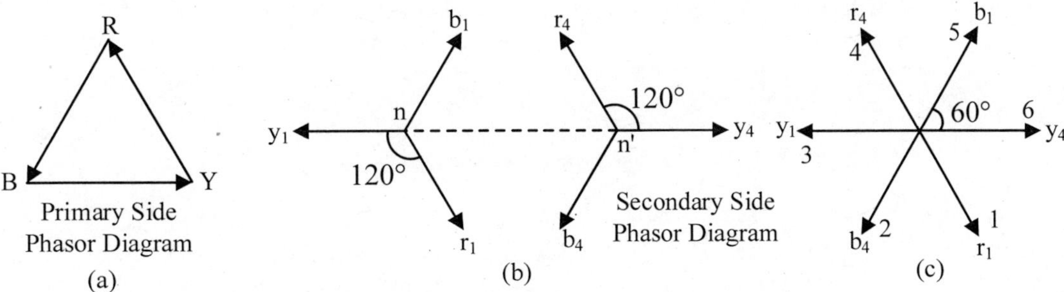

Fig. 5.83 Phasor Diagram (a) Primary side (b) Secondary side (c) Six-phase star connection

The phasor diagram of the primary side and the secondary side is shown in fig. 5.83(a) and fig. 5.83(b). Now, in order to determine the connections of the transformer terminals to the load terminals, the transformer terminal is marked 1 as shown in fig.5.83(c). The other terminals are marked 2, 3, 4, 5, 6 in the clockwise direction. In the double star connection, six voltages are obtained which are displaced by 60° from each other.

5.41.2 Double-Delta Connection

The double-delta connection of transformers for three-phase to six-phase conversion, where three identical single-phase transformers are used shown in fig. 5.84. The three primaries are connected in delta, whereas each transformer in the secondary unit splits into two equal connections. Each set of secondaries is connected in delta.

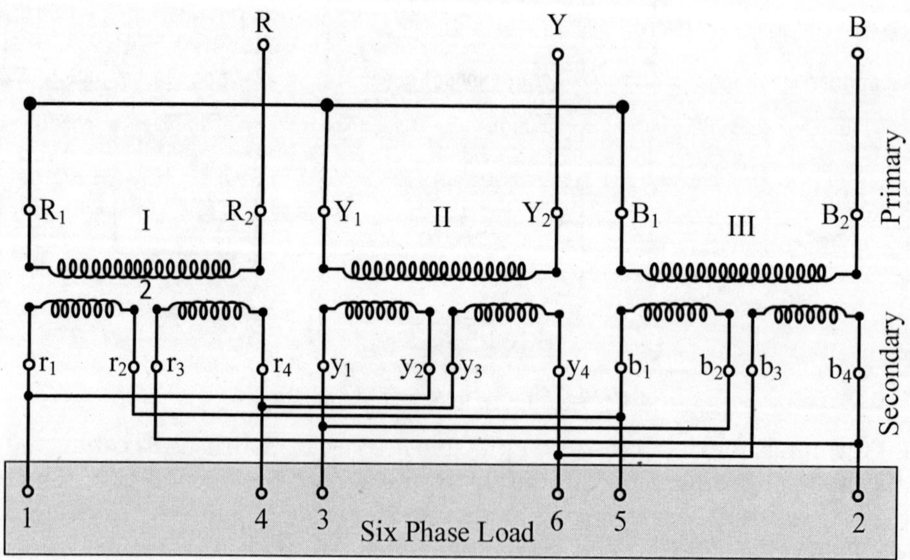

Fig. 5.84 Double-delta connection

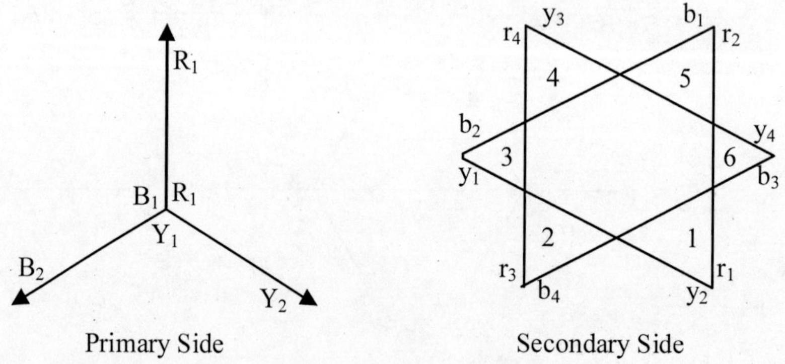

Fig. 5.85 Phasor Diagram

In the double delta connection of 3-to-6 phase transformation, three identical single-phase transformers T_1, T_2 and T_3 are used. The primary windings of the three transformers can be connected in *star* because the two secondary windings are connected in delta and provide the required path for third harmonic currents. Each transformer has its secondary winding split into two equal sections. One set of three secondary windings r_1r_2, y_1y_2 and b_1b_2 being connected in delta by connecting starting end r_1 of the first winding to the finishing end y_2 of the second winding, starting end y_1 of the second winding to the finishing end b_2 of the third winding and the starting end b_1 of the third winding to the finishing end r_2 of the first winding. The other set of three secondary windings r_3r_4, y_3y_4 and b_3b_4 being connected in *reverse* delta by connecting finishing end r_4 of the first winding to the starting end y_3 of the second winding, finishing end y_4 of the second winding to the starting end b_3 of the third winding and the finishing end b_4 of the third winding to the starting end r_3 of the first winding. The phasor diagrams of the primary and secondary windings are shown in fig. 5.85.

The main advantage of the double-delta connection is that, it has good harmonic elimination. But, the main problem associated with the double-delta connection is, it is not suitable for rectifier circuits, due to the absence of the secondary neutral. Therefore, a true 6-phase supply is obtained when the six terminals are connected to a 6-phase load.

5.41.3 Six-Phase Star Connection

The six-phase star connection of transformers for three-phase to six-phase conversion, where three identical single-phase transformers are used shown in fig. 5.86. The three primaries are connected in delta and the center tap of each secondary transformer is connected to neutral.

Here, the secondary windings have center taps which are joined together to form the neutral on the 6-phase side. The primary windings are connected in delta, but these may also be connected in star. The six-phase load terminals 1, 2, 3, 4, 5, 6 are connected to the six secondary winding terminals $r_1, b_2, y_1, r_2, b_1, y_2$ of the transformer respectively.The phasor diagrams of the primary and secondary windings are shown in fig. 5.87.

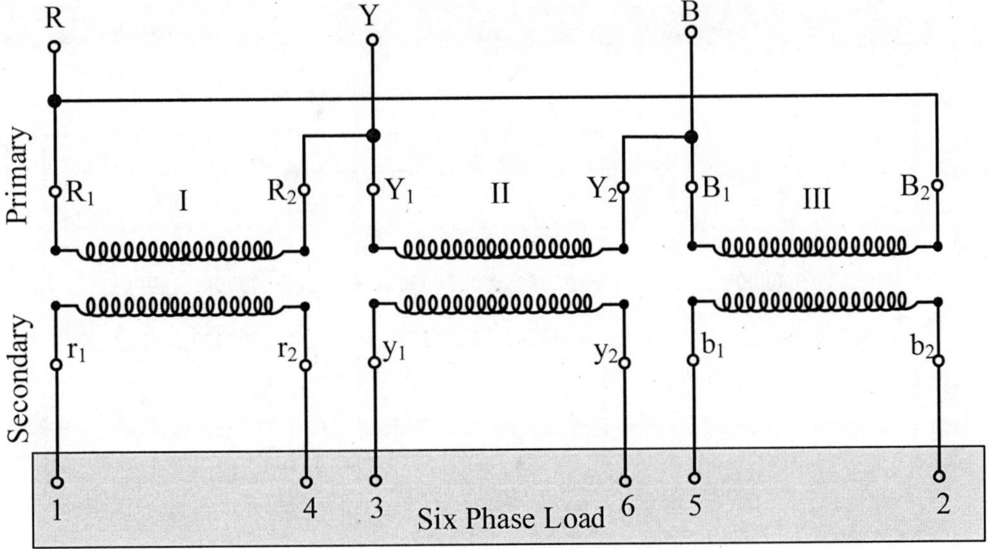

Fig. 5.86 Six-phase star connection

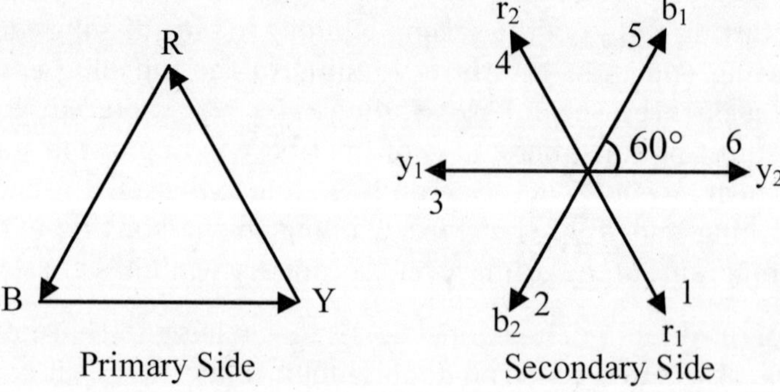

Fig. 5.87 Phasor diagram

5.41.4 Diametrical Connection

Fig 5.88 shows a diametrical connection. No center tapping is used here. Neutral connections are not required. In the case of a diametrical connection, no center tappings are required i.e. no neutral terminal is taken out. A true 6-phase supply is only obtained using the diametrical connection when the six secondary terminals are connected to the terminals of the suitable 6-phase load.

Here, the transformer terminals r_1, r_2, y_1, y_2, b_1, b_2 are connected to the load terminals 1, 4, 3, 6, 5, 2 respectively. Since the neutral terminal is not present, hence, the diametrical connection cannot be used for the rectifier circuits. The phasor diagram of the diametrical connection is shown in fig. 5.89.

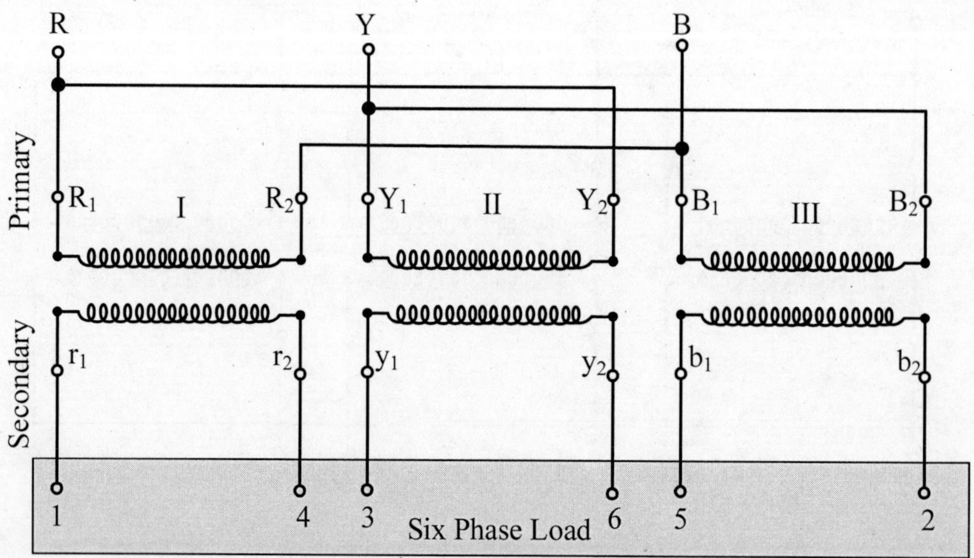

Fig.5.88 Diametrical Connection

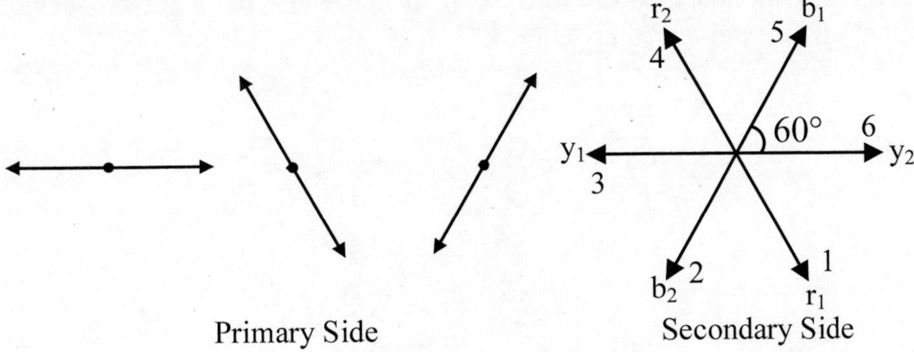

Fig. 5.89 Phasor diagram

5.42 Three-Winding Transformer

Transformers may be constructed with a third winding in addition to the primary and secondary. The third winding is called tertiary. The primary has the highest voltage rating, the tertiary has the lowest voltage rating, and the secondary has the intermediate voltage rating. In two-winding transformers, kVA ratings of both *HV* and *LV* windings are equal, but in 3-winding transformers, kVA ratings of the three windings are usually unequal.

The tertiary winding is connected in delta. The main advantage of using tertiary windings is that the delta connection suppresses any harmonic voltages that may be generated in star-connected primaries and secondaries of transformers. In case of unbalanced secondary load currents, which are reflected as unbalanced primary currents, an increased circulating current is reduced in the tertiary windings. This tends to restore both primary and secondary phase voltages to their normal phase magnitudes and angles. Thus, the secondary and primary unbalanced is reduced and the secondary load unbalance is more evenly distributed among primary phases.

Tertiary windings are also used for the following purposes:

- Tertiary windings are used to supply substation auxiliaries (for example, lights, fans and pumps) at a voltage different from those of primary and secondary windings.
- Synchronous capacitors or static high-voltage capacitors are connected across the delta-connected output of the tertiary windings for reactive power injection into the system for either power factor correction or voltage regulation or both.
- Tertiary windings are used to interconnect three supply systems operating at different voltages.
- A delta-connected tertiary reduces the impedance offered to the zero-sequence currents to allow sufficient earth fault current for the proper operation of protective devices.
- Tertiary can be used for measuring the voltage of high-voltage testing transformers.
- Supply of a single load from two sources, where the continuity of supply is important.

It is to be noted that the unbalance and third harmonic problems do not arise when one or both sets of windings are connected in delta.

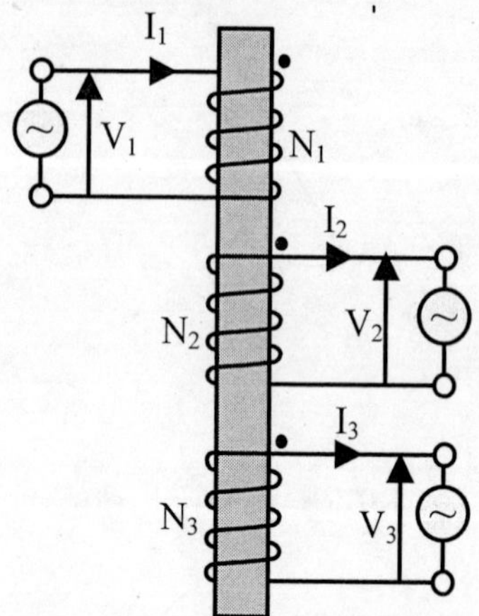

Fig. 5.90 Schematic diagram of the winding of a 3-winding transformer

Tertiary winding transformers are currently being manufactured with tertiary VA ratings up to 35 percent of the total VA rating of the transformer. The advantages of a 3-winding transformer are economy of construction and efficiency.

Fig. 5.90 shows a schematic diagram of the windings in a 3-winding transformer. The subscripts 1,2 and 3 are used to indicate primary, secondary and tertiary windings respectively.

For an ideal 3-winding transformer

$$\frac{V_2}{V_1} = \frac{N_2}{N_1} \tag{5.211}$$

$$\frac{V_3}{V_1} = \frac{N_3}{N_1} \tag{5.212}$$

And $$I_1 N_1 = I_2 N_2 + I_3 N_3 \tag{5.213}$$

Equivalent circuit of a 3-winding transformer

The equivalent circuit of a 3-winding transformer can be represented by the single-phase equivalent circuit. Each winding is represented by its equivalent resistance and reactance. Fig. 5.91 shows the equivalent circuit referred to as primary. Here terminals 1, 2, and 3 indicate primary, secondary, and tertiary terminals respectively.R_1, R_2 and R_3 are the resistances and X_1, X_2 and X_3 are the leakage reactances of primary, secondary, and tertiary windings respectively. If the exciting current is considered, then R_0 and X_0, can be connected as shown in fig. 5.91. Three external circuits are connected between terminals 1, 2, 3, and common terminal 0. Let V_1, V_2 and V_3 be the voltages and I_1, I_2 and I_3 be the currents of primary, secondary, and tertiary windings.

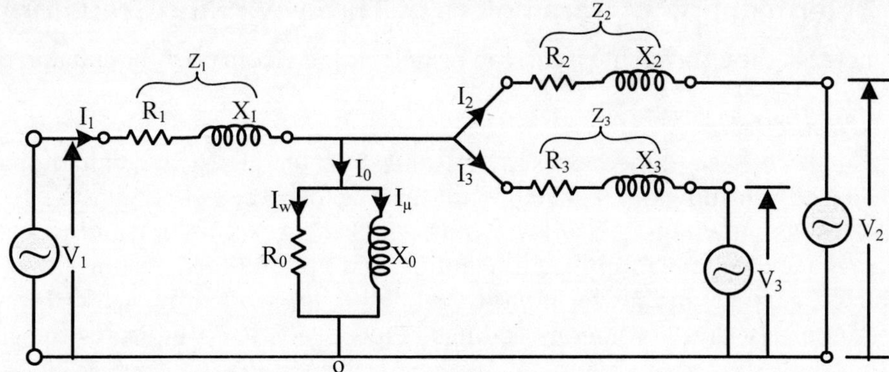

Fig. 5.91 Single phase equivalent circuit of a three-winding referred to as primary

Fig. 5.92 shows the equivalent circuit in per unit where the resistance and reactance have been converted to per unit values on the basis of common VA rating base and respective base voltages.

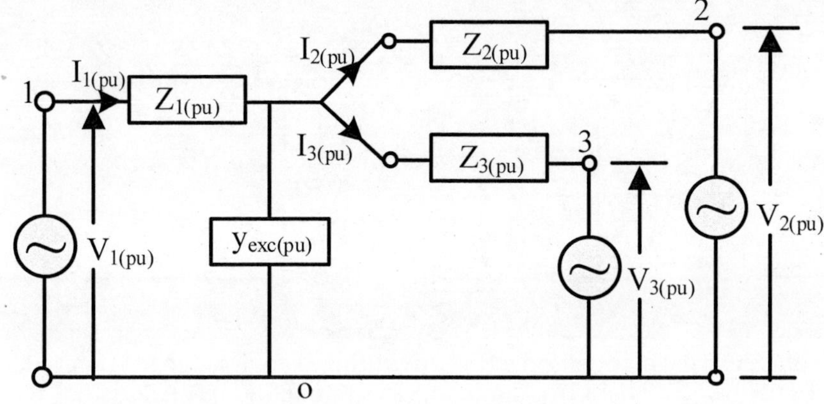

Fig. 5.92 Single phase equivalent circuit of a three-winding transformer in per unit

The equivalent circuit neglecting magnetizing admittance y_{exc} is shown in fig. 5.93.

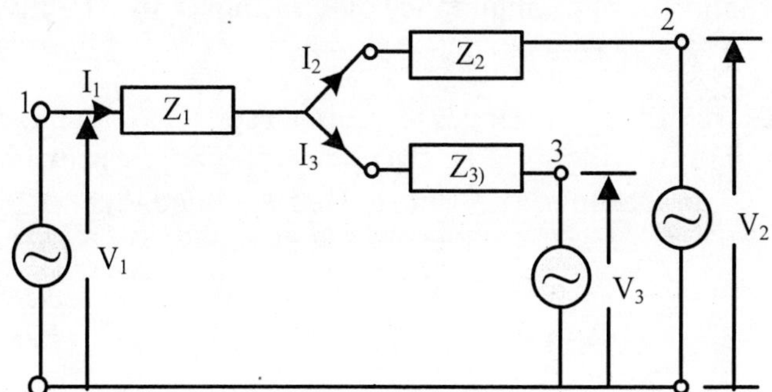

Fig. 5.93 Equivalent circuit neglecting magnetizing admittance

5.42.1 Determination of Parameters of Three-Winding Transformers

The parameter of the equivalent circuit can be determined from open-circuit and three short-circuit tests.

A. Short-Circuit Tests

The equivalent leakage impedances Z_1, Z_2 and Z_3 referred to a common base can be determined by performing three short-circuit tests as follows:

In the first test as shown in fig. 5.94(a), winding 2 is short-circuited, winding 3 is kept open circuited and a low voltage is applied to winding 1 so that full-load current flows in winding 2. The voltage, current, and power input to winding 1 are measured. Let V_1, I_1 and P_1 be the voltmeter, ammeter, and wattmeter readings respectively. If Z_{12} indicates the short-circuit impedance of windings 1 and 2 with windings 3 open, then $Z_{12} = \frac{V_1}{I_1}$. Equivalent resistance $R_{12} = \frac{P_1}{I_1^2}$, and equivalent leakage reactance $X_{12} = \sqrt{Z_{12}^2 - R_{12}^2}$. It is seen from the equivalent circuit of fig. 5.94(b) that Z_{12} is a series combination of Z_1 and Z_2.

$$Z_{12} = R_{12} + jX_{12} = Z_1 + Z_2 \tag{5.214}$$

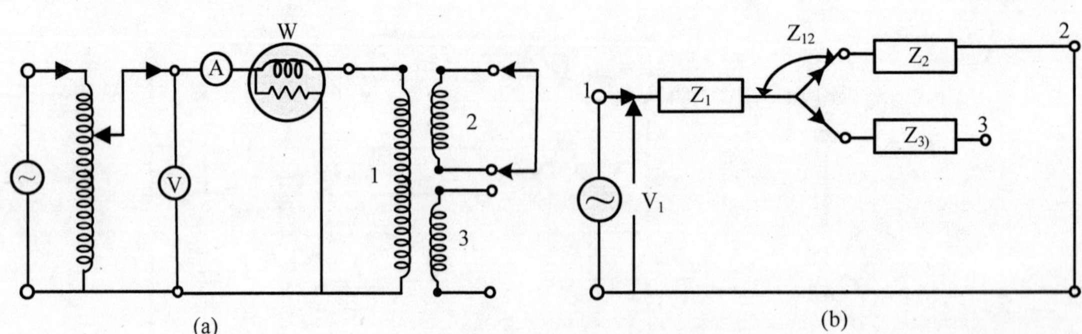

Fig.5.94 Short-circuit test on a three-winding transformer (a) Connection diagram (b) Single phase equivalent circuit

In the second short-circuit test, winding 3 is short-circuited and winding 2 is kept open. A low voltage is applied to winding 1 to circulate full-load current in winding 3. If Z_{13} represents the short-circuit impedance of windings 1 and 3 with winding 2 left open

$$Z_{13} = Z_1 + Z_3 \tag{5.215}$$

In the third short-circuit test winding 3 is short-circuited and winding 1 is kept open. A low voltage is applied to winding 2 to circulate full-load current in the short-circuited winding 3. If Z_{23} represents the short-circuit impedance of windings 2 and 3 with winding 1 open

$$Z_{23} = Z_2 + Z_3 \tag{5.216}$$

All the impedances are referred to a common base. Solving equations (5.214), (5.215) and (5.216) we get the leakage impedances Z_1, Z_2 and Z_3 all referred to primary as

$$Z_1 = \frac{1}{2}(Z_{12} + Z_{13} - Z_{23}) \tag{5.217}$$

$$Z_2 = \frac{1}{2}(Z_{23} + Z_{12} - Z_{13}) \tag{5.218}$$

$$Z_3 = \frac{1}{2}(Z_{13} + Z_{23} - Z_{12}) \tag{5.219}$$

Where,

$$Z_1 = R_1 + jX_1, \quad Z_2 = R_2 + jX_2, \quad Z_3 = R_3 + jX_3$$

It is to be noted that impedance Z_{12} and Z_{13} are referred to winding 1, because the instruments are connected in winding1. The impedance Z_{23} is referred to winding 2. Therefore, it must be referred to winding 1 $\left[= Z_{23}\left(\frac{T_1}{T_2}\right)^2\right]$ and only then the equations (5.214), (5.215) and (5.216) are used for calculating Z_1, Z_2 and Z_3. Alternatively, Z_{12}, Z_{13} and Z_{23} should be expressed in per unit and then equations (5.217), (5.218) and (5.219) are used to determine Z_1, Z_2 and Z_3.

B. *Open-Circuit Test*

The open-circuit test is made in the same manner as that for a 2-winding transformer and it gives the data for calculating the core loss. magnetizing impedance and turns ratios. Thus, magnetizing impedance may be found by exciting winding 1 with both windings 2 and 3 on open circuit. Then we have:

$$a_{12} = \frac{V_1}{V_2}, \quad a_{13} = \frac{V_1}{V_3} \tag{5.220}$$

$$a_{23} = \frac{V_2}{V_3} = \frac{V_2/V_1}{V_3/V_1} = \frac{a_{13}}{a_{12}} \tag{5.221}$$

5.43 Tap Changing Transformer

All the electrical equipment connected at the consumer's end is designed to operate satisfactorily at a particular voltage level. Therefore, it is essential to supply electrical energy to the consumers at a level that must fall within the prescribed limits. However, in the power system, due to change in load (may be seasonal or otherwise), the transformer output voltage on the consumer's terminal may change beyond the permissible limits. This can be controlled by providing tap-changing transformers. Taps can be provided either on the primary or on the secondary.

The secondary output voltage can be regulated by either changing the number of turns in the primary or secondary on the basis of following principle.

Let V_1, N_1 and V_2, N_2 be primary and secondary quantities respectively. If N_1 is decreased, the emf per turn on primary $= \left(\frac{V_1}{N_1}\right)$ increases which increases the secondary terminal voltage

$\left(V_2 = \frac{V_1}{N_1} \times N_2\right)$. On the other hand, if N_2 is increased keeping N_1 constant, the secondary terminal voltage ($V_2 \propto N_2$) still increases.

Thus, the secondary terminal voltage can be increased either by decreasing the primary turns or by increasing the secondary turns and vice-versa.

The choice between primary and secondary to provide taps to regulate the output terminal voltage of the transformer

While selecting the side to provide taps to regulate the secondary output voltage, we always try to maintain voltage per turn, as far as possible, constant. If the primary voltage per turn decreases, the core flux decreases which results in poor utilization of the core, although it reduces the core losses. On the other hand, if the primary voltage per turn increases, the core flux increases which results in magnetic saturation of the core. It also increases the core losses.

In the transformers located at the generating stations, the primary voltage has to be kept constant, consequently, the taps are provided on the secondary side. However, if transformers are energized from a variable source, as at the receiving end of a transmission line (receiving sub-stations), the taps are usually provided on the primary side.

The other factors which may also be taken into consideration are given below:

- Transformers with large turns ratio, are tapped on the *HV* side since this enables a smoother control of the output voltage. If in such transformers, the taps are provided on the *LV* side, it varies the output voltage to a large step which is usually undesirable.
- Tap-changing gear provided on the *HV* side are to handle low currents, although more insulation has to be provided.
- It is difficult to tap the *LV* winding, since it is placed next to the core. Whereas, the *HV* winding, placed outside the *LV* winding, is easily accessible and can, therefore, be tapped more easily.

5.43.1 Types of Tap-changers

There are two types of tap-changers, called

a. No-load (or off-load) tap-changers
b. On load tap-changers

The working of these tap-changers is explained below with the help of schematic diagrams.

a. No-load (or off-load) tap changer

These tap-changers are used for seasonal voltage variations. The schematic diagram of a no-load tap-changer is shown in fig. 5.95. A winding is tapped and its leads are connected to six studs marked 1 to 6. The studs are stationary and are arranged in a circle. The face plate carrying the studs can be mounted anywhere on the transformer, say on the yoke or on any other convenient place (say a separate box). The rotatable arm '*A*' is attached to a hand wheel which is kept outside the tank and can be rotated.

Sometimes, in case of large transformers, the rotatable arm is rotated with the help of a motor (With gear drive) and the controls of the motor are placed on the panel board although hand wheel is also provided for manual operation.

The active number of turns of the winding which remain in the circuit depends upon the position of the rotating arm. If the winding is tapped at 2% intervals, then at various positions of the rotatable arm '*A*', the winding in the circuit will be as under:

a) at studs 1–2: Full winding is in the circuit.
b) at studs 2–3: 98% of the winding is in the circuit.
c) at studs 3–4: 96% of the winding is in the circuit.
d) at studs 4–5: 94% of the winding is in the circuit.
e) at studs 5–6: 92% of the winding is in the circuit.

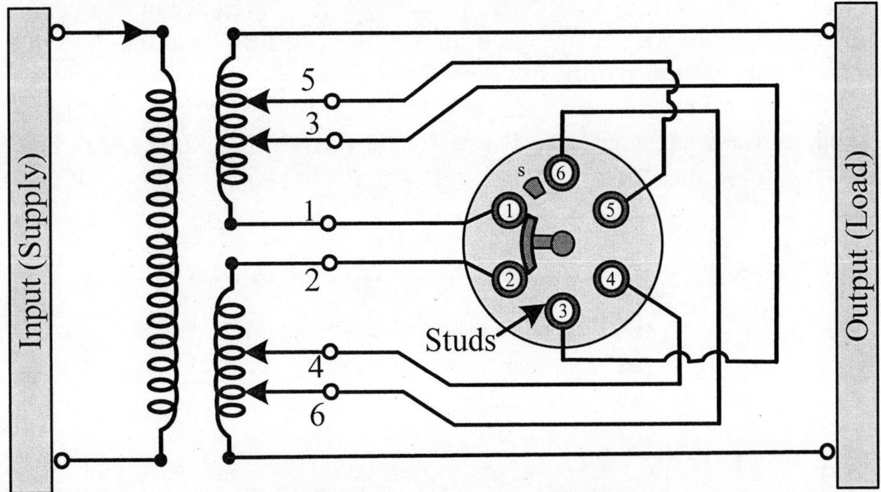

Fig. 5.95 OFF-load tap-changer

A stopper *S* is placed in between the stud 1 and 6. It fixes the final position of the arm '*A*' at stud 5–6. Moreover, it prevents the clockwise rotation of the arm '*A*' from stud-1. It prevents the connections of stud 1 and 6 through arm '*A*'. If stud 1 and 6 are connected then only the lower part of the winding is cut out of circuit which is undesirable from mechanical-stress considerations.

In this case, the tap-changing is carried out only after the transformer is disconnected from the supply. For instance, let the arm '*A*' is at stub 1 and 2 and the whole winding is in the circuit. Now, if we want to reduce the winding to 96%, we have to rotate the arm in an anticlockwise direction to bridge stud 3 and 4 through arm '*A*'. While doing so, the transformer is disconnected from the supply, arm is rotated to the desired position and then the transformer is energized.

This tap-changer is never operated on load. If it would be operated on load, there would be heavy sparking at the studs when arm '*A*' is separated from them. It may damage the tap-changer and the transformer winding.

b. *On-load tap-changer*

This tap-changer is used for daily or short-period voltage regulations. The output voltage can be regulated with the help of this tap-changer without any supply interruptions. During the operation of an on-load tap-changer, the following points must be kept in mind;

- Never open the main circuit during the operation of the tap-changer otherwise dangerous sparking will occur, and
- No part of the tapped winding should get short-circuited.

One form of an on-load tap-changer provided with a center-tapped reactor is shown in fig. 5.96. The function of the reactor is to prevent the short-circuiting of the tapped winding. The switches 1,2, 3, 4 and 5 are connected to the correspondingly marked taps.

During normal operation switch *S* is closed, switches 2, 3, 4 and 5 are opened and switch-1 is closed as shown in fig.5.96. The entire winding is in the circuit. The two halves of the reactor carry half of the total current in opposite directions. Since the whole reactor is wound in the same direction, the mmf produced by the two halves is opposite to each other. Since these mmfs are equal, therefore, the net mmf is practically zero. Hence, the reactor is almost non-inductive and the impedance offered by it is very small, consequently, the voltage drop in the center-tapped reactor is negligible.

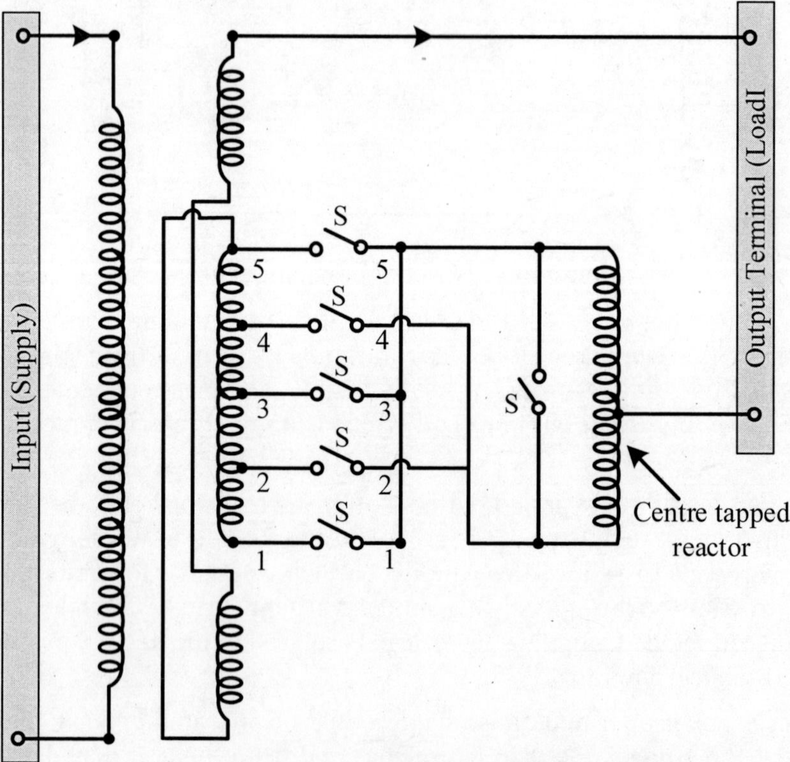

Fig. 5.96 On-load tap-changer

When a change in voltage is required, the following sequence of operations is adopted.

- **Open the switch S:** By opening this switch, total current flows through the upper half of the reactor and there is more voltage drop. Since reactor is to carry full load current momentarily, the reactor must be designed accordingly.
- **Close the switch 2:** When switch 2 is closed, the winding between taps 1 and 2 is connected across the reactor. Since the impedance offered by the reactor is high for a current flowing in only one direction, the local circulating current flowing through the reactor and tapped winding is quite small. Thus, reactor prevents the tapped winding from getting short circuited. The terminal voltage at this instant is mid-way between the potentials of tappings 1 and 2.
- **Open the switch I:** The entire current now flows through the lower half of the reactor which causes more voltage drop.
- **Close the switch S:** By closing the switch, current is divided equally in the upper and lower part of the reactor which causes almost negligible voltage drop. The same sequence of operations is repeated if the tapping is to be changed from stud 2 to 3.

Off line tap changers are preferred for distribution transformer for seasonal correction of output voltage. On load tap changing is used in power system. On load changers are also in use in electrical traction locomotives. A locomotive, faces 25 KV near feeder, as the locomotive is mid way between two feeders, the line voltage may even drop to about 17 KV. Real time, on line tap changing is required to correct the voltage on board of the locomotive.

5.44 Cooling of Transformer

When a transformer is connected to the mains, heat is produced in the transformer due to various losses. This increases its temperature. In fact, output of the transformer is limited by the rise in temperature. Therefore, some means are provided to cool down the transformer so that the rise in temperature may not go beyond the permissible limits.

According to specifications, the rise in temperature of a transformer when working at the rated output shall not exceed 45°C to 60°C.

Most of the electrical machines have rotating part, therefore, it is easy to cool down such machines by providing fan on their shaft and ducts in their construction. But in case of transformers there is no rotating part, therefore, it is difficult to cool down the inner parts of a transformer. However, depending upon the size of a transformer, various methods have been evolved to dissipate heat produced in the transformer. These are mentioned below:

1) **Natural Cooling**
    a. Air Natural cooling (*AN*)
    b. Oil immersed Natural cooling (*ON*)
    c. Oil immersed Forced oil circulation Natural cooling (*OFN*)
2) **Artificial Cooling (Air)**
    a. Oil immersed Forced oil circulation with air Blast cooling (*OFB*)
    b. Oil immersed air Blast cooling (*OB*)
    c. Air Blast cooling (*AB*)

3) **Artificial Cooling (Water)**
    a. Oil immersed Water cooling (*OW*)
    b. Oil immersed Forced oil circulation with Water cooling (*OFW*)
4) **Mixed Cooling**
    a. Oil immersed Natural cooling with alternative additional air Blast cooling (*ON/OB*)
    b. Oil-immersed Natural cooling with alternative additional Forced oil circulation (*ON/OFN*)
    c. Oil-immersed Natural cooling with alternative additional Forced oil circulation air Blast cooling (*ON/OFB*)
    d. Oil-immersed Natural cooling with alternative additional forced oil circulation and Water cooling (*ON/OFW*)

5.44.1 Methods of transformer cooling

Some of the common methods of transformer cooling are explained below:

1. **Air Natural Cooling (AN)**
   Small transformers of a few kVA rating are cooled by the Natural air surrounding the core and winding of the transformer. The heat produced in the transformer due to losses dissipated by conduction, convection and radiation.
   The transformers cooled by this method are used in laboratories and small appliances like radios, rectifiers etc.

2. **Oil Immersed Natural Cooling (ON)**
   In this case, the assembly of core and winding (assembled unit) of a transformer is placed in a steel tank filled with pure mineral/insulating oil. The cooling is affected by natural circulation (convection currents) of the oil through the cooling ducts provided between *coils* and between coils and core.
   For small transformers, the tanks are usually smooth surfaced but for large sizes the tanks are made of corrugated sheets. Sometimes fins are attached to improve cooling. In most of the pole mounted transformers up to the rating of 200 kVA number of round or elliptical tubes are provided with the tank externally. The transformer oil has large coefficient of expansion; therefore, convection currents develop which cause the oil to circulate through the external tubes. The cooling can further be improved by using a radiator in place of external cooling tubes.

3. **Oil Immersed Forced oil-circulation Natural Cooling (OFN)**
   In this type, core and coils are immersed in oil and the cooling is affected basically by forced oil circulation through motor driven oil pump as shown in fig. 5.97. This method of cooling is one of the latest to be adopted and is employed with the radiators separately attached with the main tank of the transformer. A pump forces the oil through the ducts provided between winding and back to the radiators. The transformer tank in this case is also made plain, the radiators being relied upon to give necessary cooling.

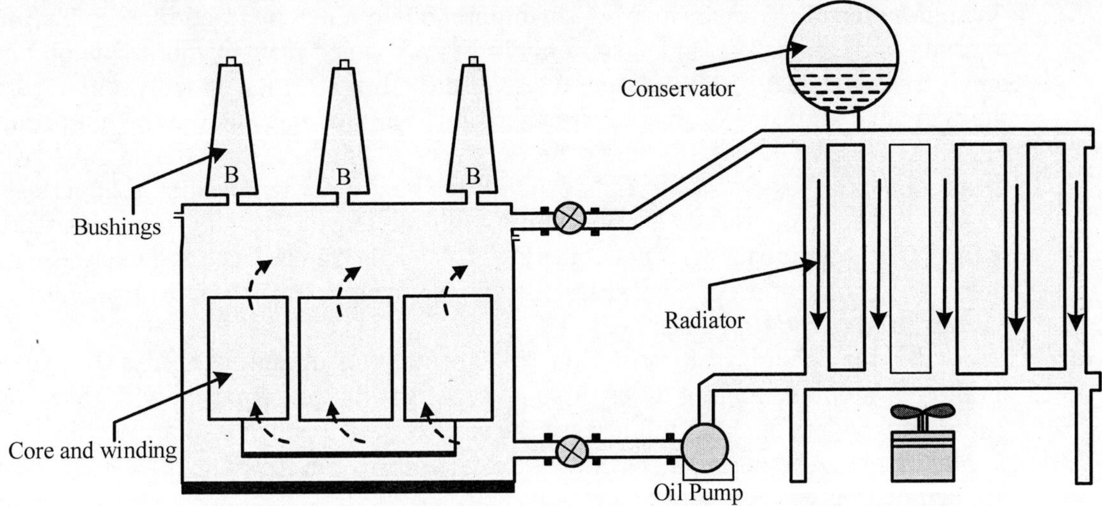

Fig. 5.97 Oil immersed forced oil circulation with air blast cooling

4. **Oil Immersed Forced Oil Circulation with Air Cooling (OFB)**
   This is achieved by means of fans blowing under the radiator as shown in fig. 5.97. The radiators may be placed at some little distance from the transformer tank. Fans are placed under the radiators in such a way as to blow air through them and thus increases the cooling.
   The method in which forced air cooling is adopted in addition to the oil-forced cooling is called *OFB* method i.e., oil immersed forced oil circulation with air blast cooling. *OFN* method rises the rating by one-third and *OFB* by another third.

5.45 Power Transformer and its Auxiliaries

The transformers used in the power system for transfer of electric power or energy from one circuit to the other are called power transformers. The rating of a transformer includes voltage, frequency and kVA. The kVA rating is the kVA output that a transformer can deliver at the rated voltage and frequency under general service conditions without exceeding the standard limit of temperature rise (usually 45° to 60°C) The power transformer has the following important parts:

1. **Magnetic circuit:** The magnetic circuit comprises of transformer core. The transformer core may be core type or shell type in construction. The power transformers used in the power system are mostly three phase transformers. In a core type three-phase transformer core has three limbs of equal area of cross-section.
2. **Electrical circuit:** In three phase transformers there are three primary (*HV*) windings and three secondary (*LV*) windings. Whole of the *LV* winding is wound over one limb next to the core, then whole of the *HV* winding is wound over the *LV* winding. In between the *LV* winding and *HV* winding and between core and *LV* winding insulation is provided.

3. **Transformer oil:** Transformer oil is a mineral oil obtained by fractional distillation of crude petroleum. The oil is used only in the oil cooled transformers. The oil not only carries the heat produced due to losses in the transformer, by convection from the windings and core to the transformer tank, but also has even more important function of insulation. When transformer delivers power, heat is produced due to the iron and copper losses in the transformer. This heat must be dissipated effectively otherwise the temperature of the winding will increase. The rise in temperature further increases the losses. Thus, the efficiency of the transformer will decrease. As there is no rotating part in the transformer, it is difficult to cool down the transformer as compared to rotating machines.
   Generally, for cooling of distribution transformers, oil immersed natural cooling method is adopted. Cooling tubes or small cooling radiators are used with the main tanks, as shown in fig. 5.98, to increase the surface area for the dissipation of heat.
4. **Tank Cover:** A number of parts are arranged on the tank cover of which most important are:
5. **Bushing:** The internal winding of the transformer is connected to the lines through copper rods or bars which are insulated from the tank cover, these are known as bushings. Up to 33 kV ordinary porcelain bushing can be used. Above this voltage oil filled bushings or condenser bushing are employed. Provision of bushings avoids strong electromagnetic forces among the high current (LV) carrying conductors

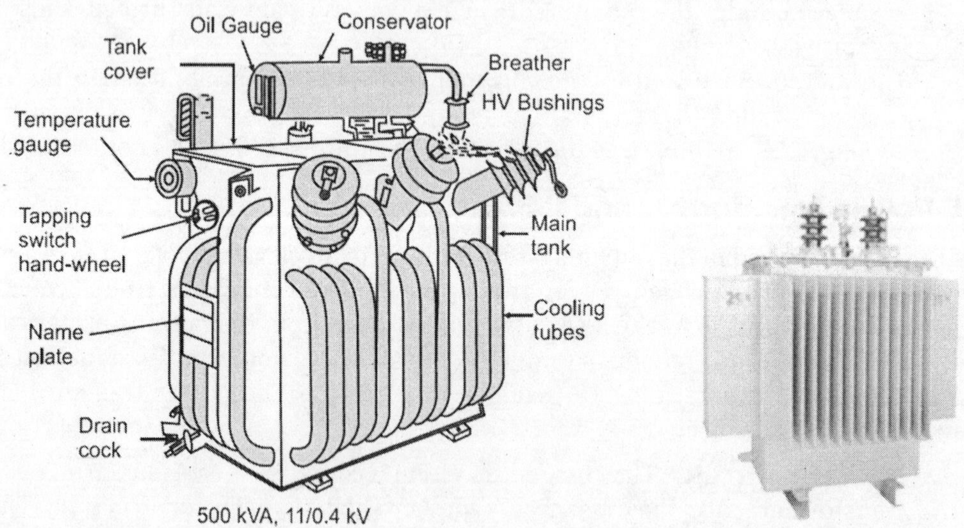

Fig. 5.98 (a) A 500kVA, 11kV/400V ONAN colled distribution transformer (Single phase transformer)

6. **Oil conservator tank:** Oil conservator is also known as an oil expansion chamber. It is a small cylindrical air tight and oil tight vessel. The oil conservator is connected with a tube to the main transformer tank at the tank cover. This tank is partially filled with oil. The expansion and contraction of oil, changes the oil level in the conservator.

7. **Breather:** The transformer oil should not be allowed to come in contact with atmospheric air, since a small amount of moisture causes a great decrease in the dielectric strength of transformer oil. All the tank fittings are made air tight. When oil level in the oil conservator tank changes due to expansion and contraction of oil because of change of oil temperature, air moves in and out of the conservator. This action is known as breathing.
   The breathed air is made to pass through an apparatus called breather to remove moisture. Breather, contains Silica gel or some other drying agent such as *calcium chloride*. This ensures that only dry air enters the transformer tank.
8. **Buchholz Relay:** This is installed in between the main tank and the oil conservator. It is a gas relay that gives warning of any fault developing inside the transformer, and if the fault is dangerous, the relay disconnects the transformer circuit. This relay is installed in the transformer having capacity more than 750 kVA.

All the important parts of a 200 kVA, 11 kV/400 V oil immersed natural cooled distribution transformer are shown in fig. 5.94.

5.46 Maintenance schedule of a Transformer

A power transformer is the most costly and essential piece of power equipment within an electrical substation. As such it is desirable to perform various preventive maintenance activities to ensure that the transformer maintains a high level of performance and a long functional life.

A power transformer requires various routine maintenance tasks including measurement and testing of different parameters of the transformer. There are two main types of maintenance of transformers. We perform one group on a routine basis (known as preventive maintenance), and the second group on an ad-hoc basis (i.e., as required).

That means for getting smooth performance from a transformer we have to perform some maintenance actions on a regular basis.

Some other types of maintenance for a transformer we perform only as they are required – known as emergency or breakdown transformer maintenance. But if one performs regular maintenance properly, this significantly reduces the changes of needs to perform such emergency maintenance. The regular checking and maintenance of transformers is also known as condition maintenance.

Hence by proper condition maintenance, one can avoid emergency and breakdown maintenance. That is why technical personnel should mainly concentrate on condition maintenance. As 100% of condition maintenance causes 0% breakdown of equipment.

There are many different preventive maintenance actions to be performed on a power transformer. They can be on a daily, monthly, yearly, quarterly, half-yearly, or yearly basis. Some transformer maintenance activities only need to be performed once in a 3 to 4 years interval.

Daily Basis Maintenance and Checking

There are three main things which are to be checked on a power transformer on a daily basis:

- Reading of MOG (Magnetic Oil Gauge) of main tank and conservator tank.
- Color of Silica Gel in breather.
- Leakage of oil from any point of a transformer.

In case of unsatisfactory oil level in the MOG, oil to be filled in transformer and also the transformer tank to be checked for oil leakage. If oil leakage is found take required action to plug the leakage. If silica gel becomes pinkish, it should be replaced.

Monthly Basis Maintenance of Transformer

Let us first discuss about the action to be taken on power transformer in monthly basis.

- The oil level in oil cap under Silica Gel breather must be checked in a one-month interval. If it is found the transformer oil inside the cup comes below the specified level, oil to be top up as per specified level.
- Breathing holes in silica gel breather should also be checked monthly and properly cleaned if required, for proper breathing action.
- If the transformer has oil filled bushing the oil level of transformer oil inside the bushing must be visually checked in the oil gauge attached to those bushing. This action also to be done on monthly basis.

If it is required, the oil to be filled in the bushing up to correct level. Oil filling to be done under shutdown condition.

Maintenance of Transformer on Half Yearly Basis

The transformer oil must be checked half yearly basis that means once in 6 months, for dielectric strength, water content, acidity, sludge content, flash point, DDA, IFT, and resistivity of transformer oil.

In the case of a distribution transformer, as they are operating light load condition all the time of day remaining peak hours, so there is no maintenance required.

Yearly Basis Transformer Maintenance Schedule

- The auto, remote, manual function of cooling system that means, oil pumps, air fans, and other items engaged in cooling system of transformer, along with their control circuit to be checked in the interval of one year. In the case of trouble, investigate control circuit and physical condition of pumps and fans.
- All the bushings of the transformer to be cleaned by soft cotton cloths yearly. During cleaning the bushing should be checked for cracking.
- Oil condition of OLTC to be examined in every year. For that, oil sample to be taken from drain valve of divertor tank, and this collected oil sample to be tested for dielectric strength (BDV) and moisture content (PPM). If BDV is low and PPM for

moisture is found high compared to recommended values, the oil inside the OLTC to be replaced or filtered.

- Mechanical inspection of Buchholz relays to be carried out on yearly basis.
- All marshalling boxes to be cleaned from inside at least once in a year. All illumination, space heaters, to be checked whether they are functioning properly or not. If not, required maintenance action to be taken. All the terminal connections of control and relay wiring to be checked and tighten at least once in a year.
- All the relays, alarms and control switches along with their circuit, in R&C panel (Relay and Control Panel) and RTCC (Remote Tap Changer Control Panel) to be cleaned by appropriate cleaning agent.
- The pockets for OTI, WTI (Oil Temperature Indicator & Winding Temperature Indicator) on the transformer top cover to be checked and if required oil to be replenished.
- The proper function of Pressure Release Device and Buchholz relay must be checked annually. For that, trip contacts and alarm contacts of the said devices are shorted by a small piece of wire, and observe whether the concerned relays in remote panel are properly working or not.
- Insulation resistance and polarization index of transformer must be checked with battery operated megger of 5 kV range.
- Resistive value of earth connection and riser must be measured annually with clamp on earth resistance meter.
- DGA or Dissolve Gas Analysis of transformer Oil should be performed, annually for 132 KV transformer, once in 2 years for the transformer below 132 KV transformer and in 2 years interval for the transformer above 132 KV transformer.

The Action to be taken once in 2 years:

- The calibration of OTI and WTI must be carried once in two years.
- Tan & delta; measurement of bushings of transformer also to be done once in two years.

5.47 Trouble Shooting a Transformer

a fault in these transformers can lead to critical problems for the system's operation. To avoid such risks of damage or malfunctioning, a proper follow-up of the maintenance schedule is recommended. This article illustrates the proper procedure for transformer troubleshooting and testing.

For any type of internal fault of the transformer, the winding insulation and bushings are tested. For this insulation test between transformer winding and core or tank is carried out. Similarly, to test insulation of bushings, insulation test between each terminal of transformer and tank is carried out. When fault is located, it is removed.

Trouble shooting of a transformer			
S.No.	Trouble	Cause	Action
1	Stop of the transformer (except OLTC)	Internal fault of main tank	Check on relay operation. Electrical tests. Measurement of insulation resistance. Measurement of winding resistance. Measurement of voltage ratio. Measurement of the exciting current. Gas analysis of transformer.
1-1	Operation of differential relay and Buchholz relay or pressure relief device operate simultaneously	1. Internal fault 2. Inrush current at time of energization 3. Internal fault 4. Faulty operation	1. Action shall be same as internal fault procedure 2. Setting values of relays. 3. Check wiring. 4. Check contacts.
1-2	Winding temperature indicator or oil temperature indicator	1. Internal fault 2. Faulty operation 3. Stop of cooling fan	1. Action shall be same as internal fault procedure. 2.1 Check wiring. 2.2 Check contacts. 3. Check cooling fan.
2	Operation of Buchholz: relay (alarm) only	1. Internal Fault 2. Trapped air or evaluation of dissolved gases	1. Action shall be same as internal fault procedure. 2.1 Analysis of gas accumulated in Buchholz relay. 2.2 Sudden variation of oil temperature.
3	Operation of relays such as over current relay, ground relay, arrestor etc.	1. Internal fault 2. External faults such as short-circuit in the system	1. Action shall be same as internal fault procedure. 2.1 Check any sign of: Heavy current circulation in the windings. 2.2 Check any sign of: On-Load tap changer interrupting heavy short-circuit current. And

			short-circuit current value, its duration time.
4	Abnormality is found in recording devices	1. Internal fault 2. External faults such as short-circuit in the system	1. Action shall be same as internal fault procedure. 2.1 Check any sign of: Heavy current circulation in the windings. 2.2 Check any sign of: On-Load tap changer interrupting heavy short-circuit current. And short-circuit current value, its duration time.
5	High-temperature rise	Period required to reach the maximum temperature and ambient temperature, load factor or type of load. Connection diagram of bank and internal connection tap voltage.	If the oil temperature is about 80° C over, the transformer must not be operated continuously.
6	Burned Coil	Date and time of burning. Installation site. Period of use (immediately after operation period). Cause of failure: Overload lighting, short circuiting (internal or external). Others: existence or absence of copper powder, operation of protective devices (primary and secondary)	
7	Local Heating	Heated point. Load factor or type of load. Smoke. Discoloration.	Check core clamping bolts, nuts fitting and remove the fault.
8	Insulation resistance failure	Insulation resistance: High voltage winding to low voltage winding and tank.	Check transformer coil. Replace or filter the oil. Dry the transformer. Check series reactor. Lightening arrestors.

		Low voltage winding to high voltage winding and tank. High voltage winding to low voltage. Temperature and humidity at which measurement is made. Installation site (indoor or outdoor). History of the transformer. Dielectric strength of the insulation oil.	
9	Noise	Operating conditions (Overload, Normal)	Check and tighten the nut-bolts and stamping.
10	Rusting/Corroding of tank	Oil might be having more acidity.	Take the acidity test of oil. Use activated ammonia.
11	Sagging	Used wrong grade oil. May be due to overheating.	Replace the present oil with the appropriate grade oil.

Solved Examples:

Q1. The maximum flux density in the core of a 250/3000 volts, 50 Hz single-phase transformer is 1.2 Wb/ m². If the e.m.f. per turn is 8 volts, determine (i) Primary and secondary turns (ii) the Area of the core

Solution: -

(i) $E_1 = N_1 \times \text{e.m.f. induced/turn}$

$$N_1 = \frac{250}{8} = 32$$

$$N_2 = \frac{3000}{8} = 375$$

(ii) We may use $E_2 = -4.44 f N_2 B_m A$

$$3000 = 4.44 \times 50 \times 375 \times 1.2 \times A$$

$$A = 0.03 m^2$$

Q2. The core of a 100 kVA, 11000/550 V, 50 Hz 1-Phase, core type transformer has a cross-section of 20cm X 20cm. Find (i) The number of H.V. and L.V. turns per phase and (ii) The e.m.f. per turn if the maximum core density is not to exceed 1.3 tesla. Assume a stacking factor of 0.9. What will happen if its primary voltage is increased by 10% on no-load?

Solution: -

(i) $B_m = 1.3\, T.\, A = (0.2 \times 0.2) \times 0.9 = 0.036 m^2$

$$11000 = 4.44 \times 50 \times N_1 \times 1.3 \times 0.036$$

$$N_1 = 1060$$

$$550 = 4.44 \times 50 \times N_2 \times 1.3 \times 0.036$$

$$N_2 = 53$$

$$N_2 = KN_1 = \left(\frac{550}{11000}\right) \times 1060 = 5$$

(ii) $\text{E.m.f./turn} = \frac{11000}{1060} = 10.4\, V$

$$550/53 = 10.4\, V$$

Q3. The core of a three-phase, 50 Hz, 11000/550 V delta/star, 300 kVA, core-type transformer operates with a flux of 0.05 Wb. Find (i) e.m.f. per turn (ii) Number of H.V. and L.V. turns per phase (iii) Full load H.V. and L.V. phase-currents.

Solution: -

Maximum value of flux has been given as 0.05 Wb

(i) E.m.f. per turn $= 4.44\, f \phi_m$

$= 4.44 \times 50 \times 0.05 = 11.1$ volts

(ii) Calculation of number of turns on two sides:

Voltage per phase on delta-connected primary winding = 11000 volts

Voltage per phase on star-connected secondary winding = 550/1.732 = 317.5 volts

T_1 = Number of turns on primary, per phase

= voltage per phase/e.m.f. per turn

= 11000/11.1 = 991

T_2 = Number of turns on secondary, per phase

= voltage per phase/e.m.f. per turn

= 317.5/11.1 = 28.6

(iii) Full load H.V. and L.V. phase currents:

Output per phase = (300/3) = 100 kVA

H.V. phase-current = (100 X 1000/11000) = 9.1 Amp

L.V. phase-current = (100 X 1000/317.5) = 315 Amp

Q4. A single-phase transformer has 500 turns on the primary and 40 turns on the secondary winding. The mean length of the magnetic path in the iron core is 150 cm and the joints are equivalent to an air gap of 0.1 mm. When a p.d. of 3,000 V is applied to the primary, maximum flux density is 1.2 Wb/m². Calculate (i) The cross-sectional area of the core (ii) No-load secondary voltage (iii) The no-load current drawn by the primary (iv) Power factor on no-load. Given that AT/cm for a flux density of 1.2 Wb/m² in iron to be 5, the corresponding iron loss to be 2 watt/kg at 50 Hz and the density of iron as 7.8 gram/cm³.

Solution: -

(i) $E = 4.44 f N_1 B_m A$

$3000 = 4.44 \times 50 \times 500 \times 1.2 \times A$

$A = 0.0225$ m² = 225 cm²

This is the net cross-sectional area. However, the gross area would be about 10% more to allow for the insulation between laminations.

(ii) $K = N_2/N_1 = 40/500 = 4/50$

N.L. secondary voltage $= KE_1 = (4/50) \times 3000 = 240\ V$

(iii) AT per cm = 5

AT for iron core = 150 X 5 = 750

AT for air-gap $= Hl = \frac{B}{\mu_0} \times l = \frac{1.2}{4\pi \times 10^{-7}} \times 0.0001 = 95.5$

Total AT for given $B_{max} = 750 + 95.5 = 845$

Max. value of magnetising current drawn by primary = 84.5/500 = 1.691 A

Assuming this current to be sinusoidal, its r.m.s value is $I_\mu = 1.691/\sqrt{2} = 1.196\ A$

Volume of iron = Length X Area = 150 X 225 = 33750 cm³

Density = 7.8 gram/cm³

Mass of iron =33750 X 7.8/1000 = 263.25 kg

Total iron loss = 263.25 X 2 = 526.5 W

Iron loss component of no-load primary current I_0 is I_w = 526.5/3000 = 0.176 A

$$I_0 = \sqrt{I_\mu^2 + I_w^2} = \sqrt{1.196^2 + 1.196^2} = 0.208 \text{ A}$$

(iv) Power factor, $\cos\theta_0 = I_w/I_o = 0.176/1.208 = 0.1457$

Q5. A 30 kVA, 2400/120 V, 50 Hz transformer has a high voltage winding resistance of 0.1 Ω and a leakage reactance of 0.22 Ω. The low voltage winding resistance is 0.035 Ω and the leakage reactance is 0.012 Ω. Find the equivalent winding resistance, reactance and impedance referred to the (i) High voltage side (ii) The low voltage side.

Solution: -

$$K = 120/2400 = 1/20$$

$$R_1 = 0.1\ \Omega, X_1 = 0.22\ \Omega, R_2 = 0.035\ \Omega \text{ and}, X_2 = 0.012\ \Omega$$

(i) Here, high-voltage side is, obviously, the primary side. Hence, values as referred to primary side are

$$R_{01} = R_1 + R_2' = R_1 + R_2/K^2 = 0.1 + 0.035/(1/20)^2 = 14.1\ \Omega$$

$$X_{01} = X_1 + X_2' = X_1 + X_2/K^2 = 0.22 + 0.12/(1/20)^2 = 5.02\ \Omega$$

$$Z_{01} = \sqrt{R_{01}^2 + X_{01}^2} = \sqrt{14.1^2 + 5.02^2} = 15\ \Omega$$

(ii) $$R_{02} = R_2 + R_1' = R_2 + K^2R_1 = 0.035 + (1/20)^2 \times 0.1 = 0.03525\ \Omega$$

$$X_{02} = X_2 + X_1' = X_2 + K^2X_1 = 0.012 + (1/20)^2 \times 0.22 = 0.01255\ \Omega$$

$$Z_{02} = \sqrt{R_{02}^2 + X_{02}^2} = \sqrt{0.0325^2 + 0.01255^2} = 0.0374\ \Omega$$

Or $Z_{02} = K^2Z_{01} = (1/20)^2 \times 15 = 0.0375\ \Omega$

Q6. The parameters of a 2300/230 V, 50 Hz transformer are given below:

$R_1 = 0.286\ \Omega$	$R_2' = 0.319\Omega$	$R_0 = 250\ \Omega$
$X_1 = 0.73\ \Omega$	$X_2' = 0.73\Omega$	$X_0 = 1250\ \Omega$

The secondary load impedance $Z_L = 0.387 + j0.29$. Solve the exact equivalent circuit with normal voltage across the primary.

Solution: -

$$K = 230/2300 = 1/10;\ Z_L = 0.387 + j0.29$$

$$Z_{L.}' = Z_L/K^2 = 100\,(0.387 + j0.29) = 38.7 + j29 = 48.4\ \angle 36.8°$$

$$Z_2' + Z_L' = (38.7 + 0.319) + j(29 + 0.73) = 39.02 + j29.73 = 49.0\ \angle 37.3°$$

$$Y_m = (0.004 - j0.0008);\ Z_m = 1/Y_m = 240 + j48 = 245\ \angle 11.3°$$

$$Z_m + (Z_2' + Z_L') = (240 + j48) + (39 + j29.7) = 290\ \angle 15.6°$$

$$I_1 = \frac{V_1}{Z_1 + \frac{Z_m\,(Z_2' + Z_L')}{Z_m\,(Z_2' + Z_L')}} = \left[\frac{2300\ \angle 0°}{0.286 + j0.73 + 41.4\ \angle 33°}\right]$$

$$\frac{2300\ \angle 0°}{42\ \angle 33.7°} = 54.8\ \angle -33.7°$$

$$I_2' = I_1 \times \frac{Z_m}{(Z_2' + Z_L') + Z_m} = 54.8\ \angle -33.7° \times \frac{245\ \angle 11.3°}{290\ \angle 15.6°}$$

$$= 54.8\ \angle -33.7° \times 0.845\ \angle -4.3° = 46.2\ \angle -38°$$

$$I_0 = I_1 \times \frac{(Z_2' + Z_L')}{Z_m\,(Z_2' + Z_L')} = 54.8\ \angle -33.7° \times \frac{49\ \angle 37.3°}{290\ \angle 15.6°}$$

$$= 54.8\ \angle -33.7° \times 0.169\ \angle -21.7° = 9.26\ \angle -12°$$

Input power factor = cos 33.7° = 0.832 lagging

Power input = $V_1 I_1 \cos\theta_1$ = 2300 X 54.8 X 0.832 = 105 kW

Power output = $46.2^2 \times 38.7$ = 82.7 kW

Primary Cu loss = $54.8^2 \times 0.286$ = 860 W

Secondary Cu loss = $46.2^2 \times 0.319$ = 680 W

Core loss = $9.26^2 \times 240$ = 20.6 kW

$\eta = (82.7/105) \times 100 = 78.8\%$

$V_2' = I_2' Z_L' = 46.2 \times 48.4 = 2240\ V$

Regulation = $\frac{2300-2240}{2240} \times 100 = 2.7\%$

Q7. The no-load test of single transformer gives the following test data: Primary voltage: 220 V; Secondary voltage: 110 V; Primary current: 0.5 A; Power input: 30 W. Find (i) The turns ratio (ii) The magnetizing component of no-load current (iii) its working (or loss) component (iv) The iron loss.
Resistance of the primary winding = 0.6 ohms.
Solution: -

(i) Turn ratio $N_1/N_2 = 220/110 = 2$

(ii) $W = V_1 I_0 \cos\theta_0$; $\cos\theta_0 = 30/220 \times 0.5 = 0.273$; $\sin\theta_0 = 0.962$

$$I_\mu = I_0 \times \sin\theta_0 = 0.5 \times 0.962 = 0.48\ A$$

(iii) $I_w = I_0 \times \cos\theta_0 = 0.5 \times 0.273 = 0.1365\ A$

(iv) Primary loss = $I_0^2 R_1 = 0.5^2 \times 0.6 = 0.15\ W$

Iron loss = 30 – 0.15 = 29.85 W

Q8. A transformer is connected to 2200 V, 40 Hz supply. The core-loss is 800 watts out of which 600 watts are due to hysteresis and the remaining, eddy current losses. Determine the core-loss if the supply voltage and frequency are 3300 V and 60 Hz respectively.

Solution: -

For constant flux density (i.e. constant V/f ratio), which is fulfilled by 2200/40 or 3300/60 figures in two cases,

Core-loss = $Af + Bf^2$

First term on the right-hand side represents hysteresis loss and the second term represents the eddy-current loss.

At 40 Hz, 800 = 600 + eddy current loss

Thus, $Af = 600$, or A =15

$Bf^2 = 200$, or B = 200/1600 = 0.125

At 60 Hz, Core-loss = 15 X 60 + 0.125 X 60^2

= 900 + 450

= 1350 watts

Q9. A 100 kVA transformer has 400 turns on the primary and 80 turns on the secondary. The primary and secondary resistance are 0.3 Ω and 0.01 Ω respectively and the corresponding leakage reactances are 1.1 and 0.035 Ω respectively. The supply voltage is 2200 V. Calculate (i) the Equivalent impedance referred to primary and (ii) The voltage regulation and the secondary terminal voltage for full load having a power factor of 0.8 leading.

Solution: -

K = 80/400 = 1/5, $R_1 = 0.3\ \Omega, R_{01} = R_1 + R_2/K^2 = 0.3 + 0.01/(1/5)^2 = 0.55\ \Omega$

$X_{01} = X_1 + X_2/K^2 = 1.1 + 0.035/(1/5)^2 = 1.975\ \Omega$

(i) $Z_{01} = 0.55 + j\,1.975 = 2.05\ \angle 74.44°$

(ii) $Z_{02} = K^2 Z_{01} = (1/5)^2\ (0.55 + j\,1.975) = (0.022 + j\,0.079)$

No-load secondary voltage $= KV_1 = (1/5) \times 2200 = 440\ V$,

$I_2 = 10 \times 10^3/440 = 227.3\ A$

Full-load voltage drop as referred to secondary

$= I_2\ (R_{02}\ \cos\theta - X_{02}\ \sin\theta\ )$

$= 227.3\ (0.022 \times 0.8 - 0.079 \times 0.6) = -6.77\ V$

% Regulation = - 6.77 X 100/440 = - 1.54

Secondary terminal voltage on load = 440 – (-6.77) = 446.77 V

Q10. A short-circuit test when performed on the h.v. side of a 10 kVA, 2000/400 V single phase transformer, gave the following data; 60 V, 4 A, 100 W. If the l.v. side is delivering full load current at 0.8 p.f. lag and at 400 V. Find the voltage applied to h.v. side.

Solution: -

Here, the test has been performed on the h.v. side i.e., primary side.

$$Z_{01} = \frac{60}{4} = 15\ \Omega;$$

$$R_{01} = 100/4^2 = 6.25\ \Omega;$$

$$X_{01} = \sqrt{15^2 - 6.25^2} = 13.63\ \Omega$$

$$\text{F.L. } I_1 = \frac{10000}{2000} = 5A$$

Total transformer voltage drop as referred to primary is

$$I_1(R_{01}\ \cos\theta + X_{01}\ \sin\theta\ ) = 5\ (6.25\ \times 0.8\ + 13.63\ \times 0.6 = 67\ V\ )$$

Hence, the primary voltage has to be raised from 2000 V to 2067 v in order to compensate for the total voltage drop in the transformer. tn that case secondary voltage on load would remain the same as on no-load.

Q11. A 10 kVA, 500/250 V, single-phase transformer has a maximum efficiency of 94% when delivering 90% of its rated output at unity p.f. Estimate its efficiency when delivering its full-load output at p.f. of 0.8 lagging.

Solution: -

Rated output at unity p.f. = 10000 W. Hence, 90% of rated output = 9000 W

Input with 94% efficiency = 9000/0.94 W

Losses = 9000 [(1/0.94) – 1] = 574 W

At maximum efficiency, variable copper-loss = constant = Core loss = 574/2 = 287 W

At rated current, Let the copper-loss = P_c watts

At 90% load with unity p.f., the copper-loss is expressed as $0.90^2\ \times P_c$

Hence, $P_c = \frac{287}{0.81} = 354W$

Output at full-load, 0.8 log p.f. = 10000 X 0.80 = 8000 W

At the corresponding load, Full-load Copper-loss = 354 W

Hence, efficiency = 8000/(8000+354+287) = 0.926 = 92.6%

Q12. A 3300/230 V, 50 kVA, transformer is found to have impedance of 4% and a Cu loss of 1.8% at full-load. Find its percentage reactance and also the ohmic values of resistance, reactance and impedance as referred to primary. What would be the value of primary short-circuit current if the primary voltage is assumed constant?

Solution: -

$$\%X = \sqrt{\%Z^2 - \%R^2} = \sqrt{4^2 - 1.8^2} = 3.57\% \ (Cu\ loss = \%R)$$

Full load $I_1 = \frac{50000}{3300} = 15.2\ A$ (Assuming 100% efficiency). Considering primary winding, we have

$$\%R = \frac{R_{01} I_1 \times 100}{V_L} = 1.8$$

$$R_{01} = \frac{1.8 \times 3300}{100 \times 15.2} = 3.91\ \Omega$$

Similarly,

$$\%X = \frac{X_{01} I_1 \times 100}{V_1} = 3.57$$

$$R_{01} = \frac{3.57 \times 3300}{100 \times 15.2} = 7.76\ \Omega$$

Similarly,

$$Z_{01} = \frac{4 \times 3300}{100 \times 15.2} = 8.7\ \Omega$$

Now,

$$\frac{Short - circuit\ current}{full\ load\ current} = \frac{100}{4}$$

Short Circuit current = 15.2 X 25 = 380 A

Q13. A transformer has copper-loss of 1.5% and reactance-drop of 3.5% when tested at full-load. Calculate its full-load regulation at (i) u.p.f. (ii) 0.8 p.f. lagging and (iii) 0.8 p.f. leading.

Solution: -

The test data at full load gives the following parameters:

p.u. resistance = 0.015, p.u. reactance = 0.035

(i) Approximate voltage – Regulation at unity p.f. full-load

$= 0.015\ \cos\theta + 0.035\ sin\ \theta$

$= 0.015$ per unit = 1.5%

(ii) Approximate voltage – Regulation at 0.80 lagging p.f.

= (0.015 X 0.8) + (0.035 X 0.6) = 0.033 per unit = 3.3%

(iii) Approximate voltage – Regulation at 0.80 leading p.f.

$= I_r\ \cos\theta - I_x\ \sin\theta$

= (0.015 X 0.8) – (0.035 X 0.6) = 0.009 per unit = -0.9%

Q14. Two 100 kVA, single-phase transformer are connected in parallel both on the primary and secondary. One transformer has an ohmic drop of 0.5% at full-load and an inductive drop of 8% at full-load current. The other has an ohmic drop of 0.75% and inductive drop of 2%. Show how will they share a load of 180 kW at 0.9 power factor.

Solution: -

A load of 180 kW at 0.9 p.f. means a kVA of 180/0.9 = 200

Load S = 200 – 25.8°

$$\frac{Z_1}{Z_1 + Z_2} = \frac{(0.5 + j8)}{(1.25 + j12)} = \frac{(0.5 + j8)(1.25 - j12)}{1.25^2 + 12^2}$$

$$= \frac{96.63 + j4}{145.6} = \frac{96.65 \angle 2.4^\circ}{145.6} = 0.664\angle 2.4^\circ$$

$$\frac{Z_2}{Z_1 + Z_2} = \frac{(0.75 + j4)(1.25 - j12)}{145.6}$$

$$= \frac{48.94 - j4}{145.6} = \frac{49.1 \angle -5^\circ}{145.6} = 0.337\angle -5^\circ$$

$$S_1 = S\frac{Z_2}{Z_1 + Z_2} = 200\angle -25.8^\circ \times 0.337 \angle -5^\circ = 67.4 \angle -30.8^\circ$$

$$kW_1 = 67.4 \times \cos 30.8^\circ = 67.4 \times 0.859 = 57.9\ kW$$

$$S_2 = 200\angle -25.8^\circ \times 0.664 \angle -2.4^\circ = 132.8 \angle -23.4^\circ$$

$$kW_2 = 132.8 \times \cos 23.4^\circ = 132.8 \times 0.915 = 121.5\ kW$$

Q15. What should be the kVA rating of each transformer in a V-V bank when the 3-phase balanced load is 40 kVA? If a third similar transformer is connected for operation, what is the rated capacity? What percentage increase in rating is affected in this way?

Solution: -

As pointed out earlier, the kVA rating of each transformer has to be 15% greater.

kVA/transformer = (40/2) X 1.15 = 23

Δ-Δ bank rating = 23 X 3 =69; Incraese = [(69-40)/40] X 100 = 72.5%

Q16. A Δ-Δ bank consisting of three 20 kVA, 2300/230 V transformers supplies a load of 40 kVA. If one transformer is removed, find the resulting V-V connection (i) kVA load carried by each transformer (ii) percent of rated load carried by each transformer (iii) total kVA rating of the V-V bank (iv) ration of V-V bank to Δ-Δ bank transformer ratings (v) percent increase in load on each transformer when bank is converted into V-V bank.

Solution: -

(i) $\frac{Total\text{ kVA load in V–V bank}}{VA/Transformer} = \sqrt{3}$

kVA load supplied by each of the two transformers = $40/\sqrt{3}$ = 23.1 kVA

Obviously, each transformer in V-V bank does not carry 50% of the original load but 57.7%

(ii) Percent of rated load = $\frac{\text{kVA load/transformer}}{\text{kVA rating/transformer}} = \frac{23.1}{20} = 115.5\%$ carried by each transformer.

Obviously, in this case, each transformer is overloaded to the extent of 15.5 percent.

(iii) kVA rating of the V-V bank = (2 X 20) X 0.866 = 34.64 kVA

(iv) $\frac{\text{V–V rating}}{\Delta\text{–}\Delta\text{ rating}} = \frac{34.64}{60} = 0.577$ or 57.7%

As seen, the rating is reduced to 57.7% of the original rating.

(v) Load supplied by each transformer in Δ-Δ bank = 40/3 = 13.33 kVA

Percentage increase in load supplied by each transformer

$$= \frac{\text{kVA load}/transformer\ in\ V - V\ bank}{\text{kVA load}/transformer\ in\ \Delta - \Delta\text{ bank}} = \frac{23.1}{13.3} = 1.732 = 173.2\%$$

It is obvious that each transformer in the Δ-Δ bank supplying 40 kVA was running underloaded (13.33 vs. 20 kVA) but runs overloaded (23.1 vs. 20 kVA) in V-V connection.

Multiple Choice Questions

Q1. A transformer transforms

e. Frequency
f. Voltage
g. Current
h. Voltage and Current

Q2. The main purpose of using core in a transformer is to

a. Decrease iron losses
b. Prevent eddy current loss
c. Eliminate magnetic hysteresis
d. Decrease reluctance of the common magnetic circuit

Q3. Transfomer cores are laminated in order to

a. Simplify its construction
b. Minimize eddy current loss
c. Reduced cost
d. Reduced hysteresis loss

Q4. Which of the following is not a basic element of a transformer?

a. Core
b. Primary winding
c. Secondary winding
d. Mutual flux

Q5. In an ideal transformer

a. Windings have no resistance
b. Core has no losses
c. Core has infinite permeability
d. All of the above

Q6. In a two-winding transformer, the e.m.f. per turn in secondary winding is always............. the induced e.m.f. power turn in primary.

a. Equal to K times
b. Equal to $1/K$ times
c. Equal to
d. Greater than

Q7. In a transformer, the leakage flux of each winding is proportional to the current in that winding because

a. Ohm's law applies to magnetic circuits
b. Leakage paths donot saturate
c. The two windings are electrically isolated
d. Mutual flux is confined to the core

Q8. The primary and secondary windings of an ordinary 2-winding transformer always have

a. Different number of turns
b. Same size of copper wire
c. A common magnetic circuit
d. Separate magnetic circuits

Q9. A transformer has negative voltage regulation when its load power factor is

a. Zero
b. Unity
c. Leading
d. Lagging

Q10. In performing the short circuit test of a transformer

a. High voltage side is usually short circuited
b. Low voltage side is usually short circuited
c. Any side is short circuited with preference
d. None of the above

Q11. No-load test on a transformer is carried out to determine

a. Copper loss
b. Magnetizing current
c. Magnetizing current and no-load loss
d. Efficiency of the transformer

Q12. The voltage applied to the h.v. side of a transformer during short-circuit test is 2% of its rated voltage. The core loss will be............ percent of the rated core loss.

a. 4
b. 0.4
c. 0.25
d. 0.04

Q13. When a 400-Hz transformer is operated at 50-Hz its kVA rating is

a. Reduced to1/8
b. Increased 8 times
c. Unaffected
d. Increased 64 times

Q14. The main purpose of performing open-circuit test on a transformer is to measure its

a. Cu loss
b. Core loss
c. Total loss
d. Insulation resistance

Q15. At relatively light loads, transformer efficiency is low because

a. Secondary output is low
b. Transformer losses are high
c. Fixed loss is high in proportion to the output
d. Cu loss is small

Q16. The maximum efficiency of a 100-kVA transformer having iron loss of 900 kW and F.L. Cu loss of 1600 W occurs at .... kVA.

a. 56.3
b. 133.3
c. 75
d. 177.7

Q17. The ordinary efficiency of a given transformer is maximum when

a. It runs at half full-load
b. It runs at full-load
c. Its Cu loss equals iron loss
d. It runs slightly overload

Q18. The saving in Cu achieved by converting a 2-winding transformer into an autotransformer is determined by

a. Voltage transformation ratio
b. Load on the secondary
c. Magnetic quality of core material
d. Size of the transformer core

Q19. The all-day efficiency of a transformer depends primarily on

a. Its copper loss
b. The amount of load
c. The duration of load
d. Both (b) and (c)

Q20. Two transformers A and B having equal outputs and voltage ratios but unequal percentage impedances of 4 and 2 are operating in parallel.transformer A will be running over-load by.........

a. 50
b. 66
c. 33
d. 25

Q21. The essential condition for parallel operation of two 1-ϕ transformers is that they should have the same

a. Polarity
b. kVA rating
c. Voltage ratio
d. Percentage impedance

Q22. A T-T transformer cannot be paralleled with .... transformer.

a. V-V
b. Y-Δ
c. Y-Y
d. Δ-Δ

Q23. Of the following statement concerning parallel operation of transformers, the one which is not correct is

a. Transformers must have equal voltage ratings
b. Transfomers must have same ratio of transformation
c. Transformers must be operated at the same frequency
d. Transformers must have equal kVA ratings

Q24. A T-T connection has higher ratio of utilization that a V-V connection only when

a. Identical transformers are used
b. Load power factor is leading
c. Load power factor is unity
d. Non-identical transformers are used

Q25. When a V-V system is converted in to a Δ-Δ system, the increase in capacity of the system is .... percent.

a. 86.6
b. 66.7
c. 73.2
d. 50

Q26. In a three-phase Y-Y transformer connection, neutral is fundamental to the

a. Suppression of harmonics
b. Passage of unbalanced currents due to unbalanced loads
c. Provision of dual electric service
d. Balancing of phase voltages with respect to line voltages

Q27. Which of the following connections is best suited for 3-phase, 4-wire service?

a. Δ-Δ
b. Y-Y
c. Δ-Y
d. Y-Δ

Q28. For supplying a balanced 3-ϕ load of 40-kVA, rating of each transformer in V-V bank should be nearly .... kVA.

a. 20
b. 23
c. 34.6
d. 25

Q29. If the load p.f. is 0.866, then the average p.f. of the V-V bank is

a. 0.866
b. 0.75
c. 0.51
d. 0.65

Q30. Before removing the ammeter from a current transformer, its secondary must be short-circuited in order to avoid

a. Excessive heating of the core
b. High secondary e.m.f.
c. Increase in iron losses
d. All of the above

Keys to multiple choice questions

1.	d	2.	d	3.	b	4.	d	5.	d	6.	c
7.	b	8.	c	9.	c	10.	b	11.	c	12.	d
13.	a	14.	b	15.	c	16.	c	17.	c	18.	a
19.	d	20.	c	21.	a	22.	b	23.	d	24.	d
25.	c	26.	a	27.	c	28.	b	29.	b	30.	d

Short type questions

Q31. What essentially is a transformer? What are the broad areas of applications of transformer?

Q32. Why does a transformer have an iron core?

Q33. Why is the transformer core laminated?

Q34. When a transformer is connected to the supply, how its windings are named?

Q35. Why Sandwiched winding arrangement is preferred in large transformers?

Q36. What do you mean by major and minor insulation used in transformer winding?

Q37. What is an ideal transformer?

Q38. Can a transformer work on DC? Justify.

Q39. A transformer is said to be analogous to mechanical gear, why?

Q40. While drawing phasor diagram of an ideal transformer, the flux vector is drawn 90° out of phase (lagging) to the supply voltage, why?

Q41. Why is the area of yoke of a transformer kept 15 to 20% more than limb?

Q42. What is the thickness of laminations used for transformer core?

Q43. Even at no-load, a transformer draws current from the mains. Why?

Q44. What is doubling effect in transformer core?

Q45. What are the ill effects of inrush current of transformer?

Q46. What do you know about reactance in a transformer?

Q47. When load current of a transformer increases, how does the input current adjust to meet the new conditions?

Q48. Does the flux in a transformer core increase with load?

Q49. What is voltage regulation of a transformer?

Q50. Why does voltage drop in a transformer?

Q51. What are no-load losses occurring in the transformer?

Q52. Why is efficiency of a transformer high as compared to other electrical machines?

Q53. Are transformers normally considered to be efficient devices?

Q54. How may the iron loss be reduced to a minimum?

Q55. Why short circuit test is performed on high voltage side of transformer?

Q56. Why are iron losses or core losses assumed to remain constant in a power transformer from no load to full-load?

Q57. How can iron loss be measured?

Q58. What is the necessity of parallel operation of the transformer?

Q59. What conditions are required to be fulfilled for parallel operation of transformers?

Q60. What are the advantages of using three-phase transformers over a bank of three one phase transformers?

Q61. Draw the connection diagram for delta-Y connection.

Q62. Star-star connected transformers are rarely used, why?

Q63. What is the necessity of parallel operation of three-phase transformers?

Q64. What are on-load tap changing transformers?

Q65. Why a tertiary winding is also called as auxiliary winding and stabilizing winding?

Q66. Give limitations of on-load tap changing transformer.

Q67. Compare delta and open delta connections.

Q68. Why Scott connections are also known as T-connections?

Q69. What is a tap-changer? Where and how they are placed with the transformers?

Q70. What is the necessity of cooling of transformer?

EXERCISES

Q1. The number of turns on the primary and secondary windings of a 1-ϕ transformer are 350 and 35 respectively. If the primary is connected to a 2.2 kV, 50- Hz supply, determine the secondary voltage on no-load.

(**Ans.** 220 V)

Q2. A 3000/200-V, 50-Hz, 1-phase transformer is to be worked at a maximum flux density of 1.2 Wb/m^2 in the core. The effective cross-sectional area of the transformer core is 150 cm^2. Calculate suitable values of primary and secondary turns.

(**Ans.** 830;58)

Q3. A double-wound, 1 phase transformer is required to step down from 1900 V to 240 V, 50-Hz. It is to have 1.5 V per turn. Calculate the required number of turns on the primary and secondary windings respectively. The peak value of flux density is required to be not more than 1.2 Wb/m^2. Calculate the required cross-sectional area of the steel core. If the output is 10 kVA, calculate the secondary current.

(**Ans.** 1267;160;56.4 cm^2;41.75 A)

Q4. The no-load voltage ratio in a 1-phase, 50 Hz, core-type transformer is 1200/440. Find the number of turns in each winding if the maximum flux is to be 0.075 Wb.

(**Ans.** 24 and 74 turns)

Q5. The no-load current of a transformer is 5.0 A at 0.3 power factor when supplied at 230 V, 50-Hz. The number of turns on the primary winding is 200. Calculate (i) The maximum value of the flux in the core (ii) The core loss (iii) The magnetizing current.

(**Ans.** 5.18 mWb; 345 W; 4.77 A)

Q6. The no-load current of a transformer is 15 A at a power factor of 0.2 when connected to a 460 V, 50-Hz supply. If the primary winding has 550 turns, calculate (i) The magnetizing component of no-load current (ii) The iron loss (iii) The maximum value of the flux in the core.

(**Ans.** 14.7 A; 1380 W; 3.77 mWb)

Q7. The primary of a certain transformer takes 1 A at a power factor of 0.4 when it is connected across a 200-V, 50 Hz supply and the secondary is on open circuit. The number of turns on the primary is twice that on the secondary. A load taking 50 A at a lagging power factor of 0.8 is now connected across the secondary. What is now the value of primary current?

(**Ans.** 25.9 A)

Q8. A 1-phase transformer is supplied at 1,600 V on the h.v. side and has a turn ratio of 8:1. The transformer supplies a load of 20 kW at a power factor of 0.8 lag and takes a magnetizing current of 2.0 A at a power factor of 0.2. Calculate the magnitude and phase of the current taken from the h.v. supply.

(**Ans.** 17.15 A; 0.753 lag)

Q9. The iron loss in a transformer core at normal flux density was measured at frequencies of 30 and 50 Hz, the results being 30 W and 54 W respectively. Calculate (i) The hysteresis loss (ii) The eddy current loss at 50 Hz.

(**Ans.** 44 W; 10 W;)

Q10. A transformer has no-load losses of 55 W with a primary voltage of 250 V at 50 Hz and 41 W with a primary voltage of 200 V at 40 Hz. Compute the hysteresis and eddy current losses at a primary voltage of 300 volts at 60 Hz of the above transformer. Neglect small amount of copper loss at no-load.

(**Ans.** 43.5 W; 27 W)

Q11. A 50 kVA, 2200/110 V transformer when tested gave the following results:

- O.C. Test (L.V. side): 400 W, 10 A, 110 V.
- S.C. Test (H.V. side): 808 W, 20.5 A, 90 V.

Compute all the parameters of the equivalent circuit referred to the H.V. side and draw the resultant circuit.

(**Ans.**Shunt branch: R_0 = 12.1 k-ohms, X_M = 4.724 k-ohms, Series branch: r = 1.923 ohms, x = 4.39 ohms)

Q12. A 200 kVA transformer has an efficiency of 98% at full-load. If the maximum efficiency occurs at three-quarters of full-load, calculate (i) Iron loss at F.L. (ii) Cu loss at F.L. (iii) Efficiency at half-load. Ignore magnetizing current and assume a p.f. of 0.8 at all loads.

(**Ans.** 1.777 kW; 2.09 kW; 97.92%)

Q13. A transformer, when tested on full-load, is found to have Cu loss 1.8% and reactance drop 3.8% . Calculate its full-load regulation (i) At unity p.f. (ii) 0.8 p.f. lagging (iii) 0.8 p.f. leading.

(**Ans.** 1.80%; 3.7%; -0.88%)

Q14. An 11000/230 V, 150 kVA, 50Hz, 1-phase transformer has a core loss of 1.4 kW and full-load Cu loss of 1.6 kW. Determine (i) The kVA load for maximum efficiency and the minimum efficiency (ii) The efficiency at half full-load at 0.8 power factor lagging.

(**Ans.** 140.33 kVA; 97.6%; 97%)

Q15. A 100- kVA, single-phase transformer has an iron loss of 600 W and copper loss of 1.5 kW at full-load current. Calculate the efficiency at (i) 100 kVA output at 0.8 p.f. lagging (ii) 50 kVA output at unity power factor.

(**Ans.** 97.44%; 98.09%)

Q16. The primary resistance of a 440/110 V transformer is 0.5 Ω and the secondary resistance is 0.04 Ω. When 440 V is applied to the primary and secondary is left open-circuited, 200 W is drawn from the supply. Find the secondary current which will give maximum efficiency and calculate this efficiency for a laod having unity power factor.

(**Ans.** 53 A; 93.58%)

Q17. Two transformers A and B are connected in parallel to supply a load having an impedance of (2 + j 1.5Ω). The equivalent impedance referred to the secondary windings are 0.15 + j 0.5Ω and 0.1 + j 0.6Ω respectively. The open-circuit e.m.f. of A is 207 V and of B is 205 V. Calculate (i) The voltage at the load (ii) The power supplied to the load (iii) The power output of each transformer and (iv) The kVA input to each transformer.

(**Ans.** 189 $\angle$-3.8°V; 11.5 Kw; 6.5 kW, 4.95 kW; 8.7 kVA, 6.87 kVA)

Q18. Two electric furnaces are supplied with 1-phase current at 80 V from 3-phase, 11000 V supply mains by means of two Scott-connected transformers with similar secondary windings. Calculate the current flowing kW respectively in each of the 3-phase lines at U.P.P. when the loads on the two transformers are 550 kW of 800 kW.

(**Ans.** With 550 Kw on teaser, line current are: 57.5 A; 78.2; 78.2 A)

Books for further Reading

1 Electric Machinery, Fitzgerald, Kingslay, Umans, Tata McGraw-Hill.

2. Electric Machinery Fundamentals, Chapman, McGraw-Hill Higher Education.

3. Electric Machines, Nagrath and Kothari, Tata McGraw-Hill.

4. Electric Machinery, P.S. Bimbhra, Khanna Publishers.

5. Electrical Machines, R. K. Srivastava, 2/e, Cengage Learning Pvt. Ltd.-2011

6. Electrical Machines, Smarajit Ghosh, 2/e, Pearson edu, 2012

7. Electrical Machines-I, D. K. Palwalia, N K Garg, P Kumar, G Jain. Ashirwad Publishers-2020

REFERENCE BOOKS:

1. Electric Machinery and Transformer, Guru, Hiziroglu, Oxford University press.

3. Basic Electric Machines, Vincent Deltoro, Prentice Hall.

3. Performance and Design of A.C. machines, M. G. Say

For further reading scan to:-

Scan for video lecture on transformers

scan for video lecture on transformers

scan for video lecture on transformers

scan for lecture notes, problems and solution on transformers

Appendix

Practicals and Lab Experiments

Experiment: 01

Analysis of open circuit characteristics (O.C.C.) of DC shunt generator.

AIM: To obtain the O.C.C of the given self excited DC shunt generator and to determine

- O.C.C at the specified speed
- Critical field resistance
- Critical speed

Name Plate Details:			
DC Generator:		**DC Motor:**	
Rated Voltage		**Rated Voltage**	
Rated Current		**Rated Current**	
Rated Speed		**Rated Speed**	
Power Rating		**Power Rating**	

Scan to perform experiment through VLAb

Apparatus required:

Sr. No.	Apparatus	Range	Type	Quantity

Theory:

A D.C. generator requires an excitation circuit to generate an induced voltage. Depending on whether the excitation circuit consumes power from the armature of the machine or from separately required power supply, the generators may be classified as self excited or separately excited generators respectively.

The induced EMF in a DC generator is given by the equation $E_g = P\phi NZ/60A$ volts. Since P,Z and A are constants the above equation can be rewritten as $E_g = K\phi N$. If the speed of the generator is also maintained constant then $Eg = K_1\phi$, but the flux is directly proportional to the field current, hence $Eg = K_2I_f$. From the above equation it is clear that the induced EMF is directly proportional to the field current when speed is maintained constant. The plot between the induced EMF and the field current is known as open circuit characteristics of the DC generator. The typical shape of this characteristic is shown in figure. The induced EMF when the field current is zero is known as residual voltage. This EMF is due to the presence of a small amount of flux retained in the field poles of the generator called residual flux. Once the O.C.C is obtained the parameters such as critical field resistance, critical speed and the maximum voltage to which the machine can build up can be determined. If required the O.C.C at a different speeds can also be obtained. Critical speed is minimum speed below which the generator shunt fails to excited.

Precautions:

1. Remove the fuse carriers before wiring and start wiring as per the circuit diagram.
2. Check the position of the various rheostats as specified below:
   - Motor field rheostat is kept at minimum resistance position.
   - Generator field rheostat is kept at maximum resistance position.
3. The DPST switch on the load side is kept open at the time of starting the experiment.
4. Fuse calculations. As this is a load test, the required fuse ratings are
   - 120% of the motor rated current for supply side DPST.
   - 120% of the generator rated current for load side DPST.
5. Replace the fuse carriers with appropriate fuse wires after the circuit connections are checked by the staff in charge.

Procedure:

1. The circuit connections are made as per the circuit diagram.
2. Keeping the motor field rheostat in its minimum position, the generator field rheostat in maximum position, and the starter in its OFF position, the main supply is switched ON to the circuit.
3. The motor is started using the 3-point starter by slowly and carefully moving the starter handle from its OFF to ON position.

4. The motor is brought to its rated speed by adjusting its field rheostat and checked with the help of the tachometer.
5. With the DPST switch open, the residual voltage is noted.
6. Now the DPST switch is closed and the Potential divider is varied in steps and at each step the field current (I_f) and the corresponding induced EMF (E_g) are recorded in the tabular column. This procedure is continued until the generator voltage reaches 120% of its rated value the speed of the machine is maintained constant.
7. After the experiment is completed the various rheostats are brought back to their original position in sequence and then main supply is switched OFF.

Table for OCC

S.No	Field current I_f (amps)	Generator Voltage (E_g) Volt
1.		
2.		
3.		
4.		
5.		

Table for Measuring field resistance (R_f)

V_f (Volts)	I_f (amps)	(R_f) Ohms

Model Calculation:

To calculate critical resistance R_C (Ohms) and critical speed N_C(RPM)

$$\frac{N_{Critical}}{N_{Rated}} = \frac{E_{Critical}}{E_{Rated}}$$

Circuit Diagram:

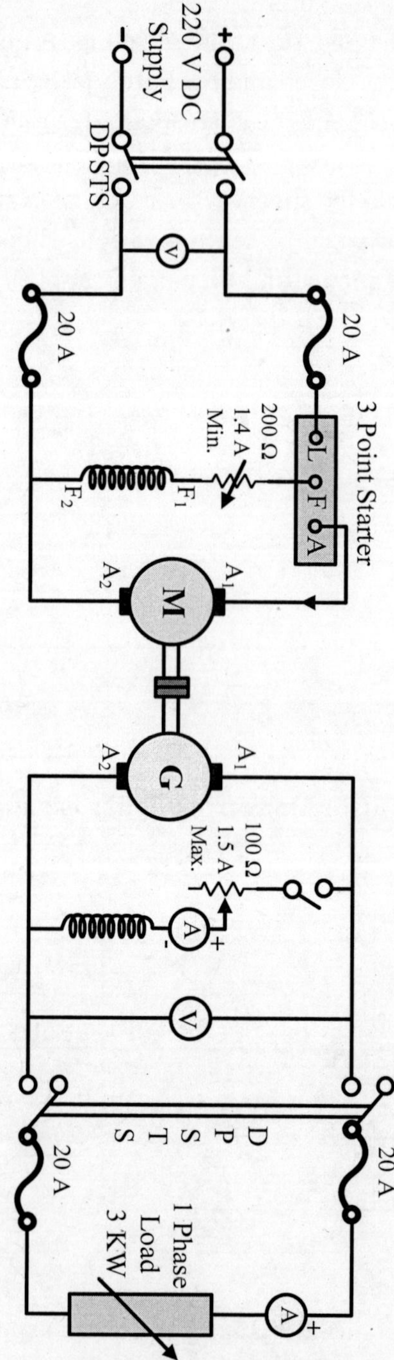

Fig.P1: OCC and Load test on self-excited DC shunt generator

Model Graph:

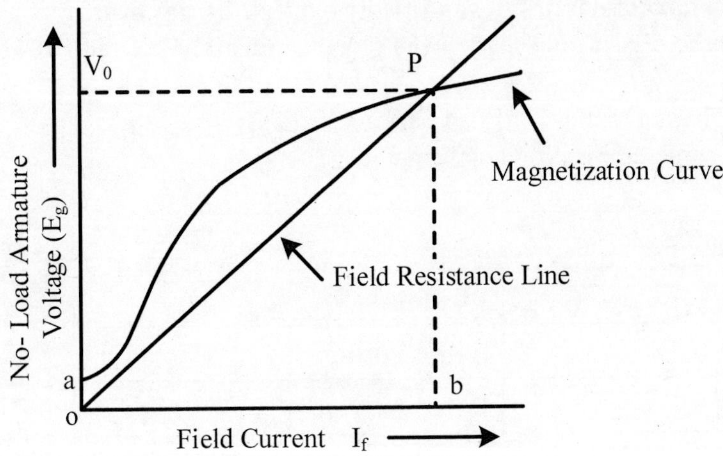

Fig.P2: Magnetization characteristics curve

Results:

The magnetization characteristics curve is drawn and the value of build-up voltage is obtained from the graph. The critical resistance and critical speed are determined from the graph.

Some viva-voce questions:

1. What is the principle of a generator?
2. What is meant by residual magnetism?
3. What is critical field resistance?
4. What is meant by saturation?
5. What is the difference between a separately excited DC generator and a shunt generator?
6. If a DC shunt generator fails to build up voltage, what may be the probable reasons?
7. What is DPST? What is its use in this experiment?
8. What is the reason for the presence of residual magnetism in the field poles?

Experiment: 02

Analysis of load characteristics of separately excited DC generator.

Aim: To conduct the direct load test on the given separately excited DC shunt generator to plot

- External Characteristics (or Load Characteristics)
- Internal Characteristics (or Total Characteristics)

Name Plate Details:			
DC Generator:		**DC Motor:**	
Rated Voltage		**Rated Voltage**	
Rated Current		**Rated Current**	
Rated Speed		**Rated Speed**	
Power Rating		**Power Rating**	

Scan to perform experiment through VLAb

Apparatus required:

Sr. No.	Apparatus	Range	Type	Quantity

Theory:

A D.C. Generator works on the principle of Faraday's law of electro magnetic induction which says that, "whenever a conductor is moved in magnetic field, an emf is generated in it". "The magnitude of induced emf is directly proportional to the rate of change of flux". The voltage equation for a DC shunt generator is given; by $V_L = E_g - I_a R_a$, Under no load condition; since I_a is negligibly small, from the above equation, the terminal voltage (V_L) is the no; load induced EMF (E_g), As the load on the generator increases, the load current and hence the armature current increases due to armature reaction the induced emf in the armature decreases. Also increased armature current causes increase in laRa drop. Hence the terminal voltage decreases with increasing load. The plot between the terminal voltage (V_L) and load current (I_L) is known as the external or load characteristics. The plot between the induced EMF (E_g) and the armature current (I_a) is known as the internal or total characteristics. The typical graph of internal and external characteristics is shown in model graph.

Precautions:

1. Remove the fuse carriers before wiring and start wiring as per the circuit diagram.
2. Check the position of the various rheostats as specified below:
   - Motor field rheostat is kept at minimum resistance position.
   - Generator field rheostat is kept at maximum resistance position.
3. The DPST switch on the load side is kept open at the time of starting the experiment.
4. Fuse calculations. As this is a load test, the required fuse ratings are
   - 120% of the motor rated current for supply side DPST.
   - 120% of the generator rated current for load side DPST.
5. Replace the fuse carriers with appropriate fuse wires after the circuit connections are checked by the staff-in-charge.

Procedure:

1. The circuit connections are made as per the circuit diagram.
2. Keeping the motor field rheostat in its minimum position, generator field rheostat in maximum position and the starter in its OFF position, the main supply is switched ON to the circuit.
3. The motor is started using the 3-point starter by slowly and carefully moving the starter handle from its OFF to ON position.
4. The motor is brought to its rated speed by adjusting its field rheostat and checked with the help of the tachometer.
5. With the DPST switch open, the generator field rheostat is slowly decreased until the generator voltage is equal to its rated value (220V). The terminal voltage and the field current are noted in the tabular column.
6. The DPST switch on the load side is now closed and the load on the generator is gradually increased in steps by switching on the lamps one by one. At each step the speed of the generator is checked and maintained constant at its rated value by adjusting the field rheostat of the motor. After satisfying this condition at each step of

loading, the terminal voltage (V_L), field current (I_f) and the load current (I_L) are noted down in the tabular column.

7. This procedure is continued until the generator is loaded to 120% of its rated value.
8. Once the experiment is completed the load on the generator is gradually decreased, the various rheostats are brought back to their original position in sequence and the main supply is switched OFF.

Sr. No.	V_L(V)	I_L(A)	I_f(A)	I_a(A)	I_aR_a(A)	$E_g = V_L + I_aR_a(V)$

Procedure for measurement of armature resistance

1. The circuit connections are made as per the circuit diagram.
2. Keeping the lamp load at the OFF position the main supply is switched ON.
3. The load is increased such that the current in the circuit is approximately adjusted to 25%, 50% and 75% of rated current of the generator and at these load conditions the armature voltage (V) and current (I) are noted in the tabular column.

Sr. No.	V_a(V)	I_a(A)	R_a(Ohms)

Calculation:

1. Determination of armature resistance (R_a):
   The armature winding resistance is calculated using ohms law R_a = V/I for each set of readings and the average of them is calculated. The effective resistance of the armature winding after taking into account the effect of temperature rise and skin effect is 1.2 times the average resistance R_a i.e. R_a (effective) = 1.2 R_a (average).
2. To plot the internal characteristics, the armature current and the induced EMF are calculated using the expressions,

$I_a = I_L + I_f$ and $E_g = V_L + I_a R_a(eff)$

3. The plots of V_L Vs I_L and E_g Vs I_a are drawn to scale in the same graph sheet

Model Graph:

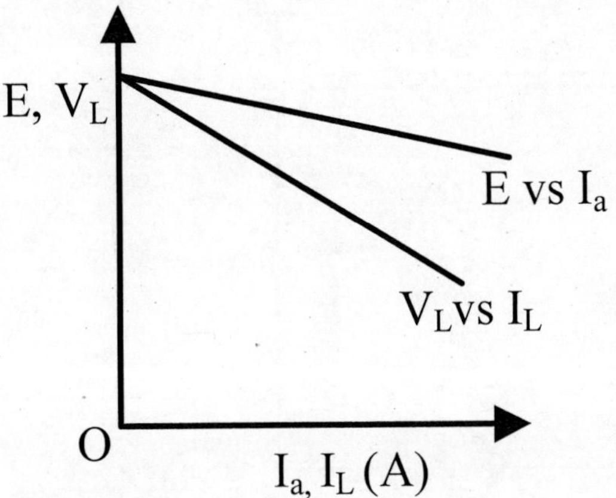

Fig.P5:

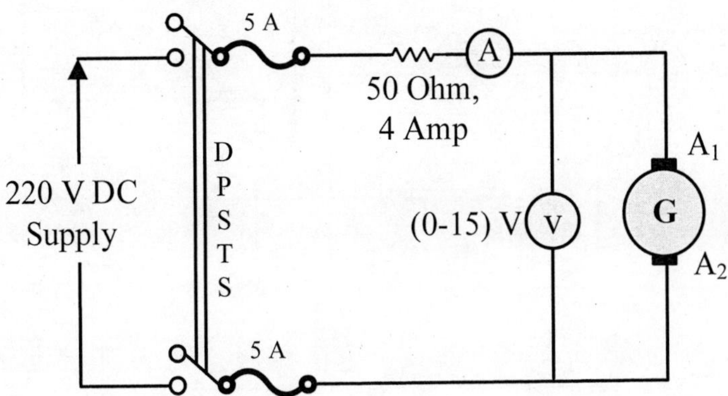

Fig.P6: Armature resistance Measurement

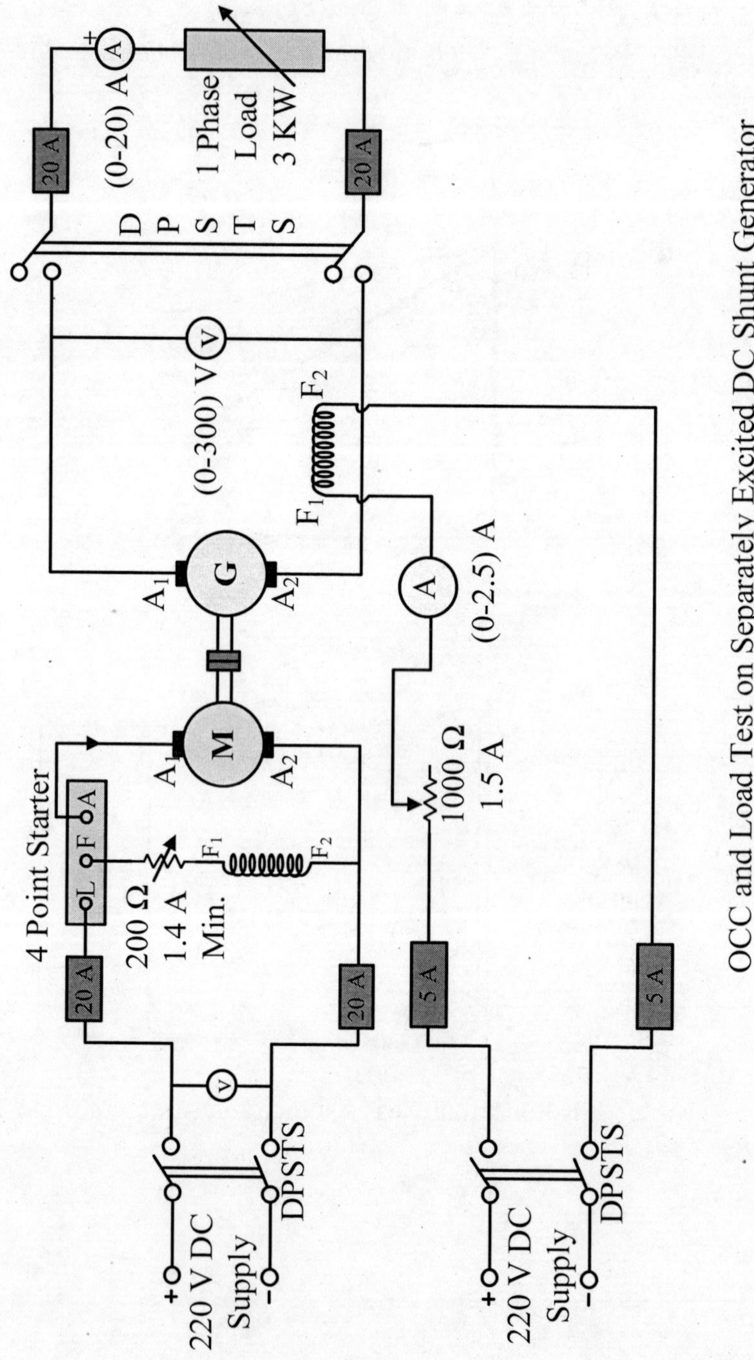

Fig.P7: Load test on separately excited DC shunt generator

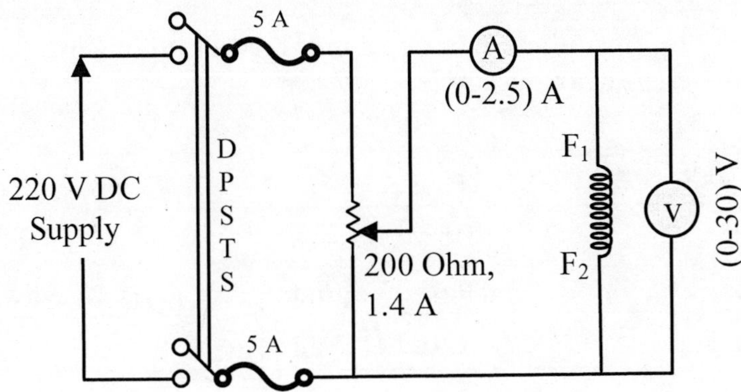

Fig.P8: Measurement of field resistance

Result:

The direct load test on the given separately excited DC generator has been conducted and the internal and external characteristics are plotted.

Some viva-voce questions:

1. What are the reasons for the drooping load characteristics?
2. Why does the terminal voltage decrease as the load current increases?
3. Why the load characteristics of dc shunt is having Drooping characteristics?
4. Why the Drooping of Dc Shunt Generator is more when compared to separately excited generator? (If independent of voltage.)
5. How can the external characteristics be drawn?
6. How can the internal characteristics be drawn from External characteristics?
7. What are the applications of shunt generator?
8. Shunt field winding of a dc machine consists of

   a. Many turns of thin wire

   b. Few turns of thick wire

Experiment: 03

Determination of efficiency of DC machine through Hopkinson's Test.

Aim: To perform Hopkinson's test on two similar DC shunt machines and hence obtain their efficiencies at various loads.

Name Plate Details:			
DC Generator:		**DC Motor:**	
Rated Voltage		**Rated Voltage**	
Rated Current		**Rated Current**	
Rated Speed		**Rated Speed**	
Power Rating		**Power Rating**	

Instruments

S. No.	Name	Type	Range	Quantity
1.	Ammeter	MC	0-30 A	2
2.	Ammeter	MC	0-2 A	2
3.	Voltmeter	MC	0-300 V	1
4.	Voltmeter	MC	0-600 V	1
5.	Rheostats	Single tube	290 Ω, 1.4 A	1
6.	Techometer	Digital	0-2000 rpm	1

Scan to perform experiment through VLAb

Theory

The connection diagram of Hopkinson's test is shown in fig.P9 In the connection diagram, the machine M acts as a motor and is started from the supply with the help of starter. The switch S is kept open. The field current of the machine M is adjusted with the help of field rheostat R_m to make the motor to run at its rated speed. Machine G acts as a generator.

As the G is driven by the machine M, hence it runs at rated speed of M. The field current of machine G is so adjusted with the help of its field rheostat R_g that the armature voltage of the generator G is somewhat higher than the supply voltage. When the voltage of the generator is equal to and of the same polarity of the busbar voltage (voltmeter across Switch S reads zero), the switch S is closed and the generator is connected to the busbar.

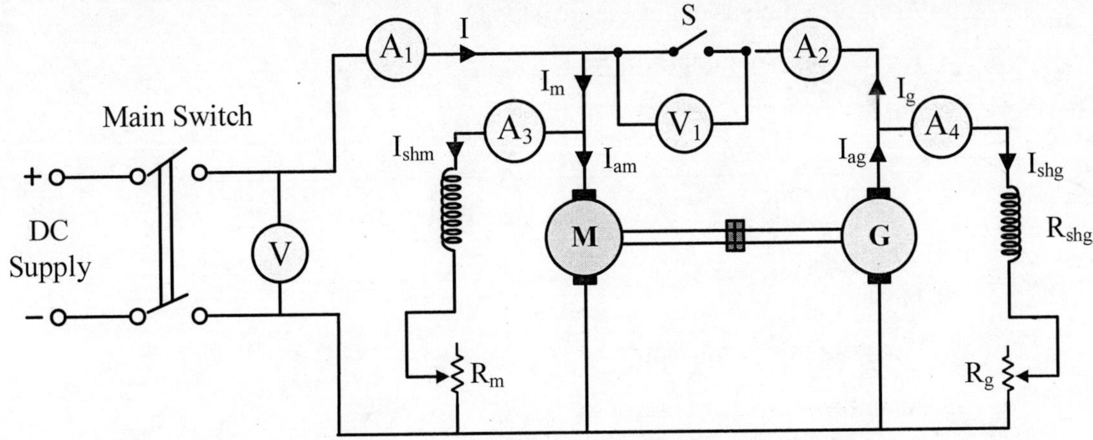

Fig.P9: Circuit arrangement for Hopkinson's test

Now, both the machines are connected in parallel across the supply voltage. Under this condition, the generator neither taking any current nor giving any current to the supply, thus it is said to float. Now, by adjusting the excitation of the machines with the help of the field rheostats, any load can be thrown on the machines.

Here, Input power from the supply = VI

This input power from the supply is equal to the total losses of both machines.

$$\text{Armature Cu loss of motor} = I_{am}^2 R_a$$

$$\text{Field Cu loss of the motor} = I_{shm}^2 R_{shm}$$

$$\text{Armature Cu loss of generator} = I_{ag}^2 R_a$$

$$\text{Field Cu loss of the generator} = I_{shg}^2 R_{shg}$$

As the two machines are identical, the constant losses of both the machines P_C are assumed to be equal and given by,

$$P_C = (Power\ input\ from\ supply) - (Armature\ and\ shunt\ cu\ losses\ of\ both\ machines)$$

$$P_C = VI - (I_{am}^2 R_a + I_{shm}^2 R_{shm} + I_{ag}^2 R_a + I_{shg}^2 R_{shg})$$

It was assumed that the constant losses are being equally divided between the two machines.

$$\therefore Constant\ loss\ per\ machine = \frac{P_C}{2}$$

Now, the efficiency of two machines can be determined as follows-

$$\text{Generator output} = VI_{ag}$$

$$\text{Constant losses for generator} = \frac{P_C}{2}$$

Armature Cu loss of generator = $I_{ag}^2 R_a$

Field Cu loss of generator = $I_{shg}^2 R_{shg}$

Therefore, the efficiency of the generator is given by,

$$\eta_g = \frac{Output}{Output + Losses} = \frac{VI_{ag}}{VI_{ag} + I_{ag}^2 R_a + I_{shg}^2 R_{shg} + \frac{P_C}{2}}$$

Efficiency of motor

Motor input = $VI_m = V\,(I_{am} + I_{shm}\,)$

Constant losses for motor = $\frac{P_C}{2}$

Armature Cu loss of motor = $I_{am}^2 R_a$

Field Cu loss of motor = $I_{shm}^2 R_{shm}$

Therefore, the efficiency of the motor given by,

$$\eta_m = \frac{Input - Losses}{Input}$$

$$= \frac{[V\,(I_{am} + I_{shm})] - \left[I_{am}^2 R_a + I_{shm}^2 R_{shm} + \frac{P_C}{2}\right]}{V\,(I_{am} + I_{shm})}$$

Hence, the efficiency of the motor and the generator at various loads can be worked out, recording the various currents and voltage during the experiment and measuring the resistance of armature of both the machines.

The major advantages of this test are as follow :

1. Power drawn from the supply is low.
2. Both the dc machine are operating under loaded conditions, as such stray load losses are taken into account. Moreover, final temperature rise of the machines can be checked.

Circuit Diagram

Fig.P9 shows the circuit diagram, in which two identical dc machines are connected in such a way that one of them is acting as a motor and another as a generator. The following instrument connected in the circuit serve the function indicated against each.

1. Ammeter A_1, A_2 – to measure the current drawn form the supply and the generator current respectively.
2. Ammeter, A_3, A_4 – to measure field current of motor and generator respectively.
3. Voltmeter, V – to measure applied voltage.
4. Voltmeter, V_1 – to check the condition for closing the switch, S.
5. Rheostat – to vary field current of motor and generator respectively.

Procedure

Connect the two dc machine, coupled mechanically as per the circuit diagram shown in fig.

1. Ensure that the switch, S in the open position.
2. Adjust the rheostat, so that the field current of the motor is maximum.
3. Adjust the rheostat, so that the field current of the generator is minimum.
4. Switch on the dc supply and start the dc motor using the starter properly.
5. Adjust the speed of the motor to rated value by varying the rheostat provided in its field circuit.
6. Adjust rheostat in the field circuit of generator, so that that generated voltage of the generators is equal to the supply voltage.
7. Check the voltage across the switch, S. In case of wrong polarity, the voltmeter will record twice the supply voltage. In such a case, switch off the mains and reverse the armature terminals of the generator. Repeat steps 5, 6 and 7. Now the voltmeter will indicate zero condition, the machine working as a generator is just floating i.e. neither drawing any current from nor giving into the lines.
8. Increase the field current of the generator or decrease the field current of the motor or by doing both adjustments, any desired load can be put on the generator. Presently , adjust the field currents of motor and generator, so that load on the generator is approximately 10 percent of its full load value. Record the readings of all the meters connected in the circuit.
9. Repeat step 9 for various values of load current, till the full load current of the generator.
10. Reduce the load on the generator and motor by varying their field currents..
11. Switch off the dc supply.
12. Measure the resistance of the armature circuit of motor and generator.

Observation : May be tabulated as follows.

S. No.	V	I_{ag}	I_{am}	I_{shm}	I_{shg}	W_c	η_m	η_G

Result:

Hopkinson's test on two similar DC shunt machines has been performed successfully and the efficiency obtained is.................... percent.

Some viva-voce questions:

1. What is the purpose of the Hopkinson back-to-back test on transformers?
2. Can you explain the basic principle behind the Hopkinson back-to-back test?
3. How are the two transformers connected in the back-to-back configuration for this test?
4. What are the main parameters that can be determined using this test?

5. How is the no-load current and losses of the transformers obtained from the back-to-back test data?
6. What measurements are needed during the test, and how are they taken?
7. What are the advantages of using the Hopkinson back-to-back test over conventional transformer testing methods?
8. Are there any limitations or potential sources of error associated with the Hopkinson back-to-back test? How can they be minimized?
9. How is the magnetizing current of the transformers accounted for in the test results?
10. What types of transformers are suitable for conducting the Hopkinson back-to-back test?

Experiment: 04

Examine the effective efficiency and, DC series motor.

1. **Aim:**
   - i. To discover the characteristic change in speed of a DC Shunt Motor as it is loaded.
   - ii. To discover the characteristic change in torque output of a DC Shunt Motor as it is loaded

Name Plate Details:			
DC Generator:		**DC Motor:**	
Rated Voltage		**Rated Voltage**	
Rated Current		**Rated Current**	
Rated Speed		**Rated Speed**	
Power Rating		**Power Rating**	

Instruments

S. No.	Name	Type	Range	Quantity

Scan to perform experiment through VLAb

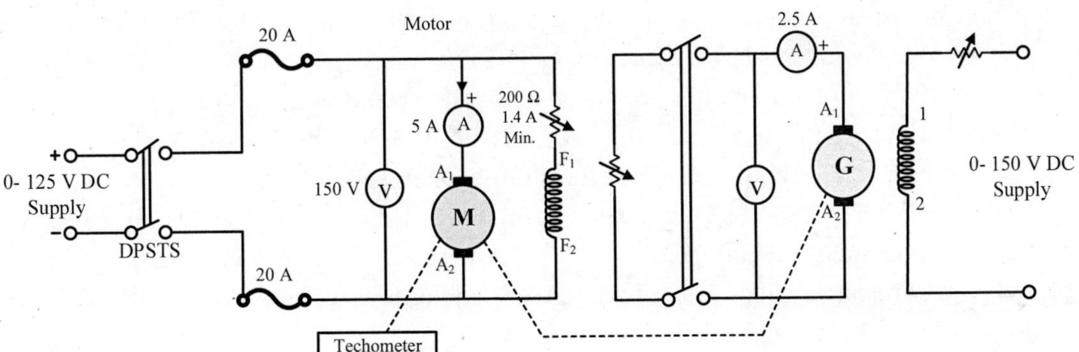

Fig.P10: Experimental setup for speed-torque characteristic of DC shunt motor

Procedure

1. Place the two machines on the bedplate with the motor on the left and the dynamometer on the right.
2. Couple the two machines tightly, using the rubber coupling. Be sure the coupling fits snugly inside both flanges. Be sure the rotor locking device has been removed from the dynamometer.
3. Clamp the machines tightiy to the bedplate. Place the coupiing guard over the coupling and the shaft guard over the motor and dynamometer shafts.
4. Connect the motor as shown in Figure. P10. Note that this is a shunt moror connection. Turn the motor's field rheostat fully countetclockwise to its minimum resistrnce position.
5. Turn the knob of the 0-125 volt supply fully counterclockwise to its zero output position. Power should remain off.
6. Connect the dymanometer as shown in Figure. P10. Note that this is a separately excited generator connection. Turn the dynamometer's field rheostat knob fully counterclockwise to its minimum resistance position.
7. Have someone check your connections to be sure they are conect. Then turn ON only the main AC and the 0-150 supply circuit breakers.
8. Temporarily connect a voltmeter across the 0-150 volt supply and adjust the knob until there is 115 volts across the dynamometel's shunt field. Then connect the voltmeter across the RL-100A Resistance Load Bank. Be sure all toggle switches are in the downward (OFF) position. Turr the dymamometer's field rheostat fully clockwise to its maximum resistance position.
9. Turn ON the 0-125 volt supply and the motor's circuit breaker switches. Start the motor by slowly increasing the output of the 0-125 volt supply to 125 volts.
10. Place the Tacho-Generator against the motor shaft and turn the knob of the field rheostat clockwise until the indicator reads 1950 rpm.
11. Use the dynamometer's field rheostat to adjust its output voltage to 115 volts.
12. Recheck the output of the 0-125 volt supply to be sure that there is still 125 volts applied to the motor armature. If not adjust the supply. Then repeat step 11.
13. In Table of TEST RESULTS, record the motor speed, armature voltage, armature current, torque, and generated voltage.
14. Switch ON resistance legs 1, 2, and 3 on the load bank.
15. Repeat Steps 11, 12, and 13.
16. Switch ON resistance legs 4, 5, and 6, on the load bank.
17. Repeat Steps 11, 12, and 13.
18. Switch ON resistance legs 7, 8, and 9 on the load benk.

19. Repeat Steps 11, 12, and 13.
20. Turn OFF all circuit breaker switches. Disconnect all leads

Test Result	No Load	Step-14	Step-16	Full Load
Armature voltage				
Armature Current				
Torque (N-m)				
Generated Voltage				
Speed				

1. On the graph provided, used the data you have recorded in Table to plot a curve showing how the speed of a DC shunt motor changes as the armatur.e current increases with increasing load. Label the curve SPEED.
2. On the same graph, use the data from Table to plot a curve showing how the output torque of a DC shunt motor changes as the arnature current increases with increasing load. Label the curve TORQUE.
3. The percent change in speed from no load to full load is called "speed regulation'. It is computed by dividing the change in speed by the full load speed.

$$\% \text{ Speed Regulation } = \frac{Speed\ (NL) - Speed\ (FL)}{Speed\ (FL)} \times 100$$

Compute the speed regulation for the DC shunt motor.

Graph between N and Ta:

2. **Aim**
   i. To discover the characteristic change in speed of a DC Series Motor as it is loaded.
   ii. To discover the characteristic change in torque output of a DC Series Motor as it is loaded

Name Plate Details:			
DC Generator:		**DC Motor:**	
Rated Voltage		**Rated Voltage**	
Rated Current		**Rated Current**	
Rated Speed		**Rated Speed**	
Power Rating		**Power Rating**	

Instruments

S. No.	Name	Type	Range	Quantity

Theory:

DC series motors are widely used in traction vehicles, such as locomotives and electric cars, because of their ability to produce a high toryue without a corresponding high armature current. However, there is a drastic change in speed when a series motor is loaded. In traction equipment, the heaviest load is at start, when torque is more important than speed.

In a DC series motor, the armature current and the field current are the same current, since the two are in series. Torque output is proportional to the armature current and field flux. The field flux, then is proportional to armature current. This makes torque proportional to the square of armature current. For light loads a series motor produces less torque for the same arrnature current; for heavy loads it produces more torque than a shunt motor.

Series motors are alwaye rigtdly connected to their loads, and never operated without some Ioad. A shunt motor can operate unloaded because a strong field flux is always present. Enough CEMF is generated to lower armature current to just the amount needed to overcome windage and friction losses. A series motor, on the other hand, will get into a nrun awayn condition, if unloaded. That is when the motor continues to increase in speed until it tears itself to pieces.

Assume a series motor is running under load and the load is suddenly removed. The torque that was driving the load is now applied to the motor shaft as accelerating torque. As the motor begins to speed up, additional CEMF is generated, reducing armature current. But this also reduces the strength of the field flux. Not eough CEMF is generated to reduce the armature current enough to stop the motor fiwm accelerating. The motor speeds up morle. Armature current continues dropping but the field keeps getting weaker. CEMF can never catch up with the armature current, Even at light loads, the speed can become excessive before armaturre current decreases enough to eliminate the accelerating torque.

Procedure

1. Place the two machines on the bedplate with the motor on the left and the dynamometer on the right.
2. Couple the two machines tightly, using the rubber coupling. Be sure the coupling fits snugly inside both flanges. Be sure the rotor locking device has been removed from the dynamometer.
3. Clamp the machines tightiy to the bedplate. Place the coupiing guard over the coupling and the shaft guard over the motor and dynamometer shafts.
4. Connect the motor as shown in Figure. P11. Note that this is a shunt moror connection. Turn the motor's field rheostat fully counter clockwise to its minimum resistance position.
5. Turn the knob of the 0-125 volt supply fully counter clockwise to its zero output position. Power should remain off.
6. Connect the dynamometer as shown in Figure. P11. Note that this is a separately excited generator connection. Note also that the shunt field rheostat is not being used.
7. Have someone check your connections to be sure they are correct. Then turn ON resistance legs 1, 2, and 3 on the load bank.
8. Turn ON the main AC and the 0-150 volt supply circuit breakers.
9. Temporarily connect a voltmeter acnoss the 0-150 volt excitation supply and adjust the knob for 125 volts. Then connect the voltmeter acrose the RJ-100A Resistance Load Bank.
10. Turn ON the 0-125 volt supply and the motor circuit breaker switches. Start the motor by slowly increasing the output of the 0-125 volt supply to 115 volts.
11. Use the 0-150 volt excitation supply knobto adjust the dynamometer's output voltage to 125 volts. If necessary, re-adjust the 0-125 volt supply to 115 volts.
12. Read, and record in Table of TEST RESULTS, the values of motor speed, armature voltage, armature current, torque, and load voltage.
13. Switch ON resistance legs 4, 5, and 6 on the load bank.
14. Repeat Steps 11 and 12.
15. Switch ON resistance leg 7, 8, and 9 on the load bank.

16. Repeat Steps 11 and 12.
17. Turn OFF all circuit breaker switches. Disconnect all loads.

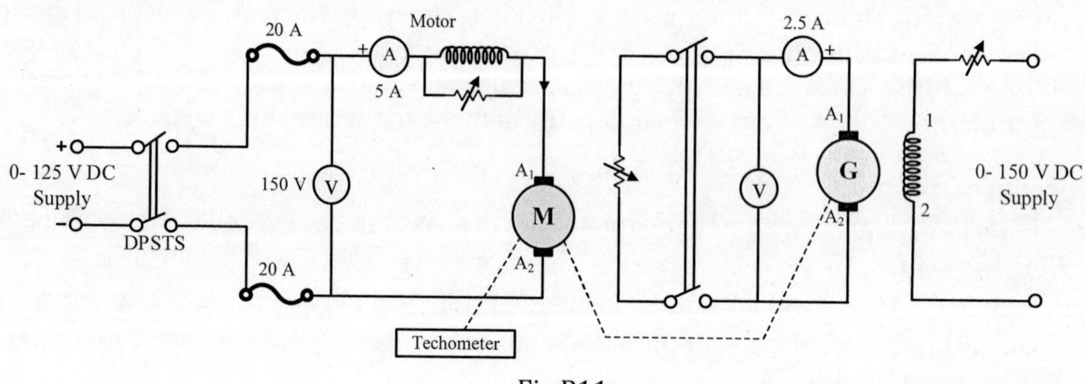

Fig.P11:

Test Result	Step-12	Step-13	Step-15
Armature voltage			
Armature Current			
Torque (N-m)			
Generated Voltage			
Speed			

1. On the graph provided, used the data you have recorded in Table to plot a curve showing how the speed of a DC series motor changes as the armature current increases with increasing load. Label the curve SPEED.
2. On the same graph, use the data from Table to plot a curve showing how the output torque of a DC shunt motor changes as the arnature current increases with increasing load. Label the curve TORQUE.

Graph between N and Ta:

Some viva-voce questions:

1. What is the purpose of conducting the speed-torque characteristic experiment on a DC shunt motor?
2. Can you explain the basic principle behind the speed-torque characteristic of a DC shunt motor?
3. How do you ensure that the motor is in a steady-state condition before taking readings?
4. How does the field winding control the speed-torque characteristic of a DC shunt motor?
5. What factors may cause deviations in the obtained speed-torque characteristic from theoretical expectations?
6. What is the purpose of conducting the speed-torque characteristic experiment on a DC series motor?
7. How is the speed-torque characteristic of a DC series motor different from that of a DC shunt motor?
8. How can the speed-torque characteristic of a DC series motor be used in practical applications or motor selection?

Experiment: 05

Determination the load characteristics of DC compound motor.

Aim:

To obtain internal and external characteristic of DC compound generator by conducting load test.

Name Plate Details:			
DC Generator:		**DC Motor:**	
Rated Voltage		**Rated Voltage**	
Rated Current		**Rated Current**	
Rated Speed		**Rated Speed**	
Power Rating		**Power Rating**	

Scan to perform experiment through VLAb

Apparatus required:

Sr. No.	**Apparatus**	**Range**	**Type**	**Quantity**

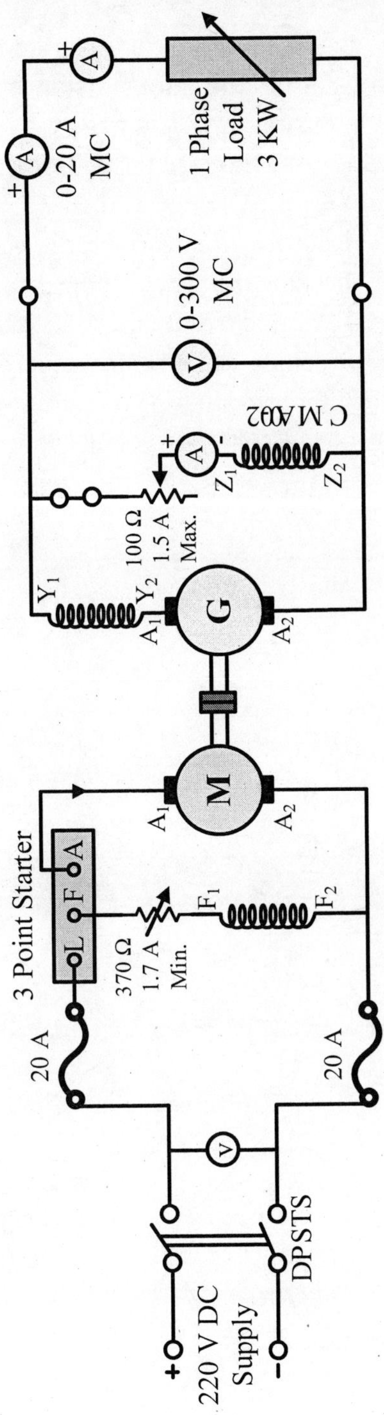

220 V DC
Supply
DPSTS
20 A
20 A
3 Point Starter
L
F
A
370 Ω
1.7 A
Min.
F_1
F_2
A_1
A_2
M
G
Y_1
Y_2
100 Ω
1.5 A
Max.
Z_1
Z_2
0-300 V
MC
0-20 A
MC
1 Phase
Load
3 KW

Fig.P12:

Procedure:

1. Make the connections as shown in the circuit diagram.
2. Keep the motor field rheostat in minimum resistance position (Resistance) and the Generator field rheostat in maximum resistance position at starting.
3. Start the MG set and bring it to the rated speed of the Generator by adjusting the motor field rheostat.
4. Adjust the terminal voltage of the generator to rated value by means of the generator field rheostat. Keep the rheostat in this position throughout the experiment as its variation changes the field circuit current and hence the generated e.m.f.
5. Put on the load and note down the values of load current IL and terminal voltage VT at the generator side, for different values of load until full load current.
6. Draw external characteristics VT vs IL & Internal characteristics Eg vs I Where Eg = V+ IaRa.

Sr. No.	V_T(V)	I_L(A)	$E_g = V + I_aR_a(V)$

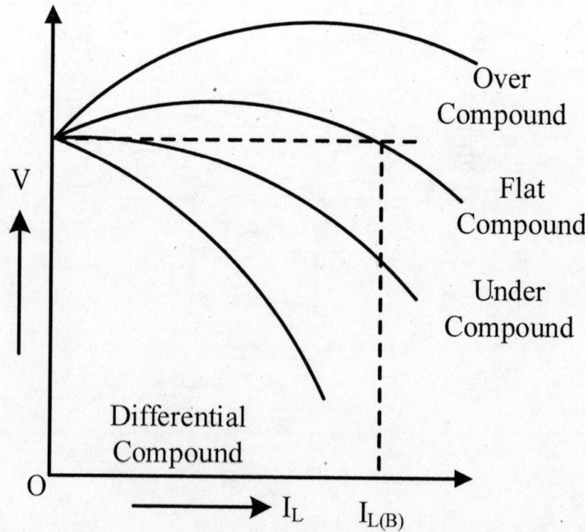

Fig.P13: Load Characteristics of DC Compound Generator

Some viva-voce questions:

1. Can you explain the concept of a DC compound motor and how it differs from a DC series or shunt motor?
2. How is the motor connected to the load during the experiment, and how is the load varied?
3. What measurements are taken during the experiment, and what instruments are used for this purpose?
4. How is the armature current varied during the experiment, and what range of currents is typically used?
5. What are the different operating modes of a compound motor, and how do they affect its performance?
6. What are the advantages and disadvantages of using a compound motor compared to a series or shunt motor?
7. How can the load characteristics of a DC compound motor be used in practical applications or motor selection?

Experiment: 06

Examine the Speed control methods of DC Shunt motor using armature control menthod and field control method.

(a) Aim: - Speed control of D.C. shunt motor by field current control method & plot the curve for speed verses field current.

Name Plate Details:	
DC Motor:	
Rated Voltage	
Rated Current	
Rated Speed	
Power Rating	

Scan to perform experiment through VLAb

S. No.	Name	Type	Range	Quantity

Theory

The back emf for a dc motor is given by,

Back emf, $E_b = \frac{P\phi NZ}{60\ A}$

The number of poles, P the armature conductors, Z and the number of parallel paths, A are constant for a particular machine.

The equation for the speed of the motor clearly indicated the following,

(i) Speed of the DC motor can be controlled below the normal range of speed by varying the resistance in the armature circuit included in the form of a rheostat as a variable resistance (armature control).

(ii) Speed of the DC motor can be controlled above the normal range of speed by decreasing the flux Ø i.e. by decreasing the current in the field circuit by including an external resistance in the form of a rheostat as variable resistance (field control).

Armature Control

Let the external resistance in the armature circuit of shunt motor be R ohms, then the speed equation modifies to,

$$N = K \frac{V - I_a\,(R_a + r)}{\phi}\ rpm$$

Hence the speed of the motor decreases with an increase in the value of external resistance R. Thus reduced speeds lower than the no load speed can be obtained by this method. However, there is an excessive wastage of power in the additional resistance, which lowers the efficiency of the motor considerably.

Field Control

The speed of the dc motor can be increased beyond the no load speed by inserting an external resistance in the shunt field circuit. The current in the external resistance is very low, hence the losses occurring in the additional resistance is quite small.

Circuit Diagram

Figure. P16 shows the circuit diagram for speed control of dc motor. Instruments used in the circuit serve the function mentioned against each.

- Rheostat (45 Ω, 5 A) - to vary the voltage applied to the armature winding of dc motor.
- Voltmeter - to measure the applied voltage across the armature winding.
- Rheostat (290 Ω, 1.8 A) - to vary the field current of dc motor.
- Ammeter - to measure the field current.

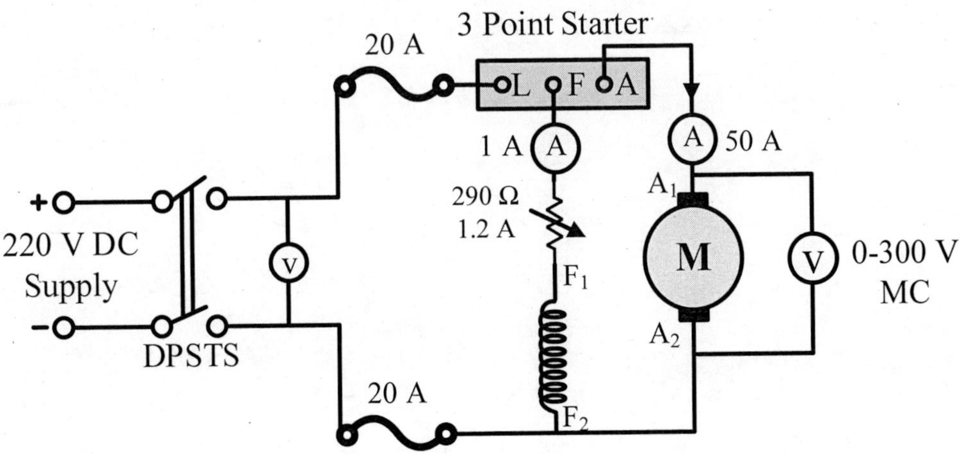

Fig.P16: Circuit diagram for speed control of dc motor

Procedure

1. Connect the dc motor as per the circuit diagram shown in figure.
2. Ensure that the external resistance in the armature circuit is maximum.
3. Ensure that the external resistance in the field circuit is minimum.
4. After ensuring step 2 and 3, switch on the dc supply, as a result motor will start running at a low speed.
5. Cut out the external resistance in the armature circuit and adjust the field current, so that the speed of the motor becomes rated speed.
6. The field current is kept constant to the above value. Vary the voltage applied to the armature by varying the external resistance in the armature circuit. Record the applied voltage and the corresponding speed.
7. Repeat step 6 for various values of applied voltage to the armature, till the rated voltage of the motor.
8. Keep the applied voltage to the armature constant at its rated value. Vary the speed of the motor by inserting external resistance in the field circuit. Record the field current and the corresponding speed of the motor.
9. Repeat step 8 for various values of field current, till the speed of the motor is about 1.4 times the rated speed of the motor. It is not advisable to increase the speed beyond 1.4 times the rated speed, otherwise mechanical stresses will be high, which may damage the motor. Hence, the field current should not be decreased to a very low value.
10. Switch off the main supply to stop the motor.

Observation: May be tabulated as follows.

Field control		
S. No.	Field Current	Speed

(b) Aim:- Speed control of D.C. shunt motor by armature voltage control method & plot the curve for speed verses armature voltage.

S. No.	Name	Type	Range	Quantity

Theory

The back emf for a dc motor is given by,

Back emf, $E_b = \dfrac{P\phi NZ}{60\,A}$

The number of poles, P the armature conductors, Z and the number of parallel paths, A are constant for a particular machine.

The equation for the speed of the motor clearly indicated the following,

(i) Speed of the dc motor can be controlled below the normal range of speed by varying the resistance in the armature circuit included in the form of a rheostat as a variable resistance (armature control).

(ii) Speed of the dc motor can be controlled above the normal range of speed by decreasing the flux φ i.e. by decreasing the current in the field circuit by including an external resistance in the form of a rheostat as variable resistance (field control).

Armature Control

Let the external resistance in the armature circuit of shunt motor be R ohms, then the speed equation modifies to,

$$N = K\,\frac{V - I_a\,(R_a + r)}{\phi}\ rpm$$

Hence the speed of the motor decreases with an increase in the value of external resistance R. Thus reduced speeds lower than the no load speed can be obtained by this method. However, there is an excessive wastage of power in the additional resistance, which lowers the efficiency of the motor considerably.

Circuit Diagram

Fig. P17 shows the circuit diagram for speed control of dc motor. Instruments used in the circuit serve the function mentioned against each.

- Rheostat (45 Ω, 5 A) - to vary the voltage applied to the armature winding of dc motor.

- Voltmeter - to measure the applied voltage across the armature winding.
- Rheostat (290 Ω, 1.8 A) - to vary the field current of dc motor.
- Ammeter - to measure the field current.

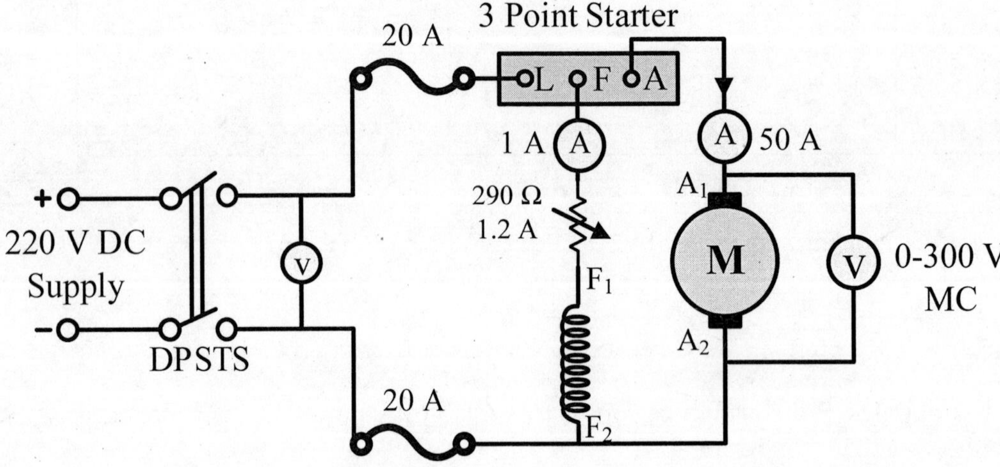

Fig.P17: Circuit diagram for speed control of dc motor.

Procedure

1) Connect the dc motor as per the circuit diagram shown in figure.
2) Ensure that the external resistance in the armature circuit is maximum.
3) Ensure that the external resistance in the field circuit is minimum.
4) After ensuring step 2 and 3, switch on the dc supply, as a result motor will start running at a low speed.
5) Cut out the external resistance in the armature circuit and adjust the field current, so that the speed of the motor becomes rated speed.
6) The field current is kept constant to the above value. Vary the voltage applied to the armature by varying the external resistance in the armature circuit. Record the applied voltage and the corresponding speed.
7) Repeat step 6 for various values of applied voltage to the armature, till the rated voltage of the motor.
8) Keep the applied voltage to the armature constant at its rated value. Vary the speed of the motor by inserting external resistance in the field circuit. Record the field current and the corresponding speed of the motor.
9) Repeat step 8 for various values of field current, till the speed of the motor is about 1.4 times the rated speed of the motor. It is not advisable to increase the speed beyond 1.4 times the rated speed, otherwise mechanical stresses will be high, which may damage the motor. Hence, the field current should not be decreased to a very low value.

10) Switch off the main supply to stop the motor.

Observation: May be tabulated as follows.

Armature control		
S. No.	Applied voltage	Speed

Some viva-voce questions:

1. What is the purpose of speed control in a DC shunt motor?
2. Can you name and briefly explain the different methods of speed control used for DC shunt motors?
3. How does the field flux control method affect the speed of the DC shunt motor? What are the advantages and disadvantages of this method?
4. Describe the armature voltage control method and its impact on the motor's speed. What are its pros and cons?
5. Can you explain the concept of field weakening in the context of DC shunt motor speed control?
6. How is the efficiency of a DC shunt motor affected when using different speed control methods?
7. Under what circumstances would you choose a particular speed control method over others for a DC shunt motor application?
8. Can you explain the concept of armature reaction and how it can affect the speed control of a DC shunt motor?
9. How does the speed-torque characteristic of a DC shunt motor vary when different speed control methods are applied?

Experiment: 07

Examine the Speed control of DC motor using ward leonard method

Aim: To perform speed control of DC motor by using Ward- Leonard Method of speed control

Scan to perform experiment through VLAb

In the ward-leonard method, the speed control of D.C. motor can be obtained by varying the applied voltage to the armature. In this method M is the main D.C. motor whose speed is to be controlled, and G is a separately excited D.C. generator which is driven by a 3-phase induction motor. The combination of ac driving motor and the dc generator is called the motor-generator set.

Let the external resistance in the armature circuit of shunt motor be R ohms, then the speed equation modifies to,

$$N = K\,\frac{V - I_a\,(R_a + r)}{\phi}\,rpm$$

Hence the speed of the motor decreases with an increase in the value of external resistance R. Thus reduced speeds lower than the no load speed can be obtained. However, there is an excessive wastage of power in the additional resistance, which lowers the efficiency of the motor considerably.

Thus, the speed of a D.C motor may be varied by either of the following adjustments:

1. Changing the flux per pole Φ, by varying the field current.
2. Changing external resistance in the armature circuit.
3. Changing the applied voltage V (Ward-Leonard speed control method).

Ward-Leonard System:

Ward Leonard system of speed control is based on principle of armature voltage control. This method was introduced in 1891. The drawbacks of the earlier methods can be overcome by this method. When large motors are to be controlled (rotating by starting and speed reversal), a separate motor-driven generator of suitable rating is used. The method is useful in rolling mills, where slab of molten metal is pressed between two rollers, which are rolled in forward and reverse directions so that a sheet metal is finally obtained. In this process the distance between rollers is also adjusted automatically. The schematic diagram of the Ward Léonard method of speed control of a DC shunt motor is shown in Fig. P18.

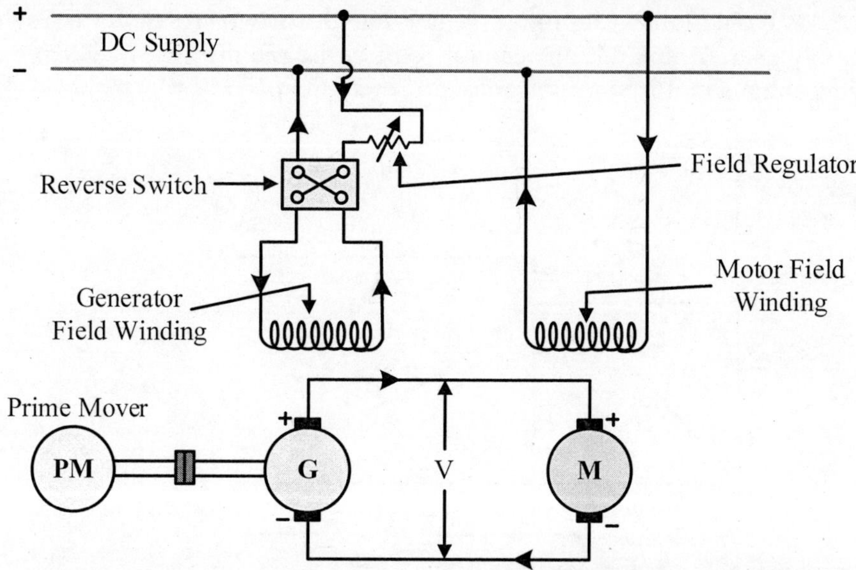

Fig. P18 Ward-Leonard method of speed control for DC motor

In this system M is the main DC motor whose speed is to be controlled, and G is a separately excited DC generator. The DC motor *M* is fed from the generator *G* and its field winding is connected directly to a constant DC supply line. The generator G is driven by a 3-phase driving motor or prime mover (PM) which may be an induction motor or a synchronous motor. The combination of AC driving motor and the DC generator is called the motor-generator (M-G) set. The field winding of the DC generator is connected to a constant voltage DC supply line through a field regulator and reversing switch.

The voltage of the generator fed to the motor, can be varied from zero to its maximum value by means of its field regulator. By reversing the direction of the field current by means of the reversing switch, the polarity of the generated voltage can be reversed at zero speed and field current is increased in reverse direction and hence the direction of rotation of motor *M*. Hence, by this method, the speed and direction of rotation both can be controlled very accurately. When speed control over a wide range is required, combination of armature voltage control and field flux control is used. In armature voltage control method constant torque and variable power drive is obtained from speed below the base speed. This is shown in Fig. P19.

As mentioned, earlier the driving AC motor can be an induction motor or synchronous motor. An induction motor operates at a lagging power factor. The synchronous motor running at fixed speed may be operated at a leading power factor by over -excitation of its field. Leading reactive power taken by over-excited synchronous motor compensates for the lagging reactive power taken by other inductive loads in the plant. Thus, the power factor of the plant is improved.

When the load is heavy and intermittent, a slip-ring induction motor is used as a prime mover. A flywheel is mounted on its shaft. This scheme is known as Ward-Leonard-Ilgener Scheme.

It prevents heavy fluctuations in supply current. When the driving AC motor is a synchronous motor, its supply current fluctuations cannot reduced by mounting a flywheel on its shaft, because a synchronous motor operates only at a constant speed.

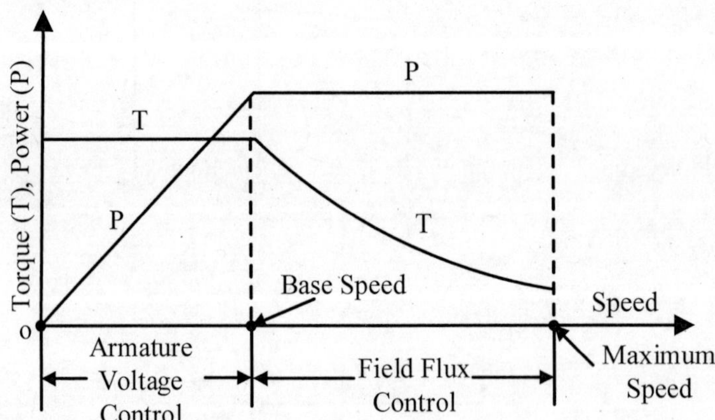

Fig. P19 Torque and power characteristic for combined armature voltage and field control

In another form of Ward- Leonard drive, non- electrical prime movers can also be used to drive the DC generator either by a diesel engine or a gas turbine. In this system regenerative braking is not possible because energy cannot flow in the reverse direction in the prime mover.

The variable voltage generator in Ward Leonard system is driven by a constant speed 3-phase induction motor.

If the constant voltage d.c power for excitation is not available otherwise, the same may be obtained from a constant voltage exciter coupled with the auxiliary motor-generator set. The direction of the field current of the variable voltage generator may be reversed by anyone of the following two methods..

- By providing a reversing switch in the field circuit.
- By connecting two potentiometer rheostats across generator field across the movable terminals.

Procedure:

STEP 1: Make connections as per the as per circuit given.

STEP 2: Check the connections.

STEP 3: If the connections are correct, then Turn “ON” the MCB Switch.

STEP 4: Then, check the Auto Transformer knob.

STEP 5: Now, Slide the Knob of the Rheostat. The readings of Voltmeter and RPM will be shown in the display.

STEP 6: Add the readings to the observation table and Graph will be created when readings are added in the table.

Advantages of using Ward-Leonard method

- It is a very smooth speed control system over a very wide range (from zero to normal speed of the motor).
- The speed can be controlled in both the direction of rotation of the motor easily.
- The motor can run with a uniform acceleration.
- Speed regulation of DC motor in this Ward Leonard system is very good.
- It has inherent regenerative braking property.

Drawbacks of Ward-Leonard method

- Higher initial cost due to use of two additional machines of the same rating as the main dc motor.
- Larger size and weight.
- Requires more floor area and costly foundation.
- Frequent maintenance is needed.
- Lower efficiency due to higher losses.
- The drive produces more noise.

Some viva-voce questions:

1. What is the purpose of conducting the experiment on DC motor speed control using the Ward Leonard method?
2. Can you explain the basic principle behind the Ward Leonard method and how it works to control the speed of a DC motor?
3. What are the main components of the Ward Leonard control system, and what is the function of each component?
4. What are the different control modes (forward, reverse, and braking) available in the Ward Leonard method?
5. Can you explain the concept of motoring and regenerative braking in the Ward Leonard system?
6. What are the advantages of using the Ward Leonard method for DC motor speed control compared to other methods?
7. Can you describe the procedure for conducting the DC motor speed control experiment using the Ward Leonard method step by step?

Experiment: 08

Experiment on DC motor *efficiency determination* using direct test or Brake test

Brake Test or Direct Test:

This test is a kind of direct method and used to determine the efficiency of comparatively small motors. In direct method, the DC machine is subjected to rated load and entire output power is wasted. The motor is loaded directly by means of a mechanical brake or by means of an eddy current brake or a calibrated air fan. Fig. P20 shows a common type of mechanical brake used in testing – the rope or belt brake. The brake is applied to a pulley mounted on the motor shaft. The load on the motor is increased by tightening the belt mounted on the pulley. The electrical connections are shown in the circuit diagram.

Let, spring balanced reading on tight side $= W_a$ Kg

Spring balanced reading on loose side $= W_b$ Kg

Motor Speed $= N$ rpm

Radius of pully $= r$ metre

Thickness of belt $= t$ metre

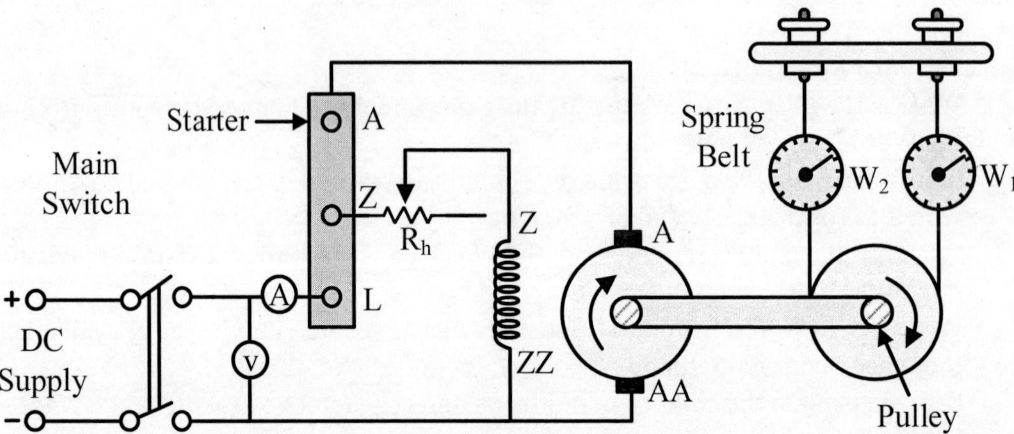

Fig.P20. Circuit arrangement for brake test

$$\text{Motor output} = T \times \frac{2\pi N}{60}$$

$$= (W_a - W_b)\left(r + \frac{t}{2}\right)\frac{2\pi N}{60}$$

Since T=effective pull ×effective radius

$$= (W_a - W_b)r \times \frac{2\pi N}{60} \text{ Kg m/s}$$

$$= (W_a - W_b)r \times \frac{2\pi N}{60} \times 9.81 \text{ Nm/s or watt}$$

If voltmeter reading is V volt and ammeter reading, is I ampere

Motor input=VI Watt

Then, efficiency of motor,

$$\eta = \frac{output}{Input}$$

$$\eta = \frac{(W_a - W_b)r \times \frac{2\pi N}{60} \times 9.81}{VI}$$

This method of measuring efficiency has the following disadvantages:

- It is not possible to measure the output power directly.
- It is not possible to use this method for determining internal losses and efficiency of large motors because such facilities of loading are not available.
- The output measured by this method is not accurate because belt is not offering a constant load, because usually belt slips over the pulley.

Belt encircles half the rotating hollow drum. Output power is wasted in rubbing of belt with rotating drum causing rise in temperature. Water is normally poured inside rotating hollow drum for cooling purpose. At high temperature and continuous rubbing action, the belt becomes smooth resulting in slippage of belt over drum. Sprinkling of sand is required for creating friction between belt and drum.

Procedure:

1. Connections are made as per the circuit diagram

2. Before starting the motor, Ensure that the field rheostat and Pot meter of Drive Control Unit are in minimum position.

3. Observing all the precautions, the motor is started using Drive Control Unit and the speed is increased until the rated armature voltage (of motor) is reached. At this instant the speed would be slightly lesser than the rated speed.

4. By adjusting the motor field rheostat, the motor is brought to rated speed.

5. The no load readings are tabulated.

6. The load is applied with the help of Mechanical Loading arrangement with Load cells, gradually in small steps and each step, take the reading of ammeters, voltmeter and load cells.

7. The motor is then brought to no load condition and field rheostat to minimum position and MCB is opened.

Some viva-voce questions:

1. What is the purpose of conducting the direct test or brake test on a DC motor?
2. Can you explain the basic principle behind the direct test or brake test and how it helps in determining the motor's efficiency?
3. How is the brake applied to the DC motor during the test, and how is the braking torque measured?
4. How are the input power and output power of the motor measured during the direct test or brake test?
5. What other parameters are recorded during the test to calculate the efficiency of the DC motor?
6. What are the sources of error that can affect the accuracy of the test results, and how can they be minimized?
7. Are there any limitations or challenges associated with implementing the direct test or brake test method for efficiency determination?
8. Can you describe the procedure for conducting the direct test or brake test on a DC motor step by step?

Experiment: 09

Predetermination of Efficiency using Swinburne's test

Aim: Pre - Determine the efficiency and constant losses of a DC shunt machine by Swinburne's method.

Name Plate Details:	
DC Motor:	
Rated Voltage	
Rated Current	
Rated Speed	
Power Rating	

Scan to perform experiment through VLAb

This method is an indirect method of testing a DC machine. It is named after Sir James Swinburne. **Swinburne's test** is the most commonly used and simplest method of testing shunt and compound wound DC machines which have constant flux. This test is performed to determine the constant losses and efficiency at any desired load. In this test, the machine is operated as a motor on no-load or as a generator. The iron and friction losses are determined by measuring the input. A voltmeter and two ammeters A_1 and A_2 are connected in the circuit as shown in fig. P21. The normal rated voltage V is applied to the motor terminals. The ammeter A_1 and A_2 measure the no-load line current I_{L0} and shunt field current I_{sh} respectively. The voltmeter measures the applied voltage. As there is no output at no-load, all the power supplied to the motor, given by the product of current I_{L0} and voltage V is being utilized to meet losses only. The speed of the machine is adjusted to the rated speed with the help of the shunt regulator R_d as shown in the figure.

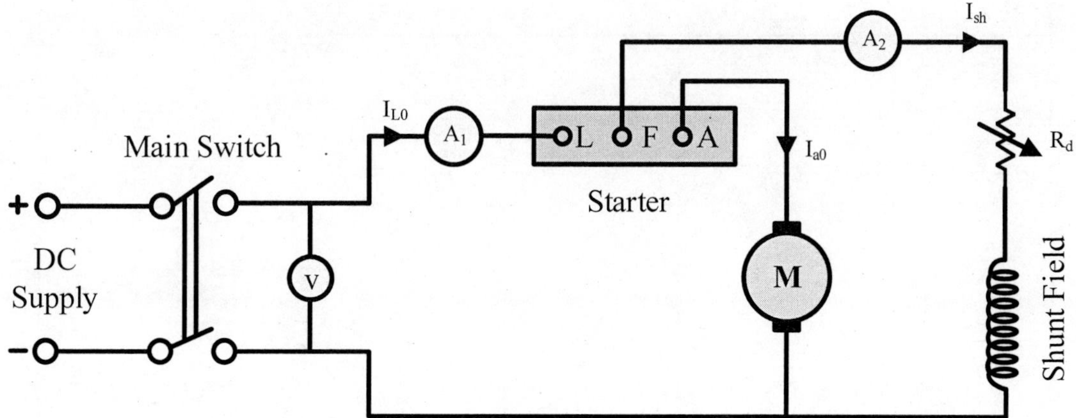

Fig.P21: Circuit arrangement for Swinburne's test

APPARATUS REQUIRED:

Sr. No.	Apparatus	Range	Type	Quantity

Procedure:

1. Choose the proper ranges of meters after noting the name plate details of the given machine and make the connections as per the circuit diagram.
2. Keep the motor field rheostat (R_{fm}) in the minimum position, start the motor by closing the switch and operating the starter slowly.
3. Run the motor at rated speed by adjusting the motor field rheostat.
4. Note down the voltage, no load current and field current.

Sr. No.	V(Volt)	I_{L0}(Amp)	$Speed$ (rpm)

Let,

V= Supply Voltage

I_{L0}= No load line current

I_{sh}= Shunt field current

The following are the losses at no-load:

a) Shunt field copper losses.
b) Armature copper losses at no-load (very small)
c) Iron losses in the core
d) Windage and friction losses at bearing and commutator.

$$\text{Armature current, } I_{a0} = I_{L0} - I_{sh}$$

$$\text{No load input power} = VI_{L0} \quad \text{Watt}$$

The resistance of armature circuit including the inter pole winding, etc., is measured by disconnecting one end of the shunt field circuit. Let its value be R_a.

$$\text{Then, variable losses} = I_{a0}^2 R_a$$

$$\text{Shunt field copper} = I_{sh}^2 R_{sh}$$

$$\text{Constant losses, } P_c = \left(VI_{L0} - I_{a0}^2 R_a - I_{sh}^2 R_{sh}\right)$$

After determining the constant losses, the efficiency of the machine, when it is working as a motor or generator can be calculated at any load, as discussed below:

Let I_L be the line current at which efficiency is to be calculated.

(i) When the machine is working as a motor:

$$\text{Armature current, } I_a = I_L - I_{sh}$$

$$\text{Variable or armature copper loss at load} = I_a^2 R_a$$

$$\text{Total losses} = P_c + I_a^2 R_a + I_{sh}^2 R_{sh}$$

$$\text{Input value} = VI_L$$

Output power = Input power – total losses

$$= VI_L - \left(P_c + I_a^2 R_a + I_{sh}^2 R_{sh}\right)$$

$$\text{Efficiency, } \eta = \frac{output}{Input} = \frac{VI_L - \left(P_c + I_a^2 R_a + I_{sh}^2 R_{sh}\right)}{VI_L}$$

(ii) When the machine is working as a generator

$$\text{Armature current, } I_a = I_L + I_{sh}$$

$$\text{Variable or armature copper loss at load} = I_a^2 R_a$$

$$\text{Shunt field copper} = I_{sh}^2 R_{sh}$$

$$\text{Total losses} = P_c + I_a^2 R_a + I_{sh}^2 R_{sh}$$

$$\text{Output power} = VI_L$$

$$\text{Input power} = \text{output power} + \text{total losses}$$

$$= VI_L + \left(P_c + I_a^2 R_a + I_{sh}^2 R_{sh}\right)$$

$$\text{Efficiency}, \eta = \frac{VI_L}{VI_L + \left(P_c + I_a^2 R_a + I_{sh}^2 R_{sh}\right)}$$

Advantages

Following are the advantages of Swinburne's test

- The power required for the testing of large machines is very small, therefore it is an economical and convenient method of testing DC machines.
- As the constant losses are known, thus the efficiency can be pre-determined at any load.

Disadvantages

The main disadvantages of Swinburne's test are

- Since Swinburne's test is performed on no-load, thus it does not indicate whether the commutation on full load is satisfactory and whether the temperature rise would be within specified limits.
- This test cannot be performed with DC series motors because at no-load series motors obtain dangerously high speeds.
- The change in iron losses is not considered from no-load to full load. At full load, due to the armature reaction, the flux is distorted which increases the iron losses.
- Stray load loss is not considered.

As a Motor:		**Rated Voltage (V_L):**		**Rated Speed (N):**	
S. No.	I_L	**Input Power** VI_L	**Total Losses** $P_c + I_a^2 R_a + I_{sh}^2 R_{sh}$	**Output Power=Input-Total losses** $VI_L - (P_c + I_a^2 R_a + I_{sh}^2 R_{sh})$	η

As a Motor:		Rated Voltage (V_L):		Rated Speed (N):	
S. No.	I_L	**Output Power** $V_L I_L$	**Total Losses** $P_c + I_a^2 R_a + I_{sh}^2 R_{sh}$	**Input Power=Output+Total losses** $V_L I - (P_c + I_a^2 R_a + I_{sh}^2 R_{sh})$	η

Model graph:

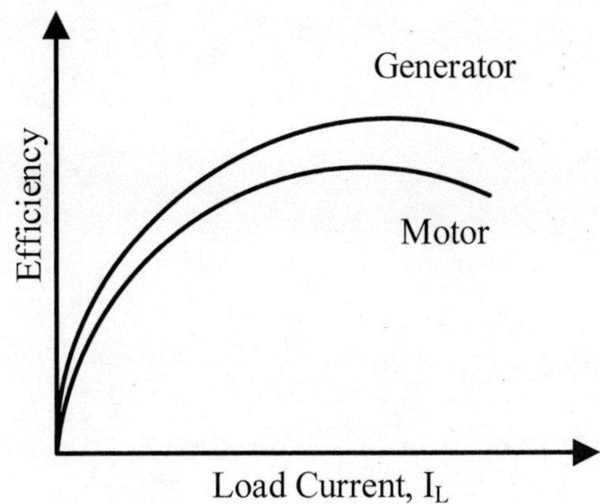

Fig.P22: Load Characterstic of DC Shunt Motor and Generator

Result:

Some viva-voce questions:

1. What is the objective of the Swinburne's test experiment in the context of a DC motor?
2. Can you explain the basic principle behind Swinburne's test for determining motor efficiency?
3. What are the main components of the experimental setup for this test?
4. How is the motor connected in the Swinburne's test setup, and what measurements are taken during the experiment?
5. How is the field current and armature voltage adjusted during the experiment, and why is it necessary to vary them?
6. What are the different losses considered in the efficiency calculation of the DC motor using Swinburne's test?
7. How is the efficiency of the DC motor calculated from the experiment data?
8. What factors may cause deviations in the obtained efficiency value from theoretical expectations?
9. Are there any limitations to Swinburne's test method?
10. How does Swinburne's test differ from other methods used to determine motor efficiency, such as the direct and indirect methods?

Experiment: 10

Determine and draw the equivalent circuit parameters of single-phase transformer.

AIM : To perform O.C. and S.C. test on a 1-phase transformer and to determine the parameters of its equivalent circuit its voltage regulation and efficiency.

Name Plate Details:	
Single phase transfromer:	
Rated Voltage	
Rated Current	
Rated frequency	
Power power (kVA)	
Number of phase	
Type: core/shell	

Scan to perform experiment through VLAb

Instruments

S. No.	Name	Type	Range	Quantity

Theory

Open Circuit Test

In this test low voltage winding (primary) is connected to a supply of normal voltage and frequency (as per the rating of transformer) and the high voltage winding (secondary) is left open as shown in figure. The primary winding draws very low current hardly 3 to 5 percent of full load current (may be upto 10 % for very small rating transformers used for laboratory purposes) under this condition. As such copper losses in the primary winding will be negligible. Thus mainly iron losses occur in the transformer under no load or open circuit condition, which are indicated by the wattmeter connected in the circuit.

Hence, total iron losses = W_0 (Reading of wattmeter)

From the observations of this test, the parameters R_0 and X_m of the parallel branch of the equivalent circuit can also be calculated, following the steps given below :

Power drawn, $W_0 = V_0 I_0 \cos\phi_0$

Thus, no load power factor, $\cos\phi_0 = \dfrac{W_0}{V_0 I_0}$

Core loss componenet of no load current, $I_w = I_0 \cos\phi_0$

And, magnetizing component of no load current, $I_m = I_0 \sin\phi_0$

Equivalent resistance representing the core loss, $R_0 = \dfrac{V}{I_w}$

Magnetizing reactance representing the magnetizing current, $X_m = \dfrac{V}{I_w}$

Short circuit test

In this test, low voltage winding is short circuited and a low voltage hardly 5 to 8 percent of the rated voltage of the high voltage winding is applied to the winding. This test is performed at rated current flowing in both the windings. Iron losses occurring in the transformer under this condition is negligible, because of very low applied voltage. Hence the total losses occurring under short circuit are mainly the copper losses of both the winding, which are indicated by the wattmeter connected in the circuit as shown in figure.

Thus total full load copper losses = W_{SC} (reading of wattmeter)

The equivalent resistance R_{eq}, and reactance X_{eq} referred to a particular winding can also be calculated from the observations of this test, following the step given below.

Equivalent resistance referred to H. V. wunding $R_{eq} = \dfrac{W_{sc}}{I_{sc}^2}$

Also, equivalent impedance referred to H. V. winding $Z_{eq} = \dfrac{V_{sc}}{I_{sc}}$

Thus equivalent reactance referred to H.V. winding $X_{eq} = \sqrt{(Z_{eq}^2 - R_{eq}^2)}$

Performance calculations

Complete performance of the transformer can be calculated based on the above observations of open-circuit and short-circuit test following the steps given by,

Efficiency at different loads :

Efficiency at full load:

Total losses at full load, $= W_0 + W_{sc}$

Let the full load out put power of the transformer in kVA be $\boldsymbol{P_0}$.

Then percentage efficiency at full load, $\eta_f = \dfrac{P_0 \times 1000 \times \cos\phi}{P_0 \times 1000 \times \cos\phi + W_0\, W_{sc}} \times 100$

Efficiency at half the full load:

Iron losses at half the full load $= W_0$ (constant)

$$\text{Total copper losses at half the full load } = \frac{1}{4} W_{sc}$$

$$\text{Output power at half full load} = \frac{1}{2} P_0\, kVA$$

% efficiency at half the full load $\eta_{\frac{1}{2}f}$

$$= \frac{\frac{1}{2} P_0 \times 1000 \times \cos\phi}{\frac{1}{2} P_0 \times 1000 \times \cos\phi + W_0 + \frac{1}{4} W_{sc}} \times 100$$

In a similar manner, efficiency at other loads can be found out and the efficiency V_s output power curve can be plotted.

Equivalent circuit:

All the parameters of the approximate equivalent circuit has been calculated above. Thus an approximate equivalent circuit of the transformer can be drawn with these values of parameters marked on it. The equivalent circuit can be solved easily for estimating the performance like terminal voltage across the secondary etc.

Regulation:

Regulation of the transformer can now be calculated based on the parameters of the equivalent circuit, using the approximate formula given below.

$$\text{Percentage regulation} = \frac{I\left(R_{eq} \cos\phi \pm X_{eq} \sin\phi\right)}{V} \times 100$$

Where, I – rated current of the winding, referred to which R_{eq} and X_{eq} have been calculated.

V – Voltage of that winding.

cos ϕ - Power factor at which regulation is to be calculated

Circuit diagram

Fig.P23 shows the circuit diagram to perform the no load test at rated voltage of the transformer. Ammeter, Voltmeter and wattmeter have been connected in the circuit to measure, no load current, rated voltage applied and the no load power drawn by the transformer respectively.

Fig.P24 shows the circuit diagram to perform the short circuit test at a reduced voltage, such that the current flowing in the windings is of rated full load value. As such a single phase variac has been included in the circuit. Ammeter, wattmeter and voltmeter connected in the circuit measure, full load short circuit, power drawn under this condition and the reduced voltage applied to the winding of the transformer respectively.

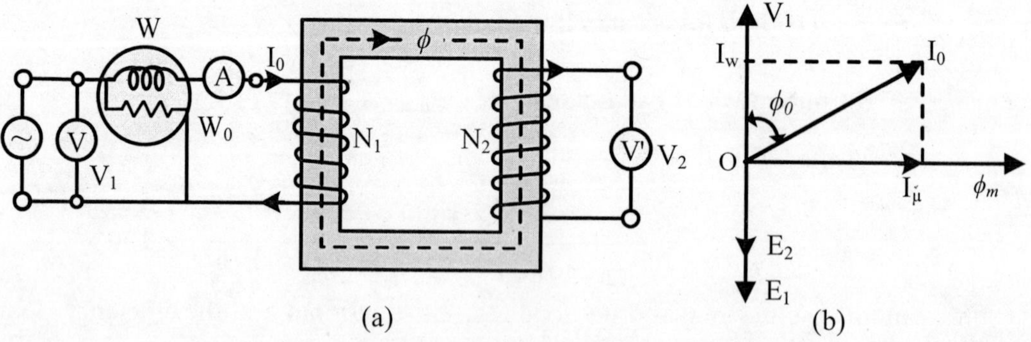

Fig.P23: Circuit diagram to perform the no load test

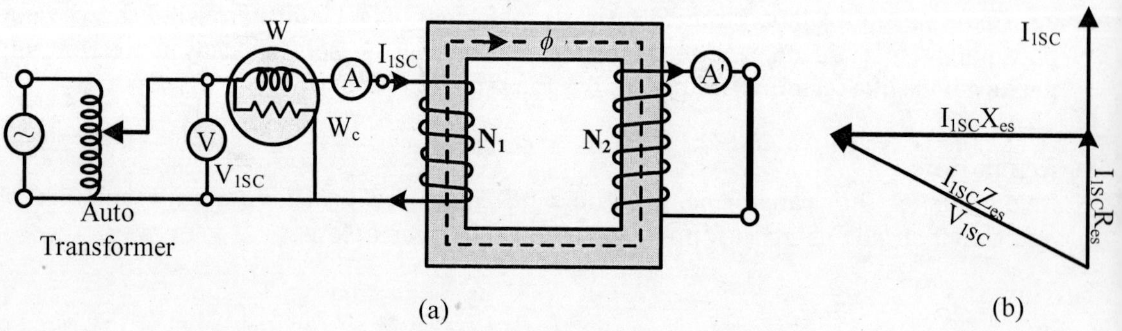

Fig.P24: Circuit diagram to perform the short circuit test

Procedure

(a) Open circuit test

1. Connect the circuit as per Fig.P23
2. Ensure that the setting of the variac is at low output voltage.

3. Switch on the supply and adjust rated voltage across the transformer circuit.
4. Record no load current, voltage applied and no load power, corresponding to the rated voltage of the transformer winding.
5. Switch off the ac supply.

(b) Short circuit test

1. Connect the circuit as per Fig.P24 for conducting short circuit test.
2. Adjust the setting of the variac, so that the output voltage is zero.
3. Switch on the ac supply to the circuit.
4. Increase the voltage applied slowly, till the current in the windings of the transformer is full load rated value.
5. Record, short circuit current, corresponding applied voltage and power with full load current flowing under short circuit conditions.

Observation: May be tabulated as follow :

S. No.	No Load Test			Short Circuit Test		
	V_0	I_0	W_0	V_{sc}	I_{sc}	W_{sc}

Calculations: May be tabulated as follows,

S. No.	Cos ϕ_0	I_w	I_m	R_0	X_m	R_{eq}	X_{eq}	Load	η	Regulation

Some viva-voce questions:

1. What is the purpose of conducting the O.C. and S.C. tests on a 1-phase transformer?
2. How is the transformer connected during the O.C. test, and what are the measured quantities?
3. How is the transformer connected during the S.C. test, and what are the measured quantities?
4. What is meant by the voltage regulation of a transformer, and how is it calculated using the O.C. and S.C. test data?
5. How can you determine the efficiency of the transformer from the O.C. and S.C. test results?
6. What are the sources of error that can affect the accuracy of the test results, and how can they be minimized?
7. How does the frequency and voltage level of the test affect the transformer's performance and test results?
8. What are the practical applications of the transformer equivalent circuit parameters, voltage regulation, and efficiency calculations?
9. Can you explain the concept of copper and iron losses in a transformer and how they are determined from the test data?
10. What are the advantages and limitations of using the O.C. and S.C. test methods compared to other transformer testing methods?

Experiment: 11

Testing of transformers using Sumpner's Test.

AIM : To perform sumpner's test on two identical 1-phase transformers and find their efficiency & parameters of the equivalent circuit.

Name Plate Details:	
Single phase transfromes	
Rated Voltage	
Rated Current	
Rated frequency	
Power power (kVA)	
Number of phase	
Type: core/shell	

Scan to perform experiment through VLAb

Apparatus required:

S.No.	Name	Range	Qty.

Theory:

To determine the maximum temperature rise, it is necessary to conduct a full-load test on a transformer. For small transformers, the full-load test is conveniently possible, but for large transformers, the full-load test is very difficult. A suitable load to absorb the full-load power of a large transformer may not be easily available. It will also be very expensive as a large amount of energy will be wasted in the load during the test. Large transformers can be tested for determining the maximum temperature rise by the back-to-back test. This test is also called the **Regenerative test or Sumpner's test.**

The **Sumpner's test** on single-phase transformers requires two identical transformers. Fig. P.25, shows the circuit diagram for the **Sumpner's test** on two identical single-phase transformers T_{r_1} and T_{r_2}. The primary windings of the two transformers are connected in parallel and supplied at rated voltage and rated frequency. A voltmeter, an ammeter, and a wattmeter are connected to the input side as shown in fig. P25.

Procedure

The secondaries are connected in series with their polarities in phase opposition, which can be checked by the voltmeter V_2. The range of this voltmeter should be double the rated voltage of a transformer secondary. To check that secondaries are connected in series opposition, any two terminals (say B and C) are joined together and the voltage is measured between the remaining terminals A and D.

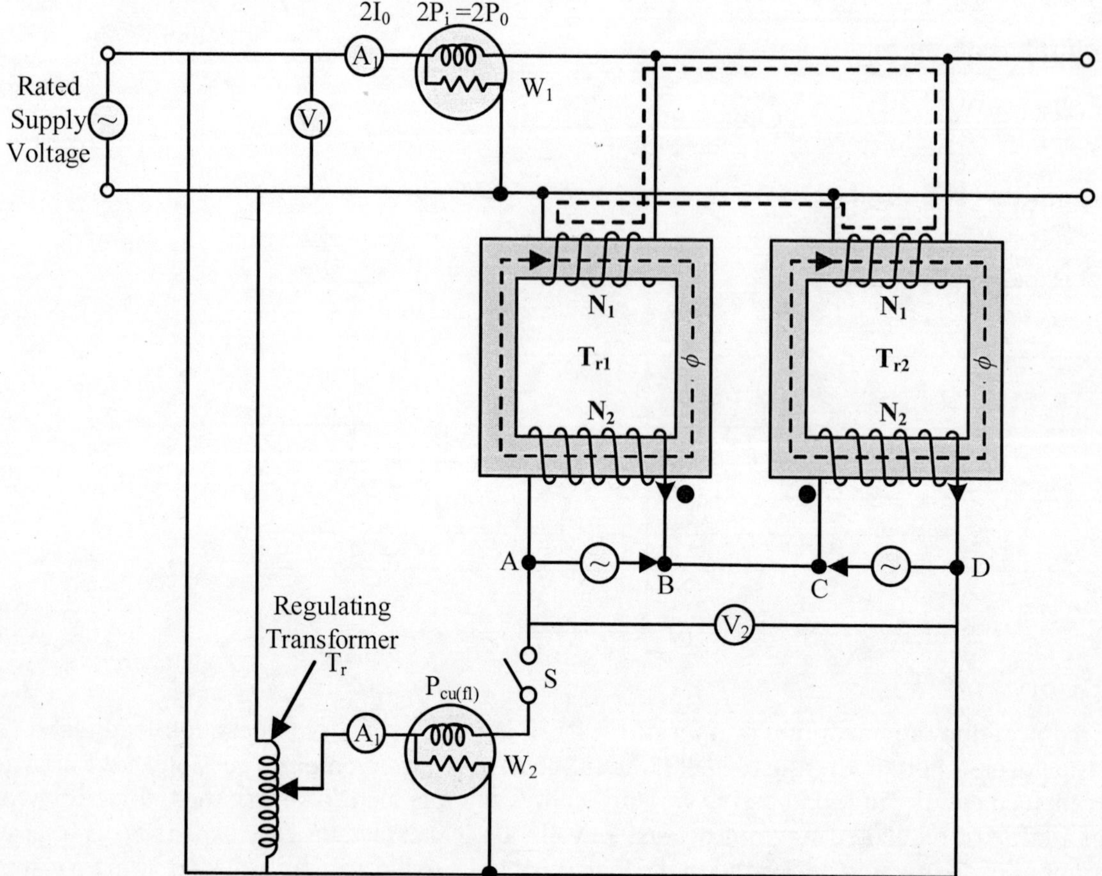

Fig.P25 Circuit diagram for the Sumpner's test on two identical single-phase transformers

If the voltmeter V_2 reads zero, the two secondaries are in series opposition, and terminals A and D are used for the test. If the voltmeter reads a value approximately equal to twice the rated secondary voltage of either transformer, then the secondaries are acting in the same direction. Then terminals A and C are joined and the terminals B and D are used for the test.

If the primary circuit is now closed, the total voltage across the two secondaries in series will be zero. There will be no current in the secondary windings. The transformers will behave as if their secondary windings are open-circuited. Hence, the reading of the Wattmeter W_1 gives the iron losses of both the transformers when rated low voltage V_1 is applied across parallel connected primaries.

A small voltage is injected into the secondary circuit by a regulating transformer T_r excited by the main supply. The magnitude of the injected voltage is adjusted till the armature A_2 reads full-load secondary current.

The secondary current produces full-load current to flow through the primary windings. This current will follow a circulatory path through the main busbars as shown by the dotted line in fig. P25. The reading of the Wattmeter W_2 will not be affected by this current. Thus, wattmeter W_2 gives the sum of full-load copper losses of the two transformers.

The armature A_1 gives the total no-load current of the two transformers. Thus, in this method, we have loaded the two transformers to full load but the power taken from the supply is that necessary to supply the losses of both transformers.

The temperature rise of the transformers can be determined by operating these transformers back-to-back for a long time, say 48 hours, and measuring the temperature of the oil at periodic intervals of time, say every hour.

OBSERVATION : May be tabulated as follows,

S. No.	Primary side				Secondary side			
	V_0	I_0	W_0	W_0O	V_{sc}	I_{sc}	W_{sc}	W_{2gc}

Result: Sumpner's test on two identical 1-phase transformers has been performed successfully.

Some viva-voce questions:

1. What is the purpose of conducting the Sumpner's Test on transformers?
2. Can you explain the basic principle behind Sumpner's Test?
3. How are the transformers connected in the Sumpner's Test, and what are the measured quantities?
4. What are the main parameters that can be determined from the Sumpner's Test data, and how are they calculated?
5. What is the significance of using transformers with different voltage ratings in the Sumpner's Test setup?
6. How can the obtained equivalent circuit parameters be used to predict the performance of the transformers under different operating conditions?
7. Can you explain the concept of copper and iron losses in transformers and how they are determined from the Sumpner's Test data?
8. What are the advantages and limitations of using the Sumpner's Test compared to other transformer testing methods?
9. How can the Sumpner's Test results be used to optimize the efficiency and performance of the transformers in real-world applications?

Experiment: 12 Study of Scott Connection of Two Single phase Transformers.

Aim: To study conversion of three-phase supply to two-phase supply using Scott-Connection.

Name Plate Details:	
Single phase transfromes	
Rated Voltage	
Rated Current	
Rated frequency	
Power power (kVA)	
Number of phase	
Type: core/shell	

Apparatus required:

S. No	Name	Type	Range	Quantity

Theory:

In some cases, we may require 2-phase power instead of 3-phase or 1-phase power. For that it is necessary to convert 3-phase to2-phase power. Scott connection is one by which 3-phase to 2-phase transformation is accomplished with the help of two identical 1-phase transformers having same current rating. One transformer has a center tap on primary side and it is known as Main transformer. It forms the horizontal member of the connection. Another transformer has 0.866 tap on primary side and known as Teaser transformer. The 50% tap point on primary side of the main transformer is joined to 86.6% tap on primary of the teaser transformer. Obviously full rating of the transformers is not at all used. Refer to the fig. The main transformer primary winding center tap point D is connected to one end of the primary of the teaser transformer on secondary side, both the main & teaser transformer turns are used (not only 86.6%). Hence the voltage per turn will be equal for both transformer.

Procedure:

1. Perform polarity test of the two single-phase transformers and find the polarities.

2. Connect the circuit as shown in the fig. P27 for three-phase to two-phase conversion.
3. The three-phase balanced supply should be given to the circuit through auto-transformer and switch S_1.
4. With switch S_2 open, close S_1 to supply rated voltage on primary side. Note down the voltages on primary and secondary sides of teaser and main transformers.
5. Start with balanced load on the secondary side (Z_{L1}=Z_{L2}). Close S_2 and take all the ammeter and voltmeter readings. (Note that for balanced load the two secondary currents are equal and the three primary currents are equal. Else, it implies that the load is unbalanced).
6. Now keep unbalanced load on secondary side and take all the voltmeter and ammeter readings. Repeat this for 3 more sets of unbalanced load.
7. Verify the results with theoretical predictions and draw Phasor diagrams.
8. For three-phase to single-phase conversion, short the negative polarity side of teaser transformer and positive polarity side of main transformer in the previous circuit. Note the load is kept across positive polarity side of teaser and negative polarity side of main transformer. (Note that the single-phase voltage is higher than the secondary voltage in two-phase conversion and connect the load accordingly. If necessary connect two loads in series to maintain the rated voltage of load greater than or equal to single-phase voltage.)
9. Keep load at some value and note down all the voltmeter and ammeter readings.
10. Verify the results with theoretical predictions and draw Phasor diagrams.

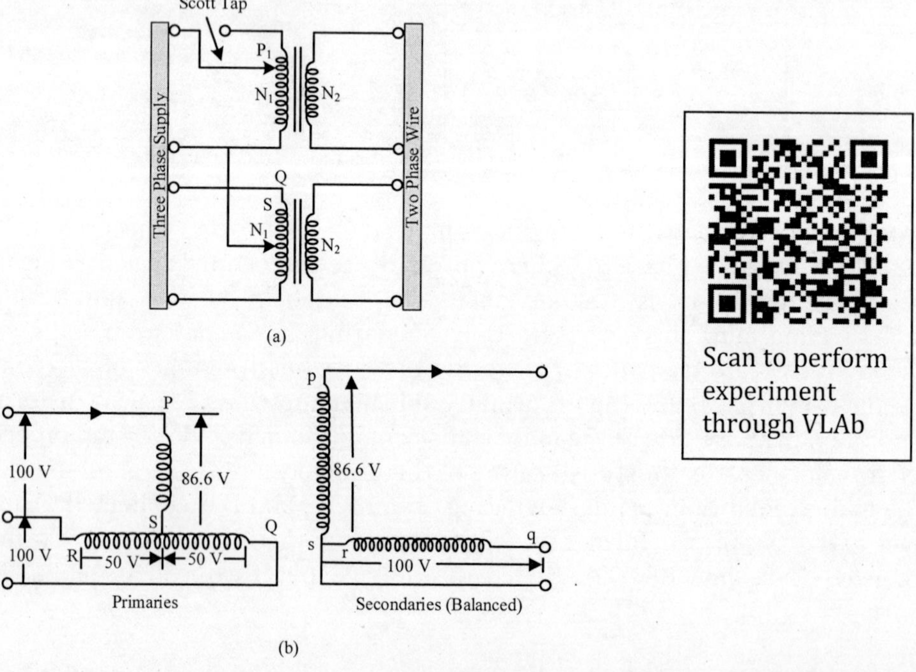

Fig. P.26 Three-phase to two-phase conversion

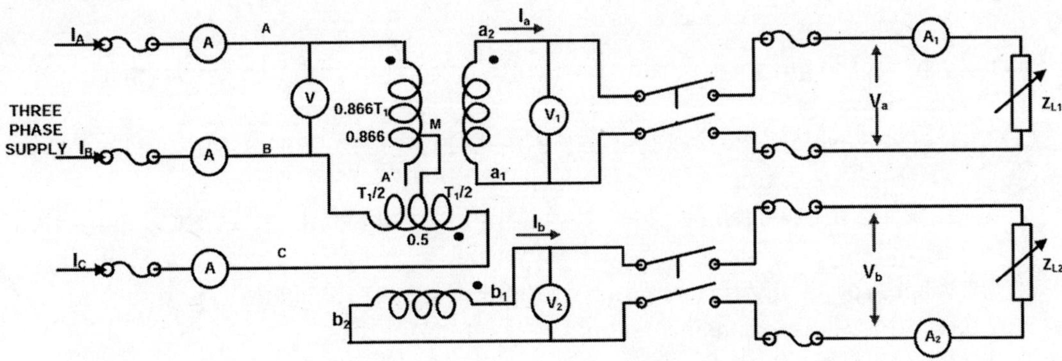

Fig.P27: Connection for load test with Scott connected transformers

The condition due to a balanced two-phase load at a lagging power factor of 0.866 has been shown in Fig.P28 (a). The three-phase side is balanced here. The main transformer rating is 15 percent greater than the rating of the teaser transformer. This is because its voltage is 15 percent greater. But its current remains the same. Fig.P28 (b) shows the condition due to an unbalanced two-phase load having different currents and power factors.

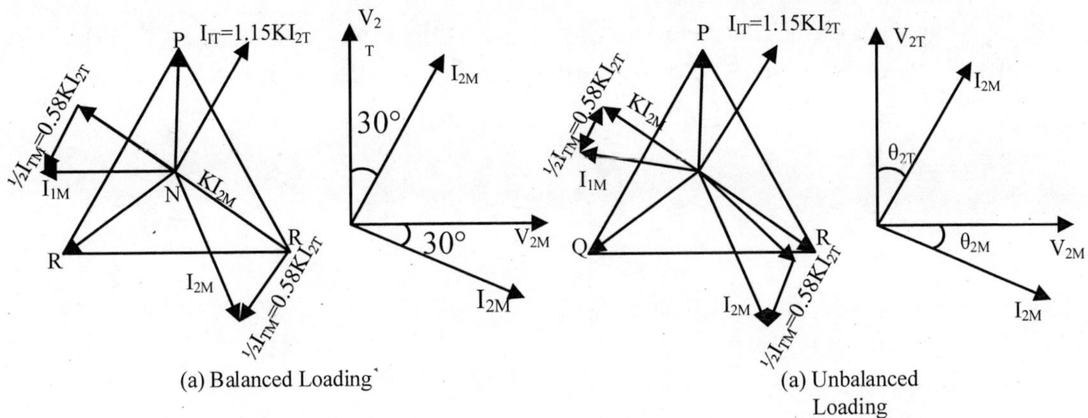

Fig.P24 Vector diagram for Balanced and Unbalanced loading

Some viva-voce questions:

1. What is the purpose of using the Scott Connection with two single-phase transformers?
2. Can you explain the basic principle behind the Scott Connection and how it converts a two-phase system to a three-phase system?
3. How are the two single-phase transformers connected in the Scott Connection setup?
4. What are the different winding configurations for the primary and secondary windings of the transformers in the Scott Connection?
5. How does the Scott Connection provide a 90-degree phase shift between the two secondary voltages?
6. What are the applications of the Scott Connection in electrical power systems?
7. What are the advantages of using the Scott Connection compared to other methods of achieving a 90-degree phase shift in a two-phase system?
8. How do you determine the turns ratio and voltage ratings of the transformers for the Scott Connection experiment?
9. How can you verify the phase shift between the secondary voltages in the Scott Connection experiment?
10. What are some alternative methods of converting a two-phase system to a three-phase system, and how do they compare to the Scott Connection?

Experiment: 13 Separation of No-load losses in single phase transformers.

Aim: To seprate the hysteresis and eddy current losses from iron loss in a single phase transformer ar normal voltage and frequency.

Name Plate Details:	
Single phase transfromes	
Rated Voltage	
Rated Current	
Rated frequency	
Power power (kVA)	
Number of phase	
Type: core/shell	

Scan to perform experiment through VLAb

Apparatus required:

S. No	Name	Type	Range	Quantity

Theory:

The component of iron loss consist of hysteresis loss and eddy current loss. Both are functions of frequency and maximum flux density in the core can be separated by finding iron losses at various frequencies and plotting the graphs Pc/N Vs N. Variable supply frequency can be obtained from an alternator. There are mainly two types of losses occurs in a transformer. They and iron loss or core loss and copper loss or winding loss. As the name indicates, the loss occurs in core is known as core loss and loss in winding is known as winding loss. The iron loss includes hysteresis losses and eddy current losses, both are functions of frequency and maximum flux density in the core. The values of these losses are independent of load current. Hence it is assumed as constant from no load to full load and named as constant loss.Hence constant loss in a transformer is given by,

Constant loss (core loss/Iron loss) = Hysteresis loss + Eddy current loss

$$P_C = W_H + W_E W_H = K_H B_m^{1.6} f W_E = K_E B_m^2 f^2$$

Where K_H and K_E are proportionality constant.

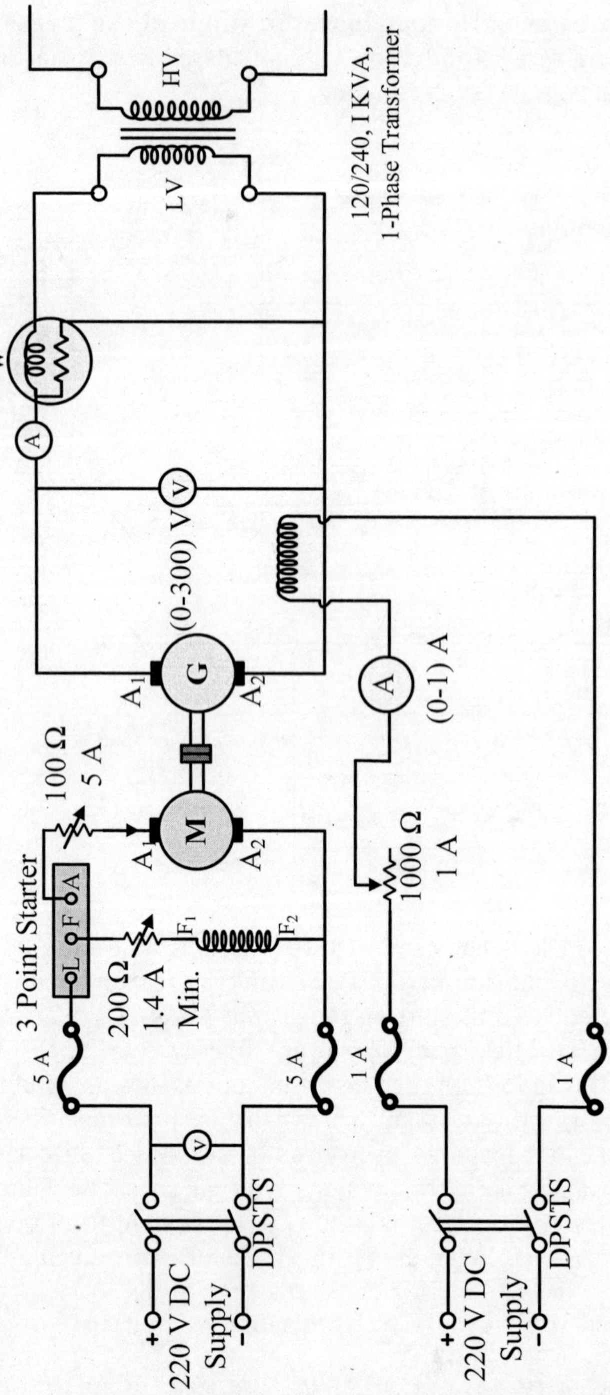

Fig.P29 Circuit Diagram Speration on Losses

Therefore,

$$P_C = K_H B_m^{1.6} f + K_E B_m^2 f^2$$

The hysteresis loss is varying linearly with the frequency while the eddy current loss varies as the square of supply frequency.
The core loss per cycle is given by

$$P_C/f = K_H B_m^{1.6} + K_E B_m^2 f$$

This shows that hysteresis loss per cycle is independent of frequency and eddy current per cycle is proportional to the frequency. For the open circuit test V and f are varied together so that V/f is a constant. Since Bm α V/f for a particular value of V/f, the equation for core loss per cycle can be written as,

$$P_C/f = K_1 + K_2 f$$

Where,

$$K_1 = K_H B_m^{1.6},\ K_2 = K_E B_m^2 f^2$$

Thus the plot of P_C/f versus f results a straight line. From the graph, value of K_1 and K_2 can be determined. Slope of the straight line gives K_2 and intercept gives K_1. Thus core loss can be separated as,

Hysteresis loss, $W_H = K_1 f$, Eddy Current Loss, $W_E = K_2 f^2$

An alternator is a three phase a.c. generator whose speed and frequency are related as N=120f/P.
Where N- Speed
f- Frequency
P- Number of poles of alternator

Procedure:

1. Connect the circuit as shown in figure.
2. Keep the field rheostat of alternator in maximum position and field rheostat of motor in minimum position. Also keep the armature rheostat of motor in maximum position.
3. Switch ON the power supply.
4. Start the motor using three point starter.
5. Cut off the starting rheostat gradually.
6. Adjust the field rheostat of motor to drive the alternator at its rated speed to get normal supply frequency (50Hz).
7. Adjust the field rheostat of alternator to supply rated voltage to the transformer.
8. Note the wattmeter reading.
9. Now the frequency is varied to different convenient values by adjusting

10. the speed of the prime mover (motor) and in each case voltage is also adjusted to keep V/f ratio constant
11. Tabulate the readings.
12. Switch OFF the power supply after bringing all the rheostats to initial positions.

Observation : May be tabulated as follows

S. No.	Frequency	Applied voltage	No load current	No load power	V/f	P_c/f

Result: Separation of no load losses in single phase transformer has been performed successfully.

Some viva-voce questions:

1. What is the purpose of conducting the experiment on the separation of no-load losses in single-phase transformers?
2. Can you explain the difference between no-load losses, core losses, and copper losses in a transformer?
3. What are the factors that contribute to no-load losses in a transformer?
4. How do you calculate the core losses of the transformer from the test results?
5. What are the challenges or sources of error in accurately measuring the no-load losses?
6. How can the obtained data be used to improve the efficiency and performance of the transformer?
7. Can you explain the concept of hysteresis losses and eddy current losses in transformers and their contribution to core losses?
8. How can this experiment aid in selecting the most efficient transformer for specific applications?
9. How does the frequency and voltage level of the test affect the transformer's no-load losses and test results?

Electrical Machine Laboratory Safety Rules:

1. Always ensure that any terminals or switches are dead before touching them.
2. It is advisable to wear shoes with rubber soles for added protection against electric shock.
3. Use a fuse wire with the appropriate rating for the circuit to prevent overloading.
4. When connecting leads, use sufficiently long ones instead of joining multiple short ones, as loose joints can be dangerous.
5. Double-check all electrical connections for correctness before switching on any circuit, as wrong connections can lead to equipment damage.
6. De-energize the circuit before making any changes to the connections.
7. In the event of an emergency or fire, immediately switch off the master switch on the main panel board.
8. Maintain a safe distance from all moving parts of the electrical machines.
9. If a fuse blows, do not replace it until you have determined the cause of the problem and rectified it.
10. Avoid touching an electric circuit with wet or bleeding hands to prevent electric shock.
11. Never disconnect a plug by pulling the flexing cable while the switch is still on.
12. In case of a fire involving electrical equipment, do not throw water on it, as water conducts electricity and can worsen the situation. Use appropriate fire extinguishing methods.
13. Avoid testing circuits with bare fingers; always use appropriate testing equipment and tools.
14. Refrain from wearing loose garments while working in the laboratory to minimize the risk of getting entangled in moving parts or equipment.
15. Open or close switches and fuses quickly and decisively, avoiding slow or hesitant actions that may cause sparking or arcing.

References for Further Learning

1. Fitzgerald, A. E., Kingsley Jr, C., & Umans, S. D. (2003). Electric Machinery (6th ed.). McGraw-Hill Education.
2. Chapman, S. J. (2005). Electric Machinery Fundamentals (4th ed.). McGraw-Hill Education.
3. Kosow, I. L. (2004). Electric Machinery and Transformers (3rd ed.). Prentice Hall.
4. Bhadra, S. K. (2007). Electrical Machines. Tata McGraw-Hill Education.
5. Wildi, T. (2013). Electrical Machines, Drives, and Power Systems (6th ed.). Pearson.
6. Guru, B. S., & Hiziroglu, H. R. (2001). Electric Machinery and Transformers (3rd ed.). Oxford University Press.
7. Hubert, G. (2009). Electrical Machines and Drives: Principles, Control, Modeling, and Simulation. Wiley-Blackwell.
8. Lyshevski, S. E. (2008). Electrical Engineering Handbook. Academic Press.
9. Narayana, P. N. (2007). Electric Machines and Drives. PHI Learning Pvt. Ltd.
10. Say, M. G. (2014). Performance and Design of Alternating Current Machines (3rd ed.). Elsevier.
11. Murthy, V. V. S., & Kumar, P. S. (2012). Electrical Machines (2nd ed.). Pearson Education India.
12. Sen, P. C. (2003). Principles of Electric Machines and Power Electronics. John Wiley & Sons.
13. Boldea, I., & Nasar, S. A. (2002). Linear Electric Actuators and Generators. Cambridge University Press.
14. Fink, D. G., & Beaty, H. W. (2006). Standard Handbook for Electrical Engineers (15th ed.). McGraw-Hill Professional.
15. Charles, D. S. (2007). Electrical Machines: Mathematical Fundamentals of Machine Topologies. Springer.
16. Xu, D. (2012). Advanced Electrical Drives: Analysis, Modeling, Control. Springer Science & Business Media.
17. Nagrath, I. J., & Kothari, D. P. (2007). Electric Machines. Tata McGraw-Hill Education.
18. Krause, P. C., Wasynczuk, O., & Sudhoff, S. D. (2013). Analysis of Electric Machinery and Drive Systems. John Wiley & Sons.
19. Gieras, J. F., & Piech, Z. J. (2004). Linear Synchronous Motors: Transportation and Automation Systems. CRC Press.
20. Slemon, G. R. (2001). Electric Machinery (2nd ed.). Wiley-Blackwell.
21. Jones, J. A. (2003). Electrical Machines and Transformers. Oxford University Press.

22. Rosenblatt, M. (2013). Electric Motors and Drives: Fundamentals, Types, and Applications (4th ed.). Newnes.
23. Bose, B. K. (2006). Power Electronics and Variable Frequency Drives: Technology and Applications. John Wiley & Sons.
24. McFarland, D. M. (2003). Electric Machines and Drives: Principles, Control, Modeling, and Simulation. John Wiley & Sons.
25. Nair, M. N. (2007). Basic Electrical Engineering. Tata McGraw-Hill Education.
26. Hughes, T. P. (2010). Electric Motors and Drives: Fundamentals, Types and Applications. Newnes.
27. Bose, B. K. (2002). Modern Power Electronics and AC Drives. Prentice Hall.
28. Begamudre, R. D. (1999). Electrical Machines. Oxford University Press.
29. Zipse, H. G. (2001). Electric Motor Handbook. McGraw-Hill Professional.
30. Vasic, V. M. (2014). Electric Drives. CRC Press.
31. Singh, M. D. (2001). Electrical Machines. Tata McGraw-Hill Education.
32. Dorf, R. C., & Bishop, R. H. (2004). Modern Control Systems (10th ed.). Pearson.
33. Siskind, C. S. (2003). Electrical Design of Through Silicon Via. IEEE Press.
34. Krause, P. C., Wasynczuk, O., & Sudhoff, S. D. (2013). Analysis of Electric Machinery and Drive Systems. John Wiley & Sons.
35. Hameyer, K., Belmans, R., & Meys, R. (2006). Magnetic Field Calculation of Electrical Machines for Analytical and Numerical Methods. Springer Science & Business Media.
36. Belmans, R., & Driesen, J. (2006). Advanced Modeling and Optimization of Manufacturing Processes: International Research and Development (Vol. 5). Springer.
37. Boldea, I., & Nasar, S. A. (2002). Linear Electric Actuators and Generators. Cambridge University Press.
38. Bose, B. K. (2002). Modern Power Electronics and AC Drives. Prentice Hall.
39. Carter, G. C., & Nielsen, J. G. (2004). Electric Energy Conversion (2nd ed.). CRC Press.
40. Drury, B. (2009). Control Techniques Drives and Controls Handbook (2nd ed.). Newnes.
41. Dubey, G. K. (2009). Fundamentals of Electrical Drives. CRC Press.
42. Fitzgerald, A. E., & Kusko, A. (1998). Electric Machinery (5th ed.). McGraw-Hill Professional.
43. Hughes, T. P. (2010). Electric Motors and Drives: Fundamentals, Types and Applications. Newnes.
44. Kingsley Jr, C., & Umans, S. D. (2002). Electric Machinery (5th ed.). McGraw-Hill Professional.

45. Krause, P. C., Wasynczuk, O., & Sudhoff, S. D. (2013). Analysis of Electric Machinery and Drive Systems. John Wiley & Sons.
46. Kundur, P. (1994). Power System Stability and Control. McGraw-Hill.
47. Miller, T. J. E. (2004). Reactive Power Control in Electric Systems. Wiley-IEEE Press.
48. Mohan, N., Undeland, T. M., & Robbins, W. P. (2003). Power Electronics: Converters, Applications, and Design. John Wiley & Sons.
49. Ned Mohan, Tore M. Undeland, William P. Robbins (2003). Power Electronics: Converters, Applications, and Design (3rd ed.). John Wiley & Sons.
50. Lyshevski, S. E. (2008). Electrical Engineering Handbook. Academic Press.
51. Fitzgerald, A. E., Kingsley Jr, C., & Umans, S. D. (2003). Electric Machinery (6th ed.). McGraw-Hill Education.
52. Chapman, S. J. (2005). Electric Machinery Fundamentals (4th ed.). McGraw-Hill Education.
53. Kosow, I. L. (2004). Electric Machinery and Transformers (3rd ed.). Prentice Hall.
54. Bhadra, S. K. (2007). Electrical Machines. Tata McGraw-Hill Education.
55. Wildi, T. (2013). Electrical Machines, Drives, and Power Systems (6th ed.). Pearson.
56. Guru, B. S., & Hiziroglu, H. R. (2001). Electric Machinery and Transformers (3rd ed.). Oxford University Press.

CO AND PO ATTAINMENT TABLE

Course outcomes (COs) for this course can be mapped with the programme outcomes (POs) after the completion of the course and a correlation can be made for the attainment of POs to analyse the gap. After proper analysis of the gap in the attainment of POs necessary measures can be taken to overcome the gaps.

Table for CO and PO attainment

Course Outcomes	Expected Mapping with Programme Outcomes *(1- Weak Correlation; 2- Medium correlation; 3- Strong Correlation)*											
	PO-1	PO-2	PO-3	PO-4	PO-5	PO-6	PO-7	PO-8	PO-9	PO-10	PO-11	PO-12
CO-1												
CO-2												
CO-3												
CO-4												

The data filled in the above table can be used for gap analysis.

Index